Detection and Estimation

Dimitri Kazakos

P. Papantoni-Kazakos
University of Virginia

COMPUTER SCIENCE PRESS
An imprint of W. H. Freeman and Company
New York

ANDERSONIAN LIBRARY

2 3. AUG 91

UNIVERSITY OF STRATHCLYDE

Library of Congress Cataloging-in-Publication Data

Kazakos, Dimitri, 1944-
  Detection and estimation / D. Kazakos, P. Papantoni-Kazakos.
    p.   cm — (Electrical engineering, communications, and signal processing)
  Includes bibliographies and index.
  ISBN 0-7167-8181-6
  1. Signal theory (Telecommunication)   2. Statistical communication theory.   3. Estimation theory.   I. Papantoni-Kazakos, P., 1945-
II. Title.   III. Series: Electrical engineering, communications, and signal processing series.
TK5602.5.K38   1990   89-15867
621.382′23—dc20   CIP

Copyright © 1990 Computer Science Press

No part of this book may be reproduced by any mechanical, photographic, or electronic process, or in the form of a phonographic recording, nor may it be stored in a retrieval system, transmitted, or otherwise copied for public or private use, without written permission from the publisher.

Printed in the United States of America

Computer Science Press

An imprint of W. H. Freeman and Company
41 Madison Avenue, New York, NY 10010
20 Beaumont Street, Oxford OX1 2NQ, England

1 2 3 4 5 6 7 8 9 0   RRD   89

Detection and Estimation

ELECTRICAL ENGINEERING, COMMUNICATIONS, AND SIGNAL PROCESSING
Raymond L. Pickholtz, Series Editor

Computer Network Architectures
Anton Meijer and Paul Peeters
Spread Spectrum Communications, Volume I
Marvin K. Simon, Jim K. Omura, Robert A. Scholtz, and Barry K. Levitt
Spread Spectrum Communications, Volume II
Marvin K. Simon, Jim K. Omura, Robert A. Scholtz, and Barry K. Levitt
Spread Spectrum Communications, Volume III
Marvin K. Simon, Jim K. Omura, Robert A. Scholtz, and Barry K. Levitt
Elements of Digital Satellite Communications: System Alternatives, Analyses and Optimization, Volume I
William W Wu
Elements of Digital Satellite Communication: Channel Coding and Integrated Services Digital Satellite Networks, Volume II
William W Wu
Current Advances in Distributed Computing and Communications
Yechiam Yemini
Digital Transmission Systems and Networks, Volume II: Applications
Michael J. Miller and Syed V. Ahamed
Transmission Analysis in Communication Systems, Volume I
Osamu Shimbo
Transmission Analysis in Communication Systems, Volume II
Osamu Shimbo
Spread Spectrum Signal Design: LPE and AJ Systems
David L. Nicholson
Digital Signal Processing Design
Andrew Bateman and Warren Yates
Probability for Engineering with Applications to Reliability
Lavon B. Page
Detection and Estimation
Dimitri Kazakos and P. Papantoni-Kazakos

OTHER WORKS OF INTEREST

Local Area and Multiple Access Networks
Raymond L. Pickholtz, Editor
Telecommunications and the Law: An Anthology
Walter Sapronov, Editor

To the memory of a unique, talented, and loving man, our father Athanasios Papantonis.

To Effie

CONTENTS

Preface ... xi

CHAPTER 1
INTRODUCTION TO DETECTION AND
ESTIMATION ... 1
  1.1 Basic Ideas and Their Physical Interpretations ... 1
  1.2 Applications to Modern Communication and
     Biological Systems ... 7
  1.3 Historical Background and Organization
     of this Book ... 10
     References ... 14

CHAPTER 2
SOME PRINCIPLES OF OPTIMIZATION THEORY ... 21
  2.1 Optimization Under Constraints ... 21
  2.2 Lagrange Multipliers ... 23
  2.3 Kuhn–Tucker Conditions ... 28
  2.4 Saddle-Point Theory ... 33
     References ... 36

CHAPTER 3
PRINCIPLES OF DETECTION AND PARAMETER
ESTIMATION THEORIES ... 37
  3.1 Detection-Theory Concepts ... 37
  3.2 Detection Schemes ... 46

3.3 Concepts in Parameter Estimation	49
3.4 Parameter Estimation Schemes	51
3.5 Applications and Examples	52

CHAPTER 4
BAYESIAN DETECTION — 61

4.1 Introduction	61
4.2 The Optimization Problem	63
4.3 Gram-Schmidt Orthogonalization Method	71
4.4 Continuous-Time Hypotheses	76
4.5 Examples	85
4.6 Appendix	107
4.7 Problems	107
References	111

CHAPTER 5
NEYMAN-PEARSON DETECTION RULE — 112

5.1 Introduction	112
5.2 The Optimization Problem	114
5.3 Uniformly Most Powerful Tests	133
5.4 Locally Most Powerful Tests	140
5.5 Examples and Applications	145
5.6 Problems	150
References	152

CHAPTER 6
MINIMAX AND ROBUST DETECTION — 154

6.1 The Game in Minimax Detection	154
6.2 The Game in Robust Detection	166
6.3 Robust Detection for Classes of Stationary and Memoryless Processes	168
6.4 Examples	185
6.5 Problems	195
References	196

CHAPTER 7
NONPARAMETRIC DETECTION — 198

7.1 Introduction	198
7.2 The Sign Test	200

7.3	Rank Tests	205
7.4	Small-Sample Results	214
7.5	An Asymptotic Comparison	216
7.6	Concluding Remarks	219
7.7	Problems	220
	References	221

CHAPTER 8
SEQUENTIAL DETECTION ... 224

8.1	Introduction	224
8.2	Wald's Procedures	225
8.3	Robust Sequential Tests	233
8.4	Tests for Detecting a Change in Distribution	235
8.5	Examples	238
8.6	Proof of Theorem 8.4.1	246
8.7	Problems	256
	References	258

CHAPTER 9
BAYESIAN AND MINIMAX PARAMETER ESTIMATION ... 260

9.1	Introduction	260
9.2	The Bayesian Optimization Problem	261
9.3	The Linear Mean-Squared Scheme	274
9.4	Unbiased Mean-Squared Estimates	280
9.5	The Minimax Optimization Problem	284
9.6	Lower Bounds in Parameter Estimation	287
9.7	Problems	295
	References	297

CHAPTER 10
MAXIMUM LIKELIHOOD PARAMETER ESTIMATION ... 299

10.1	Introduction	299
10.2	Parametrically Known Stationary and Memoryless Processes	300
10.3	Consistent Location Parameter Estimates	310
10.4	Robust Location Parameter Estimation—Memoryless Processes	318

10.5 Discussion and Examples	328
10.6 Problems	336
References	338

CHAPTER 11
STOCHASTIC DISTANCE MEASURES ... 341

11.1 Introduction	341
11.2 Properties and Relationships	345
11.3 Relation to Statistical Procedures	350
11.4 Explicit Expressions for Gaussian Processes	356
11.5 Applications and Examples	371
11.6 Problems	379
References	381

CHAPTER 12
QUALITATIVE ROBUSTNESS ... 384

12.1 Introduction	384
12.2 Parametric Prediction, Interpolation, and Filtering	388
12.3 Robust Prediction and Interpolation	397
12.4 Robust Filtering	404
12.5 Examples	412
12.6 Appendix	418
12.7 Problems	420
References	423
Index	427

PREFACE

We have written *Detection and Estimation* to fill the need for a systematic and up-to-date presentation of statistical hypothesis testing and estimation schemes. In more than ten years of teaching experience at various universities, we have discovered that the deficiencies of existing books have forced us to teach from class notes and collections of technical papers. This is unsatisfactory for the mathematically competent students who are being introduced to the subject. The existing books, on the other hand, are either outdated (ignoring important concepts such as robust statistics), or are specialized monographs that do not provide sufficient guidance for the researcher who wishes to understand and apply statistical tests. Likewise, the need for advanced statistical procedures is acute in many modern research and development disciplines, such as the analysis of physical and biological systems, the design of high-performance signal-processing algorithms, and the design of advanced communication systems. The application of hypothesis testing statistical schemes has been particularly emphasized lately, due to the evolution of digital systems.

Detection and Estimation would not have been possible without the stimulation and feedback provided by a generation of alert and competent graduate students. The book addresses both those who wish to use existing statistical methods for specific applications and those who intend to carry out further research in statistical communications. We hope to provide the latter with sufficient initial resources and stimulating conceptual skills. We have tried to present the basic concepts systematically, have provided relatively complete bibliography, and have reserved the later chapters for the presentation of more advanced concepts. The first five chapters and Chapter 9 can be included in a first-semester graduate course. The remaining chapters may comprise the material for a second-semester graduate course or seminar. Solid foundation on probability theory and stochastic processes is prerequisite.

We wish to acknowledge the professional and efficient typing work by Ms. Susan Warren and Mrs. Sharon Smalley. Some numerical computations were performed by Dr. Michael Georgiopoulos. The support of the U.S. Air Force Office of Scientific Research and the Signal Processing Branch at the Rome Air Development Center are gratefully acknowledged. In particular, we wish to acknowledge Dr. Vincent C. Vannicola and Mr. Michael C. Wicks of the Rome Air Development Center for the stimulating discussions and interactions which provided us with engineering intuition on signal processing and radar problems for our research on the subject.

Dimitri Kazakos
P. Papantoni-Kazakos

CHAPTER 1

INTRODUCTION TO DETECTION AND ESTIMATION

This chapter aims to provide a basis for a comprehensive and consistent study of the theories on statistical detection and estimation. First, the basic ideas are presented and interpreted. Then, some applications of the theories are qualitatively discussed. Finally, a historical background is provided.

1.1 BASIC IDEAS AND THEIR PHYSICAL INTERPRETATIONS

The complexity of physical phenomena necessitates the consideration of probabilistic models. A probabilistic or stochastic description usually models the effect of causes whose origin and nature are either unknown or too complex to describe deterministically. Thus, a stochastically modelled physical phenomenon is not necessarily nondeterministic by nature; its stochastic description may merely represent the best known model for its behavior. The best model of a physical phenomenon is the simplest existing model that describes the phenomenon with satisfactory accuracy, with the emphasis on simplicity.

The simplest form of a stochastic model is represented by a scalar random variable $X = X(\omega)$, where X represents the name or identity of the variable and ω represents elements on the probability space. For simplicity, ω is usually deleted from the notation, and the random variable X is fully described by its probability distribution, $F_X(x) = \Pr\{X \leq x\}$. The next simplest stochastic model is represented by a vector random variable, $X^n = [X_1, \ldots, X_n]^T$, where X_i is a scalar random variable for every i, and T means transpose. The vector random variable, X^n, is fully described by the joint probability distribution, $F_{X^n}(x^n) = \Pr\{X^n \leq x^n\}$, where x^n represents vector values. The most general form of a stochastic model is represented by a stochastic process $X(t) = X(t, \omega)$, where X is the name of the process, ω represents an element in the probability space, and t is time. For every fixed time instant t_i, $X(t_i) = X(t_i, \omega)$ is a scalar random variable. A stochastic process is completely described by the probability distributions $F_{X^n}(x^n; t_1, \ldots, t_n) = \Pr\{X(t_i) \leq x_i; 1 \leq i \leq n\}$ for every n and every set $\{t_i\}$ of time instants, where $x^n = [x_1, \ldots, x_n]^T$. The stochastic process $X(t)$ is *stationary*, if the property $F_{X^n}(x^n; t_1, \ldots, t_n) = F_{X^n}(x^n; t_1 + \tau, t_2 + \tau, \ldots, t_n + \tau)$ holds for every n, every set $\{t_i\}$ of time instants, and every τ value. The stochastic process is *discrete-time* if it is described only by its realizations on a countable set $\{t_i\}$ of time instants. Then, time is counted by the indices i in $\{t_i\}$ and we can denote $X_i^{n+i} = [X_i, X_{i+1}, \ldots, X_{n+i}]^T$. The discrete-time process is *stationary* if $F_{X_i^{n+i}}(x^n) = F_{X_j^{n+j}}(x^n)$ for every n, i, and j, and it is *memoryless* if it generates independent random variables in time; that is, if $F_{X^n}(x^n) = \prod_{i=1}^n F_{X_i}(x_i)$. The process is memoryless and stationary if $F_{X^n}(x^n) = \prod_{i=1}^n F_{X_i}(x_i)$ and $F_{X_i}(x) = F_{X_j}(x) = F_X(x)$ for every i, j, and x. Therefore, a memoryless and stationary discrete-time process generates in time independent and identically distributed (i.i.d.) random variables.

It should be clear from the above that any stochastic model can be represented by a stochastic process. Indeed, the scalar and vector random variable representations can be considered as the fixed-in-time restrictions of a stochastic process. In our general discussion, we will denote an arbitrary stochastic process $X(t)$. By $x(t)$ we will then denote a realization of the process; that is, an observed waveform generated by $X(t)$. If the stochastic process is discrete-time, $X(t)$ actually represents a sequence $X_1, X_2, \ldots$ of random variables, and $x(t)$ represents a sequence $x_1, x_2, \ldots$ of observed scalar values. It is said that the stochastic process $X(t)$ is *well known*, if the distribution $F_{X^n}(x^n; t_1, \ldots, t_n)$ is precisely known for all n, every set $\{t_i\}$, and every vector value x^n. The process is, instead, *parametrically known*, if there exists a vector parameter θ^m of finite dimensionality m such that the conditional distribution $F_{X^n}(x^n; t_1, \ldots, t_n | \theta^m)$ is precisely known for all n, $\{t_i\}$, x^n, and given vector value θ^m. The stochastic process, $X(t)$, is *nonparametrically described*, if there is no vector parameter θ^m of finite dimensionality such that the distribution $F_{X^n}(x^n; t_1, \ldots, t_n | \theta^m)$ is completely described for every given vector value θ^m and for all n, $\{t_i\}$, and x^n. We note that in the definition of the parametrically known processes, some components of the vector parameter θ^m may be deterministic functions with known support. Such a function may be the spectral density of a discrete-time stationary process, whose support is then the interval $[-\pi, \pi]$. As an example, assume that the stochastic process $X(t)$ is stationary and discrete-time, and that the random variables $\{X_i\}$ that the process generates have finite variance. Given only these facts, the process $X(t)$ is *nonparametrically described*, and it actually represents a whole class of stochastic processes. Now assume that the process $X(t)$ is also Gaussian. At this point, the process is *parametrically known*, since only its mean and spectral density functions are needed for its full description. When the latter two quantities are also provided, the process becomes *well known*; that is, it is Gaussian with known mean and spectral density functions.

The analyst who observes the behavior of a physical phenomenon searches for the stochastic process that best describes the phenomenon, assuming that deterministic modeling has been rejected. The observed behavior of the phenomenon is then represented by a realization $x(t)$ of the underlying stochastic

1.1 Basic Ideas and Their Physical Interpretations

process in some time interval $[0, T]$. The laws that govern this search as well as the search's outcome are the objectives of *statistical decision theory*. The analyst, a term that may also mean an automated system here, performs his search and subsequently makes a decision, using the following:

i. A realization $x(t)$ from the underlying stochastic process on some observation time interval $[0, T]$, where T may be controlled dynamically by the analyst.
ii. Careful experimental control to assure that the total observed outcome $x(t)$ on $[0, T]$ represents a realization from the same physical phenomenon or stochastic process, rather than a mixture of partial outcomes from different processes.
iii. A library of stochastic processes, with either finite or infinite membership, in which each stochastic process describes a different physical phenomenon.
iv. A performance criterion for evaluating decisions.
v. Possibly a probability measure determining the a priori probability with which each of the stochastic processes in the library of processes may occur.

Using these assets, the analyst formulates a generalized optimization problem; the solution of this problem is his decision. The nature of the optimization problem and the subsequent decision vary significantly with the specifics of the library of stochastic processes available, whether a known a priori probability measure on those processes exists or not, the performance criterion used, and the possibility of controlling the observation time interval $[0, T]$ dynamically. For any fixed specifications on the above issues, the decision also depends on the observed realization $x(t); t \in [0, T]$ from the underlying process. As a function of this observed realization, the decision is called a *decision rule* or test.

If the library of stochastic processes available has a finite number of members, the decision process is classified as *detection* or *hypothesis testing*. The analyst must decide in favor of one of the stochastic processes in the library; that is, he decides among a finite number of alternatives. If those alternatives are only two, then the problem is also called *single hypothesis testing*. If the alternatives or hypotheses in the detection or hypothesis testing problem are represented by well known processes, the *parametric detection* problem evolves. If, instead, the alternatives are represented by nonparametrically described processes, the *nonparametric detection* problem appears. In both the parametric and the nonparametric detection problems, the optimization problem that leads to the decision rule can take a number of forms. Those forms change as the number of assets available to the analyst decreases, and as their specifics change. We will start with the fullest possible set of assets. Let us suppose that an a priori probability measure of a finite number of alternatives is given. Then the class of Bayesian detection formulations and solutions appear. If a specific cost function that penalizes wrong decisions is provided, minimization of the expected penalty is chosen as the performance criterion for the design of the decision rule. If such a cost function is not provided, the probability of making a decision error is minimized instead. The resulting test is called an *ideal observer test*. Now suppose the a priori probability measure is unavailable. Then, if a specific cost function is provided, a least favorable a priori probability is found and the class of minimax decision rules evolves. If such a cost function is not provided and the single hypothesis problem is present, a constraint optimization problem is formulated. One of the two hypotheses is selected in advance as most important, and the performance criterion becomes maximization of the probability that this hypothesis is detected correctly, under the constraint that the probability that a wrong decision in its favor is made does not exceed a given value. The resulting test is known as the *Neyman-Pearson* test. All the above tests take a dynamic form if the observation time interval $[0, T]$ can be controlled dynamically by the analyst. The analyst then observes the realization $x(t)$ from the underlying process for as long as seems necessary, in order to make a relatively reliable decision. Thus, he decides in favor of some hypothesis after he has decided dynamically on the appropriate observation interval $[0, T]$. The resulting tests are called *sequential*.

Now suppose the library of stochastic processes available to the analyst has an infinite membership. A special case of such a library is represented by a parametrically known stochastic process $X(t)$. Then there exists a parameter vector θ^m of finite dimensionality such that the stochastic process is well known when parametrized by a given fixed value of θ^m. The infinite membership library of stochastic processes is generated, in this case, by letting the value of θ^m vary,

Table 1.1.1
Classes of Detection or Hypothesis Testing Schemes

		Nature of Hypotheses and Observation Interval			
	Available assets	Well known hypotheses		Nonparametrically determined hypotheses	Controllable $[0, T]$
A priori probability available	Cost function given	Bayesian tests	Minimization of expected penalty	Nonparametric tests	Sequential tests
A priori probability available	No cost function	Bayesian tests	Minimization of error probability Ideal observer test	Nonparametric tests	Sequential tests
No a priori probability available	Cost function given		Minimax rules	Nonparametric tests	Sequential tests
No a priori probability available	No cost function		For single hypothesis testing: Neyman-Pearson test	Nonparametric tests	Sequential tests

1.1 Basic Ideas and Their Physical Interpretations

assuming that this parameter vector can take infinitely many values. The decision then corresponds to selecting a single value for the parameter vector θ^m, and the *parametric parameter estimation* problem arises. The *nonparametric parameter estimation* problem evolves when the process parametrized by a given fixed vector parameter value is nonparametrically described. In both the parametric and the nonparametric parameter estimation procedures, the optimization problem that leads to the decision rule or test can take different forms. If an a priori probability measure on the vector parameter θ^m is given, then, as in hypothesis testing, the *Bayesian tests* appear. There, either an expected cost is minimized (if a cost function is available) or the probability of correct decision is maximized, with the possible inclusion of a constraint on the induced bias. If no a priori probability measure on the parameter vector θ^m is given, the *minimax and maximum likelihood tests* can be formalized. The first appears when a cost function is given and it develops as in the case of hypothesis testing. The latter is used when no cost function is given; it decides in favor of the value of θ^m that makes the observed process realization $x(t); t \in [0, T]$ most probable. As in hypothesis testing, *sequential parameter estimation tests* can be developed if the observation time interval $[0, T]$ can be controlled by the analyst.

The hypothesis testing and parameter estimation problems posed above require the preexistence of a library of stochastic processes among which the analyst makes a selection. If such a library does not exist, a different problem arises. The analyst must then decide on an appropriate stochastic model for

Table 1.1.2
Classes of Parameter Estimation Schemes

	Available assets	Nature of Library of Processes and Observation Interval		
		Parametrically known library	Nonparametrically described library	Controllable $[0, T]$
Parameter a priori probability available	Cost function given	Minimization of expected penalty (Bayesian tests)	Nonparameteric tests	Sequential tests
	No cost function	Maximization of Probability of correct decision (Bayesian tests)		
Parameter a priori probability unavailable	Cost function given	Minimax tests		
	No cost function	Maximum likelihood tests		

the corresponding physical phenomenon, based solely on the observation of its behavior. He can do that only if he adopts certain assumptions regarding stationarity and ergodicity. He can then make probabilistic inferences based on histograms. The study of this problem is not within the objectives of this book; some of its implications are important, however, for understanding the importance of nonparametric decision rules. In the process of making probabilistic inferences from histograms, the largest errors induced correspond to observation values whose probability of occurrence is small. These values correspond to the "tails" of the underlying distributions and rarely appear. Thus, reliable probabilistic inferences about such "distribution tails" cannot be derived. So, in practice, the developed stochastic models are described subject to uncertainty regarding distribution tails when the library of stochastic processes used in the hypothesis testing and parameter estimation problems consists of stochastic models so developed. This fact, in conjunction with the established observation that test performance may deteriorate significantly if the distribution tails are conjectured ad hoc, led to the consideration of nonparametric techniques. Such techniques are based on the assumption that each stochastic model considered is described by a class of well known stochastic processes rather than a single well known stochastic process. A special case of such a class represents uncertainty in distribution tails; its description consists of a well known stochastic process that is contaminated to a small degree by some unknown stochastic process. In general, statistical contamination of small degree is represented by a sphere of stochastic processes of relatively small radius, with a well known stochastic process as the center; the radius is measured by an appropriate stochastic distance on the space of stochastic processes. Statistical procedures that address the problem of small statistical contamination have been called *robust* rather than nonparametric. There are two reasons for this choice, one qualitative and one historical. Qualitatively speaking, a robust statistical procedure is characterized by robust performance—stable performance whether or not small statistical contamination is present. Historically, on the other hand, statistical contamination of large degree was studied first. The objective was a guaranteed minimum level of performance when very little is known about the underlying stochastic process, such as the existence of some symmetries in the induced distributions and bounded variances. The resulting statistical procedures are called *nonparametric*, in general they do not guarantee stable performance in the presence or absence of small statistical contamination. So, although both the nonparametric and robust statistical procedures are concerned with nonparametrically described stochastic models, they adopt different objectives. From another viewpoint, the nonparametric procedures are designed for nonparametric classes of stochastic processes that are much larger than the corresponding classes for the robust procedures.

In the development of all the tests we have discussed to this point, it is very crucial that the observed realization $x(t); t \in [0, T]$ be generated 'y a single stochastic process, even if the process is not well known. The nonparametric models used, then, reflect the analyst's uncertainty as to the exact description of the process acting; the process itself is still unique. We will now pose a different problem, which is related to the single hypothesis testing problem. Let it be known that a given stochastic process is acting when the analyst starts observing some realization $x(t)$. Let it be possible that at some point in time another given stochastic process may start acting instead. Then the objective of the analyst is to decide if this change has occurred. Since both the actual occurrence of the change as well as the timing are uncertain, an observation time interval $[0, T]$ determined a priori is meaningless. Instead, the realization $x(t)$ should be observed and processed sequentially in time in order to develop a decision rule. The resulting sequential tests are called *shift detection tests* or *quality control tests*. For historical reasons, the latter is the term most frequently used—these tests were originally developed for industrial control of product quality. If the two given stochastic processes are well known, the shift detection or quality control tests are parametric. If, instead, at least one of the two given processes is nonparametrically described, the tests are nonparametric. In both cases, the performance criterion used is minimization of the time delay in the detection of an occurred change, subject to a false alarm constraint. The resulting tests are the solution of a constraint optimization problem.

Let us go back to the case where the observed realization $x(t); t \in [0, T]$ is generated by a single stochastic process $X(t)$. Let it be known that the process $X(t)$ is the sum of two processes $Y(t)$ and $Z(t)$ whose realizations are respectively denoted $y(t)$ and $z(t)$. Then $x(t) = y(t) + z(t); t \in [0, T]$. Assuming that the stochastic processes $Y(t)$ and $Z(t)$ and their mutual statistical relationship are known to the analyst, and given the observed realization $x(t); t \in [0, T]$ and some performance criterion, let the problem be to estimate the realization $y(t); t \in [0, T]$ of the process $Y(t)$. This problem is known as filtering since the realization $y(t)$ is filtered out from the observed realization $x(t)$. The answer to the filtering problem is the solution of an optimization problem, whose specifics depend on the performance criterion used and on the mutual characteristics of the processes $Y(t)$ and $Z(t)$. When the stochastic processes $Y(t)$ and $Z(t)$ are well known, the *parametric filtering* problem arises. When those processes are nonparametrically described, the *nonparametric filtering* formalization appears.

In this book, we will study the hypothesis testing, parameter estimation, shift detection, and filtering problems. We will cover the more classical parametric versions of those problems as well as their robust and nonparametric forms. Since statistical tests are the solutions of certain optimization problems, we will include some principles in optimization theory. We will also include some background on stochastic distance measures; they are needed in the development and analysis of robust statistical procedures as well as in the performance evaluation of parametric statistical schemes.

1.2 APPLICATIONS TO MODERN COMMUNICATION AND BIOLOGICAL SYSTEMS

In processing communication and biological data, decisions are made at various stages; in fact, the term *processing* itself implies decision making. Such decisions include recognition of important characteristics, estimation of representative parameters, model identification, and detection of model changes. The current

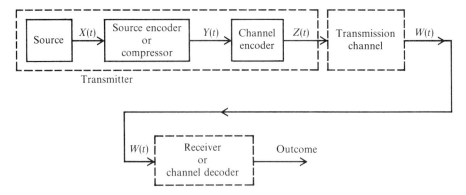

Figure 1.2.1 Single-link communication system

revolution in technology has led to the discovery of new and highly complex biological phenomena and to the design of highly sophisticated and complex communication systems. This has made the need for implementing advanced decision making procedures more acute. Procedures that are characterized by high performance are in great demand; the new high-memory and high-speed computers respond very effectively to the required operational complexity. Not only can high performance be afforded in modern times, it is also necessary, in response to the ever-increasing demand for service quality. Modern communication and biological systems require that decisions of high accuracy be made in the presence of disadvantageous conditions.

We will first present some applications of statistical decision theory to communication systems. In Figure 1.2.1, the block diagram of a single-link communication system is shown. The various blocks represent important operations. The specifics of each operation vary with the objectives of the communication system, and the characteristics of the transmission channel and the source.

The source in Figure 1.2.1 generates a waveform $X(t)$. This waveform may be continuous or discrete-time and deterministic or stochastic, depending on the nature of the source. The source encoder or compressor may or may not exist; both its existence and operational characteristics are determined by the objective of the system and the source and channel characteristics. If the source encoder exists, its objective is the extraction of characteristics relevant to the system source; this extraction is the result of a decision-theory formalization. The output from the source encoder waveform $Y(t)$ may be discrete-valued or digital, even if the source waveform $X(t)$ is continuous-valued or analog; indeed, the source encoder may perform a quantization operation. If the source waveform $X(t)$ is analog and the transmission channel is digital, a quantization or analog-to-digital source encoding operation is necessary. The channel encoder takes into consideration the nature of the transmission channel and the objective of the communication system, to perform an appropriate transformation of the waveform $Y(t)$. The objective of this transformation is to facilitate the operation performed by the receiver. If, for example, the transmission channel is digital, then the waveform $Y(t)$ is also digital, the objective of the receiver is recognition

1.2 Applications to Modern Communication and Biological Systems

of the digital sequences that represent $Y(t)$, and the channel encoder realizes an *error correcting code*. If, on the other hand, the transmission channel is represented by an analog stochastic process that is additive to the channel input $Z(t)$, then a *modulation operation* may also be included in the functions performed by the channel encoder. In every case, the receiver or channel decoder is represented by a decision-theory operation. If the transmission channel is digital, the source encoder generates one of a finite number of digital messages well known to the receiver, and the objective of the system is recognition of those messages, then the receiver performs a *hypothesis testing* or *detection* operation. If the messages are also equally probable and the digital channel is binary, memoryless, and symmetric, then the *ideal observer hypothesis test* is adopted; its form becomes then the familiar *minimum distance decoding* scheme. If the objective of the communication system is, instead, the identification of a vector parameter θ^m that characterizes the source, and if this vector parameter can take an infinite number of values, then the receiver is a *parameter estimator*. Finally, if the source is well known but not always active, the channel is represented by an analog and additive stochastic process $N(t)$, and the objective of the system is distinction between an active versus a nonactive source, then the *radar problem* arises. The Radar problem is a single hypothesis testing problem, where the decision consists of $\alpha = 0$ versus $\alpha = 1$ from the observed waveform

$$W(t) = \alpha Z(t) + N(t) \qquad t \in [0, T] \qquad (1.2.1)$$

Modern complex communication systems usually include several decision-theory operations under rigid and demanding conditions, as in the case of spread spectrum communication systems. There, low power signals are transmitted, for protection against intelligent adversaries. Those low power signals must be effectively recognized in the presence of noise and interference from other transmissions. For guaranteed high performance, the inclusion of sophisticated source and channel encoding schemes is then necessary. Both error correcting codes and frequency modulation procedures are incorporated. The signal recognition operations include hypothesis testing schemes for data decoding, as well as parameter estimation operations for phase identification (Holmes 1982).

The design of appropriate decision-theory operations becomes even more challenging when several single-link communication systems and computers are united in an interactive fashion. Then, a computer-communication network evolves. Such a network may include satellite and radio transmissions as well as computer operations, it may operate on realizations from various types of sources, and it may require different performance characteristics for different operations. For guaranteed performance of the network, automated performance monitoring techniques must be deployed. The objective of those techniques is the reliable and quick identification of deteriorating or faulty network elements, such as transmission channels, receivers, encoders, etc. The performance monitoring techniques are basically quality control tests that operate on appropriately selected network data and incorporate in their operation knowledge of the operational characteristics of the network elements [Papantoni-Kazakos (1979a and 1979b)]. Some basic concepts in communication systems that involve decision theory can be found in Van Trees (1968) and Wozencraft and Jacobs (1965).

In biological systems, cyclic phenomena are commonly observed. Such phenomena, which range from reaction enzyme cycles to predator-prey cycles, are known as *biological oscillators*. In the last twenty years they have been studied both theoretically and experimentally (Pavlidis 1974, Boiteux et al 1977). *Point biological oscillators* arise when the time evolution of lumped phenomena is studied at discrete points in space. Then, modeling via difference or differential equations is both relevant and useful. The presence of contaminating noise processes in the observed data transforms the model into an autoregressive process. In one-dimensional discrete-time form, the autoregressive process is given by

$$X_n = \sum_{k=1}^{m} \alpha_k X_{n-k} + W_n \qquad (1.2.2)$$

In (1.2.2), $\{\alpha_k\}$ are the autoregressive parameters, m is the order of the autoregressive model, and W_n is the noise or driving stochastic process. The parameter set $\{\alpha_k\}$ is considered as the model characteristic, and the objective of the analyst is the estimation of $\{\alpha_k\}$ from a number of observations. Thus, the problem is parameter estimation, where the test to be used depends on the assumed knowledge about the stochastic process W_n. If the process W_n is assumed to be well known, a parametric test evolves. If the process W_n is instead assumed to be nonparametrically described, a robust or nonparametric test is appropriate. Finally, if a transition from one model (or model class) to another is possible and the occurrence of such a transition is monitored, then a shift detection test is adopted and applied.

In this section, we have described only very few of the numerous applications of decision theory. Many more applications appear, not only in communication and biological systems, but also in the modeling and study of sociological and economic systems, in production studies, in the design of transit systems, and so on.

1.3 HISTORICAL BACKGROUND AND ORGANIZATION OF THIS BOOK

The roots of statistical decision theory are old and deep. Some of the oldest known problems included in its evolution are the needle problem, problems connected with card games, the Petersburg problem, and problems arising from the Mendelian laws of heredity. Those problems can be found in the *Encyklopädie der Mathematischen Wissenschaften*; additional material on card problems is provided by Tietze (1943). The elements of mathematical abstraction, counting, and calculating—essential in the development of statistical theory—were introduced by Aristotle (384–322 B.C.). More concrete ties with modern statistical decision theory can be traced in the work of Carl Friedrich Gauss (1777–1855)

1.3 Historical Background and Organization of this Book

and Henri Poincaré (1854–1912). Poincaré's recurrence theorem is presented by Wintner (1941); it provides the seed for the development of the ergodic theory so extensively used in the derivation of decison rules.

The 20th century can be considered to be the Golden Age of statistical decision theory. Early in the century, some optimal decision-theory formalizations were presented and analyzed, and methods for data analysis were studied. Fisher (1920) studied methods for determining confidence in observations; Neyman (1935, 1952) and Neyman and Pearson (1933, 1936, 1938) studied statistical sufficiency, statistical efficiency, and statistical bias for hypothesis testing; Cramér (1946) presented a mathematically optimal formalization for several statistical problems; Wald (1947) and Wald and Wolfowitz (1948) presented and analyzed an optimal formalization for sequential hypothesis testing; Blackwell (1947) and Savage (1947) studied sequential and unbiased parameter estimation tests; Pitman (1939) and Rao (1945) studied methods for the parametric estimation of a location parameter and analyzed the induced performance; Halmos and Savage (1948) studied sufficient statistics in relationship to the Randon-Nikodym theorem; Arrow et al (1949) studied sequential hypothesis tests with Bayesian and minimax formalizations; Fix (1949) provided tables for noncentral χ^2 distributions; Von Mises (1947) studied the asymptotic distribution of certain statistical functions; and Hoel et al (1949) studied the problem of optimal classification.

In the early work on statistical decision theory, the need to use concepts and methods from advanced probability theory and the strong relationship between decision theory and the theories of games and functions were recognized. The fundamental principles of advanced probability theory are presented by Kolmogorov (1956) and Loéve (1963). In addition to the fundamental principles, Cramér (1955), Feller (1966), Parzen (1960), and Tucker (1962), also included applications of probability theory and relationships to mathematical statistics. Strassen (1965) studied the existence of probability measures with given marginals. The fundamental role that the theories of functions and games play in several important problems was also recognized early in the 20th century. Jensen (1906) studied the importance of convexity in the presence of inequality constraints; Von Neumann et al (1944) studied the application of game theory to the analysis of economic behavior; Luce et al (1957) studied the decisions that evolve from certain games; and Karlin (1959) studied the relationships among mathematical methods for games, programming, and economics. The relationships between statistics and the theories of functions and games were first studied by Wald (1950), who first studied the behavior of certain statistical decision functions, and then presented randomized statistics from a game-theory viewpoint (1951). Dvoretzky et al (1951) studied the relationship between zero-sum two person games and the elimination of randomization in hypothesis testing; Blackwell et al (1954) presented a study that related the statistical decision and game theories; and Daniels (1954) studied some applications of saddle-point games and approximations in statistics.

The decade between 1950 and 1960 has been very fruitful for statistical decision theory. Here we will only cover some indicative advances, since it is impossible to cover all the contributions. We will attempt more complete coverage of the literature as we proceed with each different topic in this book. From 1950 to 1960, many early formalizations were further studied and further advanced. Pearson et al (1951) provided charts of power functions for test analysis; Scheffé (1959) analyzed variance; Chernoff et al (1959) and Lehman (1959) provided comprehensive presentations of statistical theory; Quenouille (1956) studied the presence of bias in parameter estimation; Agarwal (1955), Blyth (1951), Kiefer (1957), and Wolfowitz (1950) elaborated on minimax procedures; Karlin et al (1956), and Le Cam (1953) studied Bayesian decision rules; Bahadur (1954, 1958), Blackwell (1951), De Groot (1959), Le Cam (1955) and Fix et al (1959) studied important properties for various decision rules; Page (1954, 1955) introduced the quality control sequential tests; and Tuckey (1953) presented some empirical statistical tests. But perhaps the most significant advancement in the decade from 1950–1960 was the introduction and study of nonparametric tests. The first documented study on those tests was provided by Hoeffding (1951), and additional studies were carried on by Chernoff et al (1958), Fraser (1957), Hodges et al (1956), Siegel (1956), and Stein (1956).

The 1960s started a new direction in statistical decision theory. Robust statistics were formally introduced by Huber (1964, 1965, 1968, 1969). Huber's minimax formalization of robustness was not truly carried on by other investigators until the early 70s. Hampel (1971) presented a qualitative study of robustness for memoryless and stationary processes. Later, Papantoni-Kazakos and Gray (1979) extended Hampel's qualitative robustness to include stationary processes with memory. Some partial surveys of robust statistical procedures are given by Andrews (1972) and Huber (1972). A more complete coverage of statistical robustness is provided by Huber (1981). In addition, Beran (1977) studied robust estimates for location parameters, Hampel (1976) analyzed the breakdown point for such estimates, and Dempster (1975) presented a subjectivist's view point of robustness. In parallel, the studies of well-established statistical procedures have continued since 1960. To mention only a few, Romanowski et al (1965) studied some applications of modified normal distributions; Ferguson (1967) presented a theory for mathematical statistics; Shafer (1976) provided a mathematical theory of evidence; Lorden (1971) studied and analyzed an optimal formalization for Page's quality control tests; Daniel et al (1971) provided methods for fitting equations to data; Mosteller et al (1977) and Tuckey (1977) presented empirical methods for data analysis, Stone (1975) studied parametric adaptive estimates for a location parameter; James et al (1961) studied Bayesian parametric estimates for a location parameter; Chernoff et al (1967) and Beran (1974) studied the asymptotic properties of Rank nonparametric statistical tests; Hájek et al (1967) presented a theory for Rank nonparametric tests; and Miller (1964, 1974) studied the method of jackknifing in nonparametric statistics. In sequential tests, optimality and admissibility aspects were studied by Burkholder et al (1963), Chernoff (1972), and Matthens (1963); stopping rules were studied by Chow et al (1965a, 1965b); and bounds on the sample size needed for decision were provided by Hoeffding (1960) and Ray (1965). Finally, Harding and Kendall (1974) presented a theory on stochastic geometry.

1.3 Historical Background and Organization of this Book

The spectacular advances in statistics have been parallel to and frequently stimulated by the development of communication technology. The demonstration of telegraphy by Joseph Henry in 1832 and by Samuel F. B. Morse in 1838, the telephone, patented by Alexander Graham Bell in 1876, and the invention of the wave filter by G. A. Campell in 1917 were some of the major initial discoveries that led to the revolution of the communication technology and to the formulation of statistical communication theory. The roots of the theory can be traced to the beginning of our century, when the concept of information transmission was first presented by Hartley (1928). Hartley's paper was followed by the pioneering studies of Rice (1945) on the analysis of noise, Shannon (1948, 1949) and Shannon and Weaver (1949) on the development of a theory for transmission of information, and Wiener (1949) on the mathematical formalization for filtering and smoothing. By 1960 the basic principles of statistical communication theory were well established; the reader may refer to the books by Kotel'nikov (1959), Schwartz (1959), Helstrom (1960), Middleton (1960), and Fano (1961). Since then, the theory has further developed to include sophisticated source and channel encoding schemes, statistical techniques in signal processing and spread spectrum, and techniques for multiuser transmission of information in computer-communication networks.

The contributors to the development of statistics and statistical communication theory have been numerous and the existing advances are many. In this book, we will make an attempt to present some of the important advances methodically and comprehensively, and to give credit to as many of the contributors as possible. To cover all the existing advances would be a life-long task and would require the space of several books. To give credit to all the contributors is likewise impractical. Scientific accomplishments are the result of collective efforts in our times; they can not be attributed to a few, even if only a few are given the credit.

The organizational flow in the book does not necessarily follow the historical evolution of statistical decision theory. Instead, statistical schemes are presented according to their model characteristics, richness in available information, and conceptual complexity. Since statistical tests are the solutions of certain optimization or game formalizations, some principles on optimization theory and game theory are first provided. Then, the problem of distinguishing among a finite number of models—that is, the hypothesis testing or detection problem—is studied. The hypothesis testing problem is formalized and its solution is found as the information available to the analyst decreases. First to be studied are the parametric Bayesian detection schemes where the models are well known, both an a priori probability measure on the different models and a cost function are given, and the observation interval is not controlled by the analyst. Then, the Neyman-Pearson and the minimax detection schemes are presented. In both, the observation interval is fixed, the models are well known, and no a priori probability measure on the models is given. In the first, no cost function is available either; in the second such a cost function is provided. Chapter 6 includes the robust detection schemes in which the models are nonparametrically described, each representing a relatively small class of stochastic processes. Chapter 7 considers models that represent a large, nonparametrically described class of stochastic processes. Finally, Chapter 8 presents sequential detection procedures in which the observation interval is controlled dynamically by the analyst. Some advanced concepts for the performance evaluation of detection schemes are included in Chapter 11.

We introduce the problem of parameter estimation in Chapter 3. As in detection, we start with parametric Bayesian estimation schemes in which the model is parametrically known, a cost function is available, and an a priori probability measure on the parameter is given. When the model is parametrically known, but neither a cost function nor an a priori probability measure on the parameter are provided, the maximum likelihood parameter estimation schemes arise. Those schemes are studied in Chapter 10. When the model is nonparametrically described, robust parameter estimation is necessary; its qualitative formalization requires the use of stochastic distance measures. Chapter 11 presents some stochastic distance measures that are useful in the development and evaluation of both detection and parameter estimation procedures. In Chapter 12, qualitative robustness is presented and analyzed, and the important aspects of parametric and robust filtering are studied.

REFERENCES

Agarwal, O. P. (1955), "Some Minimax Invariant Procedures for Estimating a Cumulative Distribution Function," *Ann. Math. Statist.* 26, 450–463.

Andrews, D. F. et al (1972), *Robust Estimates of Location: Survey and Advances*, Princeton University Press, Princeton, NJ.

Arrow, K. J., D. Blackwell, and M. A. Girshick (1949), "Bayes and Minimax Solutions of Sequential Decision Problems," *Econometrica* 17, 213–244.

Bahadur, R. R., (1954), "Sufficiency and Statistical Decision Functions," *Ann. Math. Statist.* 25, 423–462.

——— (1958), "A Note on the Fundamental Identity of Sequential Analysis," *Ann. Math. Statist.* 29, 534–543.

Beran, R. (1974), "Asymptotically Efficient Adaptive Rank Estimates in Location Models," *Ann. Math. Statist.* 2, 63–74.

——— (1977), "Robust Location Estimates," *Ann. Math. Statist.* 5, 431–444.

Blackwell, D. (1947), "Conditional Expectation and Unbiased Sequential Estimation," *Ann. Math. Statist.* 18, 105–110.

——— (1951), "On the Translation Parameter Problem for Discrete Variables," *Ann. Math. Statist.* 22, 393–399.

Blackwell, D. and M. A. Girshick (1954), *Theory of Games and Statistical Decisions*, Wiley, New York.

References

Blyth, C. R. (1951), "On Minimax Statistical Decision Procedures and Their Admissibility," *Ann. Math. Statist.* 22, 22–42.

Boiteux et al (1977), "Oscillatory Phenomena in Biological Systems," *EMBO Proceedings* 75, 1–4.

Burkholder, D. L. and R. A. Wijsman (1963), "Optimum Properties and Admissibility of Sequential Tests," *Ann. Math. Statist.* 34, 1–17.

Chernoff, H. (1972), Sequential Analysis and Optimal Design, Regional Conference Series in Applied Mathematics, Society for Industrial and Applied Math., Philadelphia, PA.

Chernoff, H. and L. E. Moses (1959), *Elementary Decision Theory*, Wiley, New York.

Chernoff, H., J. L. Gastwirth, and M. V. Johns (1967), "Asymptotic Distribution of Linear Combinations of Functions of Order Statistics with Applications to Estimation," *Ann. Math. Statist.* 38, 52–72.

Chernoff, H. and I. R. Savage (1958), "Asymptotic Normality and Efficiency of Certain Nonparametric Test Statistics," *Ann. Math. Statist.* 20, 972–994.

Chow, Y. S. and H. Robbins (1965a), "On Optimal Stopping Rules for S_n/n," *Illinois J. Math.* 9, 444–454.

Chow, Y. S., H. Robbins, and H. Teicher (1965b), "Moments of Randomly Stoped Sums," *Ann. Math. Statist.* 36, 789–799.

Cramér, H. (1955), *The Elements of Probability Theory and Some of Its Applications*, Wiley, New York.

——— (1946), *Mathematical Methods of Statistics*, Princeton Univ. Press, Princeton, NJ.

Daniel, C. and F. S. Wood (1971), *Fitting Equations to Data*, Wiley, New York.

Daniels, H. E. (1954), "Saddle Point Approximations in Statistics," *Ann. Math. Statist.* 25, 631–650.

De Groot, M. H. (1959), "Unbiased Sequential Estimation for Binomial Populations," *Ann. Math. Statist.* 30, 80–101.

Dempster, A. P. (1975), "A Subjectivist Look at Robustness," Proc. 40th Session I.S.I., Warsaw, *Bull. Int. Statist. Inst.* 46, Book 1, 349–374.

Dvoretzky, A. A. Wald, and J. Wolfowitz (1951), "Elimination of Randomization in Certain Statistical Decision Procedures and Zero-Sum Two-Person Games," *Ann. Math. Statist.* 22, 1–21.

Encyklopädie der Mathematischen Wissenschaften, Vol. 1, Part 2, 754–765 and 1090.

Fano, R. M. (1961), *The Transmission of Information*, MIT Press and Wiley, New York.

Feller, W. (1966), *An Introduction to Probability Theory and its Applications*, Vol. II, Wiley, New York.

Ferguson, T. S. (1967), *Mathematical Statistics: A Decision Theoretic Approach*, Academic Press, New York.

Fisher, R. A. (1920), "A Mathematical Examination of the Methods of Determining the Accuracy of an Observation by the Mean Error and the Mean Square Error," *Monthly Not. Roy. Astron. Soc.* 80, 758–770.

Fix, E. (1949), "Tables of the Non-Central χ^2," *Univ. of Calif. Pub. in Statist.* 1, 15–19.

Fix, E., J. L. Hodges, and E. L. Lehmann (1959), "The Restricted χ^2 Test," in *Studies in Probability and Statistics Dedicated to Harold Cramér*. Almquist and Wiksell, Stockholm.

Fraser, D. A. S. (1957), *Nonparametric Methods in Statistics*, Wiley, New York.

Hájek, J. and Z. Šidák (1967), *Theory of Rank Tests*, Academic Press, New York.

Halmos, P. R. and L. J. Savage (1948), "Application of the Radon-Nikodym Theorem to the Theory of Sufficient Statistics," *Ann. Math. Statist.* 20, 225–241.

Hampel, F. R. (1971), "A General Qualitative Definition of Robustness," *Ann. Math. Statist.* 42, 1887–1896.

—— (1976), "On the Breakdown Point of Some Rejection Rules with Mean," *Res. Rep. No. 11, Fachgruppe für Statistik, Eidgen*, Technische Hochschule, Zurich.

Harding, E. F. and D. G. Kendall (1974), *Stochastic Geometry*, Wiley, London.

Hartley, R. V. L. (1928), "Transmission of Information," *Bell System Tech. J.* 7, No. 3, 535–563.

Helstrom, C. W. (1960), Statistical Theory of Signal Detection, Pergamon, New York.

Hodges, J. L. and E. L. Lehman (1956), "The Efficiency of Some Nonparametric Competitors to the t-test," *Ann. Math. Statist.* 27, 324–335.

Hoeffding, W. (1951), "Optimum Nonparametric Tests," *Proc. 2nd Berkeley Symp. on Math. Statist. and Prob.*, Univ. of Calif. Press, Berkeley, 83–92.

—— (1960), "Lower Bounds for the Expected Sample Size and the Average Risk of a Sequential Procedure," *Ann. Math. Statist.* 31, 352–368.

Hoel, P. G. and R. P. Peterson (1949), "A Solution to the Problem of Optimum Classification," *Ann. Math. Statist.* 20, 433–38.

Holmes, J. K. (1982), *Coherent Spread Spectrum Systems*, Wiley, New York.

References

Huber, P. J. (1964), "Robust Estimation of a Location Parameter," *Ann. Math. Statist.* 35, 73–101.

——— (1965), "A Robust Version of the Probability Ratio Test," *Ann. Math. Statist.* 36, 1753–1758.

——— (1968), "Robust Confidence Limits," *Z. Wahrscheinlichkeitstheorie Verw. Gebiete* 10, 269–278.

——— (1969), *Théorie de L'Inférence Statistique Robuste*, Presses de l' Université, Montreal.

——— (1972), "Robust Statistics: A Review," *Ann. Math. Statist.* 43, 1041–1067.

——— (1981), *Robust Statistics*, Wiley, New York.

James, W. and C. Stein (1961), "Estimation with Quadratic Loss," *Proc. 4th Berkeley Symp. on Math. Statist. and Prob.* 1, Univ. of Calif. Press, Berkeley, 361–380.

Jensen, J. L. W. V. (1906), "Sur Les Functions Convexes et Les Inégalités Entre Les Valeurs Moyennes," *Acta Math.* 30, 175–193.

Karlin, S. (1959), *Mathematical Methods and Theory in Games, Programming, and Economics*, Vols. 1 and 2, Addison-Wesley, Reading, MA.

Karlin, S. and H. Rubin (1956), "The Theory of Decision Procedures for Distributions with Monotone Likelihood Ratio," *Ann. Math. Statist.* 27, 272–299.

Kiefer, J. (1957), "Invariance, Minimax Sequential Estimation and Continuous Time Processes," *Ann. Math. Statist.* 28, 573–601.

Kolmogorov, A. N. (1956), *Foundations of the Theory of Probability*, Chelsea, New York.

Kotel'nikov, V. A. (1959), *The Theory of Optimum Noise Immunity*, McGraw-Hill, New York.

LeCam, L. (1953), "On Some Asymptotic Properties of Maximum Likelihood Estimates and Related Bayes Estimates," *Univ. Calif. Publ. Statist.* 1, 277–330.

——— (1955), "An Extension of Wald's Theory of Statistical Decision Functions," *Ann. Math. Statist.* 26, 69–81.

Lehmann, E. L. (1959), *Testing Statistical Hypotheses*, Wiley, New York.

Loéve, M. (1963), *Probability Theory*, 3rd ed., Van Nostrand, Princeton, NJ.

Lorden, G. (1971), "Procedures for Reacting to a Change in Distribution," *Ann. Math. Statist.* 42, 1897–1908.

Luce, R. D. and H. Raiffa (1957), *Games and Decisions*, Wiley, New York.

Matthens, T. K. (1963), "On the Optimality of Sequential Probability Ratio Tests," *Ann. Math. Statist.* 34, 18–21.

Middleton, D. (1960), *An Introduction to Statistical Communication Theory*, McGraw-Hill, New York.

Miller, R. (1964), "A Trustworthy Jackknife," *Ann. Math. Statist.* 35, 1594–1605.

———— (1974), "The Jackknife—A Review," *Biometrika* 61, 1–15.

Mosteller, F. and J. W. Tukey (1977), *Data Analysis and Regression*, Addison-Wesley, Reading, MA.

Neyman, J. (1935), "Su Un Teorema Concernente Le Cosiddette Statistiche Sufficienti," *Giornale dell' Instituto degli Attuari* 6, 320–334.

———— (1952), *Lectures and Conferences on Mathematical Statistics and Probability*, 2nd ed., Graduate School, U.S. Dept. of Agriculture.

Neyman, J. and E. S. Pearson (1933), "On the Problem of the Most Efficient Tests of Statistical Hypotheses," *Philos. Trans. Roy. Soc. London Series A* 231, 289–337.

———— (1936, 1938), "Contributions to the Theory of Testing Statistical Hypotheses. I. Unbiased Critical Regions of Type A and Type A_1. II. Certain Theorems on Unbiased Critical Regions of Type A. III. Unbiased Tests of Simple Statistical Hypotheses Specifying the Values of More than One Unknown Parameter," *Statist. Res. Mem.* 1, 1–37, and 2, 25–57.

Page, E. S. (1954), "Continuous Inspection Schemes," *Biometrika* 41, 100–115.

———— (1955), "A Test for a Change in a Parameter Occurring at an Unknown Point," *Biometrika* 42, 523–527.

Papantoni-Kazakos, P. (1979a), "Algorithms for Monitoring Changes in Quality of Communication Links," *IEEE Trans. Comm.*, COM-27, No. 4, 682–693.

———— (1979b), "The Potential of End-to-End Observations in Trouble Localization and Quality Control of Network Links," *IEEE Trans. Commun.*, COM-27, No. 1, 16–30.

Papantoni-Kazakos, P. and R. M. Gray (1979), "Robustness of Estimators on Stationary Observations," *Ann. Prob.* 7, 989–1002.

Parzen, E. (1960), *Modern Probability Theory and its Applications*, Wiley, New York.

Pavlidis, T. (1974), *Biological Oscillators: Their Mathematical Analysis*, Academic Press, New York.

Pearson, E. S. and H. O. Hartley (1951), "Charts of the Power Function for Analysis of Variance Tests Derived from the Noncentral F-Distribution," *Biometrika* 38, 112–130.

References

Pitman, E. J. G. (1939), "The Estimation of Location and Scale Parameters of a Continuous Population of Any Given Form," *Biometrika* 30, 391–421.

Quenouille, M. H. (1956), "Notes on Bias in Estimation," *Biometrika* 43, 353–360.

Rao, C. R. (1945), "Information and Accuracy Attainable in Estimation of Statistical Parameters," *Bull. Calcutta Math. Soc.* 37, 81–91.

Ray, S. N. (1965), "Bounds on the Maximum Sample Size of a Bayes Sequential Procedure," *Ann. Math. Statist.* 36, 859–878.

Rice, S. O. (1945), "Mathematical Analysis of Random Noise," *Bell Syst. Tech. J.* 23, 283–332, and 24, 46–156.

Romanowski, M. and E. Green (1965), "Practical Applications of the Modified Normal Distribution," *Bull. Géodésique* 76, 1–20.

Savage, L. J. (1947), "A Uniqueness Theorem for Unbiased Sequential Estimation," *Ann. Math. Statist.* 18, 295–297.

——— (1954), *The Foundations of Statistics*, Wiley, New York.

Scheffé, H. (1959), *The Analysis of Variance*, Wiley, New York.

Schwartz, M. (1959), *Information Transmission, Modulation, and Noise*, McGraw-Hill, New York.

Shafer, G. (1976), *A Mathematical Theory of Evidence*, Princeton Univ. Press, Princeton, NJ.

Shannon, C. E. (1948), "A Mathematical Theory of Communication," *Bell Syst. Tech. J.* 27, 379–423, and 623–56.

——— (1949), "Communication in the Presence of Noise," *Proc. IRE*, 37, No. 1, 10–21.

Shannon, C. E. and W. Weaver (1949), *The Mathematical Theory of Communication*, Univ. of Ill. Press, Urbana.

Siegel, S. (1956), *Nonparametric Statistics for Behavioral Sciences*, McGraw-Hill, New York.

Stein, C. (1956), "Efficient Nonparametric Testing and Estimation," *Proc. 3d Berkeley Symposium on Math. Statist. and Prob.*, Vol. 1, Univ. of Calif. Press, Berkeley.

Steinhaus, H. (1957), "The Problem of Estimation," *Ann. Math. Statist.* 28, 633–648.

Stone, C. J. (1975), "Adaptive Maximum Likelihood Estimators of a Location Parameter," *Ann. Math. Statist.* 3, 267–284.

Strassen, V. (1965), "The Existence of Probability Measures with Given Marginals," *Ann. Math. Statist.* 36, 423–439.

Tietze, H. (1943), "Über Gewisse Umordnungen Von Permutationen und Ein Zugehöriges Stabilitätskriterium," *Jahresbericht der Deutschen Mathematiker Vereinigung*, Stuttgart, Germany.

Tucker, H. G. (1962), *An Introduction to Probability and Mathematical Statistics*, Academic Press, New York.

Tukey, J. W. (1953), "Some Selected Quick and Easy Methods of Statistical Analysis," *Trans. N.Y. Acad. Sci., Series II* 16, 88–97.

——— (1977), *Exploratory Data Analysis*, Addison-Wesley, Reading, Mass.

Van Trees, H. L. (1968), *Detection, Estimation, and Modulation Theory*, Wiley, New York.

Von Mises, R. (1947), "On the Asymptotic Distribution of Differentiable Statistical Functions," *Ann. Math. Statist.* 18, 309–348.

Von Neumann, J. and O. Morgenstern (1944), *Theory of Games and Economic Behavior*, 3d ed., Princeton University Press, Princeton, NJ.

Wald, A. (1947), *Sequential Analysis*, Wiley, New York.

——— (1950), *Statistical Decision Functions*, Wiley, New York.

Wald, A. and J. Wolfowitz (1948), "Optimum Character of the Sequential Probability Test," *Ann. Math. Statist.* 19, 326–339.

——— (1951), "Two Methods of Randomization in Statistics and the Theory of Games," *Ann. Math. Statist.* 23, 581–586.

Wiener, N. (1949), *The Extrapolation, Interpolation, and Smoothing of Stationary Time Series with Engineering Applications*, Wiley, New York.

Wintner, A. (1941), *The Analytical Foundations of Celestial Mechanics*, Princeton University Press, Princeton, NJ.

Wolfowitz, J. (1950), "Minimax Estimates of the Mean of a Normal Distribution with Known Variance," *Ann. Math. Statist.* 21, 218–230.

Wozencraft, J. M. and I. M. Jacobs (1965), *Principles of Communication Engineering*, Wiley, New York.

CHAPTER 2

SOME PRINCIPLES OF OPTIMIZATION THEORY

In this chapter, we briefly review some principles of optimization theory that are extensively used in the analyses and derivations included in this book. For complete information and proofs, the reader may refer to the references at the end of the chapter.

2.1 OPTIMIZATION UNDER CONSTRAINTS

We are generally concerned with finding extreme points of functions, subject to a number of constraints. In particular, we are searching for maxima or minima within a given region of functional support, where this region is generally determined by a set of constraints. If such extrema exist, then it is said that the *constraint optimization problem* has a solution. We point out that the existence of such a solution is possible, even if maxima or minima do not exist when the constraints are eliminated.

Example 2.1.1

Consider the function $f(x) = x^2$, defined on the real line. Without constraints, this function has a unique global minimum at $x = 0$, but no maximum. Thus, the unconstrained optimization problem regarding a maximum for $f(x)$, has no solution. Let us now search for a maximum of $f(x)$, subject to the constraint, $0 \leq x \leq 1$. The latter constraint optimization problem has a solution, attained at $x = 1$.

Given a single-variable function $f(x)$, its extrema may be sought subject to either equality or inequality constraints. *Equality constraints* are generally expressed via a finite set of functions, $\{g_i(x); i = 1, \ldots, m\}$ each set equal to zero, while *inequality constraints* are expressed via a similar set of functions, $\{h_i(x);$ $i = 1, \ldots, k\}$ each required to be larger than or equal to zero. The x-regions $C_g = \{x: g_i(x) = 0; i = 1, \ldots, m\}$ and $C_h = \{x: h_i(x) \geq 0; i = 1, \ldots, k\}$ are called *convex* if $x_1 \in C_g$ and $x_2 \in C_g$ implies $[(1 - \varepsilon)x_1 + \varepsilon x_2] \in C_g$ for every ε such that $0 \leq \varepsilon \leq 1$ and if $x_1 \in C_h$ and $x_2 \in C_h$ implies $[(1 - \varepsilon)x_1 + \varepsilon x_2] \in C_h$ for every ε as above. The sets C_g and C_h are *compact* with respect to some metric if every open with respect to the same metric set that covers them has a finite open-set subcovering. The function $f(x)$ is *concave*, or *convex*, on C_g if for every ε such that $0 < \varepsilon < 1$ we have, respectively, $f[\varepsilon x_1 + (1 - \varepsilon)x_2] \geq \varepsilon f(x_1) + (1 - \varepsilon)f(x_2)$ $\forall\ x_1, x_2 \in C_g$ or $f[\varepsilon x_1 + (1 - \varepsilon)x_2] \leq \varepsilon f(x_1) + (1 - \varepsilon)f(x_2)\ \forall\ x_1, x_2 \in C_g$. Concavity or convexity of $f(x)$ on C_h is similarly defined. If $f(x)$ is concave, or convex, on C_g, and if the set C_g is convex and compact, then a unique maximum, or minimum, of $f(x)$ in C_g exists. A similar statement holds if the set C_h is substituted for the set C_g. We point out that, as in Example 2.1.1, a minimum, or maximum, of the function $f(x)$ on a compact and convex set may exist even if $f(x)$ is not respectively convex, or concave, in it.

Let us consider a two-variable function $f(x, y)$, where $x \in C_x$ and $y \in C_y$. The function is called *concave-convex* on $C_x \times C_y$ if it is concave with respect to x on C_x for every y on C_y and it is convex with respect to y on C_y for every x on C_x. The function $f(x, y)$ is *convex-concave* on $C_x \times C_y$ if it is convex with respect to x on C_x for every y on C_y and it is concave with respect to y on C_y for every x on C_x. If both the sets C_x and C_y are convex and compact, and if the function $f(x, y)$ is concave-convex, then it has a unique maximum on C_x for every given y on C_y and a unique minimum on C_y for every given x on C_x. Similar conclusions are drawn if the function is instead convex-concave. The set $C_x \times C_y$ may be determined via constraints, represented by a set of functions $\{g_i(x, y); i = 1, \ldots, l\}$. If all those functions are set equal to zero, equality constraints are active. If the functions are instead required to be larger than or equal to zero, inequality constraints are present.

Example 2.1.2

Consider the function $f(x, y) = -2x^2 + 4xy + y^2$, initially defined on the two-dimensional Euclidean space. It is easily verified that the function is concave-convex on the whole two-dimensional Euclidean space. It is also continuous with respect to both x and y and with respect to the absolute difference metric; that is, $|x_1 - x_2|$ for x, and $|y_1 - y_2|$ for y. Let us now consider the inequality constraint $g(x, y) = 1 - x^2 - y^2 \geq 0$, which represents the unit circle, including its circumference. This constraint determines a set $C_{x,y}$ of x and y values, and it is easily concluded that $C_{x,y}$ is convex with respect to both x and y. $C_{x,y}$ is also compact with respect to both x and y and with respect to the absolute difference metric for each. From the above we conclude that $f(x, y)$ has a unique maximum on $C_{x,y}$ for every given y on $C_{x,y}$ and a unique minimum on $C_{x,y}$ for every given x on $C_{x,y}$.

Let us now consider instead, the equality constraint $g(x, y) = 1 - x^2 - y^2 = 0$. This constraint does not specify a convex (x, y) set. However, the function

2.2 Lagrange Multipliers

$f(x, y)$ attains both minimum and maximum on the (x, y) set. To see that, let us express everything in polar coordinates. Writing $x = r \sin \theta$, $y = r \cos \theta$, we easily obtain,

$$f(x, y) = \frac{5r^2}{2}\left[\sin(\alpha + 2\theta) - \frac{1}{5}\right] \quad \text{where} \quad \alpha = \tan^{-1}\frac{3}{4}$$

$$g(x, y) = 1 - x^2 - y^2 = 0 \to r = 1$$

We then easily observe that subject to the constraint $r = 1$, the function $f(x, y)$ attains a minimum at $\theta_1 = (3\pi/4) - (\alpha/2)$ and $\theta_2 = \pi + (3\pi/4) - (\alpha/2)$ and it attains a maximum at $\theta_3 = (\pi/4) - (\alpha/2)$ and $\theta_4 = \pi + (\pi/4) - (\alpha/2)$. Thus, the extrema subject to the equality nonconvex constraint exist, but they are not unique.

Consider a function $f(x)$ whose every order derivatives, with respect to a proper metric, exist. Then, the function $f(x)$ is concave or convex on some convex set C_x if and only if, respectively, $f^{(2)}(x) \leq 0 \ \forall \ x \in C_x$ or $f^{(2)}(x) \geq 0 \ \forall \ x \in C_x$, where (2) denotes the second derivative. Similarly, a function $f(x, y)$ whose every order derivatives, with respect to both x and y and with respect to appropriate metrics, exist is concave-convex on a convex set $C_{x,y}$ if and only if $f_x^{(2)}(x, y) \leq 0$ $\forall \ x, y \in C_{x,y}$ and $f_y^{(2)}(x, y) \geq 0 \ \forall \ x, y \in C_{x,y}$, where $f_x^{(2)}$ and $f_y^{(2)}$ denote second order derivatives, with respect to x versus y. For differentiable functions as above, the calculus of variations applies. Specifically, let x^* be a unique extreme point of $f(x)$ on C_x, where C_x is convex, and let $x \in C_x$. Then, given any ε such that $0 \leq \varepsilon \leq 1$, we conclude that $[(1 - \varepsilon)x^* + \varepsilon x] \in C_x$. Denoting by $f_\varepsilon^{(1)}$ the first order derivative with respect to ε, the calculus of variations gives, $f_\varepsilon^{(1)}[(1 - \varepsilon)x^* + \varepsilon x]|_{\varepsilon=0} = 0 \ \forall \ x \in C_x$. The same results hold for unique extrema (x^*, y^*) of $f(x, y)$ on $C_{x,y}$. In conclusion, if $f(x)$ is strictly either concave or convex and multiple differentiable on C_x, and if C_x is convex and compact, then the existence of a unique extremum x^* on C_x is guaranteed. This extremum can be then found via the application of the calculus of variations. That is, x^* is such that $f_\varepsilon^{(1)}[(1 - \varepsilon)x^* + \varepsilon x]|_{\varepsilon=0} = 0 \ \forall \ x \in C_x$. Similar conclusions are drawn for the function $f(x, y)$.

2.2 LAGRANGE MULTIPLIERS

In the previous section we briefly and qualitatively discussed optimization, subject to either equality or inequality constraints. Here, we will discuss a methodology which allows the incorporation of the constraints into an overall optimization functional. Without losing generality, we will consider two-variable functions $f(x, y)$ and search for minima. We will also assume, in substance, equality constraints. We immediately state, without proof, the following theorem.

THEOREM 2.2.1: Let the function $f(x, y)$ have a minimum on the set of points determined by the constraints $\{g_i(x, y) = 0; i = 1, \ldots, k\}$ and let the minimum be attained at the point (x^*, y^*). Also let $f(x, y)$ and $g_i(x, y)$; $i = 1, \ldots, k$ all be differentiable with respect to both x and y and with respect to an appropriate metric. Then there exists a set $\{\lambda_i; i = 1, \ldots, k\}$ of multipliers, such that, if

$$F(x, y) = f(x, y) + \sum_{i=1}^{k} \lambda_i g_i(x, y) \qquad (2.2.1)$$

F_x, F_y denote first order derivatives of $F(x, y)$ with respect to x and y

F_{xx}, F_{yy}, F_{xy} denote second order derivatives with respect to x, y, and x and y.

then

$$F_x(x^*, y^*) = 0 \qquad F_y(x^*, y^*) = 0 \qquad (2.2.2)$$
$$h^2 F_{xx}(x^*, y^*) + k^2 F_{yy}(x^*, y^*) + 2hk F_{xy}(x^*, y^*) \geq 0 \qquad \forall\, (h, k) \neq (0, 0) \qquad (2.2.3)$$

The set $\{\lambda_i\}$ is called *Lagrange multipliers*, and the function in (2.2.1) is called the *Lagrange function*. ∎

Due to Theorem 2.2.1, we conclude that if the existence of a local extremum subject to some equality constraints is guaranteed, then this extremum can be found by setting the first order derivatives of the Lagrange function equal to zero. The extreme points are then found as functions of the Lagrange multipliers, whose values are in turn found by substitution in the equality constraints. The Lagrange multipliers methodology is better exhibited via some examples.

Example 2.2.1

Let us consider the function $f(x, y)$ in Example 2.1.2, and let us search for its minimum and maximum, subject to the constraint $g(x, y) = 1 - x^2 - y^2 = 0$. As we saw in that example, those extrema exist. Thus, the Lagrange multiplier methodology in Theorem 2.2.1 can be used, where the Lagrange function $F(x, y)$ is

$$F(x, y) = -2x^2 + 4xy + y^2 + \lambda(1 - x^2 - y^2) \qquad (2.2.4)$$

The derivatives with respect to the absolute difference metric on the Euclidean space are

$$F_x(x, y) = 2[-(2 + \lambda)x + 2y] \qquad F_y(x, y) = 2[2x + (1 - \lambda)y]$$
$$F_{xy}(x, y) = 4 \qquad F_{xx}(x, y) = -2(2 + \lambda) \qquad F_{yy}(x, y) = 2(1 - \lambda)$$

Setting $F_x(x, y) = F_y(x, y) = 0$, we find,

$$-(2 + \lambda)x + 2y = 0 \quad \text{and} \quad 2x + (1 - \lambda)y = 0 \qquad (2.2.5)$$

2.2 Lagrange Multipliers

The system in (2.2.5) is satisfied if $2/(2 + \lambda) = (\lambda - 1)/2$ or if either $\lambda = 2$ or $\lambda = -3$

i. If $\lambda = 2$, (2.2.5) gives $x = y/2$. Substituting the latter equality in the constraint $x^2 + y^2 = 1$ we obtain

$$\text{either} \quad x^* = \frac{1}{\sqrt{5}} \quad y^* = \frac{2}{\sqrt{5}}$$

$$\text{or} \quad x^* = -\frac{1}{\sqrt{5}} \quad y^* = -\frac{2}{\sqrt{5}}$$

Both the pairs (x^*, y^*) give

$$h^2 F_{xx}(x^*, y^*) + k^2 F_{yy}(x^*, y^*) + 2hk F_{xy}(x^*, y^*)$$
$$= -2(k - 2h)^2 \leq 0 \qquad \forall\, (k, h) \neq (0, 0)$$

Thus, the above pairs both correspond to maxima of the function $f(x, y)$.

ii. If $\lambda = -3$, (2.2.5) gives $x = -2y$. Substituting the latter equality in the constraint $x^2 + y^2 = 1$ we obtain

$$\text{either} \quad x^* = -\frac{2}{\sqrt{5}} \quad y^* = \frac{1}{\sqrt{5}}$$

$$\text{or} \quad x^* = \frac{2}{\sqrt{5}} \quad y^* = -\frac{1}{\sqrt{5}}$$

where both the above pairs give

$$h^2 F_{xx}(x^*, y^*) + k^2 F_{yy}(x^*, y^*) + 2hk F_{xy}(x^*, y^*)$$
$$= 2(h + 2k)^2 \geq 0 \qquad \forall\, (h, k) \neq (0, 0)$$

Thus, the above pairs both correspond to minima of the function $f(x, y)$.

As we pointed out in Example 2.1.2, the extrema in both cases are not unique.

Example 2.2.2

Consider density functions defined on the real line R. Let $\mathscr{F}$ be the class of all such first order density functions whose mean is zero and whose variance is σ^2. Given f in $\mathscr{F}$, its entropy $H(f)$ is defined as follows.

$$H(f) = -\int_R dx\, f(x) \log f(x)$$

It is easily verified that the class $\mathscr{F}$ is convex. That is, if $f_1 \in \mathscr{F}$ and $f_2 \in \mathscr{F}$ then $[(1 - \varepsilon)f_1 + \varepsilon f_2] \in \mathscr{F}$ for every ε such that $0 \leq \varepsilon \leq 1$. Also, the entropy

$H(f)$ is strictly concave. That is, $H[\varepsilon f_1 + (1-\varepsilon)f_2] \geq \varepsilon H(f_1) + (1-\varepsilon)H(f_2)$ for all $f_1, f_2 \in \mathscr{F}$ and for every ε such that, $0 \leq \varepsilon \leq 1$. In addition, $H(f)$ is differentiable with respect to f and with respect to the metric, $\int_R |f_1(x) - f_2(x)| dx$. The constraints imposed by the class $\mathscr{F}$ are:

For every f in $\mathscr{F}$

$$\int_R f(x)\,dx = 1 \qquad \int_R xf(x)\,dx = 0 \qquad \int x^2 f(x)\,dx = \sigma^2$$

$$f(x) \geq 0 \qquad \forall\, x \text{ in } R. \tag{2.2.6}$$

Let us temporarily ignore the inequality constraint $f(x) \geq 0$. The remaining constraints are all differentiable with respect to f and with respect to the metric $\int_R |f_1(x) - f_2(x)|\,dx$. The entropy $H(f)$ has a maximum f^* on $\mathscr{F}$, while both the Lagrange multipliers methodology and the calculus of variations apply. The Lagrange function $F(f)$ is (ignoring the inequality constraint):

$$F(f) = H(f) + \lambda_1 \int_R f(x)\,dx + \lambda_2 \int_R xf(x)\,dx + \lambda_3 \int_R x^2 f(x)\,dx \tag{2.2.7}$$

The extremum f^* is found by setting $(\partial/\partial\varepsilon)F[(1-\varepsilon)f^* + \varepsilon f]|_{\varepsilon=0} = 0$, for every f in $\mathscr{F}$.

Given f in $\mathscr{F}$, let us define, $h(x) = f(x) - f^*(x)$. Then we have $\int_R h(x)\,dx = \int_R f(x)\,dx - \int_R f^*(x)\,dx = 0$, and from (2.2.7) we obtain

$$\frac{\partial}{\partial\varepsilon}F[(1-\varepsilon)f^* + \varepsilon f]\bigg|_{\varepsilon=0} = \frac{\partial}{\partial\varepsilon}F(f^* + \varepsilon h)\bigg|_{\varepsilon=0}$$

$$= \int_R dx\, h(x)[\log f^*(x) + 1 - \lambda_1 - \lambda_2 x - \lambda_3 x^2] = 0 \tag{2.2.8}$$

We wish to have (2.2.8) satisfied for every $h(x)$ such that $\int_R h(x)\,dx = 0$. Thus, we wish that

$$\log f^*(x) + 1 - \lambda_1 - \lambda_2 x - \lambda_3 x^2 = c \qquad \forall\, x \text{ in } R \tag{2.2.9}$$

where c is some constant. The solution of (2.2.9) is

$$f^*(x) = d\,\exp\{\lambda_3(x + a)^2\} \tag{2.2.10}$$

where

$$a = \lambda_2/2\lambda_3 \qquad d = \exp\{c + \lambda_1 - 1 - (\lambda_2^2/4\lambda_3^2)\}$$

Since $d > 0$, the inequality constraint $f(x) \geq 0\ \forall\, x$ in R is also satisfied. To satisfy the equality constraints in (2.2.6), we easily find by substitution that

$$f^*(x) = (2\pi)^{-1/2}\sigma^{-1}\exp\{-(x^2/2\sigma^2)\} \tag{2.2.11}$$

2.2 Lagrange Multipliers

The highest entropy density in class $\mathscr{F}$ is thus the Gaussian density, and it is unique, due to the strict concavity of the entropy.

Example 2.2.3

Let us consider the function $f(x, y)$ in Example 2.2.1. As we have previously seen, this function is concave-convex on $R \times R$ and everywhere differentiable with respect to both x and y and with respect to the absolute difference metric for both. Let us seek maximization and minimization of $f(x, y)$ subject to the convex and compact constraint $g(x, y) = 1 - x^2 - y^2 \geq 0$. To transform the latter constraint into a temporarily equality constraint, let us select some r such that $0 < r \leq 1$ and consider instead the constraint $r^2 - x^2 - y^2 = 0$. Then, the Lagrange function $F(x, y)$ is

$$F(x, y) = -2x^2 + 4xy + y^2 + \lambda(r^2 - x^2 - y^2)$$

where, $F_x(x, y)$, $F_y(x, y)$, $F_{xx}(x, y)$, $F_{yy}(x, y)$, and $F_{xy}(x, y)$ are exactly as in Example 2.2.1. Thus, we reach again the system in (2.2.5) and we conclude that either $\lambda = 2$, or $\lambda = -3$. Then

i. If $\lambda = 2$, the system in (2.2.5) gives, $x = y/2$. Substituting the latter in the constraint $r^2 - x^2 - y^2 = 0$, we obtain

$$\text{either} \quad x^* = \frac{r}{\sqrt{5}} \quad y^* = \frac{2r}{\sqrt{5}}$$
$$\text{or} \quad x^* = -\frac{r}{\sqrt{5}} \quad y^* = -\frac{2r}{\sqrt{5}} \quad (2.2.12)$$

Both the above pairs correspond to maxima, as in part i of Example 2.2.1. In addition, they both give $f(x^*, y^*) = 2r^2$. Allowing now r to vary in $[0, 1]$, we conclude that the maximum of $f(x, y)$ subject to the constraint $1 - x^2 - y^2 \geq 0$ is attained by the pairs in (2.2.12), with $r = 1$.

ii. If $\lambda = -3$, we conclude from (2.2.5) that $x = -2y$, which, substituted in the constraint $r^2 - x^2 - y^2 = 0$, gives

$$\text{either} \quad x^* = -\frac{2r}{\sqrt{5}} \quad y^* = \frac{r}{\sqrt{5}}$$
$$\text{or} \quad x^* = \frac{2r}{\sqrt{5}} \quad y^* = -\frac{r}{\sqrt{5}} \quad (2.2.13)$$

Both the pairs in (2.2.13) correspond to minima, as in part ii of Example 2.2.1, and they both give $f(x^*, y^*) = -\frac{11}{5}r^2$. For r varying in $[0, 1]$, the value $r = 1$ attains the minimum of $f(x^*, y^*)$. Thus the minimum of $f(x, y)$, subject to the constraint $1 - x^2 - y^2 \geq 0$, is attained by the pairs in (2.2.13), with $r = 1$.

We note that temporarily assuming equality constraint, we computed extreme $f(x, y)$ values that were monotone with respect to r. This latter property allowed us to find the solution of the problem, subject to the initial inequality constraint.

2.3 KUHN–TUCKER CONDITIONS

In Section 2.2, we reviewed the principles of the Lagrangian approach for optimization subject to equality constraints. In this section, we review some basic theorems for optimization subject to mixed equality and inequality constraints. We first consider n-variable scalar real functions $f(x^n)$ where $x^n = \{x_i; i = 1, \ldots, n\}$, optimized subject to constraints determined by a set $\{g_i(x^n)\}$ of scalar, real, n-variable functions. We then denote by ∇ the gradient vector, whose components are $\{\partial/\partial x_i; i = 1, \ldots, n\}$, and we directly state without proof the Kuhn–Tucker theorem.

THEOREM 2.3.1: Let us assume that the vector value x_0^n minimizes the function $f(x^n)$ locally on the set $\mathscr{S}$ of vector values determined by the following constraints.

$$g_i(x^n) \leq 0 \quad i = 1, \ldots, p$$
$$g_i(x^n) = 0 \quad i = p+1, \ldots, m$$

Let the gradients, $\nabla f(x_0^n)$, $\nabla g_i(x_0^n)$; $i = 1, \ldots, m$, at x_0^n exist. Then, there exist multipliers $\lambda_1, \lambda_2, \ldots, \lambda_m$ such that

$$\lambda_i > 0 \quad i = 1, \ldots, p \quad \text{with} \quad \lambda_j = 0 \quad \text{if} \quad g_j(x_0^n) < 0 \quad (2.3.1)$$

and such that

$$\nabla F(x_0^n) = 0 \quad \text{where} \quad F(x^n) = f(x^n) + \sum_{i=1}^{m} \lambda_i g_i(x^n) \quad (2.3.2)$$

In addition,

$$F(x^n) \leq f(x^n) \quad \text{on} \quad \mathscr{S} \quad F(x_0^n) = f(x_0^n) \quad (2.3.3)$$

∎

The function $F(x^n)$ in (2.3.2) is the Lagrangian. We now revisit Example 2.2.3, applying Theorem 2.3.1 above.

Example 2.3.1

Let us consider minimization and maximization of the function $f(x, y) = -2x^2 + 4xy + y^2$, subject to the constraint, $x^2 + y^2 - 1 \leq 0$.

2.3 Kuhn–Tucker Conditions

i. For minimization, and applying Theorem 2.3.1, we form the Lagrangian

$$F(x, y) = -2x^2 + 4xy + y^2 + \lambda(x^2 + y^2 - 1)$$

where due to (2.3.1), $\lambda \geq 0$. We find

$$\nabla F(x, y) = \begin{bmatrix} \dfrac{\partial}{\partial x} F(x, y) \\ \dfrac{\partial}{\partial y} F(x, y) \end{bmatrix} = 2 \begin{bmatrix} (\lambda - 2)x + 2y \\ 2x + (\lambda + 1)y \end{bmatrix}$$

Setting the above gradient equal to zero, we first find

$$\frac{\lambda - 2}{2} = -\frac{2}{\lambda + 1} \quad \text{or} \quad \lambda^2 - \lambda - 6 = 0$$

whose roots are 3 and -2. Due to the requirement $\lambda \geq 0$, we select the root $\lambda = 3$. Then, from $\nabla F(x, y) = 0$, we find $x^* = -2y^*$ where (x^*, y^*) is the vector point that attains minimum. Substituting the equality $x^* = -2y^*$ in $F(x, y)$ and $f(x, y)$, and for $\lambda = 3$, we find

$$F(x^*, y^*) = -y^{*2} - 3 \qquad f(x^*, y^*) = -16y^{*2} \qquad (2.3.4)$$

But due to (2.3.3), we must also have $F(x^*, y^*) = f(x^*, y^*)$. Due to the latter, and (2.3.4), we find $y^* = \pm 1/\sqrt{5}$, where $x^* = -2y^*$. Thus, $f(x, y)$ is minimized for

$$\text{either} \qquad x^* = -\frac{2}{\sqrt{5}} \qquad y^* = \frac{1}{\sqrt{5}}$$

$$\text{or} \qquad x^* = \frac{2}{\sqrt{5}} \qquad y^* = -\frac{1}{\sqrt{5}}$$

ii. For maximization of $f(x, y)$ subject to the constraint, $x^2 + y^2 - 1 \leq 0$, we consider instead minimization of $-f(x, y)$, subject to the same constraint. The Lagrangian, with $\lambda \geq 0$, is then

$$F(x, y) = 2x^2 - 4xy - y^2 + \lambda(x^2 + y^2 - 1)$$

and

$$\nabla F(x, y) = 2 \begin{bmatrix} (\lambda + 2)x - 2y \\ -2x + (\lambda - 1)y \end{bmatrix} = 0$$

implies $(\lambda + 2)/2 = 2/(\lambda - 1)$, which gives either $\lambda = -3$, or $\lambda = 2$. Requiring $\lambda \geq 0$, we select $\lambda = 2$. Then, $\nabla F(x, y) = 0$ gives $y^* = 2x^*$. For such y^*

and x^* and for $\lambda = 2$ we obtain $F(x^*, y^*) = -2$ and $f(x^*, y^*) = -10x^{*2}$. Setting $F(x^*, y^*) = f(x^*, y^*)$, due to (2.3.3), we then conclude $x^* = \pm 1/\sqrt{5}$, where $y^* = 2x^*$. The function $f(x, y)$ is thus maximized for

$$\text{either} \quad x^* = \frac{1}{\sqrt{5}} \quad y^* = \frac{2}{\sqrt{5}}$$

$$\text{or} \quad x^* = -\frac{1}{\sqrt{5}} \quad y^* = -\frac{2}{\sqrt{5}}$$

For our problems and applications in this book, an extension of the Kuhn–Tucker theorem for optimization of functionals is of interest. Then, the minimization of some functional $G(f)$ is generally considered, with respect to some function $f(x)$. The constraints are then focused on the function $f(x)$. Some of the constraints may be in the form of functionals, while the values of $f(x)$ for every x may be constrained as well. Let $G_i(f); i = 1, \ldots, m$ be a set of real and scalar functionals defined on the space $\mathscr{S}$ of scalar and real functions $f(x)$ whose support is some subspace $\mathscr{P}$ of the real line. Let $g_0(x)$ be some given function on $\mathscr{S}$ and let the set $\mathscr{F}$ of functions f be defined as follows.

$$\mathscr{F} \subset \mathscr{S}$$

$$f \in F \text{ implies} \quad G_i(f) \leq 0 \quad i = 1, \ldots, p$$
$$G_i(f) = 0; \quad i = p+1, \ldots, m \quad (2.3.5)$$
$$f(x) \leq g_0(x) \quad \text{for all} \quad x \text{ in } \mathscr{P}$$

Let the set $\mathscr{F}$ of functions be convex; let for any $f, h \in \mathscr{F}$ and any ε such that $0 < \varepsilon < 1$ the derivatives $(\partial/\partial\varepsilon)G_i[(1-\varepsilon)f + \varepsilon h]; i = 1, \ldots, m$ exist; and let $G(f)$ be a scalar real functional defined for f in $\mathscr{F}$, and such that the derivative $(\partial/\partial\varepsilon)G[(1-\varepsilon)f + \varepsilon h]$ exists for all f, h in $\mathscr{F}$ and all ε such that $0 < \varepsilon < 1$. Then, combining Theorem 2.3.1 with the calculus of variations, we can express the following theorem.

THEOREM 2.3.2: Let f^* minimize the functional $G(f)$ in $\mathscr{F}$, where the set $\mathscr{F}$ of functions is as above. Then, multipliers $\lambda_1, \ldots, \lambda_m$ and a multiplier function $\mu(x)$, defined on $\mathscr{P}$, exist, such that

$$\begin{aligned} \lambda_i &> 0 \quad i = 1, \ldots, p \quad \text{with} \quad \lambda_j = 0 \text{ if } G_j(f^*) < 0 \\ \mu(x) &\geq 0 \quad \text{for all} \quad x \text{ in } \mathscr{P} \end{aligned} \quad (2.3.6)$$

and such that

$$\left. \frac{\partial}{\partial \varepsilon} F[(1-\varepsilon)f^* + \varepsilon h] \right|_{\varepsilon=0} = 0 \quad \text{for all} \quad h \text{ in } \mathscr{F}$$

2.3 Kuhn–Tucker Conditions

where

$$F(f) = G(f) + \sum_{i=1}^{m} \lambda_i G_i(f) + \int_{\mathscr{P}} \mu(x)[f(x) - g_0(x)]\,dx \qquad (2.3.7)$$

In addition

$$F(f) \leq G(f) \quad \text{on} \quad \mathscr{F} \qquad F(f^*) = G(f^*) \qquad (2.3.8)$$

∎

We now proceed with an example that represents an application of Theorem 2.3.2 and completes this section.

Example 2.3.2

We consider asymptotic linear prediction, for zero mean, scalar, real, discrete-time, stationary stochastic processes. Let $f(\lambda)$ be the power spectral density of such a process. Then $f(\lambda)$ is defined on $[-\pi, \pi]$ and as we will discuss in Chapter 12, the mean squared error $e_p(f)$ induced by the asymptotic optimal linear mean squared predictor is then given by the following expression, where ln denotes natural logarithm.

$$e_p(f) = \exp\left\{(2\pi)^{-1} \int_{-\pi}^{\pi} \ln[2\pi f(\lambda)]\,d\lambda\right\} \qquad (2.3.9)$$

Let $f_0(\lambda);\ \lambda \in [-\pi, \pi]$ be a given spectral density, such that $(2\pi)^{-1}\int_{-\pi}^{\pi} f_0(\lambda)\,d\lambda = 1$, and let δ be a given number in $(0, 1)$. Let $\mathscr{F}$ be the class of spectral density functions, defined as:

$$\mathscr{F} = \left\{f: f(\lambda) \text{ defined on } [-\pi, \pi]; f(\lambda) \geq (1-\delta)f_0(\lambda)\ \forall\ \lambda \right.$$

$$\left. \text{in } [-\pi, \pi]; (2\pi)^{-1}\int_{-\pi}^{\pi} f(\lambda)\,d\lambda = 1\right\}$$

We wish to find the f in $\mathscr{F}$ that maximizes the error in (2.3.9). We first observe that $\mathscr{F}$ is a convex class and that $e_p(f)$ has a maximum in $\mathscr{F}$, since it is strictly concave with respect to f. The optimization problem can be expressed as follows.

Minimize $-\ln e_p(f) = -(2\pi)^{-1}\int_{-\pi}^{\pi} \ln[2\pi f(\lambda)]\,d\lambda = G(f)$ subject to the constraints $(2\pi)^{-1}\int_{-\pi}^{\pi} f(\lambda)\,d\lambda - 1 = G_1(f) = 0;\ (1-\delta)f_0(\lambda) - f(\lambda) \leq 0\ \forall\ \lambda$ in $[-\pi, \pi]$.

Applying Theorem 2.3.2, we first form the Lagrangian

$$F(f) = G(f) + \lambda_1 G(f) + (2\pi)^{-1} \int_{-\pi}^{\pi} \mu(\lambda)[(1-\delta)f_0(\lambda) - f(\lambda)]\,d\lambda$$

$$= -(2\pi)^{-1} \int_{-\pi}^{\pi} \ln[2\pi f(\lambda)]\,d\lambda + \lambda_1\left[(2\pi)^{-1}\int_{-\pi}^{\pi} f(\lambda)\,d\lambda - 1\right]$$

$$+ (2\pi)^{-1}\int_{-\pi}^{\pi} \mu(\lambda)[(1-\delta)f_0(\lambda) - f(\lambda)]\,d\lambda \qquad (2.3.10)$$

where

$$\mu(\lambda) \geq 0 \qquad \forall \lambda \text{ in } [-\pi, \pi] \qquad (2.3.11)$$

If f^* is the power spectral density that attains the maximum $e_p(f)$ on $\mathscr{F}$, and if f is some arbitrary member of $\mathscr{F}$, we write $h(\lambda) = f(\lambda) - f^*(\lambda)$, where $\int_{-\pi}^{\pi} h(\lambda) d\lambda = 0$ and $h(\lambda)$ is arbitrary otherwise. From (2.3.10), and for some ε such that $0 < \varepsilon < 1$, we then have

$$\begin{aligned} F[(1-\varepsilon)f^* + \varepsilon f] &= F(f^* + \varepsilon h) \\ &= -(2\pi)^{-1} \int_{-\pi}^{\pi} \ln[2\pi f^*(\lambda) + \varepsilon 2\pi h(\lambda)] d\lambda \\ &\quad + \lambda_1 \left[(2\pi)^{-1} \int_{-\pi}^{\pi} [f^*(\lambda) + \varepsilon h(\lambda)] d\lambda - 1 \right] \\ &\quad + (2\pi)^{-1} \int_{-\pi}^{\pi} \mu(\lambda)[(1-\delta)f_0(\lambda) - f^*(\lambda) - \varepsilon h(\lambda)] d\lambda \end{aligned}$$

$$\begin{aligned} \frac{\partial}{\partial \varepsilon} F[(1-\varepsilon)f^* + \varepsilon f] &= \frac{\partial}{\partial \varepsilon} F(f^* + \varepsilon h) \\ &= -\int_{-\pi}^{\pi} h(\lambda)[2\pi f^*(\lambda) + \varepsilon 2\pi h(\lambda)]^{-1} d\lambda \\ &\quad - (2\pi)^{-1} \int_{-\pi}^{\pi} \mu(\lambda) h(\lambda) d\lambda \end{aligned}$$

$$\begin{aligned} \left. \frac{\partial}{\partial \varepsilon} F[(1-\varepsilon)f^* + \varepsilon f] \right|_{\varepsilon=0} &= \left. \frac{\partial}{\partial \varepsilon} F(f^* + \varepsilon h) \right|_{\varepsilon=0} \\ &= -(2\pi)^{-1} \int_{-\pi}^{\pi} h(\lambda)\{[f^*(\lambda)]^{-1} + \mu(\lambda)\} d\lambda \qquad (2.3.12) \end{aligned}$$

Via Theorem 2.3.2, we want the expression in (2.3.12) to be equal to zero for all $h(\lambda)$ such that $\int_{-\pi}^{\pi} h(\lambda) d\lambda = 0$. This requirement is equivalent to

$$[f^*(\lambda)]^{-1} + \mu(\lambda) = c \qquad \text{for all } \lambda \text{ in } [-\pi, \pi] \qquad (2.3.13)$$

where c is a constant, or

$$\mu(\lambda) = c - [f^*(\lambda)]^{-1} \qquad \text{for all } \lambda \text{ in } [-\pi, \pi] \qquad (2.3.14)$$

where, due to (2.3.11), we have

$$c - [f^*(\lambda)]^{-1} \geq 0 \qquad \text{for all } \lambda \text{ in } [-\pi, \pi]. \qquad (2.3.15)$$

Due to (2.3.8) in Theorem 2.3.2, we also require that $F(f^*) = G(f^*)$. Substituting in (2.3.10), this requirement gives

$$\lambda_1 \left[(2\pi)^{-1} \int_{-\pi}^{\pi} f^*(\lambda) d\lambda - 1 \right] + (2\pi)^{-1} \int_{-\pi}^{\pi} \mu(\lambda)[(1-\delta)f_0(\lambda) - f^*(\lambda)] d\lambda = 0 \qquad (2.3.16)$$

Substituting (2.3.14) in (2.3.16), and due to the constraint $(2\pi)^{-1} \int_{-\pi}^{\pi} f^*(\lambda) d\lambda = 1$, we obtain

$$(2\pi)^{-1} \int_{-\pi}^{\pi} \{c - [f^*(\lambda)]^{-1}\}\{(1 - \delta)f_0(\lambda) - f^*(\lambda)\} d\lambda = 0 \quad (2.3.17)$$

This last expression determines the power spectral density $f^*(\lambda)$. Indeed, since f^* is in $\mathscr{F}$, then $(1 - \delta)f_0(\lambda) - f^*(\lambda) \leq 0$, for every λ in $[-\pi, \pi]$. Also, due to (2.3.15), $c - [f^*(\lambda)]^{-1} \geq 0$ for every λ in $[-\pi, \pi]$. That is, the integrand in (2.3.17) is nonpositive for every λ. Therefore, (2.3.17) can be satisfied only if

$$\{c - [f^*(\lambda)]^{-1}\}\{(1 - \delta)f_0(\lambda) - f^*(\lambda)\} = 0 \quad \text{for all } \lambda \text{ in } [-\pi, \pi] \quad (2.3.18)$$

where

$$f^*(\lambda) \geq (1 - \delta)f_0(\lambda)$$
$$c \geq [f^*(\lambda)]^{-1} \quad \text{or} \quad f^*(\lambda) \geq c^{-1} \quad (2.3.19)$$
$$(2\pi)^{-1} \int_{-\pi}^{\pi} f^*(\lambda) d\lambda = 1$$

The combination of (2.3.18) and (2.3.19) obviously gives

$$f^*(\lambda) = \max[(1 - \delta)f_0(\lambda), c^{-1}] \quad \lambda \in [-\pi, \pi] \quad (2.3.20)$$

with c such that $c \geq 0$ and

$$g(c^{-1}) = (2\pi)^{-1} \int_{-\pi}^{\pi} \max[(1 - \delta)f_0(\lambda), c^{-1}] d\lambda = 1 \quad (2.3.21)$$

It can be easily seen that $g(c^{-1})$ is monotonically increasing with increasing c^{-1}, with $g(0) = 1 - \delta$ and $g(\infty) = \infty$. Thus, (2.3.21) has a unique solution $(c^*)^{-1}$, which implies that the spectral density $f^*(\lambda)$ in (2.3.20) is unique as well.

2.4 SADDLE-POINT THEORY

In this section, we discuss a class of optimization problems termed *saddle-point* or *minimax games*. We first present the principle of those games on two-variable, real, scalar functions, defined on subsets of the two-dimensional Euclidean space.

Consider two players. One of the players selects some value x from a subset A of the real line. The other player selects some value y from another subset B of the real line. The two selections are done independently, and they are then both announced. Upon this announcement, the first player pays the second player a penalty equal to $f(x, y)$, where $f(\cdot, \cdot)$ is a function known to both players called the *payoff function* of the games. The subsets A and B are known to both players. The objective of the first player is to minimize his loss, while the objective of the second player is to maximize his gain. A pair (x^*, y^*) that represents a satisfactory compromise to the objectives of both the players, would be such that,

$$\text{for every } y \text{ in } B, f(x^*, y) \le f(x^*, y^*) \le f(x, y^*)$$
$$\text{for every } x \text{ in } A \text{ and } x^* \in A, y^* \in B \qquad (2.4.1)$$

If a pair, (x^*, y^*), that satisfies (2.4.1) exists, it is called the *saddle point* of the game, while $f(x^*, y^*)$ is then called the *saddle value* of the game. If (x^*, y^*) exists, then

$$f(x^*, y^*) = \inf_{x \in A} \sup_{y \in B} f(x, y) = \sup_{y \in B} \inf_{x \in A} f(x, y) \qquad (2.4.2)$$

We now express a theorem, that states some conditions under which the saddle-point game has a unique solution. Its proof can be found in Rockafellar (1969).

THEOREM 2.4.1: Consider the real, scalar function $f(x, y)$, and the subsets A and B of the real line. Consider the absolute difference metric on the real line, and let at least one of the subsets A, B be compact and the function $f(x, y)$ be continuous in x and y with respect to that metric. Also let both A and B be convex and let $f(x, y)$ be convex-concave on $A \times B$. Then, the game in (2.4.1) has a unique solution (x^*, y^*). ∎

We now proceed with a simple example.

Example 2.4.1

Consider a saddle-point game with payoff function $f(x, y) = x^2 - 7xy - y^2$, where x is such that $|x| \le 1$ and where y is such that $y > \frac{2}{7}$. We observe that the set $|x| \le 1$ is convex and compact, while the set $y > \frac{2}{7}$ is convex but not compact. The function $f(x, y)$ is continuous with respect to both x and y and differentiable, where $(\partial^2/\partial x^2)f(x, y) = 2$, and $(\partial^2/\partial y^2)f(x, y) = -2$. The function $f(x, y)$ is thus strictly convex-concave. So, via Theorem 2.4.1, the game in (2.4.1) has here a unique solution. To find this solution, we write

$$f(x, y) = \left(x - \frac{7}{2}y\right)^2 - \frac{53}{4}y^2$$

Given any y such that $y > \frac{2}{7}$, the function $f(x, y)$ is clearly minimized in $|x| \le 1$ at $x = 1$. That is

$$\inf_{x: |x| \le 1} f(x, y) = \min_{x: |x| \le 1} f(x, y) = \left(1 - \frac{7}{2}y\right)^2 - \frac{53}{4}y^2 = 1 - 7y - y^2 \qquad (2.4.3)$$

2.4 Saddle-Point Theory

The function in (2.4.3) has a unique supremum in $y > \frac{2}{7}$ at $y^* = \frac{2}{7}$. This supremum is not attained; thus, it is not a maximum. Therefore, we have

$$\sup_{y: y > 2/7} \inf_{x: |x| \leq 1} f(x, y) = 1 - 7 \cdot \frac{2}{7} - \left(\frac{2}{7}\right)^2 = -\frac{53}{49} = f\left(1, \frac{2}{7}\right)$$

Since we know in advance that the game has a unique solution, we also have

$$\inf_{x: |x| \leq 1} \sup_{y: y > 2/7} f(x, y) = \sup_{y: y > 2/7} \inf_{x: |x| \leq 1} f(x, y) = f\left(1, \frac{2}{7}\right)$$

The unique saddle-point solution is thus $(x^*, y^*) = (1, \frac{2}{7})$, and the unique saddle value is $f(1, \frac{2}{7}) = -\frac{53}{49}$.

The principle of the saddle-point games can be extended to include payoff functions that are either functionals or stochastic distances, and sets A and B of more abstract nature. Theorem 2.4.1 can then be generalized, with compactness and continuity defined through appropriate metrics. Such metrics can be defined either on functional spaces or on stochastic processes, as will be the case in Chapter 12. We will complete this section with an example of a saddle-point game on functional spaces.

Example 2.4.2

Let us consider the class $\mathcal{M}$ of scalar, real, zero mean, discrete-time stationary processes, whose power spectral densities lie within the class $\mathcal{F}$ in Example 2.3.2. Let us consider the class $\mathcal{O}$ of real asymptotic linear operations that predict the datum X_0 from a process in $\mathcal{M}$, from all the past data x_i; $i \leq -1$. Given some process μ in $\mathcal{M}$ and given some operation g in $\mathcal{O}$, let the payoff function be the mean squared error $e(\mu, g)$ induced by the operation g when X_0 is predicted from x_i; $i \leq -1$ and when all the data are generated by the process μ. Let the power spectral density of the process μ be $f(\lambda)$, $\lambda \in [-\pi, \pi]$. Let us change the game, by considering instead a payoff function $e(f, g)$ that is only a function of the power spectral density of the process, and by considering the power spectral set $\mathcal{F}$ and the class of all asymptotic real linear operations $\mathcal{O}$. Let us adopt the metric $\int |h_1(x) - h_2(x)| \, dx$ on $\mathcal{O}$ and $\mathcal{F}$. Then, it can be seen that both $\mathcal{O}$ and $\mathcal{F}$ are convex, that $\mathcal{F}$ is compact with respect to the latter metric, and that $\mathcal{O}$ is not compact with respect to the same metric. Also, $e(f, g)$ is concave-convex on $\mathcal{F} \times \mathcal{O}$ and continuous in f and g with respect to the above metric. Thus, the generalization of Theorem 2.4.1 guarantees that there exists a unique pair (f^*, g^*) such that

$$\text{for all } f \text{ in } \mathcal{F} \quad e(f, g^*) \leq e(f^*, g^*) \leq e(f^*, g) \quad \text{for all } g \text{ in } \mathcal{O} \quad (2.4.4)$$

The pair (f^*, g^*) can then be uniquely determined by $\sup_{f \in \mathcal{F}} \inf_{g \in \mathcal{O}} e(f, g)$. But as we mentioned in Example 2.3.2, given f, there exists a unique g in $\mathcal{O}$ that attains the infimum $\inf_{g \in \mathcal{O}} e(f, g)$. In particular, this infimum is given by expression (2.3.9) in Example 2.3.2. That is,

$$\inf_{g \in \mathcal{O}} e(f, g) = \exp\left\{ (2\pi)^{-1} \int_{-\pi}^{\pi} \ln[2\pi f(\lambda)] \, d\lambda \right\} \qquad (2.4.5)$$

We thus only have to find the supremum of the expression in (2.4.5) on $\mathcal{F}$. That was done in Example 2.3.2, where f^* is given by (2.3.20) and (2.3.21). Thus, the unique solution of the game in (2.4.4) is comprised of the power spectral density f^* and the unique asymptotic linear mean squared predictor that corresponds to f^*. The game in $\mathcal{M} \times \mathcal{O}$ now has no unique solution. Every process in $\mathcal{M}$ whose power spectral density is f^* satisfies it. Therefore, the game on $\mathcal{M} \times \mathcal{O}$ has infinite solutions. This is so because the set $\mathcal{M}$ is not compact with respect to some appropriate metric.

REFERENCES

Adby, P. and M. Dempster (1974), *Introduction to Optimization Methods*, Halsted Press, Wiley, New York.

Akhiezer, N. (1962), *The Calculus of Variations*, Ginn (Blaisdell), Boston, MA.

Aoki, M. (1971), *Introduction to Optimization Techniques*, Macmillan, New York.

Bellman, R. (1957), *Dynamic Programming*, Princeton Univ. Press, Princeton, NJ.

——— (ed.), (1963), *Mathematical Optimization Techniques*, Univ. of Calif. Press, Berkeley, CA.

Bliss, G. (1925), *Calculus of Variations*, Open Court Publishing Co., Chicago, IL.

Clegg, J. (1968), *Calculus of Variations*, Oliver and Boyd, Edinburgh.

Courant, R. and D. Hilbert (1953), *Methods of Mathematical Physics*, Vol. 1, Interscience, New York.

Fletcher, R. (ed.), (1969), *Optimization*, Academic Press, New York.

Hestenes, M. (1966), *Calculus of Variation and Optimal Control Theory*, Wiley, New York.

——— (1975), *Optimization Theory, the Finite Dimensional Case*, Wiley, New York.

Rockafellar, R. (1969), *Convex Analysis*, Princeton Univ. Press, Princeton, NJ.

CHAPTER 3

PRINCIPLES OF DETECTION AND PARAMETER ESTIMATION THEORIES

3.1 DETECTION-THEORY CONCEPTS

In Chapter 1 we introduced the basic notions in statistical decision theory and we described qualitatively the induced generalized optimization problems. In this section, we concentrate on the part of statistical decision theory that has been termed *hypothesis testing* or *detection*. As we saw in Chapter 1, the basic ingredient here is that the library of stochastic processes available to the analyst has a finite membership. Those stochastic processes are distinct, each represents a different stochastic model, and they can be well known, parametrically known, or nonparametrically described. Let the number of available stochastic models be M. Those models are then indexed and they are called the *hypotheses*. The ith indexed model is hypothesis i, usually denoted H_i; $1 \leq i \leq M$.

An important assumption in hypothesis testing is that there exists some time interval $[0, T]$ during which *one* of the M stochastic processes is continuously active. The analyst knows this time interval precisely, he observes a realization $x(t)$ on it, and his objective is to decide which one of the M stochastic processes or hypotheses has been active. To make this decision, the analyst uses his knowledge about the M stochastic models, the realization $x(t)$, possibly some additional assets, and a performance criterion. He then formulates an optimization problem, whose solution is the *decision rule*. The decision rule is a function of the observed realization $x(t)$ on $[0, T]$, where T may be either fixed in advance or controlled dynamically by the analyst. If T is controlled dynamically, the optimization problem formulated also incorporates a decision as to when the realization from the active stochastic process will stop being observed, at which point the identity of the active stochastic process will be decided upon. The former decision is termed the *stopping rule*, and the hypothesis testing procedure is then called *sequential*.

To quantify our presentation, we will first present some notation. We will consider discrete-time stochastic processes and will denote by X_i the random variable that represents the ith datum generated by the process in time; i is thus the time index. We will denote by X_i^j, $j > i$ the sequence $\{X_i, X_{i+1}, \ldots, X_j\}$ of consecutive random variables generated by the process, and by x_i^j we will denote some realization of this sequence. An infinite-length realization from the process will be denoted x, while X^n and x^n will denote respectively a sequence of n consecutive random variables generated by the process, and some realization of this sequence. In the discrete-time model, an observation time interval $[0, T]$ with T fixed in advance corresponds to operating on realizations x^n of fixed length n. A dynamically controlled T then corresponds to a dynamically controlled length n of the observation sequence x^n. The analyst observes some realization x^n and his objective is to design a decision rule. If n is dynamically controlled, the analyst must also design a stopping rule. Both the decision and stopping rules are functions of the specific realizations they operate upon, and they have a specific form. Below, we define those two rules precisely.

Stopping Rule: $\Delta_n(x)$ Given an infinite-length realization x, the probability that the analyst will decide to stop observing the realization after the nth datum.

(3.1.1)

Decision Rule: $\delta_i(x^n)$ Given the realization x^n, the probability that the
$i = 1, \ldots, M$ analyst will decide that hypothesis H_i has been active.

(3.1.2)

We observe that the stopping and decision rules are both probabilities, conditioned on the observed realizations or data sequences. For a given infinite-length realization x, the sum $\sum_n \Delta_n(x)$ is equal to one; that is, with probability one, data will stop being observed at some point. For given n and given realization x^n, the sum $\sum_{i=1}^{M} \delta_i(x^n)$ is equal to one; that is, for any realization x^n the decision is restricted among the M hypotheses. Let Γ^n be the space where the sequences x^n take their values. Let Γ^∞ be the space where the infinite-length sequences x take their values. A stopping rule $\Delta_n(x)$ basically subdivides the space Γ^∞ into subspaces $\{\Gamma_n^\infty\}$, such that if some sequence x lies in the subspace Γ_n^∞, then stopping after the nth datum is decided. If the subspaces $\{\Gamma_n^\infty\}$ are disjoint, every given sequence x is associated with a single n; that is, exactly n observation data are collected, with probability one. The stopping rule is then called *nonrandomized*. If some subspaces $\Gamma_i^\infty, \Gamma_j^\infty; i \neq j$ overlap, and x is a sequence in their intersection, there are nonzero probabilities that either i or j data will be collected from x. The stopping rule is then called *randomized*. A decision rule $\delta_i(x^n); i = 1, \ldots, M$ subdivides the space Γ^n into M subspaces $\{\Gamma_j^n\}$. If the subspaces $\{\Gamma_j^n\}$ are all disjoint, and some sequence x^n lies in Γ_j^n, then hypothesis H_j is decided with probability one. Each sequence x^n is then associated with a single hypothesis, and the decision rule is called *nonrandomized*. If some sets Γ_j^n and Γ_k^n; $k \neq j$ overlap, and some sequence x^n lies in $\Gamma_j^n \cap \Gamma_k^n$, then

3.1 Detection-Theory Concepts

either one of the hypotheses H_j and H_k will be decided with nonzero probabilities. That is, given x^n, the hypothesis H_j will be decided with some probability $q_j > 0$, and the hypothesis H_k will be decided with probability $q_k > 0$, where $q_j + q_k = 1$. The decision rule is then called *randomized*. In Figure 3.1.1, we exhibit graphically a randomized and a nonrandomized decision rule.

As we saw, the objective of the analyst is to design a decision rule, and possibly a stopping rule as well. As a result of such a design, for every given infinite-length data sequence x some stopping time n, will exist, such that the decision rule will be based upon the subsequence x^n of x. Also, for every given subsequence x^n either a unique hypothesis will be decided with probability one, or the decision will be randomized among a number of hypotheses. The former is the result of a nonrandomized decision rule, while the latter is the result of a randomized decision rule. Both the stopping and decision rules are solutions of optimization problems, whose formalization is based on some performance criterion. Thus, those optimization problems dictate the randomized or nonrandomized nature of the two rules, as well as the appropriate subdivision of the spaces Γ^∞ and Γ^n. Qualitatively speaking, the stopping rule decides how large a portion from an observed sequence x is needed, to make a reliable decision regarding the active hypothesis. The stopping rule is thus highly dependent on the decision rule. Therefore, decision rules must be fully explored before stopping rules are even considered.

For the design of a decision rule, the minimum information available to the analyst consists of the M hypotheses and an observed realization x^n of length n. If the hypotheses are nonparametrically described stochastic processes, then each represents a class of processes, rather than a single well known or parametrically known process. This last case is studied in Chapters 6 and 7. To avoid complications in our presentation at this point, we will assume that each process is well known. Then we will denote by $F_i(x^n)$ the n-dimensional cumulative distribution induced by the hypotheses H_i at the vector point x^n. We will also denote $dF_i(x^n) = f_i(x^n) dx^n$, assuming that the density function $f_i(x^n)$ exists. Let Γ^n be the space where the sequences x^n take their values, and let $\delta = \{\delta_j(x^n); j = 1, \ldots, M\}$ be some decision rule. Given some realization x^n, it is in general possible that while this realization has been generated by hypothesis H_i, the decision rule δ will conclude that it has been generated by hypothesis H_k, where H_k may be different from H_i. Let us denote by $P_{ki}(\delta)$ the conditional

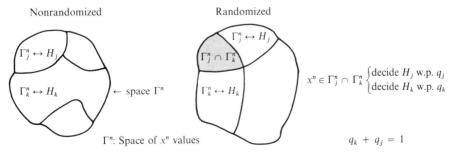

Figure 3.1.1 Nonrandomized and randomized decision rules

probability that H_k is decided given that H_i is true, as induced by the decision rule δ. Then, $P_{ki}(\delta)$ represents the probability mass of those sequences x^n in Γ^n that lead to such decision, and it is given by the following expression.

$$P_{ki}(\delta) = \Pr\{H_k \text{ decided by rule } \delta|_{H_i} \text{ true}\}$$

$$= \int_{\Gamma^n} d\Pr\{H_k \text{ decided by rule } \delta, \text{ and realization } x^n \text{ observed}|_{H_i} \text{ true}\}$$

$$= \int_{\Gamma^n} \Pr\{H_k \text{ decided by rule } \delta|_{x^n}\} d\Pr\{\text{realization } x^n \text{ observed}|_{H_i} \text{ true}\}$$

$$= \int_{\Gamma^n} \delta_k(x^n) dF_i(x^n) = \int_{\Gamma^n} \delta_k(x^n) f_i(x^n) dx^n \qquad (3.1.3)$$

We note that in the derivation of (3.1.3) we have used the theorem of total probability, the Bayes rule, and the fact that the decision induced by the decision rule is independent of the true hypothesis. We also note that, since the decision rule δ consists of probabilities such that $\sum_{j=1}^{M} \delta_j(x^n) = 1 \; \forall \; x^n \in \Gamma^n$, then the following equation is also true.

$$\sum_{k=1}^{M} P_{ki}(\delta) = \int_{\Gamma^n} \left[\sum_{k=1}^{M} \delta_k(x^n)\right] f_i(x^n) dx^n = \int_{\Gamma^n} f_i(x^n) dx^n = 1 \qquad (3.1.4)$$

That is, given the true hypothesis H_i, the decision induced by the decision rule δ will be restricted among the M hypotheses.

Let us now suppose that a set $\{p_i; i = 1, \ldots, M\}$ of a priori probabilities is available. Then p_i denotes the probability that hypothesis H_i is active throughout the duration of the observation time length n. Clearly the set $\{p_i; i = 1, \ldots, M\}$ is such that $\sum_{i=1}^{M} p_i = 1$, and a joint probability $Q_{ki}(\delta)$ can then be defined. $Q_{ki}(\delta)$ denotes the probability that H_k is decided and H_i is truly active, as induced by the decision rule δ. Then, the probability $Q_{ki}(\delta)$ can be expressed as a function of the probabilities p_i and $P_{ki}(\delta)$.

$$Q_{ki}(\delta) = p_i P_{ki}(\delta) = p_i \int_{\Gamma^n} \delta_k(x^n) f_i(x^n) dx^n \qquad (3.1.5)$$

Due to (3.1.4), we can now express the following two equations.

$$\sum_{k=1}^{M} Q_{ki}(\delta) = p_i \sum_{k=1}^{M} P_{ki}(\delta) = p_i \qquad (3.1.6)$$

$$\sum_{i=1}^{M} \sum_{k=1}^{M} Q_{ki}(\delta) = \sum_{i=1}^{M} p_i = 1 \qquad (3.1.7)$$

Expressions (3.1.6) and (3.1.7) state that the decision induced by the decision rule δ is among the M hypotheses, and that only one of those M hypotheses may be active throughout the observation time length n.

To this point, we have assumed that each of the M hypotheses is well known. We will now assume that each such process is parametrically known, instead.

3.1 Detection-Theory Concepts

In particular, let the M hypotheses be determined through a single parametrically known stochastic process, and M disjoint subdivisions of the parameter space. Let Θ^m be a parameter vector of finite dimensionality m and let $\mathscr{E}^m$ be the space where this vector takes its values. Let this space be subdivided into M disjoint subspaces $\mathscr{E}_i^m$; $i = 1, \ldots, M$, such that $\bigcup_{i=1}^{M} \mathscr{E}_i^m = \mathscr{E}^m$. Let $F_{\theta^m}(x^n)$ be the n-dimensional cumulative distribution induced by the parametrically known process for $\Theta^m = \theta^m$, at the vector point x^n, and let $f_{\theta^m}(x^n)$ denote the density function of $F_{\theta^m}(x^n)$. Let a density function $p(\theta^m)$ be given on the space $\mathscr{E}^m$, and let the hypothesis H_i be defined by the parametrically known process for θ^m values in the $\mathscr{E}_i^m$ subspace. Let us suppose that some decision rule δ is available, and let us denote by $P_{k,\theta^m}(\delta)$ the conditional probability that the H_k hypothesis is decided, given that the value of the vector parameter is θ^m. Then, $P_{k,\theta^m}(\delta)$ represents the probability mass of those x^n vectors in Γ^n that lead to such decision, and it is given by

$$P_{k,\theta^m}(\delta) = \int_{\Gamma^n} \delta_k(x^n) f_{\theta^m}(x^n) \, dx^n \qquad (3.1.8)$$

where

$$\sum_{k=1}^{M} P_{k,\theta^m}(\delta) = 1 \qquad \forall \, \theta^m \in \mathscr{E}^m$$

The density function $p(\theta^m)$ on $\mathscr{E}^m$ determines the probability masses that correspond to the a priori probabilities $\{p_i; i = 1, \ldots, M\}$ of the M hypotheses. It also determines the conditional probabilities $P_{ki}(\delta)$; $k, i = 1, \ldots, M$ parallel to those in (3.1.3), and the joint probabilities $Q_{ki}(\delta)$; $i = 1, \ldots, M$ parallel to those in (3.1.5). Indeed, we obtain then in a straightforward fashion the following expressions.

$$p_i = \int_{\mathscr{E}_i^m} p(\theta^m) \, d\theta^m \qquad (3.1.9)$$

$$P_{ki}(\delta) = p_i^{-1} \int_{\mathscr{E}_i^m} P_{k,\theta^m}(\delta) p(\theta^m) \, d\theta^m$$

$$= \left[\int_{\mathscr{E}_i^m} d\theta^m p(\theta^m) \right]^{-1} \int_{\Gamma^n} \delta_k(x^n) d \left[\int_{\mathscr{E}_i^m} F_{\theta^m}(x^n) p(\theta^m) \, d\theta^m \right]$$

$$= \left[\int_{\mathscr{E}_i^m} p(\theta^m) \, d\theta^m \right]^{-1} \int_{\Gamma^n} dx^n \, \delta_k(x^n) \int_{\mathscr{E}_i^m} f_{\theta^m}(x^n) p(\theta^m) \, d\theta^m \qquad (3.1.10)$$

$$Q_{ki}(\delta) = \int_{\mathscr{E}_i^m} P_{k,\theta^m}(\delta) p(\theta^m) \, d\theta^m = \int_{\Gamma^n} dx^n \, \delta_k(x^n) \int_{\mathscr{E}_i^m} f_{\theta^m}(x^n) p(\theta^m) \, d\theta^m \qquad (3.1.11)$$

Let us now consider the case where the hypotheses are two, and they are determined through a single parametrically known stochastic process and two disjoint subdivisions of the parameter space $\mathscr{E}^m$. Let us denote those two hypotheses H_0 and H_1. Let $\delta_k(x^n)$; $k = 0, 1$; $x^n \in \Gamma^n$, be some decision rule, where $\delta_1(x^n) + \delta_0(x^n) = 1 \, \forall \, x^n \in \Gamma^n$. We are then presented with a special case of *single hypothesis testing.* As we mentioned in Chapter 1, it is then customary that one of the two hypotheses be given more emphasis. Let this hypothesis be H_1. Then, since $\delta_0(x^n) = 1 - \delta_1(x^n)$; for every x^n in Γ^n, we can drop the indices in the decision rule, and we can denote by $\delta(x^n)$ the probability $\delta_1(x^n)$. We can also then denote by $P_{\theta^m}(\delta)$ the conditional probability $P_{1,\theta^m}(\delta)$ in (3.1.8), where,

$$P_{\theta^m}(\delta) = \int_{\Gamma^n} \delta(x^n) f_{\theta^m}(x^n) \, dx^n \tag{3.1.12}$$

The expression in (3.1.12) represents the probability that the more emphatic hypothesis H_1 is decided, conditioned on the value θ^m of the vector parameter. This probability is determined by the decision rule δ and is a function of the parameter value θ^m. As such a function, the probability $P_{\theta^m}(\delta)$ is called *the power function of the decision rule* δ, since it provides the probability with which the emphatic hypothesis is decided for each fixed parameter value θ^m.

Given M parametrically known hypotheses, their a priori probabilities, and some decision rule δ: $\{\delta_k(x^n); k = 1, \ldots, M\}$, we denote by $P_e(\delta)$ the probability that the decision induced by the decision rule δ is erroneous; that is H_k is decided and $H_i; i \neq k$ is true for $k = 1, \ldots, M$. We denote by $P_d(\delta)$ the probability that the decision induced by δ is correct; that is H_k is decided and H_k is true for $k = 1, \ldots, M$. Then, we have

$$P_e(\delta) = \sum_{k \neq i} Q_{ki}(\delta) \tag{3.1.13}$$

$$P_d(\delta) = 1 - P_e(\delta) = \sum_{k=1}^{M} Q_{kk}(\delta) \tag{3.1.14}$$

where, depending on the specific parametric hypothesis model considered, $Q_{ki}(\delta)$ is the joint probability in either (3.1.5) or (3.1.11).

The probability $P_e(\delta)$ is called the *probability of error* induced by the decision rule δ, and it equals the probability mass of those sequences x^n in Γ^n that lead to erroneous decision through the mapping performed by δ. The probability $P_d(\delta)$ is called the *probability of correct decision* or *detection* induced by δ. Either one of the two probabilities $P_e(\delta)$ and $P_d(\delta)$ can be adopted as a performance measure for the decision rule δ. Qualitatively speaking, a "good" decision rule δ should then induce a low probability of error $P_e(\delta)$, or, equivalently, a high probability of correct detection $P_d(\delta)$. If the above performance measure is adopted and there exists a whole class $\mathscr{D}$ of decision rules δ, then the "best" rule δ^* in $\mathscr{D}$ is the one that induces the lowest $P_e(\delta)$ or equivalently the highest $P_d(\delta)$ in $\mathscr{D}$. That is $P_e(\delta^*) \leq P_e(\delta) \ \forall \ \delta \in \mathscr{D}$, and equivalently $P_d(\delta^*) \geq P_d(\delta) \ \forall \ \delta \in \mathscr{D}$, where, in general, there may be more than one "best" rule, all performing equally well.

In addition to the M parametrically known hypotheses and their a priori probabilities, the analyst is sometimes also equipped with a set of real *penalty coefficients* or *functions*. A set $\{c_{ki}; k, i = 1, \ldots, M\}$ of penalty coefficients is such that $c_{ki} \geq 0 \ \forall \ k, i$, where c_{ki} is the penalty paid when hypothesis H_k is decided and hypothesis H_i is true. The implication behind the condition that each coefficient c_{ki} is nonnegative is that there is no gain associated with any

3.1 Detection-Theory Concepts

decision, thus the term penalty. If the M parametrically known hypotheses are determined through a single parametrically known stochastic process and M disjoint subdivisions of a parameter space $\mathscr{E}^m$, then a set $\{c_k(\theta^m); k = 1, \ldots, M\}$ of penalty functions may be given instead. The domain of each function $c_k(\theta^m)$ is the space $\mathscr{E}^m$, the function is nonnegative, and it signifies the penalty paid when hypothesis H_k is decided and the parameter value θ^m is true. If $p(\theta^m)$ is a given density function defined on the space $\mathscr{E}^m$, then for any given decision rule $\delta: \{\delta_k(x^n); k = 1, \ldots M\}$ the induced *expected penalty* $C(\delta)$ can be expressed as follows.

$$C(\delta) = \sum_{k=1}^{M} \int_{\mathscr{E}^m} c_k(\theta^m) P_{k,\theta^m}(\delta) p(\theta^m) \, d\theta^m$$

$$= \int_{\Gamma^n} dx^n \sum_{k=1}^{M} \delta_k(x^n) \int_{\mathscr{E}^m} c_k(\theta^m) f_{\theta^m}(x^n) p(\theta^m) \, d\theta^m \quad (3.1.15)$$

where the conditional probability $P_{k,\theta^m}(\delta)$ is given by (3.1.8).

If a set of penalty coefficients $\{c_{ki}; k, i = 1, \ldots, M\}$ is provided, the M parametrically known hypotheses are arbitrary, and some decision rule δ is given, then the induced expected penalty $C(\delta)$ is given by

$$C(\delta) = \sum_{k=1}^{M} \sum_{i=1}^{M} c_{ki} Q_{ki}(\delta) \quad (3.1.16)$$

where, depending on the particular parametric hypothesis model, $Q_{ki}(\delta)$ is the joint probability in either (3.1.5) or (3.1.11).

If available, the expected penalty $C(\delta)$ can be used as a performance measure for the decision rule δ. Then, the "best" decision rule δ^* within a whole class $\mathscr{D}$ of decision rules is such that $C(\delta^*) \leq C(\delta) \,\forall\, \delta \in \mathscr{D}$. Again, there may be, in general, more than one "best" rule, all performing equally well.

When a set $\{c_k(\theta^m); k = 1, \ldots, M\}$ of penalty functions defined on $\mathscr{E}^m$ is given, but no density function $p(\theta^m)$ on the space $\mathscr{E}^m$ is available, then for some given decision rule $\delta: \{\delta_k(x^n): k = 1, \ldots, M\}$ a *conditional expected penalty* $C(\delta, \theta^m)$; $\theta^m \in \mathscr{E}^m$ can be expressed. This conditional penalty is a function of the parameter value θ^m given by

$$C(\delta, \theta^m) = \sum_{k=1}^{M} c_k(\theta^m) P_{k,\theta^m}(\delta)$$

$$= \int_{\Gamma^n} dx^n \sum_{k=1}^{M} \delta_k(x^n) c_k(\theta^m) f_{\theta^m}(x^n) \quad (3.1.17)$$

where the conditional probability $P_{k,\theta^m}(\delta)$ is given by (3.1.8).

The conditional expected penalty function $C(\delta, \theta^m)$ can be used as a performance measure for the decision rule δ. A "good" rule would induce relatively low $C(\delta, \theta^m)$ values for all θ^m in $\mathscr{E}^m$. In fact, if there exist two decision rules $\delta^{(1)}$ and $\delta^{(2)}$ whose induced conditional expected functions are such that $C(\delta^{(1)}, \theta^m) \leq C(\delta^{(2)}, \theta^m) \,\forall\, \theta^m \in \mathscr{E}^m$, then rule $\delta^{(2)}$ should be rejected in the presence of rule $\delta^{(1)}$.

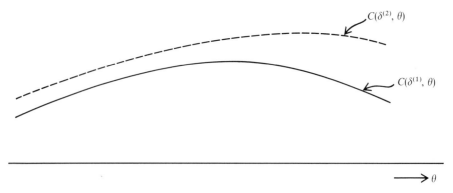

Figure 3.1.2 Decision rules $\delta^{(1)}$ and $\delta^{(2)}$–conditional expected penalty
Reject $\delta^{(2)}$ in the presence of $\delta^{(1)}$

This is shown in Figure 3.1.2, for $\mathscr{E}^m$ the real line. Now let there be two decision rules $\delta^{(1)}$ and $\delta^{(2)}$ such that $C(\delta^{(1)}, \theta^m) < C(\delta^{(2)}, \theta^m)$ for some θ^m in $\mathscr{E}^m$, and $C(\delta^{(1)}, \theta^m) > C(\delta^{(2)}, \theta^m)$ for the remaining θ^m values. Then, neither rule is uniformly (for every θ^m in $\mathscr{E}^m$) worse than the other. Thus, to select one of the two rules, we must devise a criterion other than uniform superiority. For example, we may say that we will prefer rule $\delta^{(1)}$ over rule $\delta^{(2)}$, if $\sup_{\theta^m \in \mathscr{E}^m} C(\delta^{(1)}, \theta^m) < \sup_{\theta^m \in \mathscr{E}^m} C(\delta^{(2)}, \theta^m)$. This criterion corresponds to selecting the rule that induces the minimum maximum penalty (across $\mathscr{E}^m$); such a case is exhibited in Figure 3.1.3, for $\mathscr{E}^m$ the real line. Thus, the "best" rule δ^* within a class of rules will then be such that $\sup_{\theta^m \in \mathscr{E}^m} C(\delta^*, \theta^m) \leq \sup_{\theta^m \in \mathscr{E}^m} C(\delta, \theta^m) \ \forall \ \delta \in \mathscr{D}$, and, in general, there may be more than one "best" such rule.

If we are presented with the single hypothesis problem, the two hypotheses H_0 and H_1 are determined through a single parametrically known process and two disjoint subdivisions of the parameter space $\mathscr{E}^m$, the emphatic hypothesis is H_1, and neither a penalty function nor an a priori distribution on the parameter space are available, then the performance measure for any given decision

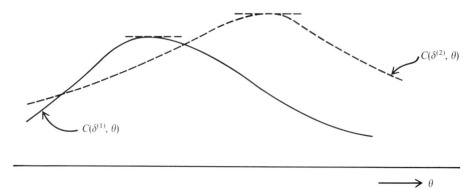

Figure 3.1.3 Decision rules $\delta^{(1)}$ and $\delta^{(2)}$–conditional expected penalty
Select rule $\delta^{(1)}$

3.1 Detection-Theory Concepts

rule δ must be based solely on the power function $P_{\theta^m}(\delta)$ in (3.1.12). Qualitatively speaking, it is then desirable that the function's values at any θ^m in $\mathscr{E}_1^m$ be high as compared to its values at any θ^m in $\mathscr{E}_0^m$, where $\mathscr{E}_1^m$ is the parameter subspace that corresponds to hypothesis H_1 and the subspace $\mathscr{E}_0^m = \mathscr{E}^m - \mathscr{E}_1^m$ corresponds to hypothesis H_0. Indeed, the probability with which H_1 is decided when H_1 is true is then high as compared to the probability that the same decision is made while H_0 is true. In Figure 3.1.4, we exhibit graphically the power functions of two decision rules $\delta^{(1)}$ and $\delta^{(2)}$ for a parameter space corresponding to the real line R (thus, scalar parameter). Decision rule $\delta^{(1)}$ performs well, while decision rule $\delta^{(2)}$ performs worse than $\delta^{(1)}$ for every θ on the real line. In the presence of $\delta^{(1)}$, $\delta^{(2)}$ should be thus rejected. In general, given some decision rule δ, the supremum $\sup_{\theta^m \in \mathscr{E}_0^m} P_{\theta^m}(\delta)$ is called the *false alarm* induced by δ. For some θ^m value in $\mathscr{E}_1^m$, the corresponding value $P_{\theta^m}(\delta)$ of the power function is then called the *power* induced by the decision rule δ at θ^m. If the subspaces $\mathscr{E}_0^m$ and $\mathscr{E}_1^m$ are fixed and there is a fixed vector value θ_1^m in $\mathscr{E}_1^m$, then the "best" decision rule δ^* among all the rules within a class $\mathscr{D}$ is the one that induces the highest power at θ_1^m, subject to a false alarm constraint. That is, $P_{\theta_1^m}(\delta^*) \geq P_{\theta_1^m}(\delta) \ \forall \ \delta \in \mathscr{D}$, where $\sup_{\theta^m \in \mathscr{E}_0^m} P_{\theta^m}(\delta) \leq \alpha \ \forall \ \delta \in \mathscr{D}$, for some given positive constant α. In general, there may be more than one "best" rule, all performing equally well.

To this point, we have basically stated various performance criteria for the evaluation of decision rules. The appropriate performance criterion will be selected according to the assets available to the analyst, where those assets themselves reflect the nature of the particular physical phenomenon considered. Through the selected performance criterion, an "optimal" decision rule can be designed. Such an optimal decision rule will "best" satisfy the performance criterion, within some class $\mathscr{D}$ of decision rules. If a number of decision rules in $\mathscr{D}$, all "best" satisfy the performance criterion, then those rules are initially called *admissible*, and the analyst searches for the one among them that is the simplest to implement. The remaining rules in $\mathscr{D}$ are then called *inadmissible* and are automatically rejected.

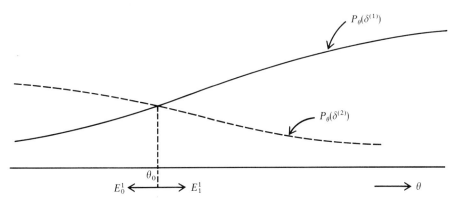

Figure 3.1.4 Decision rules $\delta^{(1)}$ and $\delta^{(2)}$–power functions Reject $\delta^{(2)}$ in the presence of $\delta^{(1)}$

As we have already stated, if the analyst is controlling the observation interval, then, in addition to the decision rule, he must also design a stopping rule. The stopping rule is designed after the decision rule has been selected, and it depends directly on the explicit form of the latter. Thus, we will state and analyze stopping rules after we have explicitly stated and analyzed the various decision rules. This will be done in Chapter 8.

In the next section, we will state the various existing detection or hypothesis testing schemes, as they evolve naturally according to the assets available to the analyst and the resulting performance criterion.

3.2 DETECTION SCHEMES

In this section, we will concentrate strictly on the various decision rules, as they evolve from the assets available to the analyst and the subsequently adopted performance criterion. Thus, we will assume a fixed-length observation interval, which in the discrete-time case is signified by a fixed number n of observations. We will then use the same notation as in Section 3.1. Below, we will list the richest possible set of assets available to the analyst. We will be subtracting some of those assets as the various hypothesis testing schemes evolve. As in Section 3.1, we will consider parametrically known hypotheses only. The case of nonparametrically described hypotheses will be dealt with in Chapters 6 and 7. There, the hypothesis testing schemes are adaptations of the parallel schemes for parametrically known hypotheses.

Let us consider the following list of all the assets available to the analyst.

i. A library of M distinct, parametrically known, discrete-time hypotheses $H_i; i = 1, \ldots, M$.
ii. A realization x^n from the underlying hypothesis.
iii. Either a set $p_i; i = 1, \ldots, M$ of a priori probabilities on the M hypotheses or a density function $p(\theta^m)$ on the parameter space $\mathscr{E}^m$, if the hypotheses are determined through a single parametrically known stochastic process, and M disjoint subdivisions of the parameter space $\mathscr{E}^m$.
iv. Either a set $\{c_{ki}; k, i = 1, \ldots, M\}$ of penalty coefficients or a set $\{c_k(\theta^m); k = 1, \ldots, M\}$ of penalty functions on $\mathscr{E}^m$, if the M hypotheses are determined through a single parametrically known stochastic process, and M disjoint subdivisions of the parametric space $\mathscr{E}^m$.
v. A performance criterion that the analyst may adopt, according to the available assets.

Some of the assets listed above may not in general be available. The minimum set of assets that is always available consists of those in i and ii. The performance criterion suffers limitations as the number of the remaining available assets decreases.

Let us first assume that all assets in i to iv are available. The existence of asset iv means that there exist specific costs paid by the system, when wrong

3.2 Detection Schemes

decisions are made. The analyst should then select the expected penalty $C(\delta)$ in either (3.1.15) or (3.1.16) as the performance criterion per decision rule $\delta: \{\delta_k(x^n); k = 1, \ldots, M\}$. Given a class $\mathscr{D}$ of decision rules, an "optimal" rule δ^* minimizes the expected penalty $C(\delta)$ in $\mathscr{D}$. This rule guarantees a minimum average cost due to wrong decisions, and it may not be unique. If uniqueness is not satisfied, there exist a number of admissible rules. Among them, the decision rule that is simplest to implement will be selected, to reduce system complexity. Given a class $\mathscr{D}$ of decision rules, an "optimal" rule δ^* in $\mathscr{D}$ is then such that $C(\delta^*) \leq C(\delta) \ \forall \ \delta \in \mathscr{D}$, where this rule may not be unique. If uniqueness is not satisfied, there exist a number of admissible rules, and among them, the one that is simplest to implement will be selected.

If assets i to iii are available, and asset iv is either unavailable or the analyst decides to ignore it, then each decision rule δ should be evaluated on the basis of the induced probability of error $P_e(\delta)$ in (3.1.13). Given a class $\mathscr{D}$ of decision rules, an "optimal" rule δ^* minimizes the probability of error $P_e(\delta)$ or equivalently maximizes the probability of correct detection $P_d(\delta)$ in $\mathscr{D}$. Again, there may not exist a unique optimal rule, but a whole class of admissible rules, instead. Then, among the latter, the simplest to implement will be selected.

When at least the assets i to iii are available and fully utilized, the "optimal" rules are the solutions of optimization problems on some class $\mathscr{D}$ of decision rules, where the optimization function is either $P_e(\delta)$ or $C(\delta)$; $\delta \in \mathscr{D}$, and where the objective is minimization of that function. The hypothesis testing problems that are based on the above formalization are called *Bayesian*, and so are the resulting "optimal" decision rules. The basic ingredient in Bayesian formalizations is thus the availability and use of either a set of a priori probabilities on the hypotheses, or a density function $p(\theta^m)$ on the appropriate parameter space $\mathscr{E}^m$—that is, asset iii. Bayesian problems and their solutions are fully stated and analyzed in Chapter 4.

Now let assets i, ii, and iv be available; let asset iii be unavailable. Then, utilizing the available assets fully, the analyst will evaluate some decision rule δ using the induced conditional expected penalty $C(\delta, \theta^m)$ in (3.1.17). As we saw in Section 3.1, this penalty corresponds to hypotheses determined through a single parametrically known stochastic process and M disjoint subdivisions of the parameter space $\mathscr{E}^m$. We note here that if the M hypotheses are well known instead, then the conditional expected penalties take the simple form $C(\delta, H_i) = \sum_{k=1}^{M} c_{ki} P_{ki}(\delta)$, where the probability $P_{ki}(\delta)$ is given by (3.1.3). Thus, in more general terms, we may consider only conditional expected penalties of the form $C(\delta, \theta^m)$. The analyst uses $C(\delta, \theta^m)$ as the optimization function on some class $\mathscr{D}$ of decision rules. Then he first rejects the uniformly inadmissible rules in $\mathscr{D}$, where some rule $\delta^{(2)}$ in $\mathscr{D}$ is uniformly inadmissible if there exists some other rule $\delta^{(1)}$ in $\mathscr{D}$ such that $C(\delta^{(1)}, \theta^m) \leq C(\delta^{(2)}, \theta^m) \ \forall \ \theta^m \in \mathscr{E}^m$. Among the remaining rules, the analyst searches for those that are "optimal," where some rule δ^* in $\mathscr{D}$ is optimal if $\sup_{\theta^m \in \mathscr{E}^m} C(\delta^*, \theta^m) \leq \sup_{\theta^m \in \mathscr{E}^m} C(\delta, \theta^m) \ \forall \ \delta \in \mathscr{D}$. The latter rules comprise the final set of admissible rules in $\mathscr{D}$. Among those, the analyst will finally select the rule that is the simplest to implement. The described procedure eventually isolates some decision rule in $\mathscr{D}$ that induces the minimum maximum value of the conditional expected penalty and protects the system against the most costly case. The formalization and the resulting "optimal" rules have been termed *minimax*. The minimax formalizations and resulting decision rules will be fully analyzed in Chapter 6. As we will see there, the procedure corresponds to a search for a *least favorable* parameter density function $p(\theta^m)$, which basically assigns mass equal to one at the parameter value θ_1^m that realizes the supremum of the penalty $C(\delta^*, \theta^m)$ for the optimal rule δ^*.

Let us now suppose that only the minimum set of assets, i and ii, is available. Then we will limit ourselves to the case of single hypothesis testing, and we will denote the two hypotheses by H_0 and H_1. We will further assume that hypothesis H_1 has been identified as the more emphatic, and that both hypotheses can be described by a single parametrically known stochastic process and two disjoint subdivisions of the parameter space $\mathscr{E}^m$. Then the analyst first formulates the power function $P_{\theta^m}(\delta)$ in (3.1.12) for any decision rule δ in some class $\mathscr{D}$, and rejects the uniformly inadmissible rules in $\mathscr{D}$. If $\mathscr{E}_i^m$; $i = 0, 1$ is the parameter subspace that determines hypothesis H_i; $i = 0, 1$, a rule $\delta^{(2)}$ in $\mathscr{D}$ is uniformly inadmissible if there exists some other rule $\delta^{(1)}$ in $\mathscr{D}$ such that $P_{\theta^m}(\delta^{(1)}) > P_{\theta^m}(\delta^{(2)}) \, \forall \, \theta^m \in \mathscr{E}_1^m$ and $P_{\theta^m}(\delta^{(1)}) < P_{\theta^m}(\delta^{(2)}) \, \forall \, \theta^m \in \mathscr{E}_0^m$. Then the analyst selects some parameter value θ_1^m in $\mathscr{E}_1^m$ and searches among the nonuniformly inadmissible rules in $\mathscr{D}$ for a rule that induces the highest power at θ_1^m, subject to a false alarm constraint. The false alarm constraint is thus an additional asset with which the analyst must be provided; it is specified by a positive constant $\alpha < 1$ called the *false alarm rate*. The problem is formulated as a constraint optimization problem, and an optimal rule δ^* in $\mathscr{D}$ is such that $P_{\theta_1^m}(\delta^*) \geq P_{\theta_1^m}(\delta) \, \forall \, \delta \in \mathscr{D}$, where $\sup_{\theta^m \in \mathscr{E}_0^m} P_{\theta^m}(\delta) < \alpha \, \forall \, \delta \in \mathscr{D}$. Any set of rules in $\mathscr{D}$ that satis-

Table 3.2.1
Classes of Hypothesis Testing Schemes for Parametrically Known Hypotheses,
Given a Class of Decision Rules $\mathscr{D}$

Assets Used	Optimization Function	Optimal Rule δ^*	Classification
i, ii, iii, iv	$C(\delta) \, \forall \, \delta \in \mathscr{D}$	$\delta^*: C(\delta^*) \leq C(\delta)$ $\forall \, \delta \in \mathscr{D}$	Bayesian
i, ii, iii	$P_e(\delta) \, \forall \, \delta \in \mathscr{D}$	$\delta^*: P_e(\delta^*) \leq P_e(\delta)$ $\forall \, \delta \in \mathscr{D}$	Bayesian
i, ii, iv	$C(\delta, \theta^m) \, \forall \, \delta \in \mathscr{D}$	$\delta^*: \sup_{\theta^m \in \mathscr{E}^m} C(\delta^*, \theta^m)$ $\leq \sup_{\theta^m \in \mathscr{E}^m} C(\delta, \theta^m)$ $\forall \, \delta \in \mathscr{D}$	Minimax
i, ii, false alarm rate α single hypothesis	Given θ_1^m in $\mathscr{E}_1^m$, $P_{\theta_1^m}(\delta)$, subject to $\sup_{\theta^m \in \mathscr{E}_0^m} P_{\theta^m}(\delta) \leq \alpha$ $\forall \, \delta \in \mathscr{D}$	$\delta^*: P_{\theta_1^m}(\delta^*) \geq P_{\theta_1^m}(\delta)$ $\forall \, \delta \in \mathscr{D}$ and $\sup_{\theta^m \in \mathscr{E}_0^m} P_{\theta^m}(\delta) \leq \alpha$ $\forall \, \delta \in \mathscr{D}$	Neyman-Pearson

fies the above conditions is a set of admissible rules; among those the rule that is simplest to implement will be selected. The present formalization and the resulting "solution" are termed *Neyman-Pearson* after the statisticians who first proposed and analyzed it. The *Neyman-Pearson rule* guarantees the highest possible recognition power to the emphatic hypothesis, while restricting false recognition within acceptable limits. We note that to formulate and solve the problem, we assumed that some specific vector value θ_1^m in $\mathscr{E}_1^m$ was selected a priori. This was done because in general there is no unique decision rule that will induce the highest power in $\mathscr{D}$ for every θ^m in $\mathscr{E}_1^m$. If such a rule exists, it is called *uniformly most powerful*. The full statement and analysis of the Neyman-Pearson formalization, and the search for uniformly most powerful rules are provided in Chapter 5.

We close the present section with Table 3.2.1, where we list all the hypothesis testing schemes discussed above. In the last section of this chapter, we will present some applications and examples.

3.3 CONCEPTS IN PARAMETER ESTIMATION

In Chapter 1, we introduced the basic notions pertaining to statistical parameter estimation theory. As we saw, the basic ingredient that distinguishes the parameter estimation concept from the hypothesis testing principle, is the presence of an infinite number of alternatives in the former. Those alternatives are then represented by some m-dimensional vector parameter Θ^m that takes values in some m-dimensional space, $\mathscr{E}^m$. The decision pertinent to the problem refers to the acting value θ^m of the parameter vector Θ^m; and its outcome is termed the *parameter estimate*.

The basic elements in the parameter estimation model are the vector parameter Θ^m the space $\mathscr{E}^m$ where Θ^m takes its values, and a stochastic process $X(t)$ which is parametized by $\theta^m \in \mathscr{E}^m$, and which is either parametrically known or nonparametrically described. If the latter process is parametrically known, it is well known for each fixed value θ^m of the vector parameter Θ^m and the *parametric parameter estimation* schemes arise. If the process is, instead, nonparametrically described, then given $\theta^m \in \mathscr{E}^m$ the process is a member of some class $\mathscr{F}_{\theta^m}$ of processes, and the *nonparametric* or *robust parameter estimation* schemes evolve. In all cases, the main assumption is that the true acting process (and thus the true value of the parameter vector as well) remains unchanged throughout the observation interval. As in the hypothesis testing formalizations, the observation interval is represented by the observation vector x^n of length n in the case of discrete-time processes. If n is fixed, the *nonsequential parameter estimation* schemes arise. If, instead, the length n can be controlled by the analyst, the *sequential parameter estimation* schemes evolve. Given n and x^n, the parameter estimate is a function of the observation vector x^n denoted $\hat{\theta}^m(x^n)$. In the case of continuous-time processes, the observation interval is $[0, T]$. Given T and an observed waveform $x(t); t \in [0, T]$, the parameter estimate is denoted $\hat{\theta}^m(x(t); t \in [0, T])$.

Given discrete-time processes and an observation vector x^n the parameter estimate $\hat{\theta}^m(x^n)$ is, in general, the solution of some optimization problem. The specific formalization of the optimization problem and its solution vary with the assets available to the analyst. For different assets, different parameter estimation schemes evolve; these will be presented in the next section. Given a specific parameter estimate $\hat{\theta}^m(x^n)$, given θ^m in $\mathscr{E}^m$, and given the stochastic process that generates the observation vector x^n, the statistical behavior of the stochastic vector, $\hat{\theta}^m(X^n)$, is well defined. If the parameter vector, Θ^m takes real values only, then given θ^m, the Euclidean norm $\|\theta^m - E\{\hat{\theta}^m(X^n)\}\|^{1/2}$ is then called the *bias* of the estimate $\hat{\theta}^m(x^n)$ at the process. The bias clearly measures the distance between the expected value of the estimate and the true value θ^m of the parameter. If the bias is zero for all θ^m in $\mathscr{E}^m$, then the estimate $\hat{\theta}^m(x^n)$ is called *unbiased* at the process. The bias of the estimate $\hat{\theta}^m(x^n)$ and the conditional variance $E\{\|\hat{\theta}^m(X^n) - E\{\hat{\theta}^m(X^n)\}\|^2|\theta^m\}$ generally present a tradeoff. Indeed, an unbiased estimate may induce relatively large variance. On the other hand, the introduction of some low-level bias, may then result in a significant reduction of the induced variance. In general, the bias versus variance tradeoff should be studied carefully for the correct evaluation of any given parameter estimate. A parameter estimate $\hat{\theta}^m(x^n)$ is called *efficient* at the process, if the conditional variance $E\{\|\theta^m - \hat{\theta}^m(X^n)\|^2|\theta^m\}$ equals a lower bound known as the Rao-Cramèr bound (see Chapter 9).

Let $\mathscr{E}^m$ denote the m-dimensional space where the parameter vector θ^m takes its values. Given $\theta^m \in \mathscr{E}^m$, a data sequence, x^n, and a parameter estimate $\hat{\theta}^m(x^n)$, a penalty function $c[\theta^m, \hat{\theta}^m(x^n)]$ is a scalar nonnegative function whose values vary, as θ^m varies in $\mathscr{E}^m$, and as the sequence x^n takes different values in the n-dimensional space, Γ^n. Given θ^m in $\mathscr{E}^m$ and some well known stochastic process, the conditional expected penalty $c(\theta^m, \hat{\theta}^m)$ induced by the parameter estimate $\hat{\theta}^m(x^n)$ and the penalty function $c[\theta^m, \hat{\theta}^m(x^n)]$ at the process are defined as follows,

$$c(\theta^m, \hat{\theta}^m) = E\{c[\theta^m, \hat{\theta}^m(X^n)]|\theta^m\} = \int_{\Gamma^n} c[\theta^m, \hat{\theta}^m(x^n)] f_{\theta^m}(x^n) dx^n \quad (3.3.1)$$

where $f_{\theta^m}(x^n)$ denotes the n-dimensional density function induced by the process at θ^m, at the vector point x^n.

If an a priori density function $p(\theta^m)$ is available, defined on the parameter space $\mathscr{E}^m$ then, given a parameter estimate $\hat{\theta}^m(x^n)$, a penalty function, $c[\theta^m, \hat{\theta}^m(x^n)]$, and the process, the expected penalty $c(\hat{\theta}^m, p)$ is defined as

$$\begin{aligned} c(\hat{\theta}^m, p) &= \int_{\mathscr{E}^m} c(\theta^m, \hat{\theta}^m) p(\theta^m) d\theta^m \\ &= \int_{\Gamma^n} \int_{\mathscr{E}^m} dx^n c[\theta^m, \hat{\theta}^m(x^n)] f_{\theta^m}(x^n) p(\theta^m) d\theta^m \end{aligned} \quad (3.3.2)$$

The various existing parameter estimation schemes evolve as the solutions of optimization problems, whose objective function is either the conditional expected penalty in (3.3.2) or the conditional density function $f_{\theta^m}(x^n)$. We will outline the various estimation schemes and their evolution in the next section.

3.4 PARAMETER ESTIMATION SCHEMES

As with the detection procedures, the various parameter estimation schemes evolve from the assets available to the analyst, and the subsequently adopted performance criteria. The nonsequential estimation schemes imply the use of an observation vector x^n of fixed length n, while the sequential schemes evolve via appropriate modifications of the former, when control of the length n is permissible. In this section, we will restrict ourselves to the consideration of nonsequential parameter estimation schemes only, and without lack in generality we will only consider discrete-time processes. The fullest possible list of assets available to the analyst is then:

i. A parametrically known, discrete-time stochastic process parametized by θ^m, where θ^m is a well defined parameter vector of finite dimensionality m.
ii. A realization x^n from the underlying active process, where the implied assumption is that this process remains unchanged, throughout the overall observation interval.
iii. A parameter space $\mathscr{E}^m$.
iv. A density function, $p(\theta^m)$, defined on the parameter space, $\mathscr{E}^m$.
v. For each data sequence x^n, parameter vector θ^m, parameter estimate $\hat{\theta}^m(x^n)$, a penalty scalar function $c[\theta^m, \hat{\theta}^m(x^n)]$ is given.
vi. A performance criterion, consistent with the available assets.

The assets actually available may be a subset of the assets i–v listed above. When all the assets, i–v are available, the *Bayesian* parameter estimation schemes evolve. The performance criterion is then the minimization of the expected penalty $c(\hat{\theta}^m, p)$ in (3.3.2) with respect to the estimate $\hat{\theta}^m(x^n)$ for all x^n in the observation space Γ^n. The estimate $\hat{\theta}_0^m$ that minimizes the expected penalty $c(\hat{\theta}^m, p)$ is called the *optimal Bayesian estimate at p*. An estimate $\hat{\theta}^m$ is called admissible if it is the optimal Bayesian estimate at some stochastic process and some a priori parameter density function p.

When the assets, i–iii and v, are available, and asset iv is not, the *minimax* parameter schemes evolve. Those schemes are solutions of saddle-point game formalizations, with payoff function the expected penalty $c(\hat{\theta}^m, p)$ and with variables the parameter estimate $\hat{\theta}^m$ and the a priori parameter density function p. If a minimax estimate $\hat{\theta}_0^m$ exists, it is an optimal Bayesian estimate, at some least favorable a priori distribution p_0; thus, it is *admissible*.

In the case where only the assets i–iii are available to the analyst, an appropriate performance criterion is the maximization of the conditional density function, $f_{\theta^m}(x^n)$, with respect to the parameter value θ^m in $\mathscr{E}^m$. Given the observation sequence x^n, the parameter value that maximizes $f_{\theta^m}(x^n)$ is, in general, a function of x^n denoted $\hat{\theta}^m(x^n)$ and called the *maximum likelihood estimate* at the process represented by the density function $f_{\theta^m}(x^n)$.

The Bayesian, minimax, and maximum likelihood estimation schemes, described above, comprise the class of the parametric parameter estimation procedures. The common characteristic of those procedures is the availability of

Table 3.4.1
Classes of Parameter Estimation Schemes

Assets Used	Optimization Function	Optimal Estimate	Classification
i–v	$c(\hat{\theta}^m, p)$	$\hat{\theta}_0^m: c(\hat{\theta}_0^m, p) \leq c(\hat{\theta}^m, p^m) \; \forall \; \hat{\theta}^m$	Bayesian
i–iii, v	$c(\hat{\theta}^m, p)$	$\hat{\theta}_0^m: \exists \; p_0: c(\hat{\theta}_0^m, p) \leq c(\hat{\theta}_0^m, p_0)$ $\leq c(\hat{\theta}^m, p_0) \; \forall$ $\hat{\theta}^m$ and $\forall \; p$	Minimax
i–iii	$f_{\hat{\theta}^m}(x^n)$	$\hat{\theta}^m: f_{\hat{\theta}^m}(x^n) \geq f_{\theta^m}(x^n) \; \forall \; \theta^m \in \mathscr{E}^m$	Maximum likelihood
ii, iii, nonparametrically described process $\forall \; \theta^m \in \mathscr{E}^m$	Originating from $f_{\theta^m}(x^n)$	Solution of appropriate saddle-point game.	Robust

some parametrically known stochastic process that generates the observation sequence x^n. When for every given parameter value θ^m the stochastic process that generates x^n is nonparametrically described, the *nonparametric* or *robust parameter estimation* schemes arise. The latter schemes may evolve as the solutions of certain saddle-point games, whose payoff function originates from the parametric maximum likelihood formalizations. It is then assumed that, in addition to the nonparametrically described data-generating process, only the assets ii and iii are available to the analyst. The *qualitative robustness*, in parameter estimation, corresponds to local performance stability, for small deviations from a nominal, parametrically known, data-generating process.

We will present and analyze the Bayesian and minimax parameter estimation schemes in Chapter 9. The maximum likelihood parameter estimation procedures are discussed, in Chapter 10, while Chapter 12 is dedicated to the presentation of qualitative robustness. Chapter 11 presents some useful concepts pertinent to the parameter estimation principles. We conclude this section with Table 3.4.1, where we list the parameter estimation schemes just discussed.

3.5 APPLICATIONS AND EXAMPLES

The applications of the hypothesis testing and parameter estimation schemes are numerous. Some of those applications have been familiar since almost the beginning of this century; others arise with the peculiarities and performance requirements that characterize modern complex systems. In this section, we discuss a few of those applications and the induced statistical formalizations. We will discuss and analyze additional applications in the chapters dedicated to the various detection and parameter estimation schemes. Here, we present our applications in the form of examples.

3.5 Applications and Examples

Example 3.5.1

It is perhaps appropriate to start with one of the oldest applications of hypothesis testing: the *radar problem*. There, a signal or a number of signals may be transmitted through the atmosphere. The objective of the radar is to detect their presence, where the nature of those signals as well as the time period during which they may be transmitted are assumed to be known. The transmission channel is the atmosphere, and it is assumed to be statistically well known. In particular, the atmospheric channel is usually modeled as an additive, stationary, and Gaussian channel with zero mean and known auto-covariance function. The Gaussian assumption is not precise enough, as we will discuss in Chapter 7, but this is beside the point here. Some signal $s(t)$ transmitted through the atmospheric channel is then received by the radar as

$$x(t) = s(t) + n(t) \tag{3.5.1}$$

where $n(t)$ is the zero mean Gaussian process that describes the atmospheric channel.

Let the signal $s(t)$ be transmitted for a time period $[0, T]$, and let this correspond to n observations in discrete-time form. Then, in discrete-time, (3.5.1) takes the form

$$x^n = s^n + n^n \tag{3.5.2}$$

where n^n is an n-dimensional vector of jointly Gaussian random variables with common mean zero and autocovariance matrix R_n. The autocovariance matrix R_n is the discrete-time form of the autocovariance function that describes the Gaussian atmospheric channel.

Let the signal s^n be deterministic and well known. Then the vector x^n in (3.5.2) is a vector of jointly Gaussian random variables with well known joint density function $f_{s^n}(x^n)$ given by

$$f_{s^n}(x^n) = (2\pi)^{-n/2}|R_n|^{-1/2} \exp\left\{-\frac{1}{2}[x^n - s^n]^T R_n^{-1}[x^n - s^n]\right\} \tag{3.5.3}$$

where $|R_n|$ is the determinant of the matrix R_n, R_n^{-1} is the inverse of the same matrix, x^n and s^n are column vectors, and $[x^n - s^n]^T$ is the transpose of the vector $x^n - s^n$.

If during the time period signified by the number n of observations, the possibilities are that exactly one of $M - 1$ deterministic signals s_i^n; $i = 1, \ldots, M - 1$ is transmitted, or no signal is transmitted, the resulting M hypotheses are described by a set of density functions as follows:

$$H_0: f_0(x^n) = (2\pi)^{-n/2}|R_n|^{-1/2} \exp\left\{-\frac{1}{2}x^{nT} R_n^{-1} x^n\right\} \tag{3.5.4}$$

$$H_i: f_{s_i^n}(x^n) = (2\pi)^{-n/2}|R_n|^{-1/2} \exp\left\{-\frac{1}{2}[x^n - s_i^n]^T R_n^{-1}[x^n - s_i^n]\right\}$$

$$i = 1, \ldots, M - 1 \tag{3.5.5}$$

where the hypothesis H_0 corresponds to no signal transmitted. The M hypotheses described above are all well known. Given some decision rule $\delta: \{\delta_k(x^n); k = 0, 1, \ldots, M - 1\}$, the conditional probabilities $P_{ki}(\delta)$ in (3.1.3) then take the form

$$P_{ki}(\delta) = \int_{\Gamma^n} \delta_k(x^n) f_{s_i^n}(x^n) \, dx^n \quad \begin{cases} i = 1, \ldots, M - 1 \\ k = 0, 1, \ldots, M - 1 \end{cases}$$

$$P_{k0}(\delta) = \int_{\Gamma^n} \delta_k(x^n) f_0(x^n) \, dx^n \quad k = 0, 1, \ldots, M - 1 \tag{3.5.6}$$

where $f_{s_i^n}(x^n)$ and $f_0(x^n)$ are the density functions in (3.5.5) and (3.5.4), respectively, and Γ^n is here the Euclidean space R^n.

If both a set $\{c_{ki}; k, i = 0, 1, \ldots, M - 1\}$ of penalty coefficients and a set $\{p_i; i = 0, 1, \ldots, M - 1\}$ of a priori probabilities on the hypotheses are available and used, the optimization function $C(\delta)$ for some decision rule δ takes the form [as in (3.1.16)]

$$C(\delta) = \sum_{k=0}^{M-1} \sum_{i=0}^{M-1} c_{ki} p_i P_{ki}(\delta)$$

$$= \sum_{k=0}^{M-1} c_{k0} p_0 \int_{\Gamma^n} \delta_k(x^n) f_0(x^n) \, dx^n + \sum_{k=0}^{M-1} \sum_{i=1}^{M-1} c_{ki} p_i \int_{\Gamma^n} \delta_k(x^n) f_{s_i^n}(x^n) \, dx^n$$

$$= \int_{\Gamma^n} \sum_{k=0}^{M-1} \delta_k(x^n) \left[c_{k0} p_0 f_0(x^n) + \sum_{i=1}^{M-1} c_{ki} p_i f_{s_i^n}(x^n) \right] dx^n \tag{3.5.7}$$

If only the set $\{p_i; i = 0, 1, \ldots, M - 1\}$ of a priori probabilities is available and used, the optimization function for some decision rule δ is the probability $P_e(\delta)$ in (3.1.13); it takes here the form

$$P_e(\delta) = 1 - \sum_{k=0}^{M-1} p_k P_{kk}(\delta)$$

$$= 1 - \int_{\Gamma^n} \left[p_0 \delta_0(x^n) f_0(x^n) + \sum_{k=1}^{M-1} \delta_k(x^n) f_{s_k^n}(x^n) p_k \right] dx^n \tag{3.5.8}$$

Let us now suppose that only a single deterministic and well known signal s^n may be present. Then $M - 1 = 1$ and we are presented with the problem of single hypothesis testing. Furthermore, in the radar problem correctly detecting the presence of the signal is more crucial than correctly detecting its absence. Thus, hypothesis H_1 is more emphatic; it is described by the density function $f_{s^n}(x^n)$ in (3.5.3). Let no a priori probabilities on the hypotheses and no penalty coefficients be available. Then the Neyman-Pearson formalization will be used. Let α be a given false alarm rate. Then, for some decision rule δ, the optimization function is

$$P_{11}(\delta) = \int_{\Gamma^n} \delta(x^n) f_{s^n}(x^n) \, dx^n$$

3.5 Applications and Examples

subject to (3.5.9)

$$P_{10}(\delta) = \int_{\Gamma^n} \delta(x^n) f_0(x^n) \, dx^n \leq \alpha$$

where $P_{ki}(\delta)$ is given by (3.1.3), $f_0(x^n)$ is given by (3.3.4), and $\delta(x^n)$ denotes the decision probability $\delta_1(x^n)$.

In the last case, the power function induced by the rule δ consists of the two points $P_{11}(\delta)$ and $P_{10}(\delta)$.

Example 3.5.2

With the current emphasis on digital systems, it is only appropriate that we present a simple application of hypothesis testing for such systems. Let us consider a digital computer. The computer transforms all inputs into binary sequences, and its processing channels are also binary. Let us first describe a simple and widely used form of such a binary channel. It is termed a binary, memoryless, and symmetric channel and it is completely described as follows:

The binary, memoryless, and symmetric channel accepts only binary sequences as inputs, and it gives binary sequences as outputs. Given an input binary sequence $x = \{\ldots, x_{-1}, x_0, x_1, \ldots\}$, the channel processes each bit separately and in a memoryless fashion. Also, the channel transmits each bit x_i correctly with probability q; it transmits the bit incorrectly with probability $1 - q$. That is, the transition per bit performed by the channel is graphically described as below, where the channel is termed symmetric due to the symmetry in the transition probabilities.

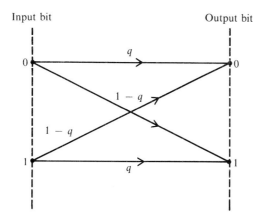

A useful binary, memoryless, and symmetric channel should transmit correctly with higher probability than the probability of transmitting incorrectly. Thus, q should be larger than .5. A given binary input sequence x^n is then transmitted by the channel as another binary sequence y^n, with probability

$$\Pr\{y^n|x^n\} = \prod_{i=1}^{n} \Pr\{y_i|x_i\} = \prod_{i=1}^{n} q^{1-(y_i \oplus x_i)}(1-q)^{y_i \oplus x_i}$$

$$= q^n \left(\frac{1-q}{q}\right)^{\sum_{i=1}^{n}(x_i \oplus y_i)} \tag{3.5.10}$$

where $x_i = 0, 1$ and $y_i = 0, 1 \ \forall \ i$ and $\oplus$ signifies binary sum.

Let us suppose that one of the binary channels is memoryless and symmetric, described as above, with $q > .5$. Let us suppose that M well known binary sequences $x^n(j); j = 1, \ldots, M$ are transmitted alternatively through this channel. Let us assume that in a period of n observations only one of the above sequences is transmitted. Our objective is to decide which one of the M message sequences has been transmitted, given an observed output binary sequence y^n. Given some decision rule $\delta: \{\delta_k(y^n); k = 1, \ldots, M\}$, the conditional probabilities $P_{ki}(\delta)$ in (3.1.3) take here the form

$$P_{ki}(\delta) = \sum_{\Gamma^n} \delta_k(y^n) \Pr\{y^n|x^n(i)\}$$

$$= q^n \sum_{\Gamma^n} \delta_k(y^n) \left(\frac{1-q}{q}\right)^{\sum_{j=1}^{n} x_j(i) \oplus y_j} \tag{3.5.11}$$

where Γ^n covers here all the 2^n values that the sequence y^n may take. If a set $\{p_i; i = 1, \ldots, M\}$ of a priori probabilities on the M messages $x^n(j); j = 1, \ldots, M$ or hypotheses is available, the optimization function for some decision rule δ is the probability of error $P_e(\delta)$ in (3.1.13), and it takes here the form

$$P_e(\delta) = 1 - \sum_{k=1}^{M} p_k P_{kk}(\delta)$$

$$= 1 - q^n \sum_{\Gamma^n} \sum_{k=1}^{M} p_k \delta_k(y^n) \left(\frac{1-q}{q}\right)^{\sum_{j=1}^{n} x_j(k) \oplus y_j} \tag{3.5.12}$$

If $p_i = M^{-1} \ \forall \ i$, the optimal rule that results from the optimization function $P_e(\delta)$ in (3.5.12) over the class of all possible rules is the minimum distance decoding in information theory. We will show that in Chapter 4.

Example 3.5.3

In Section 1.2 we presented an autoregressive model that arises from point biological oscillators, in the presence of observation noise. The model is described by

$$X_n = \sum_{k=1}^{m} \alpha_k X_{n-k} + W_n \tag{3.5.13}$$

3.5 Applications and Examples

where m is the order of the autoregressive model, $\{\alpha_k\}$ are the autoregressive parameters, and W_n is a sample from the noise or driving stochastic process.

Let us assume that the noise process is zero mean, Gaussian, and white, with variance one. That is, the variables W_n; $n = \ldots, -1, 0, 1, \ldots$ are Gaussian, zero mean, variance one, and independent. Let us suppose that either one of M well known m-order autoregressive models may be active. Then, each model or hypothesis H_j is completely described by a set $\{\alpha_k(j)\}$ of autoregressive parameters, and the objective of the analyst is to decide, within a period of n observations, which model is active. The analyst observes a realization x^n from the active autoregressive process, and formulates the problem as a hypothesis testing scheme. Let $x^n = \{x_i; i = 1, \ldots, n\}$, and let hypothesis H_j be active. Let $f_j(x_n - \sum_{k=1}^{m} \alpha_k(j)x_{n-k})$ be the density function induced at the point $x_n - \sum_{k=1}^{m} \alpha_k(j)x_{n-k}$. Then, through the above description of the model, we easily conclude:

$$f_j\left(x_n - \sum_{k=1}^{m} \alpha_k(j)x_{n-k}\right) = (2\pi)^{-1/2} \exp\left\{-\frac{1}{2}\left[x_n - \sum_{k=1}^{m} \alpha_k(j)x_{n-k}\right]^2\right\} \quad (3.5.14)$$

Thus, given the observation sequence x^n and some decision rule δ: $\{\delta_k(x^n); k = 1, \ldots, M\}$ on the M hypotheses, we can express the following form of the conditional probability $P_{kj}(\delta)$ in (3.1.3).

$$P_{kj}(\delta) = \int_{\Gamma^n} \delta_k(x^n) \prod_{i=1}^{n} f_j\left(x_i - \sum_{k=1}^{m} \alpha_k(j)x_{i-k}\right) dx^n$$

$$= (2\pi)^{-n/2} \int_{\Gamma^n} \delta_k(x^n) \exp\left\{-\frac{1}{2}\sum_{i=1}^{n}\left[x_i - \sum_{k=1}^{m} \alpha_k(j)x_{i-k}\right]^2\right\} dx^n \quad (3.5.15)$$

where Γ^n is here the Euclidean space R^n, and where the x_l data with negative or zero indices are set equal to zero.

If a set $\{p_i; i = 1, \ldots, M\}$ of a priori probabilities on the M hypotheses is available, then the optimization function for some rule δ is the probability of error $P_e(\delta)$ in (3.1.13). Its specific form here is

$$P_e(\delta) = 1 - (2\pi)^{-n/2} \int_{\Gamma^n} \sum_{j=1}^{M} p_j \delta_j(x^n)$$

$$\cdot \exp\left\{-\frac{1}{2}\sum_{i=1}^{n}\left[x_i - \sum_{k=1}^{m} \alpha_k(j)x_{i-k}\right]^2\right\} dx^n \quad (3.5.16)$$

A special form of the error $P_e(\delta)$ above is obtained when $p_j = M^{-1} \,\forall\, j$. Then the resulting optimal rule is equivalent to the maximum likelihood scheme, as we will see in the progress of this book.

Example 3.5.4

The most extensively studied parameter estimation problem is the one known as the estimation of a location parameter. The problem basically corresponds to the evaluation of some unknown scalar parameter Θ that is added to a sequence, $Y^n = \{Y_i; 1 \leq i \leq n\}$, of random variables. The observed random sequence, $X^n = \{X_i; 1 \leq i \leq n\}$, is then such that, $X_i = \Theta + Y_i; 1 \leq i \leq n$. The objective is to estimate the value of the parameter Θ, based on some observed realization x^n of the random sequence X^n. This objective is attained via the selection of some performance criterion, and the subsequent solution of the corresponding optimization problem. As discussed in Section 3.4, the formalization and solution of the optimization problem varies with the assets available to the analyst.

Let, $f(y^n)$, denote the density function of the random sequence, Y^n at the vector point y^n. Given some value θ of the location parameter, let $g_\theta(x^n)$ denote the density function of the random sequence $X^n = \{X_i = \theta + Y_i; 1 \leq i \leq n\}$ at the vector point x^n. Then the following equality clearly holds, where $s_I^n = \{s_i = 1; 1 \leq i \leq n\}$.

$$g_\theta(x^n) = f(x^n - \theta s_I^n) \tag{3.5.17}$$

If some penalty function $c[\theta, \hat{\theta}(x^n)]$ is available, for every given value θ of the parameter, and every estimate $\hat{\theta}(x^n)$ for a given x^n, then the conditional expected penalty $c(\theta, \hat{\theta})$ is expressed as

$$c(\theta, \hat{\theta}) = \int_{\Gamma^n} c[\theta, \hat{\theta}(x^n)] f(x^n - \theta s_I^n) \, dx^n \tag{3.5.18}$$

where Γ^n denotes the space where the observation vector $x^n = \theta s_I^n + y^n$, takes its values.

Let it be known that the parameter Θ takes its values on the real axis R. Then, if $p(\theta)$ is some given density function defined for each θ value in R, the expected penalty $c(\hat{\theta}, p)$ is given by

$$c(\hat{\theta}, p) = \int_R c(\theta, \hat{\theta}) p(\theta) \, d\theta$$
$$= \int_{\Gamma^n} \int_R dx^n c[\theta, \hat{\theta}(x^n)] f(x^n - \theta s_I^n) p(\theta) \, d\theta \tag{3.5.19}$$

Let the sequence $\{Y_i\}$ be a sequence of independent, zero mean, and unit variance, Gaussian random variables. Then, if $\phi(x)$ denotes the density function of the zero mean, unit variance, Gaussian random variable at the scalar point x, (3.5.17), (3.5.18), and (3.5.19) take the forms

$$g_\theta(x^n) = \prod_{i=1}^n \phi(x_i - \theta) = (2\pi)^{-n/2} \exp\left\{-\frac{1}{2} \sum_{i=1}^n (x_i - \theta)^2\right\}$$
$$= (2\pi)^{-n/2} \exp\left\{-\frac{1}{2} \sum_{i=1}^n x_i^2\right\} \exp\left\{\theta n \left[n^{-1} \sum_{i=1}^n x_i - \frac{\theta}{2}\right]\right\} \tag{3.5.20}$$

$$c(\theta, \hat{\theta}) = (2\pi)^{-n/2} \int_{\Gamma^n} c[\theta, \hat{\theta}(x^n)] \exp\left\{-\frac{1}{2} \sum_{i=1}^n x_i^2\right\}$$

3.5 Applications and Examples

$$\cdot \exp\left\{\theta n\left[n^{-1}\sum_{i=1}^{n}x_i - \frac{\theta}{2}\right]\right\}dx^n \tag{3.5.21}$$

$$c(\hat{\theta}, p) = (2\pi)^{-n/2}\int_{\Gamma^n}dx^n \exp\left\{-\frac{1}{2}\sum_{i=1}^{n}x_i^2\right\}\int_R c[\theta, \hat{\theta}(x^n)]$$

$$\cdot \exp\left\{\theta n\left[n^{-1}\sum_{i=1}^{n}x_i - \frac{\theta}{2}\right]\right\}p(\theta)\,d\theta \tag{3.5.22}$$

The above Gaussian example corresponds to the estimation of a constant signal, in the presence of additive, white, and Gaussian noise.

Example 3.5.5

Let us revisit the autoregressive model first presented in Section 1.2. This model is described by

$$X_n = \sum_{k=1}^{m} \alpha_k X_{n-k} + W_n \tag{3.5.23}$$

where m is the order of the model $\{\alpha_k\}$ are the autoregressive parameters, and $\{W_n\}$ is a sequence of random variables that are generated by the noise or driving stochastic process.

Let us adopt the widely used assumption that the noise or driving process in (3.5.23) is stationary and memoryless, and let $f(x)$ denote the density function of the random variable W_n at the scalar point x. Given the set $\{\alpha_k\}$ of the autoregressive parameters, and an n-dimensional realization $x_{l+1-m}^{l+n} = \{x_i; l+1-m \leq i \leq l+n\}$ for some l from the autoregressive process, the density function $g_{\{\alpha_k\}}(x_{l+1}^{l+n})$ of the random sequence x_{l+1}^{n+l} at the vector point x_{l+1}^{n+l} is then given by

$$g_{\{\alpha_k\}}(x_{l+1}^{l+n}) = \prod_{i=1}^{n} f\left(x_{l+i} - \sum_{k=1}^{m}\alpha_k x_{l+i-k}\right) \tag{3.5.24}$$

Let the set $\{\alpha_k\}$ of the autoregressive parameters be unknown. Let it be known only that $\alpha_k \in \mathscr{E}_k$; $1 \leq k \leq n$, where $\mathscr{E}_k$ is some known subset of the real line, and let us denote $\mathscr{E}^m = \mathscr{E}_1 \times \cdots \times \mathscr{E}_m$. Let the objective be to estimate the set $\{\alpha_k\}$ based on the observation vector x_{l+1-m}^{l+n}. If no penalty function and no a priori distribution of the set $\{\alpha_k\}$ on $\mathscr{E}^m$ are available, and if the density function f is well known, the *maximum likelihood criterion* is adopted. Then, the estimated set $\{\hat{\alpha}_k\}$ of autoregressive parameters is that which maximizes the expression in (3.5.24) on $\mathscr{E}^m$. That is,

$$\{\hat{\alpha}_k\}: \prod_{i=1}^{n} f\left(x_{l+i} - \sum_{k=1}^{m}\hat{\alpha}_k x_{l+i-k}\right) \geq \prod_{n=1}^{n} f\left(x_{l+i} - \sum_{k=1}^{n}\alpha_k x_{l+i-k}\right) \quad \forall\, \{\alpha_k\} \in \mathscr{E}^m \tag{3.5.25}$$

If f is the zero mean, unit variance Gaussian density function, then the maximum likelihood estimate $\{\hat{\alpha}_k\}$ in (3.5.25) takes the form:

$$\{\hat{\alpha}_k\}: \sum_{i=1}^{n}\left(x_{l+i} - \sum_{k=1}^{m} \hat{\alpha}_k x_{l+i-k}\right)^2 = \inf_{\{\alpha_k\}\in\mathscr{E}^m} \sum_{i=1}^{n}\left(x_{l+i} - \sum_{k=1}^{n} \alpha_k x_{l+i-k}\right)^2 \quad (3.5.26)$$

If the density function f is nonparametrically described, then robustification of the maximum likelihood criterion may in general be feasible. This robustification incorporates the class of distributions within which the density function f lies, and the solution of the latter formalization is then the robust estimate of the set $\{\alpha_k\}$.

CHAPTER 4

BAYESIAN DETECTION

4.1 INTRODUCTION

In Chapter 3, we presented an outline of the various detection schemes and their evolution as functions of the assets available to the analyst. In this chapter, we will focus on the Bayesian detection schemes, which require the maximum number of available assets. We will formulate the corresponding optimization problem, and we will find its solution for both the single hypothesis and multiple hypotheses problems, and for both discrete-time and continuous-time observed realizations. The solutions for continuous-time realizations evolve from the corresponding solutions for discrete-time realizations, so we will study the latter first. We will also use the quantities defined in Chapter 3, and in our general formalizations we will consider parametrically known hypotheses. The case of well known hypotheses evolves as a special case of the latter formalizations, where its exact form will be explained in the process. We thus assume that there exists a parameter vector Θ^m of finite dimensionality m and M stochastic processes, such that for every fixed value θ^m of the parameter vector Θ^m, the conditional (conditioned on θ^m) distributions $F_{i,\theta^m}(x^n)$; $i = 1, \ldots, M$ are well known, for all vector values x^n and all n. The M stochastic processes comprise the M parametrically known hypotheses, and $\mathscr{E}^m$ denotes the space where the vector parameter Θ^m takes its values. As in Chapter 3, we assume that density functions exist and we denote by $f_{i,\theta^m}(x^n)$ the n-dimensional conditional (conditioned on θ^m) density function induced by the ith process or hypothesis at the vector point x^n. We also denote by $p_i(\theta^m)$ the density function of the vector parameter Θ^m at the vector point θ^m, conditioned on the ith hypothesis H_i. We denote by $r_i(\theta^m)$ the unconditional density function of the vector parameter Θ^m at the point θ^m when the hypothesis H_i is active. Finally, we denote by $c_{ki}(\theta^m)$ the penalty paid when hypothesis H_k is decided, the value of the vector parameter is θ^m, and the true hypothesis is H_i. Let now the M stochastic processes be discrete-time, let Γ^n be the space where the sequences x^n take then their values, and let $\delta = \{\delta_j(x^n); j = 1, \ldots, M\}$; $x^n \in \Gamma^n$ be some decision rule. Let $C(\delta)$ be the expected penalty induced by the decision rule δ, by the penalty functions $c_{ki}(\theta^m)$; $k, i = 1, \ldots, M$, $\theta^m \in \mathscr{E}^m$, and by the density functions $f_{i,\theta^m}(x^n)$ and $r_i(\theta^m)$. Then, we easily find

$$C(\delta) = \int_{\Gamma^n} \sum_k \delta_k(x^n) \, dx^n \int_{\mathscr{E}^m} \sum_i c_{ki}(\theta^m) f_{i,\theta^m}(x^n) r_i(\theta^m) \, d\theta^m \quad (4.1.1)$$

In the formalization of the Bayesian detection schemes, it is assumed that the quantities $c_{ki}(\theta^m)$, $f_{i,\theta^m}(x^n)$, and $r_i(\theta^m)$ are available for all $k, i = 1, \ldots, M$, $\theta^m \in \mathscr{E}^m$, $x^n \in \Gamma^n$, and that they are fully utilized by the analyst. The performance criterion is the minimization of the expected penalty $C(\delta)$ in (4.1.1), where it is assumed that the penalties $c_{ki}(\theta^m)$ are nonnegative for all k, i, and $\theta^m \in \mathscr{E}^m$. We note that the case of well known hypotheses evolves when the parameter vector Θ^m is eliminated in the above formalization. Then, the quantities $c_{ki}(\theta^m)$ and $f_{i,\theta^m}(x^n)$ reduce respectively to c_{ki} and $f_i(x^n)$, as in Chapter 3. The quantity $r_i(\theta^m)$ reduces then to the a priori probability p_i of the hypothesis H_i. If, in addition, the set $\{c_{ki}; k, i = 1, \ldots, M\}$ of penalties is such that $c_{ki} = \begin{cases} 1; i \neq k \\ 0; i = k \end{cases}$, then the expected penalty $C(\delta)$ in (4.1.1) becomes the probability of error $P_e(\delta)$ in Chapter 3. When the M parametrically known hypotheses are determined through a single parametrically known stochastic process and M disjoint subdivisions of the parameter space $\mathscr{E}^m$, then the conditional density function $p_i(\theta^m)$ introduced earlier in this section reduces to $p(\theta^m)$ and so does the quantity $r_i(\theta^m)$. The quantities $c_{ki}(\theta^m)$ and $f_{i,\theta^m}(x^n)$ reduce then respectively to $c_k(\theta^m)$ and $f_{\theta^m}(x^n)$, and the expected penalty in (4.1.1) takes the form in (3.1.15).

In the Bayesian formalization, if the M stochastic processes are continuous-time, then the available observations consist of continuous waveforms, $x(t)$, in a time interval $[0, T]$. A decision rule is then denoted by $\delta = \{\delta_j(x(t); t \in [0, T])$; $j = 1, \ldots, M\}$, where $\delta_j(x(t); t \in [0, T])$ is defined as the probability that the hypothesis H_j is decided, conditioned on the observed waveform $x(t); t \in [0, T]$. The expected penalty $C(\delta)$ can be then expressed via the use of appropriate series expansions. That will be explained in the next section.

We conclude this section by emphasizing that no sequential decision schemes are considered in this chapter. Thus, it is uniformly assumed that the observation time interval is fixed, and it is not controlled by the analyst. This assumption corresponds to a fixed length n of observed data sequences x^n in the case of discrete-time processes, and to a fixed time interval $[0, T]$ of observed waveforms $x(t)$ in the case of continuous-time processes. It is also uniformly assumed that all the observed data are generated by the same underlying stochastic process (or hypothesis), and that those data are used in the decision making process.

4.2 THE OPTIMIZATION PROBLEM

Let us consider Bayesian detection for M hypotheses defined by M discrete-time, parametrically known stochastic processes. Let us use the same notation as in Section 4.1. Then, for every observed data sequence x^n a decision is made, regarding the acting underlying process or hypothesis. The performance criterion used is the expected penalty $C(\delta)$ in (4.1.1). In particular, if $\mathscr{D}$ is the class of all the legitimate decision rules, the minimization of the expected penalty $C(\delta)$ on $\mathscr{D}$ is sought. This defines a constraint optimization problem that, as we will show, has some solution. Considering the general formalization in Section 4.1, let us define

$$g_k(x^n) = \int_{\mathscr{E}^m} \sum_i c_{ki}(\theta^m) f_{i,\theta^m}(x^n) r_i(\theta^m) \, d\theta^m \quad k = 1, \ldots, M, \, x^n \in \Gamma^n \quad (4.2.1)$$

where we assume that $g_k(x^n)$ exists for all x^n in Γ^n and for all $k = 1, \ldots, M$.

Substituting (4.2.1) in (4.1.1), we obtain the following form of the expected penalty $C(\delta)$.

$$C(\delta) = \int_{\Gamma^n} dx^n \sum_k \delta_k(x^n) g_k(x^n) \quad (4.2.2)$$

We note that if the stochastic processes that determine the M hypotheses are all amplitude-discrete, then in (4.2.2) the integral is substituted by a sum. The functions $g_k(x^n)$ in (4.2.1) are clearly deterministic, known, and nonnegative for all x^n in Γ^n and for all k. This is due to the assumption that the quantities $c_{ki}(\theta^m)$, $f_{i,\theta^m}(x^n)$, and $r_i(\theta^m)$ are given, and that $c_{ki}(\theta^m)$ is nonnegative, for all $k, i, \theta^m \in \mathscr{E}^m$. The quantities $f_{i,\theta^m}(x^n)$ and $r_i(\theta^m)$ are nonnegative for all $i, k, \theta^m \in \mathscr{E}^m$, since $f_{i,\theta^m}(x^n)$ and $r_i(\theta^m)$ are density functions. For any vector point x^n in Γ^n, and for any decision rule δ in $\mathscr{D}$, we have $\delta_k(x^n) \geq 0$; $k = 1, \ldots, M$, and $\sum_k \delta_k(x^n) = 1$. The reasons for the latter are fully explained in Chapter 3. We now state the following constraint optimization problem:

$$\text{Minimize} \quad C(\delta) = \int_{\Gamma^n} dx^n \sum_k \delta_k(x^n) g_k(x^n) \quad \text{in } \mathscr{D}(\delta \in \mathscr{D})$$

subject to the constraints (4.2.3)

$$\delta_k(x^n) \geq 0 \quad k = 1, \ldots, M \quad \forall \, x^n \in \Gamma^n$$
$$\sum_k \delta_k(x^n) = 1 \quad \forall \, x^n \in \Gamma^n$$

where the functions $g_k(x^n)$; $k = 1, \ldots, M$ are given and are nonnegative for all x^n in Γ^n.

The constraint optimization problem in (4.2.3) does not in general, have, a unique solution. Any solution of the problem provides a valid Bayesian detection scheme, which consists of a specific decision rule. We state the general solution of the constraint optimization problem in a theorem.

THEOREM 4.2.1: Consider the constraint optimization problem in (4.2.3), and define

$$f(x^n) = \min_k g_k(x^n) \qquad x^n \in \Gamma^n \qquad (4.2.4)$$

For some fixed vector value x^n in Γ^n, let there be $j = j(x^n)$ indices $k_1(x^n), \ldots, k_j(x^n)$ that satisfy the minimum in (4.2.4). That is,

$$g_{k_1(x^n)}(x^n) = \cdots = g_{k_j(x^n)}(x^n) = f(x^n)$$

Then, each decision rule δ^* such that

$$\begin{aligned} \delta^*_{k_1(x^n)}(x^n) + \cdots + \delta^*_{k_j(x^n)}(x^n) &= 1 \qquad \forall\, x^n \in \Gamma^n \\ \delta^*_{k_1(x^n)}(x^n) \geq 0, \ldots, \delta^*_{k_j(x^n)}(x^n) &\geq 0 \qquad \forall\, x^n \in \Gamma^n \end{aligned} \qquad (4.2.5)$$

minimizes the expected penalty in (4.2.3). In particular, if δ_1^* and δ_2^* are two decision rules that satisfy the conditions in (4.2.5), then,

$$C(\delta_1^*) = C(\delta_2^*) \leq C(\delta) \qquad \forall\, \delta \in \mathscr{D}$$

Proof: Let δ^* be some decision rule that satisfies the conditions in (4.2.5). Then, we clearly have

$$C(\delta^*) = \int_{\Gamma^n} f(x^n)\, dx^n$$

Let δ be any decision rule in $\mathscr{D}$. Then, clearly,

$$C(\delta) = \int_{\Gamma^n} dx^n \sum_k \delta_k(x^n) g_k(x^n) \geq \int_{\Gamma^n} dx^n \sum_k \delta_k(x^n) f(x^n) = \int_{\Gamma^n} dx^n f(x^n) = C(\delta^*) \quad \blacksquare$$

From Theorem 4.2.1, it is clear that, in general, there exists a whole class $\mathscr{D}^*$ of equivalent decision rules, that are all solutions of the Bayesian detection problem. It is also clear that the class $\mathscr{D}^*$ includes both randomized and nonrandomized decision rules. But, as we emphasized in Chapter 3, when there is a choice among several equivalent (in terms of performance) decision rules, the emphasis should be on simplicity in implementation. Thus, the randomized decision rules in class $\mathscr{D}^*$ should be then excluded. In particular, among all the equivalent decision rules in class $\mathscr{D}^*$, the following simple nonrandomized rule should be selected:

Given the observation vector x^n, compute the function $f(x^n)$ in (4.2.4). Select a *single* $k = k(x^n)$ such that

$$f(x^n) = g_{k(x^n)}(x^n) \qquad (4.2.6)$$

4.2 The Optimization Problem

Then, for each vector point x^n in Γ^n select

$$\delta_j^*(x^n) = \begin{cases} 1 & j = k(x^n) \\ 0 & j \neq k(x^n) \end{cases}$$

From the above discussion, we conclude that the decision rule in (4.2.6) can be considered as the "optimal" Bayesian detection scheme. We note that the rule in (4.2.6) is still not unique, due to the free selection of the $k(x^n)$ index, among the $g_k(x^n)$ functions that satisfy the minimum $f(x^n)$ (in the case that for given x^n, more than one such index exists). In general, we can safely state that there is no benefit in adopting randomized decision rules for Bayesian detection schemes. Only nonrandomized decision rules should be considered. The function $f(x^n)$ together with the index $k(x^n)$ are called *the test* and they clearly are random variables, since X^n is a random vector. The statistical behavior of the pair $[f(X^n), k(X^n)]$ is called the *test statistics*. In most specific problems, the test can be reduced to a simplified form. Such a form results in simplified test statistics as well. The maximum possible simplification of a test should always be sought; this also induces simplicity in implementation. We will now present some examples that will perhaps further clarify some issues and concepts.

Example 4.2.1

Let us consider M well known stochastic processes. Let $f_i(x^n)$; $i = 0, 1, \ldots, M - 1$ denote the n-dimensional density function induced by the ith process at the vector point x^n. Let $\{p_i; i = 0, 1, \ldots, M - 1\}$; $p_i > 0 \, \forall \, i$ be the set of a priori probabilities corresponding to the M hypotheses. Let $\{c_{ki}; i, k = 0, 1, \ldots, M-1\}$ be the set of the penalty coefficients. Then, the functions $g_k(x^n)$; $k = 0, 1, \ldots, M - 1$ and $f(x^n)$, in (4.2.1) and (4.2.4) respectively, take the form

$$g_k(x^n) = \sum_i c_{ki} f_i(x^n) p_i \qquad (4.2.7)$$

$$f(x^n) = \min_k \sum_i c_{ki} p_i f_i(x^n) \qquad (4.2.8)$$

Given the observation vector x^n, the search for some index $k(x^n)$ that satisfies the minimum in (4.2.8) can be realized via the differences

$$\sum_i c_{ki} p_i f_i(x^n) - \sum_i c_{li} p_i f_i(x^n) = \sum_i (c_{ki} - c_{li}) p_i f_i(x^n) \qquad k \neq l \qquad (4.2.9)$$

The index $k(x^n)$ will then be some k^* such that

$$\sum_i (c_{k^*i} - c_{li}) p_i f_i(x^n) \leq 0 \qquad \forall \, l: l = 0, 1, \ldots, M - 1 \qquad (4.2.10)$$

Let x^n be such that $f_0(x^n) > 0$. Then, (4.2.10) can be written equivalently in the form

$$\sum_i (c_{k^*i} - c_{li}) \frac{p_i \, f_i(x^n)}{p_0 \, f_0(x^n)} \leq 0 \qquad \forall \, l: l = 0, 1, \ldots, M - 1 \qquad (4.2.11)$$

The expression in (4.2.11) provides an "optimal" index $k(x^n) = k^*$, and the "optimal" Bayesian decision rule $\delta_k^*(x^n)$

$$\delta_k^*(x^n) = \begin{cases} 1 & k = k^* \\ 0 & k \neq k^* \end{cases} \qquad (4.2.12)$$

The expression (4.2.11) is the *test* for the derivation of a Bayesian decision rule. The ratios $f_i(x^n)/f_0(x^n)$; $i = 0, 1, \ldots, M - 1$ in (4.2.11) are ratios of density functions traditionally called *likelihood ratios*. The test in (4.2.11) thus consists of a weighted sum of likelihood ratios, where the weights are $(c_{k^*i} - c_{li})(p_i/p_0)$; $i = 0, 1, \ldots, M - 1$. The weighted sum is a function of the observed data sequence x^n and is usually called the *test function*. In the test, the test function is tested against the number zero. The latter is not a function of the observed data vector and is usually called the *threshold* of the test. In general, therefore, the test function is a function of the observations and is tested against a threshold, which is a constant.

Let now the penalty coefficients be such that

$$c_{ki} = \begin{cases} 1 & k \neq i \\ 0 & k = i \end{cases} \qquad (4.2.13)$$

As we observed in Section 4.1, the expected penalty $C(\delta)$ becomes identical to the probability of error $P_e(\delta)$ in Chapter 3. A Bayesian decision rule is also called the *ideal observer rule* in engineering terminology, then. Substituting the penalty coefficients of (4.2.13) in (4.2.11), we trivially obtain the simplified test

$$k^* = k(x^n) : p_{k^*} f_{k^*}(x^n) \geq p_l f_l(x^n) \quad \forall l : l = 0, 1, \ldots, M - 1 \qquad (4.2.14)$$

or

$$k^* = k(x^n) : \frac{f_{k^*}(x^n)}{f_l(x^n)} \geq \frac{p_l}{p_{k^*}} \quad \forall l : l = 0, 1, \ldots, M - 1 \qquad (4.2.15)$$

As compared to the test in (4.2.11), the test in (4.2.15) is largely simplified. It consists of a set of single likelihood ratios (or test functions), tested against a set of thresholds. The latter are ratios of a priori probabilities. If, in addition, the M hypotheses are equally probable, then $p_i = M^{-1}$; $i = 0, 1, \ldots, M - 1$, and the set reduces to a single threshold, which is equal to one. Now let the M equally probable hypotheses correspond to zero mean Gaussian processes, let R_{in} denote the autocovariance matrix of the ith Gaussian process, and let the matrices be distinct. Then, the test in (4.2.15) takes the form

$$k^* = k(x^n) : \frac{\exp\{-\tfrac{1}{2} x^{nT} R_{k^*n}^{-1} x^n\} \, |R_{ln}|^{1/2}}{\exp\{-\tfrac{1}{2} x^{nT} R_{ln}^{-1} x^n\} \, |R_{k^*n}|^{1/2}}$$

$$\geq 1 \quad \forall l : l = 0, 1, \ldots, M - 1 \qquad (4.2.16)$$

4.2 The Optimization Problem

where exp $\{\cdot\}$ is the exponential function, x^{n^T} signifies transpose, and $R_{ln}^{-1}, |R_{ln}|$ denote respectively the inverse and the determinant of the matrix R_{ln}.

Due to the strict monotonicity of the logarithmic function, we can derive an equivalent test, from (4.2.16) by taking the logarithm of both parts in the inequality.

$$k^* = k(x^n): x^{n^T}[R_{ln}^{-1} - R_{k*n}^{-1}]x^n \geq \ln \frac{|R_{k*n}|}{|R_{ln}|} \quad \forall\, l: l = 0, 1, \ldots, M-1 \quad (4.2.17)$$

where ln in $\ln(|R_{k*n}|/|R_{ln}|)$ denotes natural logarithm.

In (4.2.17), the test function is a weighted inner product of x^n tested against a set of thresholds that are functions of the determinants of the covariance matrices. If, in addition, the M Gaussian processes are also memoryless, and if σ_i^2 denotes the per datum variance of the ith Gaussian process, (4.2.17) is further simplified to

$$k^* = k(x^n): \left(\frac{\sigma_{k*}^2}{\sigma_l^2} - 1\right) n^{-1} x^{n^T} x^n$$

$$\geq 2\sigma_{k*}^2 \ln \frac{\sigma_{k*}}{\sigma_l} \quad \forall\, l: l = 0, 1, \ldots, M-1 \quad (4.2.18)$$

The test in (4.2.18) can be also expressed as

$$k^* = k(x^n): \begin{cases} n^{-1} x^{n^T} x^n \geq 2(\sigma_{k*}^2 \sigma_l^2/(\sigma_{k*}^2 - \sigma_l^2)) \ln(\sigma_{k*}/\sigma_l) = T_{k*l} \\ \text{for } l: \sigma_l < \sigma_{k*} \\ n^{-1} x^{n^T} x^n \leq 2(\sigma_{k*}^2 \sigma_l^2/(\sigma_{k*}^2 - \sigma_l^2)) \ln(\sigma_{k*}/\sigma_l) = T_{k*l} \\ \text{for } l: \sigma_l > \sigma_{k*} \end{cases} \quad (4.2.19)$$

In the test in (4.2.19), the thresholds T_{kl}; $k, l = 0, 1, \ldots, M-1$ are computed in advance. Then, for every observed data vector $x^n = \{x_i; i = 1, \ldots, n\}$, the inner product $n^{-1} x^{n^T} x^n = n^{-1} \sum_{i=1}^n x_i^2$ is computed. The hypothesis H_{k*} decided is such that $n^{-1} \sum_{i=1}^n x_i^2 \leq T_{k*l}$ for l such that $\sigma_l > \sigma_{k*}$, and $n^{-1} \sum_{i=1}^n x_i^2 \geq T_{k*l}$ for l such that $\sigma_l < \sigma_{k*}$. The test statistics of (4.2.19) is controlled by the statistics of the inner product function $n^{-1} \sum_{i=1}^n x_i^2$. Since under any hypothesis the data x_i; $i = 1, \ldots, n$ are independent and identically distributed Gaussian random variables, the distribution of the quantity $n^{-1} \sum_{i=1}^n x_i^2$ is χ^2. Thus, the statistical behavior of the test in (4.2.19) can be fully studied, based on familiar properties of the χ^2 distributions.

Example 4.2.2

Let us assume M equally probable well known hypotheses H_i; $i = 0, 1, \ldots, M-1$ and penalty coefficients as in (4.2.13). Then, as we saw in Example 4.2.1, the Bayesian or ideal observer test, reduces [as in (4.2.15) with $p_l = M^{-1}$; $l = 0, 1, \ldots, M$] to the form

Given the observation sequence x^n, decide in favor of the hypothesis H_{k*}; $k* = k(x^n)$, such that

$$\frac{f_{k*}(x^n)}{f_l(x^n)} \geq 1 \qquad \forall\, l: l = 0, 1, \ldots, M-1 \tag{4.2.20}$$

Now let the M hypotheses be all described by a Gaussian process whose n-dimensional covariance matrix is denoted R_n. Let the means of the M hypotheses be distinct, and let the n-dimensional mean of hypothesis H_i be denoted s_i^n. The n-dimensional density function $f_i(x^n)$ that describes H_i is then given by

$$f_i(x^n) = (2\pi)^{-n/2} |R_n|^{-1/2} \exp\left\{-\frac{1}{2}[x^n - s_i^n]^T R_n^{-1} [x^n - s_i^n]\right\}$$

$$= (2\pi)^{-n/2} |R_n|^{-1/2} \exp\left\{-\frac{1}{2} x^{n^T} R_n^{-1} x^n\right\}$$

$$\cdot \exp\left\{s_i^{n^T} R_n^{-1} x^n - \frac{1}{2} s_i^{n^T} R_n^{-1} s_i^n\right\} \tag{4.2.21}$$

where T denotes transpose, (-1) denotes inverse, and $|\ |$ denotes determinant.

From (4.2.21), we observe that the only part of the density $f_i(x^n)$ that involves explicitly both H_i and x^n is the quadratic form $s_i^{n^T} R_n^{-1} x^n$. The vector $R_n^{-1} x^n$ is then called the *sufficient statistics* of the problem, a concept that will be further discussed in this section. Due to (4.2.21), the test in (4.2.20) can be expressed as

$$k* = k(x^n) = \frac{\exp\{s_{k*}^{n^T} R_n^{-1} x^n - 2^{-1} s_{k*}^{n^T} R_n^{-1} s_{k*}^n\}}{\exp\{s_l^{n^T} R_n^{-1} x^n - 2^{-1} s_l^{n^T} R_n^{-1} s_l^n\}} \geq 1$$

$$\forall\, l: l = 0, 1, \ldots, M-1 \tag{4.2.22}$$

Taking the logarithm in (4.2.22) we can easily obtain a simple form of the test:

$$k* = k(x^n): s_{k*}^{n^T} R_n^{-1} x^n - s_l^{n^T} R_n^{-1} x^n \geq \frac{1}{2} [s_{k*}^{n^T} R_n^{-1} s_{k*}^n - s_l^{n^T} R_n^{-1} s_l^n]$$

$$\forall\, l: l = 0, 1, \ldots, M-1 \tag{4.2.23}$$

We observe that in the test function in (4.2.23), the observation vector appears strictly in the form of the sufficient statistics $R_n^{-1} x^n$. The test consists of the test functions $\{[s_i^n - s_l^n]^T R_n^{-1} x^n\}$ and the set $\{\frac{1}{2}[s_i^{n^T} R_n^{-1} s_i^n - s_l^{n^T} R_n^{-1} s_l^n]\}$ of thresholds. The problem in this example corresponds to recognizing which of M distinct deterministic signals s_i^n; $i = 0, 1, \ldots, M-1$ has been transmitted, when the transmission channel is represented by additive, zero mean, colored Gaussian noise whose n-dimensional autocovariance matrix is R_n. The quadratic quantities $\{n^{-1} s_i^{n^T} R_n^{-1} s_i^n\}$ represent the generalized signal-to-noise ratios.

4.2 The Optimization Problem

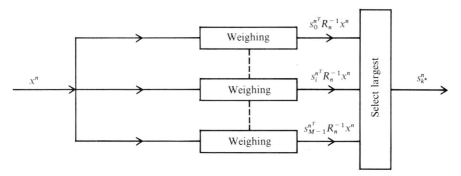

Figure 4.2.1 Matched filter with $s_i^{n^T} R_n^{-1} s_i^n = c$, for all i

Let the signals $\{s_i^n\}$ be designed to attain identical generalized signal-to-noise ratio values. That is, let $n^{-1} s_i^{n^T} R_n^{-1} s_i^n = c$; $i = 0, 1, \ldots, M - 1$. The test in (4.2.23) then takes the form

Given the observation vector x^n, decide that the signal s_{k*}^n has been transmitted if

$$s_{k*}^{n^T} R_n^{-1} x^n \geq s_l^{n^T} R_n^{-1} x^n \quad \forall\, l\colon l = 0, 1, \ldots, M - 1 \quad (4.2.24)$$

The test in (4.2.24) is known as the *matched filter*, in the engineering literature, and it represents the operations performed by a receiver. The objective of the receiver is to decide which one of the M distinct signals, $\{s_i^n\}$ has been transmitted, when the transmission is through additive, zero mean, colored Gaussian noise with n-dimensional autocovariance matrix R_n, and when the generalized signal-to-noise ratios, $\{n^{-1} s_i^{n^T} R_n^{-1} s_i^n\}$, are identical. The assumption is that the receiver knows the signals and the autocovariance matrix R_n. It then maintains in memory the precomputed row vectors $\{s_i^{n^T} R_n^{-1}\}$. Upon collection of the n data in x^n, the receiver computes the M inner products $\{s_i^{n^T} R_n^{-1} x^n\}$ and selects the largest. The signal s_{k*}^n that corresponds to this largest inner product is the receiver's decision. We exhibit the operation of the matched filter in Figure 4.2.1. If the inner products $\{s_i^{n^T} R_n^{-1} s_i^n\}$ are not identical, a matched filter-type receiver still results. Its operations are trivially concluded from the test in (4.2.23), and are exhibited in Figure 4.2.2. If the Gaussian noise is white, with power σ^2, the autocovariance matrix R_n equals $\sigma^2 I_n$, where I_n denotes the n-dimensional identity matrix. The quadratic expressions $\{s_i^{n^T} R_n^{-1} s_i^n\}$ then take the form $\{\sigma^{-2} s_i^{n^T} s_i^n\}$. If in addition the M signals have identical powers $c^2 = n^{-1} s_i^{n^T} s_i^n$; $i = 0, 1, \ldots, M - 1$ then the matched filter in (4.2.24) is simplified as

Decide in favor of the signal s_{k*}^n, if

$$s_{k*}^{n^T} x^n \geq s_l^{n^T} x^n \quad \forall\, l\colon l = 0, 1, \ldots, M - 1 \quad (4.2.25)$$

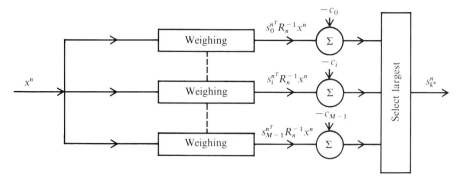

Figure 4.2.2 Matched filter with $c_i = 2^{-1} s_i^{n^T} R_n^{-1} s_i^n$, $i = 0, 1, \ldots, M-1$

An interesting special case of the matched filter in (4.2.25) arises, when the signals are orthogonal. We study this case in the form of a separate example, in Section 4.4.

In Example 4.2.2 the concept of sufficient statistics was mentioned. This concept can be used to simplify the search for optimal statistical procedures; it thus deserves special attention. Let us consider the general description of the M parametric hypotheses in Section 4.1, and let us then focus on the conditional density function $f_{i,\theta^m}(x^n)$. Let $T(x^n)$ be some generally vector function of the observation sequence x^n that does not include the hypothesis index i and the vector parameter θ^m. Let in addition $T(x^n)$ be such that, if its value is fixed, then, given fixed i and θ^m, the density function $f_{i,\theta^m}[x^n | T(x^n)]$ conditioned on this value is only a function of the observation vector x^n. $T(x^n)$ is then called the *sufficient statistics* of the hypothesis problem. As a special useful case, let us consider the possibility where the density function $f_{i,\theta^m}(x^n)$ can be factorized in the form

$$f_{i,\theta^m}(x^n) = g[i, \theta^m, T(x^n)] h(x^n) \qquad \forall\, i, \theta^m \tag{4.2.26}$$

where the functions $T(x^n)$ and $h(x^n)$ do not explicitly include i and θ^m. It is then said that the *factorization criterion* is satisfied, and $T(x^n)$ is the sufficient statistics of the problem. If (4.2.26) is true, the function $g_k(x^n)$ in (4.2.1) can be written as

$$g_k(x^n) = h(x^n) \int_{\mathscr{E}^m} \sum_i c_{ki}(\theta^m) g[i, \theta^m, T(x^n)] r_i(\theta^m)\, d\theta^m \tag{4.2.27}$$

Given x^n the minimum $f(x^n) = \min_k g_k(x^n)$ includes the sequence x^n strictly in the form of the sufficient statistics $T(x^n)$. Thus, from Theorem 4.2.1 we conclude that if the factorization criterion in (4.2.26) is satisfied, then the optimal Bayesian test involves the observed sequence x^n strictly in the form $T(x^n)$. Such is the case in Example 4.2.2, where the matched filter is strictly a function of the sufficient statistics, $R_n^{-1} x^n$.

4.3 GRAM-SCHMIDT ORTHOGONALIZATION METHOD

The concept of sufficient statistics will be used quite extensively from now on, and its presentation completes Section 4.2. In the next section, we will present, analyze, and discuss a method that facilitates the derivation and analysis of some statistical tests on discrete-time hypotheses.

4.3 GRAM-SCHMIDT ORTHOGONALIZATION METHOD

The Gram-Schmidt orthogonalization method is basically a convenient method for matrix inversion. Its use in statistical decision theory may best be shown by a simple example.

Example 4.3.1

Consider a single hypothesis problem consisting of two equally probable hypotheses, and with penalty coefficients as in (4.2.13) in Example 4.2.1. Let the two hypotheses be specified by two well known, discrete-time Gaussian processes. In particular, let the observation interval correspond to n observed data, and let the hypothesis H_0 be determined by a real, zero mean, Gaussian process, whose autocovariance matrix is R_n. Let the H_1 hypothesis be specified by a Gaussian process whose mean vector is the real constant vector s^n and whose autocovariance matrix is R_n (the same as in H_0). Let the observation vector be denoted by x^n. Then, given x^n and directly from (4.2.15) in Example 4.2.1, we conclude that the "optimal" Bayesian decision rule is

$$\text{Decide in favor of } H_1, \text{ iff } f_1(x^n)/f_0(x^n) \geq 1 \\ \text{Otherwise, decide in favor of } H_0. \tag{4.3.1}$$

where $f_1(x^n)$ and $f_0(x^n)$ are the density functions of the vector X^n at the point x^n under H_1 and H_0 respectively.

The test in (4.3.1) consists of a likelihood ratio tested against the threshold one. The densities $f_1(x^n)$ and $f_0(x^n)$ are given by

$$f_1(x^n) = \frac{\exp\{-\frac{1}{2}[x^n - s^n]^T R_n^{-1}[x^n - s^n]\}}{(2\pi)^{n/2}|R_n|^{1/2}}$$

$$= (2\pi)^{-n/2}|R_n|^{-1/2}$$

$$\cdot \exp\left\{-\frac{1}{2}x^{n^T} R_n^{-1} x^n\right\} \exp\left\{s^{n^T} R_n^{-1} x^n - \frac{1}{2}s^{n^T} R_n^{-1} s^n\right\} \tag{4.3.2}$$

$$f_0(x^n) = \frac{\exp\{-\frac{1}{2}x^{n^T} R_n^{-1} x^n\}}{(2\pi)^{n/2}|R_n|^{1/2}} \tag{4.3.3}$$

where x^n and s^n have been defined as column vectors, R_n^{-1} denotes the inverse of the matrix R_n, $|R_n|$ denotes the determinant of R_n, and T denotes transpose, and where in (4.3.2) the fact that both the quantities $s^{n^T} R_n^{-1} x^n$ and $x^{n^T} R_n^{-1} s^n$

are scalar was used, in which case

$$x^{n^T} R_n^{-1} s^n = (x^{n^T} R_n^{-1} s^n)^T = s^{n^T} (R_n^{-1})^T x^n = s^{n^T} R_n^{-1} x^n \qquad (4.3.4)$$

From (4.3.2) and (4.3.3), we note that the factorization criterion, as in (4.2.26), holds, and that the sufficient statistics is $s^{n^T} R_n^{-1} x^n$. We then obtain

$$\frac{f_1(x^n)}{f_0(x^n)} = \exp\left\{ s^{n^T} R_n^{-1} x^n - \frac{1}{2} s^{n^T} R_n^{-1} s^n \right\} \qquad (4.3.5)$$

We note that, in our example, the stochastic process that specifies H_0 is real. Thus, R_n is then symmetric, and so is its inverse R_n^{-1}. Therefore, $(R_n^{-1})^T = R_n^{-1}$. Substituting (4.3.5) in the test inequality in (4.3.1), and taking the natural logarithm of both parts (which induces an equivalent test, since the logarithm is a monotone function), we obtain the following simplified form of the test in (4.3.1).

Given x^n, decide in favor of H_1 iff

$$s^{n^T} R_n^{-1} x^n - \frac{1}{2} s^{n^T} R_n^{-1} s^n \geq 0$$

or, equivalently, iff

$$s^{n^T} R_n^{-1} x^n \geq \frac{1}{2} s^{n^T} R_n^{-1} s^n = D \qquad (4.3.6)$$

Otherwise, decide in favor of H_0.

In the test in (4.3.6), the quadratic form $s^{n^T} R_n^{-1} x^n$ is the test function, and the quadratic form $2^{-1} s^{n^T} R_n^{-1} s^n$ is the threshold. As a function of the stochastic observation vector X^n the test function is a Gaussian random variable under both the hypotheses. Indeed, under both H_0 and H_1, the random vector X^n is Gaussian. Furthermore, the scalar $s^{n^T} R_n^{-1} X^n$ is basically then a linear transformation of jointly Gaussian random variables; thus, $s^{n^T} R_n^{-1} X^n$ is Gaussian. The mean of the Gaussian variable $s^{n^T} R_n^{-1} X^n$ is zero under H_0 and equal to $s^{n^T} R_n^{-1} s^n$ under H_1. The variance σ^2 of $s^{n^T} R_n^{-1} X^n$ is the same under both hypotheses; it is derived by

$$\sigma^2 = E_{H_0}\{s^{n^T} R_n^{-1} X^n X^{n^T} R_n^{-1} s^n\} = E_{H_1}\{s^{n^T} R_n^{-1} [X^n - s^n][X^n - s^n]^T R_n^{-1} s^n\}$$
$$= s^{n^T} R_n^{-1} E_{H_0}\{X^n X^{n^T}\} R_n^{-1} s^n = s^{n^T} R_n^{-1} R_n R_n^{-1} s^n = s^{n^T} R_n^{-1} s^n = 2D \qquad (4.3.7)$$

where, by definition

$$R_n = E_{H_0}\{X^n X^{n^T}\} \qquad (4.3.8)$$

4.3 Gram-Schmidt Orthogonalization Method

The probability of correct detection P_d induced by the test in (4.3.6) is now derived by

$$P_d = \frac{1}{2} P_{H_1}\{s^{n^T} R_n^{-1} X^n \geq D\} + \frac{1}{2} P_{H_0}\{s^{n^T} R_n^{-1} X^n < D\}$$

$$= \frac{1}{2} - \frac{1}{2} P_{H_1}\{s^{n^T} R_n^{-1} X^n < D\} + \frac{1}{2} P_{H_0}\{s^{n^T} R_n^{-1} X^n < D\}$$

$$= \frac{1}{2} - \frac{1}{2}\Phi\left(\frac{D - 2D}{(2D)^{1/2}}\right) + \frac{1}{2}\Phi\left(\frac{D}{(2D)^{1/2}}\right) = \Phi\left(\left[\frac{D}{2}\right]^{1/2}\right)$$

$$= \Phi(2^{-1}[s^{n^T} R_n^{-1} s^n]^{1/2}) \qquad (4.3.9)$$

The probability P_e of error, induced by the test in (4.3.6), is clearly given by

$$P_e = 1 - P_d = 1 - \Phi(2^{-1}[s^{n^T} R_n^{-1} s^n]^{1/2}) = \Phi(-2^{-1}[s^{n^T} R_n^{-1} s^n]^{1/2}) \quad (4.3.10)$$

where $\Phi(x)$ denotes the cumulative distribution of the zero mean, unit variance, Gaussian random variable.

Remember that the Bayesian formalization in this example basically uses the probability of error $P_e(\delta)$ as the objective function in the constraint optimization problem. Thus, the test in (4.3.6) minimizes the error $P_e(\delta)$, within the class of all the legitimate decision rules. As a result, the error P_e in (4.3.10) cannot be improved upon when the vector s^n and the matrix R_n are fixed, so (4.3.10) provides the exact value of the minimum possible error.

Now consider Example 4.3.1. In principle, the corresponding single hypothesis test has been fully analyzed. The problem is that implementing and evaluating the test in (4.3.6), as exhibited by the probability P_e in (4.3.10), involves inverting the covariance matrix R_n. Such an inversion can be a horrendous task when n is large. Thus, efficient matrix inversion methods should be sought. The application of such methods is crucial to the realization of statistical tests, such as the one exhibited by Example 4.3.1. In this section, we will present only one method of matrix inversion—the *Gram-Schmidt orthogonalization method*. Presenting other methods is beyond the scope of this book. Furthermore, other methods usually apply to matrices with more specialized properties than those assumed by the Gram-Schmidt method, such as those that evolve, when symmetric Toeplitz matrices are considered. The autocovariance matrices of real, discrete-time, stationary processes are Toeplitz matrices. For the interested reader, we include some references on the subject at the end of this chapter.

The Gram-Schmidt orthogonalization method applies to nonsingular, semipositive definite square matrices, where the terms are assumed to be familiar to the reader. Let us denote by R_n such an $n \times n$ matrix. It is known [Bellman (1970)] that a discrete-time, zero mean stochastic process $X(t)$ and a lower triangular, nonsingular $n \times n$ matrix W_n exist such that,

$$R_n = E\{X^n X^{n*T}\} \qquad (4.3.11)$$

$$R_n^{-1} = W_n^{*T} W_n \qquad (4.3.12)$$

where * denotes conjugate, T denotes transpose, -1 denotes inverse and X^n is a column vector.

Let us denote by I_n the $n \times n$ identity matrix. Then, from (4.3.11) and (4.3.12) we obtain

$$R_n = W_n^{-1}(W_n^{*T})^{-1} = E\{X^n X^{n*T}\} \qquad (4.3.13)$$

$$W_n R_n W_n^{*T} = I_n = W_n E\{X^n X^{n*T}\} W_n^{*T} \qquad (4.3.14)$$

$$E\{(W_n X^n)(W_n X^n)^{*T}\} = I_n \qquad (4.3.15)$$

Let us define a column stochastic vector Z^n as

$$Z^n = W_n X^n \qquad (4.3.16)$$

From (4.3.15) we observe that the process $X(t)$ induces an orthogonal stochastic vector Z^n. In fact, this orthogonality is induced by any discrete-time process whose n-dimensional autocovariance matrix is R_n. The above observation allows us to construct the stochastic vector Z^n directly, without initially computing the lower triangular matrix W_n. (Note that the direct computation of W_n implies the direct inversion of R_n, which should be avoided.) Let us denote by Z_i the ith component of Z^n, where $i = 1, \ldots, n$. From (4.3.16) and the fact that W_n is lower triangular, we observe that the random variable Z_k is a linear combination of at most the components $X_i; i = 1, \ldots, k$ of the stochastic vector X^n. Constructing the random variables Z_i recursively, with increasing i index, so that each newly created variable has variance equal to one and is uncorrelated from all the previously constructed variables, we obtain

$$Z_1 = E^{-1/2}\{X_1 X_1^*\} \cdot X_1 \qquad (4.3.17)$$

$$Z_k = E^{-1/2}\left\{\left\|X_k - \sum_{i=1}^{k-1} E\{Z_i X_k^*\} Z_i\right\|^2\right\}$$

$$\cdot \left[X_k - \sum_{i=1}^{k-1} E\{Z_i X_k^*\} Z_i\right] \quad k > 1 \qquad (4.3.18)$$

where $\|x\|^2$ means xx^*.

From (4.3.17) and (4.3.18) we observe that in the derivation of the Z^n components, the components of R_n are used, rather than those of R_n^{-1}. Indeed, the components of the former matrix have the form $E\{X_i X_j^*\}; i, j = 1, \ldots, n$. Using the Gram-Schmidt method in (4.3.6), we compute the inner product $s^{nT} R_n^{-1} s^n$ by first constructing the column vector $a^n = W_n s^n$ as in (4.3.17) and (4.3.18) and then computing the inner product $a^{n*T} a^n$. Given an observation vector x^n, we

4.3 Gram-Schmidt Orthogonalization Method

also first compute the column vector $z^n = W_n x^n$ as in (4.3.17) and (4.3.18). Then $s^{n^T} R_n^{-1} x^n$ is computed as the inner product $a^{n*T} z^n$. We will complete this section with an example.

Example 4.3.2

Consider Example 4.3.1, and let the autocovariance matrix $R_n = \{r_{ik}; i, k = 1, \ldots, n\}$ be such that,

$$r_{ik} = \sigma^2 + \alpha \cdot \min(i-1, k-1) \tag{4.3.19}$$

where σ^2 and α are given constants.

Let N^n be some stochastic, zero mean, column, n-dimensional vector, whose autocovariance matrix is R_n. Let W_n be a lower triangular $n \times n$ matrix, and let Z^n be a stochastic, column, n-dimensional vector such that

$$R_n = W_n^{-1}(W_n^{*T})^{-1} = E\{N^n N^{n*T}\} \tag{4.3.20}$$

$$Z^n = W_n N^n \tag{4.3.21}$$

where $Z^n = \{Z_i; i = 1, \ldots, n\}$, $\{N^n = N_i; i = 1, \ldots, n\}$.

Then $r_{ik} = E\{N_i N_k^*\}$ and, substituting the vector N^n for the vector X^n in (4.3.17) and (4.3.18), we obtain

$$Z_1 = E^{-1/2}\{N_1 N_1^*\} \cdot N_1 \tag{4.3.22}$$

$$Z_k = E^{-1/2}\left\{\left\|N_k - \sum_{i=1}^{k-1} E\{Z_i N_k^*\} Z_i\right\|^2\right\}$$

$$\cdot \left[N_k - \sum_{i=1}^{k-1} E\{Z_i N_k^*\} Z_i\right] \quad k > 1 \tag{4.3.23}$$

Substituting (4.3.19) in (4.3.22) and (4.3.23), we obtain in a recursive fashion

$$Z_1 = \sigma^{-1} N_1 \tag{4.3.24}$$

$$Z_k = \alpha^{-1/2}[N_k - N_{k-1}] \quad k \geq 2 \tag{4.3.25}$$

From (4.3.21), (4.3.24), and (4.3.25), we easily conclude that the lower triangular matrix W_n has the form

$$W_n = \begin{bmatrix} \sigma^{-1} & & & & \\ & \alpha^{-1/2} & & & 0 \\ -\alpha^{-1/2} & & \ddots & & \\ 0 & & & \ddots & \\ & & & \alpha^{-1/2} & \alpha^{-1/2} \end{bmatrix} \tag{4.3.26}$$

That is, in W_n only the diagonal and the first parallel to the diagonal are nonzero. The components of the latter are all equal to $-\alpha^{-1/2}$. The first component of the former is equal to σ^{-1}, while the remaining components are all equal to $\alpha^{-1/2}$. We note that in the present example, the matrix W_n is easily and analytically computed via the Gram-Schmidt method. However, the direct computation of the inverse R_n^{-1} is still hard. In the test in (4.3.6), let us now define

$$a^n = W_n s^n \quad \begin{cases} s^n = \{s_i; i = 1, \ldots, n\} \\ a^n = \{a_i; i = 1, \ldots, n\} \end{cases} \quad (4.3.27)$$

$$y^n = W_n x^n \quad \begin{cases} x^n = \{x_i; i = 1, \ldots, n\} \\ y^n = \{y_i; i = 1, \ldots, n\} \end{cases} \quad (4.3.28)$$

where W_n is as in (4.3.26).

Then, we find by direct substitution,

$$a_1 = \sigma^{-1} s_1 \quad a_k = \alpha^{-1/2}[s_k - s_{k-1}] \quad k > 1 \quad (4.3.29)$$

$$y_1 = \sigma^{-1} x_1 \quad y_k = \alpha^{-1/2}[x_k - x_{k-1}] \quad k > 1 \quad (4.3.30)$$

$$s^{n^T} R_n^{-1} s^n = a^{n^T} a^n \quad (4.3.31)$$

$$s^{n^T} R_n^{-1} x^n = a^{n^T} y^n \quad (4.3.32)$$

Finally, substituting the vector components of (4.3.29) and (4.3.30) into (4.3.31) and (4.3.32), gives

$$s^{n^T} R_n^{-1} s^n = \sigma^{-2} s_1^2 + \alpha^{-1} \sum_{k=2}^{n} (s_k - s_{k-1})^2 \quad (4.3.33)$$

$$s^{n^T} R_n^{-1} x^n = \sigma^{-2} s_1 x_1 + \alpha^{-1} \sum_{k=2}^{n} (s_k - s_{k-1})(x_k - x_{k-1}) \quad (4.3.34)$$

From (4.3.33), we observe that, in the present case, both the threshold and the test function in the test in (4.3.6) can be computed easily and directly by using the Gram-Schmidt method. In those computations, only subtractions of consecutive vector components are involved. Such operations are simple and fast, while the direct inversion of the matrix R_n is complicated, requiring lengthy computational effort. In addition, the complexity of the direct R_n inversion also induces relatively large computational errors.

4.4 CONTINUOUS-TIME HYPOTHESES

Our formalization and solution of the Bayesian detection problem, in Section 4.2, has been general. It includes both single hypothesis and multiple hypotheses testing, for arbitrary, discrete-time, not necessarily stationary, parametrically known stochastic processes. Indeed, in the Bayesian detection problem, there are no conceptual differences between single hypothesis and multiple hypotheses

4.4 Continuous-Time Hypotheses

testing, or between stationary and nonstationary parametrically known processes. Differences appear only in the final form of the tests and in the test statistics. We believe that the latter are best exhibited by the study of specific examples, such as those included in Sections 4.2 and 4.3. We will present and fully analyze additional examples, in Section 4.5.

In Section 4.1, we mentioned that the Bayesian detection problem for continuous-time, parametrically known processes can be intermediately transformed into a discrete data form by series expansion. At this point, we will qualify the above statement. Let us consider a continuous-time, zero-mean stochastic process $X(t)$ with autocovariance function $R(t_1, t_2)$. If $X(t)$ denotes waveforms generated by the process, then

$$R(t_1, t_2) = E\{X(t_1)X^*(t_2)\} \tag{4.4.1}$$

where E denotes expected value, and $*$ denotes conjugate.

Given the process $X(t)$ above, let $[0, T]$ be a given observation time interval and let there exist an orthonormal set $\{\phi_i(t)\}$ of deterministic time functions defined on $[0, T]$ such that for every function $\phi_i(t)$ in the set, there exists a positive constant λ_i that satisfies the equation

$$\int_0^T R(t_1, t_2)\phi_i(t_2)\,dt_2 = \lambda_i\phi_i(t_1) \qquad \forall\, t_1 \in [0, T] \tag{4.4.2}$$

Expression (4.4.2) implies that the members of $\{\phi_i(t)\}$ are solutions of a set of integral equations that incorporate the autocovariance function of $X(t)$. Thus, (4.4.2) induces, in general, a nonuniversal or noncomplete set $\{\phi_i(t)\}$ of functions orthonormal in $[0, T]$. Instead, $\{\phi_i(t)\}$ is matched to a given stochastic process. The orthonormality of $\{\phi_i(t)\}$ in $[0, T]$ implies that the following condition is satisfied, where Δ_{ij} is the Kronecker delta.

$$\int_0^T \phi_i(t)\phi_j^*(t)\,dt = \Delta_{ij} \tag{4.4.3}$$

Given the process $X(t)$ and the interval $[0, T]$, let us assume that a set $\{\phi_i(t)\}$ satisfying (4.4.2) and (4.4.3) exists (the existence is not, in general, guaranteed). Let us then define the following set $\{\hat{X}_i\}$ of random variables

$$\{\hat{X}_i\}: \hat{X}_i = \int_0^T X(t)\phi_i^*(t)\,dt \tag{4.4.4}$$

Let us also define

$$\hat{X}(t) = \sum_i \hat{X}_i \phi_i(t) \tag{4.4.5}$$

Then $\hat{X}(t)$ is called the Karhunen-Loéve expansion of the process $X(t)$ in $[0, T]$ via the set $\{\phi_i(t)\}$. The rigorous analysis of the Karhunen-Loéve expansion can be found in books on stochastic processes. Here, we will only state some of its properties. Their proofs are relatively easy, and we challenge the reader to perform them. If the Karhunen-Loéve expansion exists, then the set $\{\lambda_i\}$ in (4.4.2), the set $\{\hat{X}_i\}$ in (4.4.4), the stochastic waveform $\hat{X}(t)$ in (4.4.5), the waveforms $X(t)$ generated by the stochastic process, the autocovariance function $R(t_1, t_2)$ of the process, and the orthonormal set $\{\phi_i(t)\}$ of deterministic functions, satisfy these conditions:

$$R(t_1, t_2) = \sum_i \lambda_i \phi_i(t_1) \phi_i^*(t_2) \quad \forall\, t_1, t_2 \in [0, T] \quad (4.4.6)$$

$$\int_0^T R(t, t)\, dt = \sum_i \lambda_i$$

$$E\{X(t) - \hat{X}(t)\}^2 = 0 \quad \forall\, t \in [0, T] \quad (4.4.7)$$

$$E\{\hat{X}_i \hat{X}_j^*\} = \lambda_i \Delta_{ij} \quad (4.4.8)$$

where Δ_{ij} is the Kronecker delta.

Due to (4.4.7), we say that if the Karhunen-Loéve expansion exists, it approximates the process in $[0, T]$ in the mean-squared sense. Due to (4.4.8), we conclude that the random variables $\{\hat{X}_i\}$ in the Karhunen-Loéve expansion in (4.4.5) are uncorrelated. The variance of the variable $\hat{X}_i$ is equal to the constant λ_i. If the stochastic process $X(t)$ is Gaussian, clearly the random variables $\{\hat{X}_i\}$ are also Gaussian (see (4.4.4) in conjunction with the fact that any linear transformation of a Gaussian process is also Gaussian). Then, zero correlation implies independence; thus, the random variables $\{\hat{X}_i\}$ are also independent.

If it exists, the Karhunen-Loéve expansion can be used as an intermediate step for the formulation and solution of Bayesian detection schemes for continuous-time processes. Indeed, let us approximate such a process in $[0, T]$ by its Karhunen-Loéve expansion in (4.4.5). As we saw, this approximation is in the mean-squared sense. Given the set $\{\phi_i(t)\}$ of orthonormal deterministic functions, the total stochastic information of the Karhunen-Loéve process $\hat{X}(t)$ in $[0, T]$ is included in the set of random variables $\{\hat{X}_i\}$. Thus, if a realization $\hat{x}(t);\ t \in [0, T]$ of the process $\hat{X}(t)$ were observed, it could be equivalently represented by the corresponding realizations $\{\hat{x}_i\}$ of $\{\hat{X}_i\}$. Then, the Bayesian formalization could be performed as previously, via the set $\{\hat{x}_i\}$. The resulting reduced Bayesian test should then be transformed back to a continuous-in-time form. The optimality of the latter test is accurate in the mean-squared sense, since the Karhunen-Loéve expansion approximates the original process in the same sense. Given a finite number of continuous-time processes or hypotheses, the above process can be realized only if there exists a single set $\{\phi_i(t)\};\ t \in [0, T]\}$ of orthonormal deterministic functions that provides Karhunen-Loéve expansions in $[0, T]$ for *all* the considered processes.

Let us now consider a deterministic waveform $s(t)$ observed in the time interval $[0, T]$. Let $\{\phi_i(t)\}$ be a set of deterministic time functions orthonormal in $[0, T]$, and let us seek representation of $s(t)$ in $[0, T]$ via $\{\phi_i(t)\}$. Let

$$s_i = \int_0^T s(t) \phi_i^*(t)\, dt \quad \forall\, i \quad (4.4.9)$$

$$\hat{s}(t) = \sum_i s_i \phi_i(t) \quad \forall\, t \in [0, T] \quad (4.4.10)$$

4.4 Continuous-Time Hypotheses

Then let $\{\phi_i(t)\}$ be such that

$$\int_0^T [s(t) - \hat{s}(t)][s(t) - \hat{s}(t)]^* \, dt = \int_0^T s(t)s^*(t) \, dt - \sum_i s_i s_i^* = 0 \qquad (4.4.11)$$

for every waveform $s(t)$ such that $\int_0^T s(t)s^*(t) \, dt < \infty$.

The set $\{\phi_i(t)\}$ is then called *complete in* $[0, T]$, for the class of deterministic waveforms with finite energy [see, for example, Van Trees (1968)]. The set $\{\phi_i(t)\}$ then approximates the waveform $s(t)$ via the representation $\hat{s}(t)$ in (4.4.10) in the mean-squared sense represented by (4.4.11). The coefficients $\{s_i\}$ in (4.4.9) are constants and include all the characteristics of the particular waveforms. The representation $\hat{s}(t)$ in (4.4.10) decouples the waveform characteristics, represented by $\{s_i\}$, from time, represented by $\{\phi_i(t)\}$.

Let us consider a zero mean stochastic process $X(t)$ defined in $[0, T]$ whose autocovariance function is $R(t_1, t_2)$. Let $\{\phi_i(t)\}$ be an orthonormal set complete in $[0, T]$ for the class of deterministic waveforms with finite energy. Let $s(t)$ be such a waveform, and let $\{\phi_i(t)\}$ also satisfy the integral equations in (4.4.2). Then from the discussion in this section, we conclude that $\{\phi_i(t)\}$ provides mean-squared approximations for both $s(t)$ and $X(t)$. The mean-squared approximation for $s(t)$ is in the sense provided by (4.4.11). The mean-squared approximation for $X(t)$ is in the expected value sense exhibited by (4.4.7). To clarify all the above statements, we will proceed with an example.

Example 4.4.1

Let us consider a single hypothesis problem. Let the two hypotheses be well known, let them be equally probable, and let the penalty coefficients c_{kj} be such that

$$c_{kj} = \begin{cases} 0 & k = j \\ 1 & k \neq j \end{cases} \qquad k, j = 0, 1 \qquad (4.4.12)$$

As we have seen before, given a decision rule δ the expected penalty $C(\delta)$ is then identical to the probability of error $P_e(\delta)$. Let the H_0 hypothesis be represented by a zero mean, continuous-time, scalar stationary process with autocovariance function $R(t_1, t_2)$. Let us denote by $N(t)$ stochastic waveforms generated by this process. Let us denote by $n(t)$ realizations of $N(t)$. Let the hypothesis H_1 be represented by a given deterministic scalar, finite energy, and real waveform $s(t)$ added to the stochastic waveform $N(t)$. Let us denote by $V(t)$ stochastic waveforms generated by H_1 and let $u(t)$ denote realizations of those waveforms. Then

$$V(t) = s(t) + N(t)$$
$$u(t) = s(t) + n(t) \qquad (4.4.13)$$

Let the observation interval be $[0, T]$ and let there exist an orthonormal and complete set $\{\phi_i(t)\}$ of deterministic functions for the class of deterministic and finite energy waveforms. Let $\{\phi_i(t)\}$ also provide a Karhunen-Loéve expansion for $N(t)$ in $[0, T]$. Then $\{\phi_i(t)\}$ satisfies the conditions

$$\int_0^T \phi_i(t)\phi_j^*(t)\,dt = \Delta_{ij} \tag{4.4.14}$$

$$\int_0^T R(t_1, t_2)\phi_i(t_2)\,dt_2 = \lambda_i \phi_i(t_1) \quad \begin{cases} \text{for all} & t_1 \in [0, T] \\ \text{for some} & \lambda_i > 0 \\ \text{for all} & i \end{cases} \tag{4.4.15}$$

and it provides mean squared approximations $\hat{s}(t)$ and $\hat{N}(t)$ of the respectively deterministic and stochastic waveforms $s(t)$ and $N(t)$ as

$$\hat{s}(t) = \sum_i s_i \phi_i(t) \qquad t \in [0, T] \tag{4.4.16}$$

for

$$s_i = \int_0^T s(t)\phi_i^*(t)\,dt \qquad \forall\, i \tag{4.4.17}$$

$$\hat{N}(t) = \sum_i \hat{N}_i \phi_i(t) \qquad t \in [0, T] \quad \hat{n}(t) \text{ realization of } \hat{N}(t) \tag{4.4.18}$$

for

$$\hat{N}_i = \int_0^T N(t)\phi_i^*(t)\,dt \qquad \forall\, i \quad \hat{n}_i \text{ realization of } \hat{N}_i \tag{4.4.19}$$

The set of constants $\{s_i\}$ in (4.4.17) and the set of random variables $\{\hat{N}_i\}$ in (4.4.19) represent respectively the waveform $s(t)$ and the stochastic process $N(t)$ in $[0, T]$ in the mean-squared sense. Let us denote by $v(t); t \in [0, T]$ an observed realization that may be generated by either of the hypotheses H_0 and H_1. If H_0 is active, $v(t) = n(t)$. If H_1 is active, $v(t) = s(t) + n(t)$. Let us use the above approximations. Then $v(t)$ is approximated in $[0, T]$ by $\hat{v}(t)$, where

$$\hat{v}(t) = \begin{cases} \hat{n}(t) & \text{under hypothesis } H_0 \\ \hat{s}(t) + \hat{n}(t) & \text{under hypothesis } H_1 \end{cases} \tag{4.4.20}$$

Due to the earlier discussion, given the set $\{\phi_i(t)\}$ the realization $\hat{v}(t)$ in (4.4.20) is precisely represented by the sets $\{\hat{n}_i\}$ and $\{s_i + \hat{n}_i\}$, under respectively the H_0 and the H_1 hypotheses. We can thus formulate the Bayesian detection problem for the approximate (in the mean-squared sense) hypotheses via the discrete set $\{\hat{v}_i\}$ of random variables, where

$$\hat{v}_i = \begin{cases} \hat{n}_i & \text{under hypothesis } H_0 \\ s_i + \hat{n}_i & \text{under hypothesis } H_1 \end{cases}$$
$$\hat{v}(t) = \sum_i \hat{v}_i \phi_i(t) \qquad t \in [0, T] \tag{4.4.21}$$

The joint distribution of the members in $\{\hat{v}_i\}$ is in principle known, due to (4.4.17) and (4.4.19), and due to the assumption that $s(t)$ is given, $N(t)$ is well

4.4 Continuous-Time Hypotheses

known, and $\{\phi_i(t)\}$ has been found. Let us now assume that $N(t)$ is real and Gaussian. Then, as stated earlier, the random variables $\{\hat{N}_i\}$ are zero mean, Gaussian and independent, and the variance of $\hat{N}_i$ equals λ_i; where λ_i is given by (4.4.15). The random variables in $\{s_i + \hat{N}_i\}$ are then also Gaussian and independent. In addition, the mean of $s_i + \hat{N}_i$ equals s_i and its variance still equals λ_i. Now consider the Bayesian single hypothesis test for realizations $\{\hat{v}_i\}$ under the above Gaussian assumption. Denote by $f_1(\{\hat{v}_i\})$ the density function of $\{\hat{v}_i\}$ under H_1. Denote by $f_0(\{\hat{v}_i\})$ the corresponding density, under H_0. Then, directly from (4.2.15) in Example 4.2.1, and for $p_j = .5$; $j = 0, 1$, we conclude that the optimal decision rule is

Given $\{\hat{v}_i\}$, decide in favor of H_1 iff $\dfrac{f_1(\{\hat{v}_i\})}{f_0(\{\hat{v}_i\})} \geq 1$ (4.4.22)

Otherwise, decide in favor of H_0

Due to the established statistical behavior of the variables in $\{\hat{v}_i\}$ under the two hypotheses, we also then have

$$f_1(\{\hat{v}_i\}) = \prod_i \frac{\exp\left\{-\frac{1}{2}\frac{(\hat{v}_i - s_i)^2}{\lambda_i}\right\}}{(2\pi\lambda_i)^{1/2}} = \frac{\exp\left\{-\frac{1}{2}\sum_i \frac{(\hat{v}_i - s_i)^2}{\lambda_i}\right\}}{\prod_i (2\pi\lambda_i)^{1/2}}$$

$$= \left[\prod_i (2\pi\lambda_i)^{1/2}\right]^{-1} \exp\left\{-\frac{1}{2}\sum_i \frac{\hat{v}_i^2}{\lambda_i}\right\} \exp\left\{\frac{1}{2}\sum_i \frac{s_i}{\lambda_i}(2\hat{v}_i - s_i)\right\} \quad (4.4.23)$$

$$f_0(\{\hat{v}_i\}) = \left[\prod_i (2\pi\lambda_i)^{1/2}\right]^{-1} \exp\left\{-\frac{1}{2}\sum_i \frac{\hat{v}_i^2}{\lambda_i}\right\} \quad (4.4.24)$$

where $\prod$ denotes product.

The sufficient statistics evolving from (4.4.23) and (4.4.24) is $\sum_i s_i \lambda_i^{-1} \hat{v}_i$, and the following Bayesian single hypothesis test evolves.

Given $\{\hat{v}_i\}$, decide in favor of H_1 iff:

$$\exp\left\{\frac{1}{2}\sum_i \frac{s_i}{\lambda_i}(2\hat{v}_i - s_i)\right\} \geq 1 \quad (4.4.25)$$

Otherwise, decide in favor of H_0.

As in Example 4.2.1, we simplify the test by taking the natural logarithm of both parts of the inequality in (4.4.25). We then conclude

Given $\{\hat{v}_i\}$, decide in favor of H_1 iff:

$$\sum_i q_i \hat{v}_i \geq 2^{-1} \sum_i q_i s_i \quad (4.4.26)$$

Otherwise, decide in favor of H_0

where we have defined

$$q_i = \frac{s_i}{\lambda_i} \tag{4.4.27}$$

In the test in (4.4.26), the quantities q_i and s_i are deterministic for all i's. The sequence $\{\hat{v}_i\}$ is, instead, a realization of a sequence of random variables, and it fully controls the test statistics. Let us now define

$$\hat{q}(t) = \sum_i q_i \phi_i(t) \qquad t \in [0, T] \tag{4.4.28}$$

Let the sequence in (4.4.28) exist for every t in $[0, T]$. Then let $\hat{q}(t)$ be the mean-squared approximation [in the sense in (4.4.11)] in $[0, T]$, of the finite energy deterministic waveform $q(t)$. Then, the coefficient q_i is given by

$$q_i = \int_0^T q(t)\phi_i^*(t)\,dt \tag{4.4.29}$$

From (4.4.27) and (4.4.29), we conclude

$$s_i = \lambda_i \int_0^T q(t)\phi_i^*(t)\,dt \tag{4.4.30}$$

Directly from (4.4.30), we also find

$$\sum_i s_i \phi_i(t) = \int_0^T q(\tau) \sum_i \lambda_i \phi_i(t)\phi_i^*(\tau)\,d\tau \qquad t \in [0, T] \tag{4.4.31}$$

where from (4.4.6) we also have

$$R(t, \tau) = \sum_i \lambda_i \phi_i(t)\phi_i^*(\tau) \tag{4.4.32}$$

To derive (4.4.31), we interchanged the order of integration and an infinite summation. Under the assumptions that (4.4.32) is valid [or equivalently that the set $\{\phi_i(t)\}$ provides a Karhunen-Loéve expansion in $[0, T]$ for the process $N(t)$] and the function in (4.4.28) exists, this interchange is legitimate. Substituting (4.4.16) and (4.4.32) in (4.4.31) gives

$$\hat{s}(t) = \int_0^T q(\tau) R(t, \tau)\,d\tau \qquad t \in [0, T] \tag{4.4.33}$$

Let us now focus on the quantities $\sum_i q_i \hat{v}_i$ and $2^{-1} \sum_i q_i s_i$ in the test in (4.4.26). From (4.4.17), (4.4.21), and (4.4.29) we obtain by substitution, and again valid interchange of integrations and summations,

$$\sum_i q_i \hat{v}_i = \sum_i \hat{v}_i \int_0^T q(t)\phi_i^*(t)\,dt = \int_0^T q(t) \sum_i \hat{v}_i \phi_i^*(t)\,dt = \int_0^T q(t)\hat{v}(t)\,dt \tag{4.4.34}$$

$$2^{-1} \sum_i q_i s_i = 2^{-1} \sum_i s_i \int_0^T q(t)\phi_i^*(t)\,dt = 2^{-1} \int_0^T q(t) \sum_i s_i \phi_i^*(t)\,dt$$

$$= 2^{-1} \int_0^T q(t)\hat{s}(t)\,dt \tag{4.4.35}$$

4.4 Continuous-Time Hypotheses

Due to (4.4.33), (4.4.34), and (4.4.35), we finally conclude that the test in (4.4.26) takes the form:

Given the realization $\hat{v}(t)$ in $[0, T]$, decide in favor of H_1, iff:

$$\int_0^T q(t)\hat{v}(t)\,dt \geq 2^{-1} \int_0^T q(t)\hat{s}(t)\,dt \quad (4.4.36)$$

where $q(t); t \in [0, T]$ is a solution of

$$\hat{s}(t) = \int_0^T q(\tau)R(t, \tau)\,d\tau \quad t \in [0, T] \quad (4.4.37)$$

Otherwise, decide in favor of H_0.

Since the realizations $v(t); t \in [0, T]$ and $s(t); t \in [0, T]$ are available, rather than the realizations $\hat{v}(t)$ and $\hat{s}(t)$ in $[0, T]$, we substitute $v(t)$ for $\hat{v}(t)$ and $s(t)$ for $\hat{s}(t)$, in (4.4.36) and (4.4.37). The test "optimality" is then in the mean-squared sense. In the test, the integral $D = 2^{-1} \int_0^T q(t)s(t)\,dt$, is not a function of the observed realization, and it acts as a threshold. The integral $\int_0^T q(t)v(t)\,dt$ is instead a function of the observed waveform $v(t)$, and it controls the test statistics. In fact, in the present example, where both the hypotheses are determined by Gaussian stochastic processes, the integral $\int_0^T q(t)V(t)\,dt$ is a Gaussian random variable, under both H_0 and H_1. This is so because any linear deterministic transformation of a Gaussian waveform induces Gaussian statistics as well. Let us denote by m_0 and m_1 the expected value of the random variable, $\int_0^T q(t)V(t)\,dt$, under H_0 and H_1 respectively. Let us denote by σ_0^2 and σ_1^2 the corresponding variances. Then, we easily obtain

$$E_{H_0}\{V(t)\} = E_{H_0}\{N(t)\} = 0$$

$$E_{H_1}\{V(t)\} = E_{H_1}\{s(t) + N(t)\} = s(t) + E_{H_0}\{N(t)\} = s(t)$$

$$E_{H_0}\{V(t_1)V^*(t_2)\} = E_{H_0}\{N(t_1)N^*(t_2)\} = R(t_1, t_2)$$

$$E_{H_1}\{[V(t_1) - s(t_1)][V^*(t_2) - s^*(t_2)]\} = E_{H_0}\{N(t_1)N^*(t_2)\} = R(t_1, t_2)$$

$$m_0 = \int_0^T q(t)E_{H_0}\{V(t)\}\,dt = 0$$

$$m_1 = \int_0^T q(t)E_{H_1}\{V(t)\}\,dt = \int_0^T q(t)s(t)\,dt = 2D$$

$$\sigma_1^2 = \sigma_0^2 = \int\int_0^T q(t)q^*(\tau)E_{H_0}\{V(t)V^*(\tau)\}\,dt\,d\tau = \int\int_0^T q(t)q^*(\tau)R(t, \tau)\,dt\,d\tau$$

$$= \int_0^T q(t)s(t)\,dt = 2D$$

Due to the above, we conclude that the random variable $\int_0^T q(t)V(t)\,dt$ is $G(0, 2D)$ under H_0, and $G(2D, 2D)$ under H_1, where $G(x, y)$ denotes a Gaussian random variable with mean x and variance y. Let us now denote by P_e the probability of making a wrong decision and by P_d the probability of making a correct decision, using the same test. Then, we easily obtain

$$P_e = 1 - P_d$$

$$P_d = \frac{1}{2} P_{H_1}\left\{\int_0^T q(t)V(t)\,dt \geq D\right\} + \frac{1}{2} P_{H_0}\left\{\int_0^T q(t)V(t)\,dt < D\right\}$$

$$= \frac{1}{2}\int_D^\infty \frac{\exp\left\{-\frac{1}{4D}(u-2D)^2\right\}}{2(\pi D)^{1/2}}\,du + \frac{1}{2}\int_{-\infty}^D \frac{\exp\left\{-\frac{u^2}{4D}\right\}}{2(\pi D)^{1/2}}\,du$$

$$= \frac{1}{2} - \int_{-\infty}^D \frac{1}{2}\frac{\exp\left\{-\frac{1}{4D}(u-2D)^2\right\}}{2(\pi D)^{1/2}}\,du + \frac{1}{2}\int_{-\infty}^D \frac{\exp\left\{-\frac{u^2}{4D}\right\}}{2(\pi D)^{1/2}}\,du$$

$$= \frac{1}{2} - \frac{1}{2}\Phi\left(\frac{D-2D}{(2D)^{1/2}}\right) + \frac{1}{2}\Phi\left(\frac{D}{(2D)^{1/2}}\right) = \Phi\left(\left[\frac{D}{2}\right]^{1/2}\right)$$

where $D \equiv 2^{-1}\int_0^T q(t)s(t)\,dt$; $\Phi(x)$ denotes the distribution function of the unit variance, zero mean, Gaussian random variable at the point x; and P_{H_1}, P_{H_0} denote conditional probabilities, conditioned respectively on the H_1 versus H_0.

We conclude the example by emphasizing that the key assumption in its analysis has been the existence of a set $\{\phi_i(t)\}$ that provides a mean-squared approximation [as in (4.4.11)] of the deterministic waveform $s(t)$ in $[0, T]$, and which simultaneously provides a Karhunen-Loéve expansion in $[0, T]$ for the stochastic process $N(t)$. We point out that if the waveform $s(t)$ is squared integrable in $[0, T]$, then a valid set $\{\phi_i(t)\}$ for its mean squared expansion is the Fourier set $\{e^{jkw}\}$; $w = 2\pi/T$, where $j = \sqrt{-1}$. In some cases, the Fourier set also provides a valid Karhunen-Loéve expansion. This is fully determined by the autocovariance function $R(t_1, t_2)$ of the process. One such case, for example, evolves when the process is uncorrelated; that is when $R(t_1, t_2) = \sigma^2\delta(t_1 - t_2)$,

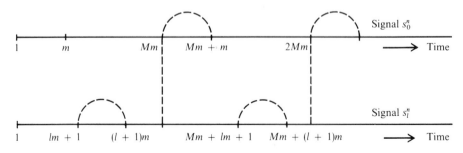

Figure 4.5.1

4.5 Examples

where $\delta(\cdot)$ denotes the delta function. Then (4.4.2) is satisfied by any set $\{\phi_i(t)\}$, thus it is also satisfied by the Fourier set. Following the analysis of the example, it is clear that, as previously claimed, the series expansions via an orthonormal set $\{\phi_i(t)\}$ are only used as an intermediate step in the derivation of Bayesian decision rules for continuous-time hypotheses.

4.5 EXAMPLES

In this section, we will solve and analyze some additional examples of Bayesian detection schemes. We will start with further discussions of some of the examples presented in this chapter. Then, we will analyze Examples 2 and 3, outlined in Section 3.3. Finally, we will present, solve, and analyze some additional problems.

Example 4.5.1

Let us consider the case where one of M distinct and equiprobable vector signals s_i^n; $i = 0, 1, \ldots, M - 1$ is transmitted in additive, white, zero mean, variance σ^2, Gaussian noise. Let the signals have identical powers, $n^{-1} s_i^{n^T} s_i$; $i = 0, 1, \ldots, M - 1$. Then, via Example 4.2.2 the optimal Bayesian, ideal observer test is a matched filter of the form in (4.2.25):

Decide in favor of the signal s_k^n, if

$$s_k^{n^T} x^n \geq s_l^{n^T} x^n \qquad \forall\, l\colon l = 0, 1, \ldots, M - 1 \qquad (4.5.1)$$

Let us denote the components of the vector signal s_l^n as s_{lj}; $1 \leq j \leq n$ and let the signals be periodic and orthogonal. In particular, for some positive integer m, each signal has period equal to mM. In addition, the only nonzero components of the signal s_l^n within its first period are the ones corresponding to the indices $lm + 1, lm + 2, \ldots, (l + 1)m$ (see Figure 4.5.1). Let us now select the observation length n in (4.5.1) equal to some multiple K of the period mM—that is, $n = KmM$. Then, we easily obtain the following expression, where $x^n = [x_1, \ldots, x_n]^T$.

$$s_l^{n^T} x^n = \sum_{i=1}^{K} \sum_{j=1}^{m} s_{l, lm+j} x_{i(lm+j)} \qquad n = KmM \qquad (4.5.2)$$

Let us now assume that when the vector x^n was observed, signal s_k^n was transmitted, and let us then denote by n_j the jth component of the additive noise. Then

$$\begin{aligned} x_{i(km+j)} &= s_{k, km+j} + n_{i(km+j)} & 1 \leq j \leq m \quad 1 \leq i \leq K \\ x_{i(lm+j)} &= n_{i(lm+j)} & 1 \leq j \leq m \quad 1 \leq i \leq K \quad l \neq k \end{aligned} \qquad (4.5.3)$$

Due to (4.5.2) and (4.5.3), we conclude that if the signal s_k^n is transmitted when x^n is observed, then

$$s_k^{n^T} x^n = \sum_{i=1}^{K} \sum_{j=1}^{m} s_{k,km+j}[s_{k,km+j} + n_{i(km+j)}] \qquad n = KmM$$

$$s_l^{n^T} x^n = \sum_{i=1}^{K} \sum_{j=1}^{m} s_{l,lm+j} n_{i(lm+j)} \qquad l \neq k \quad n = KmM$$
(4.5.4)

Since the signals are orthogonal and the noise process is memoryless, the random variables $\{s_l^{n^T} X^n\}$ in (4.5.4) are independent. In addition, the variables, $\sum_{j=1}^{m} s_{k,km+j}[s_{k,km+j} + N_{i(km+j)}]; 1 \leq i \leq K$ and $\sum_{j=1}^{m} s_{l,lm+j} N_{i(lm+j)}; 1 \leq i \leq K$ in (4.5.4) are mutually independent and each variable is Gaussian, being a linear transformation of Gaussian variables. Let us denote

$$Y_{ij} = s_{k,km+j}[s_{k,km+j} + N_{i(km+j)}]$$
$$W_{ij} = s_{l,lm+j} N_{i(lm+j)}$$
(4.5.5)

Then, the variable Y_{ij} is Gaussian, with mean $s_{k,km+j}^2$ and with variance, $\sigma^2 s_{k,km+j}^2$—that is Y_{ij} is $G(s_{k,km+j}^2, \sigma s_{k,km+j})$. The variable W_{ij} is $G(0, \sigma s_{l,lm+j})$. Due to that and the earlier discussion, we conclude that conditioned on the hypothesis H_k (signal s_k^n transmitted), the random variables $s_l^{n^T} X^n; l = 0, 1, \ldots, M - 1$ are mutually independent and Gaussian and such that

$$s_k^{n^T} X^n \text{ is } G\left(K \sum_{j=1}^{m} s_{k,km+j}^2, \sigma \left[K \sum_{j=1}^{m} s_{k,km+j}^2\right]^{1/2}\right)$$

and

$$s_l^{n^T} X^n \text{ is } G\left(0, \sigma \left[K \sum_{j=1}^{m} s_{l,lm+j}^2\right]^{1/2}\right)$$

for every $l \neq k$.

At the same time, $(nM)^{-1} \sum_{j=1}^{m} s_{l,lm+j}^2$ is the power of the signal s_l^n. Given that this latter power is identical for all signals, and denoting it by c^2, we conclude that, conditioned on the hypothesis H_k, the random variables $s_l^{n^T} X^n; l = 0, 1, \ldots, M - 1$ are mutually independent and such that,

$$s_k^{n^T} X^n \text{ is } G(nc^2, \sigma c \sqrt{n})$$
$$s_l^{n^T} X^n \text{ is } G(0, \sigma c \sqrt{n}) \quad \forall l \neq k$$

Denoting then by $\Phi(x)$ the distribution function of the zero mean, unit variance, Gaussian random variable, at the point x we obtain

$$\Pr\{s_l^{n^T} X^n \leq x; \forall l \neq k | s_k^{n^T} X^n = x, H_k\}$$
$$= \Pr\{s_l^{n^T} X^n \leq x \forall l \neq k | H_k\}$$
$$= \Phi^{M-1}\left(\frac{x}{\sigma c \sqrt{n}}\right)$$

4.5 Examples

$$\Pr\{x \leq s_k^{n^T} X^n \leq x + dx, s_l^{n^T} X^n \leq x \ \forall \ l \neq k | H_k\}$$
$$= \Pr\{x \leq s_k^{n^T} X^n \leq x + dx | H_k\} \Pr\{s_l^{n^T} X^n \leq x \ \forall \ l \neq k | H_k\}$$
$$= dx \frac{\exp\left\{-\frac{1}{2} \frac{(x-nc^2)^2}{\sigma^2 c^2 n}\right\}}{\sqrt{2\pi} \sigma c \sqrt{n}} \Phi^{M-1}\left(\frac{x}{\sigma c \sqrt{n}}\right)$$

$$\Pr\{s_k^{n^T} X^n \geq s_l^{n^T} X^n \ \forall \ l | H_k\}$$
$$= \int_{-\infty}^{\infty} \Pr\{x \leq s_k^{n^T} X^n \leq x + dx, s_l^{n^T} X^n \leq x \ \forall \ l \neq k | H_k\}$$
$$= \int_{-\infty}^{\infty} dx \frac{\exp\left\{-\frac{1}{2} \frac{(x-nc^2)^2}{\sigma^2 c^2 n}\right\}}{\sqrt{2\pi} \sigma c \sqrt{n}} \Phi^{M-1}\left(\frac{x}{\sigma c \sqrt{n}}\right) \quad (4.5.6)$$

Let us define

$$\rho = c\sigma^{-1}$$

Then, via change of variables, we obtain from (4.5.6)

$$P_d(n) = \sum_{k=0}^{M-1} \Pr\{s_k^{n^T} X^n \geq s_l^{n^T} X^n, \ \forall \ l | H_k\} \Pr(H_k)$$
$$= \Pr\{s_k^{n^T} X^n \geq s_l^{n^T} X^n, \ \forall \ l | H_k\}$$
$$= \int_{-\infty}^{\infty} dy \frac{\exp\{-\frac{1}{2}(y - \rho\sqrt{n})^2\}}{\sqrt{2\pi}} \Phi^{M-1}(y) \quad (4.5.7)$$

The expression in (4.5.7) is the probability of correct detection induced by the matched filter, given as a function of the observation length, n. The probability of error $P_e(n)$ is then equal to $1 - P_d(n)$. In Figure 4.5.2, we plot the probability of error $P_e(n)$ as a function of n and the signal-to-noise ratio ρ for $M = 2$.

Example 4.5.2

Let us revisit Example 4.3.1, which models the classic radar problem, in the discrete-time case. Hypothesis H_1 corresponds to the presence of some known signal, while hypothesis H_0 corresponds to the absence of the signal, in the presence of atmospheric, Gaussian, and in general colored noise, whose autocovariance matrix is signified by R_n. Let us temporarily assume that the atmospheric noise is white, at level σ^2. That implies that the matrix R_n equals σ^2, times the $n \times n$ identity matrix I_n. Then, it is also true that $R_n^{-1} = \sigma^{-2} \cdot I_n$, and

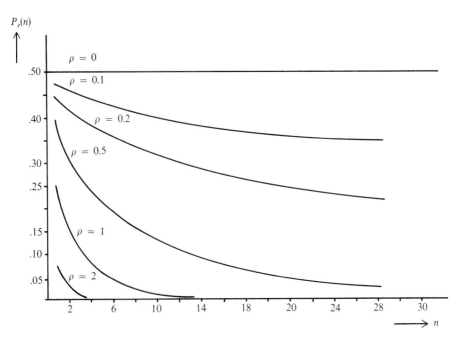

Figure 4.5.2 Probability of error for $M = 2$

substituting in expression (4.3.10) gives

$$P_e = 1 - \Phi\left(\frac{[s^{n^T}s^n]^{1/2}}{2\sigma}\right) = 1 - \Phi\left(\sqrt{n}\,\frac{[n^{-1}s^{n^T}s^n]^{1/2}}{2\sigma}\right) \qquad (4.5.8)$$

where s^n is the column vector that represents the radar signal, and $\Phi(x)$ signifies the distribution function of the zero mean, unit variance Gaussian random variable at point x.

The inner product $s^{n^T}s^n$ in (4.5.8) represents the signal energy, while σ^2 is the power of the atmospheric noise. The ratio $s^{n^T}s^n/n\sigma^2$ is called the signal-to-noise ratio, in the statistical communications literature. Given n, it is easy to see that as the signal-to-noise ratio increases from zero to infinity, the probability of error P_e in (4.5.8) decreases monotonically from .5 to 0. Thus, for fixed n, the higher the signal-to-noise ratio, the lower the probability of erroneous detection that the optimal Bayesian test induces. For a fixed signal-to-noise ratio, on the other hand, the probability of error decreases monotonically with increasing observation length n, converging to zero for $n \to \infty$. Similar conclusions can be drawn when the matrix R_n is as in Example 4.3.2, and the parameter α in (4.3.19) remains unchanged, while the noise power is represented by σ^2. In Figure 4.5.3, we plot the probability of error in (4.5.8) as a function of the signal-to-noise ratio. The behavior of the probability of error as a function of n is similar.

4.5 Examples

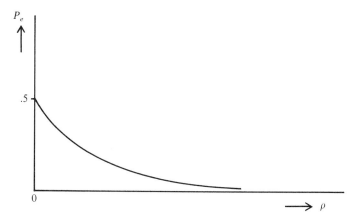

Figure 4.5.3 Probability of error $\rho = \sigma^{-1}(n^{-1}s^{nT}s^n)^{1/2}$

Example 4.5.3

Let us consider the problem in Example 3.5.2 and adopt a binary, symmetric, and memoryless channel whose probability of correct transmission is q. Let one of M binary, n-length, known sequences $x^n(j); j = 1, \ldots, M$ be transmitted through the channel. Let the output to the channel binary sequence be denoted by $y^n = \{y_i; i = 1, \ldots, n\}$, and let the ith digit of the binary sequence $x^n(j)$ be denoted by $x_i(j)$. The known sequences $x^n(j); j = 1, \ldots, M$ are called *codewords*, and we will assume that, throughout the period of n observations, each one of the M codewords may occur with probability M^{-1}. It is thus assumed that an observed sequence y^n at the output of the channel, represents only one of the M codewords. The objective is to decide which one of the M codewords has been transmitted, based on the observation vector y^n, and with performance criterion the minimization of the error probability. The optimal Bayesian test is then as in (4.2.15), Example 4.2.1, with $p_l = M^{-1}; l = 1, \ldots, M$. The likelihood ratios take here the form, due to (3.5.10),

$$\frac{\Pr\{y^n | x^n(k)\}}{\Pr\{y^n | x^n(i)\}} = \left(\frac{1-q}{q}\right)^{\sum_{j=1}^{n}[x_j(k) \oplus y_j] - \sum_{j=1}^{n}[x_j(i) \oplus y_j]} \quad (4.5.9)$$

where $\oplus$ signifies binary addition, that is, $0 \oplus 0 = 1 \oplus 1 = 0$, and $0 \oplus 1 = 1 \oplus 0 = 1$.

Given two binary sequences $x^n = \{x_i; i = 1, \ldots, n\}$ and $y^n = \{y_i; i = 1, \ldots, n\}$, the quantity $n^{-1}\sum_{i=1}^{n} x_i \oplus y_i$ is called the *Hamming distance* between the two sequences; it is strictly a distance, and it clearly equals the number of mismatching, equal order bits among the sequences x^n and y^n, divided by the sequence length n. Considering now (4.5.9), and the optimal Bayesian test in (4.2.15) for $p_l = M^{-1} \, \forall \, l$, we conclude that the optimal Bayesian test, in the present case, can be expressed as

Given the sequence y^n, decide that the codeword $x^n(k)$ has been transmitted, iff

$$\left(\frac{1-q}{q}\right)^{\sum_{j=1}^{n}[x_j(k)\oplus y_j]-\sum_{j=1}^{n}[x_j(i)\oplus y_j]} \geq 1 \quad \forall\, i=1,\ldots,M \quad (4.5.10)$$

Taking the logarithm of both parts in (4.5.10), we derive the following equivalent test, where log signifies logarithm via any base.

Given y^n, decide in favor of codeword $x^n(k)$, iff

$$\left\{\sum_{j=1}^{n}[x_j(k)\oplus y_j] - \sum_{j=1}^{n}[x_j(i)\oplus y_j]\right\}\cdot\log\frac{1-q}{q} \geq 0 \quad \forall\, i=1,\ldots,M$$
(4.5.11)

We can now discriminate between two cases.

CASE 1: Let $q > .5$, which signifies a transmission channel, that transmits a bit correctly, with probability higher than the probability of incorrect transmission. Then, $q|(1-q) > 1$, and $\log[(1-q)/q] < 0$. Then, the test in (4.5.11) reduces to the following test,

Given y^n, decide in favor of codeword $x^n(k)$, iff:

$$n^{-1}\sum_{j=1}^{n}[x_j(k)\oplus y_j] \leq n^{-1}\sum_{j=1}^{n}[x_j(i)\oplus y_j] \quad \forall\, i=1,\ldots,M \quad (4.5.12)$$

The test in (4.5.12), clearly decides in favor of this codeword, whose Hamming distance from the observed sequence y^n is the minimum. This optimal Bayesian decision rule is also then called the *minimum distance decoding scheme*. Let now the M codewords be designed so that the Hamming distance between any two such codewords equals $n^{-1}(2d+1)$, where d is some natural number. That is, let

$$n^{-1}\sum_{j=1}^{n}[x_j(k)\oplus x_j(l)] = n^{-1}(2d+1) \quad \forall\, k\neq l \quad k,l=1,\ldots,M \quad (4.5.13)$$

where d is such that

$$M\sum_{i=0}^{d}\binom{n}{i} \leq 2^n \quad (4.5.14)$$

Then, via the minimum decoding scheme in (4.5.12), if the codeword $x^n(k)$ is transmitted and if the Hamming distance between $x^n(k)$ and the observed sequence y^n is at most $n^{-1}d$, then $x^n(k)$ will be correctly recognized by the receiver. That is, if $P_{d,x^n(k)}$ denotes the probability of correct transmission, conditioned

4.5 Examples

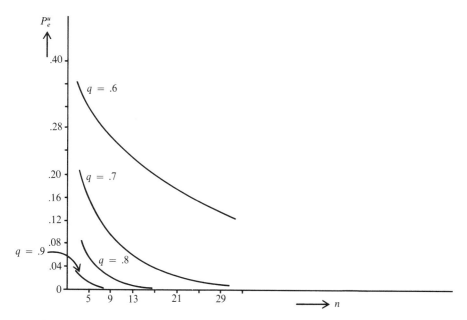

Figure 4.5.4 Upper bound to the probability of error $M = 2$ and $d = 2^{-1}(m - 1)$

on $x^n(k)$, then

$$P_{d,x^n(k)} \geq \sum_{i=0}^{d} \binom{n}{i} q^{n-i}(1-q)^i$$

$$P_d = \sum_{k=1}^{M} M^{-1} P_{d,x^n(k)} \geq \sum_{i=0}^{d} \binom{n}{i} q^{n-i}(1-q)^i$$

$$P_e = 1 - P_d \leq 1 - \sum_{i=0}^{d} \binom{n}{i} q^{n-i}(1-q)^i = P_e^u \quad (4.5.15)$$

The expression in (4.5.15) provides an upper bound on the probability of error P_e induced by the scheme in (4.5.12), if (4.5.13) and (4.5.14) are satisfied. In Figure 4.5.4, we plot the upper bound in (4.5.15) as a function of n for odd n values, for $M = 2$, for $d = 2^{-1}(n - 1)$, and for various choices of the probability q. ∎

CASE 2: If, instead, $q < .5$, then $\log[(1 - q)/q] > 0$, and the Bayesian detection scheme in (4.5.11) decides in favor of the codeword whose Hamming distance from the observed sequence is the maximum. This is intuitively not surprising, since if $q < .5$, then with higher than .5 probability, more than half of the codeword bits are changed in the transmission. ∎

Example 4.5.4

Let us consider the autoregressive model, in example 3.5.3. Let the variables $W_n; n = \ldots, -1, 0, 1, \ldots$, in (3.3.13) be zero mean, unit variance, independent, and Gaussian. Let there be M possible distinct autoregressive models, each one being completely described by a known set $\{\alpha_k(j); k = 1, \ldots, m, j = 1, \ldots, M\}$ of autoregressive parameters. Let it be assumed that, throughout n observations $x_\rho^{n+\rho-1} = \{x_i; i = \rho, \ldots, n + \rho - 1\}$ one of the M autoregressive models is active. Let each of the M models initially occur with probability M^{-1}. Utilizing the observation vector x^n the objective is to decide which one of the M autoregressive models is active. The performance criterion used is the minimization of the error probability. The problem at hand is a Bayesian detection problem, where the optimal Bayesian decision rule is as in (4.2.15), example 4.2.1, with $p_l = M^{-1}; l = 1, \ldots, M$. Let $f_j^n(x_\rho^{n+\rho-1})$ denote the conditional density function of the observed sequence x^n given the hypothesis H_j, or equivalently given the set of autoregressive parameters $\{\alpha_k(j)\}$. From the adopted assumptions on the variables $\{W_n\}$ and from the model in (3.5.13) it is easy to show recursively that the stochastic vector X^n is Gaussian and zero mean under each one of the M hypotheses. However, given some arbitrary set $\{\alpha_k\}$ of autoregressive coefficients, the autoregressive process in (3.5.13) is not, in general, stationary. In fact, setting $X_j = 0 \,\forall\, j \leq 0$ in (3.5.13) and starting at time $n = 1$, the autoregressive process is nonstationary for every set of autoregressive coefficients. If the autoregressive coefficients $\{\alpha_k; k = 1, \ldots, m\}$ are such that the z polynomial $f(z) = z^m - \sum_{k=1}^m \alpha_k z^{m-k}$ has roots that are all inside the unit circle, then the autoregressive process is asymptotically $[n \to \infty$, in (3.3.13)$]$ stationary and ergodic, with spectral density $\mathscr{S}(\omega)$ given by the expression $\mathscr{S}(\omega) = \|1 - \sum_{k=1}^m \alpha_k e^{j\omega k}\|^{-2}; \omega \in [-\pi, \pi]$, where $j = \sqrt{-1}$, and $\|\cdot\|$ means magnitude. In general, the density $f_j^n(x_\rho^{n+\rho-1})$ under H_j can be expressed as

$$f_j^n(x_\rho^{n+\rho-1}) = \prod_{l=\rho}^{n+\rho-1} (2\pi)^{-1/2} \exp\left\{-\frac{1}{2}\left[x_l - \sum_{k=1}^m \alpha_k(j)x_{l-k}\right]^2\right\}$$

$$= (2\pi)^{-n/2} \exp\left\{-\frac{1}{2}\sum_{l=\rho}^{n+\rho-1}\left[x_l - \sum_{k=1}^m \alpha_k(j)x_{l-k}\right]^2\right\} \quad (4.5.16)$$

where we set $x_i = 0$ for all nonpositive indices i.

Due to (4.5.16) we also obtain,

$$\frac{f_j^n(x_\rho^{n+\rho-1})}{f_i^n(x_\rho^{n+\rho-1})} = \exp\left\{\frac{1}{2}\sum_{l=\rho}^{n+\rho-1}\left[x_l - \sum_{k=1}^m \alpha_k(i)x_{l-k}\right]^2 \right.$$

$$\left. - \frac{1}{2}\sum_{l=\rho}^{n+\rho-1}\left[x_l - \sum_{k=1}^m \alpha_k(j)x_{l-k}\right]^2\right\}$$

$$= \exp\left\{\frac{1}{2}\sum_{l=\rho}^{n+\rho-1}\left(2x_l - \sum_{k=1}^m [\alpha_k(i) + \alpha_k(j)]x_{l-k}\right)\right.$$

$$\left. \times \left(\sum_{k=1}^m [\alpha_k(j) - \alpha_k(i)]x_{l-k}\right)\right\} \quad (4.5.17)$$

Setting $p_l = M^{-1}; l = 1, \ldots, M$ in (4.2.15), substituting it in (4.5.17), and taking the natural logarithm of both parts of the inequality, we obtain the following reduced form of the optimal Bayesian test,

4.5 Examples

Given the observation sequence $x_\rho^{n+\rho-1}$, decide in favor of the autoregressive model $\{\alpha_k(j)\}$, iff

$$\sum_{l=\rho}^{n+\rho-1}\left(2x_l - \sum_{k=1}^{m}[\alpha_k(i) + \alpha_k(j)]x_{l-k}\right)\left(\sum_{k=1}^{m}[\alpha_k(j) - \alpha_k(i)]x_{l-k}\right)$$
$$\geq 0 \qquad \forall\, i = 1, \ldots, M \qquad (4.5.18)$$

where $x_i = 0\ \forall\, i \leq 0$.

In the test in (4.5.18), the test function is the expression on the left hand side of the inequality. The test statistics have a generalized χ^2 form, and the threshold is zero. For arbitrary observation length n the probability of error induced by the test can be found numerically. For $n \to \infty$, and for autoregressive models whose polynomials $z^m - \sum_{k=1}^{m}\alpha_k(j)z^{m-k}$ have all their roots inside the unit circle, the probability of error converges asymptotically to zero. Let us now consider the special case, where there are two autoregressive models of order one. Let us then denote by $\alpha(1)$ and $\alpha(0)$ the autoregressive scalar parameters of those two models, and let us assume that the parameters $\alpha(1)$ and $\alpha(0)$ satisfy

$$\alpha(1) = -\alpha(0) \qquad 1 > \alpha = \alpha(1) > 0 \qquad (4.5.19)$$

The test in (4.5.18) is then simplified as

$$\text{Given } x_\rho^{n+\rho-1} \begin{cases} \text{Decide } \alpha(1),\ \text{iff}\ \sum_{l=\rho}^{n+\rho-1} x_{l-1}x_l \geq 0 \\ \text{Decide } \alpha(0),\ \text{iff}\ \sum_{l=\rho}^{n+\rho-1} x_{l-1}x_l \leq 0 \end{cases} \qquad (4.5.20)$$

The statistics of the random variable $\sum_{l=\rho}^{n+\rho-1} X_{l-1}X_l$ determine the probability of error P_e induced by the test in (4.5.20), where due to the symmetry of the hypotheses, we have

$$P_e = \Pr\left\{\sum_{l=\rho}^{n+\rho-1} X_{l-1}X_l > 0 \,\bigg|\, \alpha(0)\right\} \qquad (4.5.21)$$

For finite n values, the statistics of the random variable $\sum_{l=\rho}^{n+\rho-1} X_{l-1}X_l$ cannot be found in closed form. Thus, the probability of error P_e in (4.5.21) cannot be found directly. Instead, we will compute an upper bound on P_e using the Chernoff bound (see Section 4.6). Directly from the Chernoff bound, we obtain

$$P_e = \Pr\left\{\sum_{l=\rho}^{n+\rho-1} X_{l-1}X_l > 0 \,\bigg|\, \alpha(0)\right\} \leq E\left(\exp\left\{\alpha \sum_{l=\rho}^{n+\rho-1} X_{l-1}X_l\right\} \,\bigg|\, \alpha(0)\right) \qquad (4.5.22)$$

where α is equal to $\alpha(1) > 0$, and $\alpha(0) = -\alpha$.

From (4.5.22) and the characteristics of the autoregressive processes, the first order Markov property in particular, and denoting by $\phi(x)$ the zero mean, unit variance Gaussian density function at the point x, we obtain

$$P_e \leq \int_{-\infty}^{\infty} dx_p^{n+p-1} \exp\left\{\alpha \sum_{l=p}^{n+p-1} x_l x_{l-1}\right\}$$

$$\times \prod_{l=p}^{n+p-1} \phi[x_l - \alpha(0)x_{l-1}]\phi(x_{p-1})(1-\alpha^2)^{-1/2}$$

$$= (1-\alpha^2)^{-1/2} \int_{-\infty}^{\infty} dx_p^{n+p-1} (2\pi)^{-(n+1)/2} \exp\left\{\alpha \sum_{l=p}^{n+p-1} x_l x_{l-1} - \frac{1}{2} \sum_{l=p-1}^{n+p-1} x_l^2\right.$$

$$\left. -\frac{\alpha^2(0)}{2} \sum_{l=p-1}^{n+p-2} x_l^2 + \alpha(0) \sum_{l=p}^{n+p-1} x_l x_{l-1}\right\}$$

$$= (\text{since } \alpha(0) = -\alpha) \frac{1}{\sqrt{1-\alpha^2}} \int_{-\infty}^{\infty} dx_p^{n+p-1} (2\pi)^{-(n+1)/2}$$

$$\times \exp\left\{-\frac{1+\alpha^2}{2} \sum_{l=p-1}^{n+p-2} x_l^2 - \frac{1}{2} x_{n+p-1}^2\right\} \frac{1}{\sqrt{1-\alpha^2}} (1+\alpha^2)^{-(n-1)/2} = P_e^u$$

(4.5.23)

The bound in (4.5.23) converges exponentially to zero, as the number n of the observed data increases. We plot this bound in Figure 4.5.5, for various α values less than one.

Example 4.5.5

Let us consider the problem where one of M distinct, real, continuous-time, finite energy, and deterministic signal waveforms $\{s_i(t)\}$ is transmitted through an additive, zero mean Gaussian noise. Let the autocovariance function of the Gaussian noise process $N(t)$ be $R(t_1, t_2)$, and let $[0, T]$ be the observation time interval. Let it be known that the M signals are equally probable, and let the objective be to decide which one of the M signals has been transmitted, given the observed waveform $v(t)$; $t \in [0, T]$ and using the ideal observer detection scheme.

Let there exist a set, $\{\phi_i(t)\}$, of functions that is orthonormal and complete in $[0, T]$ for the class of finite energy deterministic waveforms, and that also provides a Karhunen-Loéve expansion of the process $N(t)$. Then, as discussed in Section 4.4, there exists a set $\{\lambda_i\}$ of positive numbers such that

$$\int_0^T R(t_1, t_2)\phi_i(t_2) dt_2 = \lambda_i \phi_i(t_1) \qquad \forall \, t_1 \in [0, T] \qquad (4.5.24)$$

Also, the random variables $\{\hat{N}_i\}$ such that,

$$\hat{N}_i = \int_0^T N(t)\phi_i^*(t) dt \qquad \hat{n}_i \text{ a realization of } \hat{N}_i$$

4.5 Examples

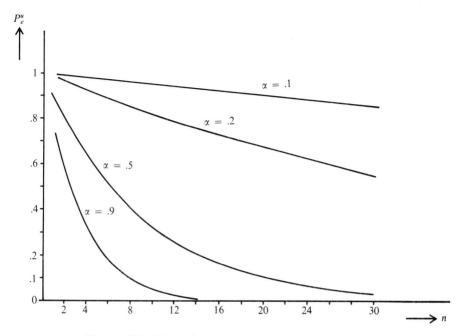

Figure 4.5.5 Upper bound on P_e from the test in (4.5.20)

are then zero mean, Gaussian, and mutually independent, and the variance of the variable $\hat{N}_i$ equals λ_i. Denoting,

$$\hat{N}(t) = \sum_i \hat{N}_i \phi_i(t) \quad t \in [0, T] \quad \hat{n}(t) \text{ a realization of } \hat{N}(t)$$

$$s_{li} = \int_0^T s_l(t) \phi_i^*(t)\, dt$$

$$\hat{s}_l(t) = \sum_i s_{li} \phi_i(t) \quad t \in [0, T]$$

then $\hat{v}_i = \hat{n}_i + s_{li}$, $\hat{v}(t) = \hat{n}(t) + \hat{s}_l(t)$. If the signal $s_l(t)$ is transmitted in $[0, T]$ we conclude directly from (4.2.23) in Example 4.2.2, that if the set $\{\hat{v}_i\}$ were observed, then the ideal observer scheme would give

Decide that $\{s_{ki}\}$ is present, iff:

$$\sum_i \frac{s_{ki}\hat{v}_i}{\lambda_i} - \frac{1}{2}\sum_i \frac{s_{ki}^2}{\lambda_i} \geq \sum_i \frac{s_{li}\hat{v}_i}{\lambda_i} - \frac{1}{2}\sum_i \frac{s_{li}^2}{\lambda_i} \quad \forall\, l\colon l = 0, 1, \ldots, M - 1 \quad (4.5.25)$$

As in Example 4.4.1, let us define

$$q_{li} = \frac{s_{li}}{\lambda_i} \quad \hat{q}_l(t) = \sum_i q_{li} \phi_i(t) \quad t \in [0, T]$$

$$q_l(t)\colon q_{li} = \int_0^T q_l(t) \phi_i^*(t)\, dt$$

where also

$$R(t_1, t_2) = \sum_i \lambda_i \phi_i(t_1) \phi_i^*(t_2)$$

Then we can write

$$\sum_i \frac{s_{li}\hat{v}_i}{\lambda_i} - \frac{1}{2}\sum_i \frac{s_{li}^2}{\lambda_i} = \int_0^T q_l(t)\hat{v}(t)\,dt - \frac{1}{2}\int_0^T q_l(t)\hat{s}_l(t)\,dt \quad (4.5.26)$$

where as in (4.4.33) in Example 4.4.1, we have

$$\hat{s}_l(t) = \int_0^T q_l(\tau) R(t, \tau)\,d\tau \quad t \in [0, T] \quad (4.5.27)$$

Following the derivations in Example 4.4.1, and due to (4.5.25) and (4.5.27), we finally conclude that the ideal observer scheme gives:

Given the observed waveform $v(t)$ in $[0, T]$, decide in favor of the signal $s_k(t); t \in [0, T]$, iff

$$\int_0^T q_k(t)v(t)\,dt - \frac{1}{2}\int_0^T q_k(t)s_k(t)\,dt \geq \int_0^T q_l(t)v(t)\,dt - \frac{1}{2}\int_0^T q_l(t)s_l(t)\,dt \quad \forall l$$

(4.5.28)

where

$$s_l(t) = \int_0^T q_l(\tau) R(t, \tau)\,d\tau$$

The test in (4.5.28) is the continuous-time version of the *matched filter* in Example 4.2.2. Now let the Gaussian noise be white with variance σ^2. Then $R(t_1, t_2) = \sigma^2 \delta(t_1 - t_2)$, where $\delta(\cdot)$ denotes the delta function, and it is easily concluded that (4.5.24) is satisfied with $\lambda_i = \sigma^2 \; \forall \; i$, for every set $\{\phi_i(t)\}$. In this latter case, the test in (4.5.28) results via any set $\{\phi_i(t)\}$ that is orthonormal and complete in $[0, T]$ for the class of finite energy deterministic waveforms, such as the Fourier set $\{e^{j\omega i}\}; \omega = 2\pi/T$. In addition, $q_l(t) = \sigma^{-2} s_l(t); t \in [0, T], \; \forall \; l$, and if the signal energies $E^2 = \int_0^T s_l^2(t)\,dt$ are also identical for all l, the test in (4.5.28) is simplified as follows.

Given the observed waveform $v(t)$ in $[0, T]$, decide in favor of the signal, $s_k(t); t \in [0, T]$, iff

$$\int_0^T s_k(t)v(t)\,dt \geq \int_0^T s_l(t)v(t)\,dt \quad \forall \; l: l = 0, 1, \ldots, M-1 \quad (4.5.29)$$

If the signals $\{s_k(t)\}$ are also orthogonal in $[0, T]$, as in Example 4.5.1, then the probability of error P_e induced by the test in (4.5.29) is exactly as computed

4.5 Examples

in the latter example, with $E^2 = \int_0^T s_l^2(t)\,dt \ \forall\ l$ substituted for nc^2. That is,

$$P_e = 1 - \int_{-\infty}^{\infty} dx \, \frac{\exp\left\{-\frac{1}{2}\frac{(x-E^2)^2}{\sigma^2 E^2}\right\}}{\sqrt{2\pi}\,\sigma E} \Phi^{M-1}\left(\frac{x}{\sigma E}\right)$$

$$= 1 - \int_{-\infty}^{\infty} dx \, \frac{\exp\left\{-\frac{1}{2}(x-\rho\sqrt{T})^2\right\}}{\sqrt{2\pi}} \Phi^{M-1}(x) \qquad (4.5.30)$$

where $\rho = \sigma^{-1}[T^{-1}\int_0^T s_l^2(t)\,dt]^{1/2} \ \forall\ l$ and $T^{-1}\int_0^T s_l^2(t)\,dt$ is the common signal power.

For $M = 2$, the behavior of the probability of error P_e as a function of the observation time length T is exactly as in Figure 4.5.2 in Example 4.5.1. The operations of the matched filter in (4.5.29) are as in Figure 4.2.1, where $v(t)$; $t \in [0, T]$ and $\int_0^T s_l(t)v(t)\,dt$ are substituted for, respectively, x^n and $s_l^{nT}R_n^{-1}x^n$.

In Figure 4.5.6, we plot the probability of error P_e in (4.5.30), as a function of the observation length T for $\rho = 1$, and for various M values.

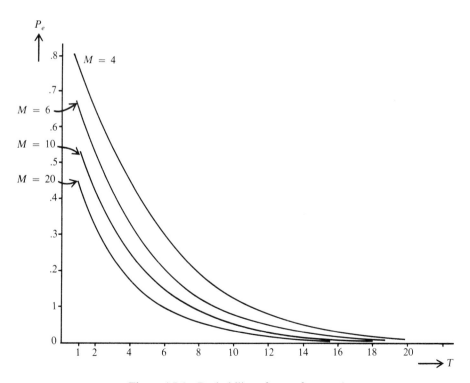

Figure 4.5.6 Probability of error for $\rho = 1$

Example 4.5.6

Let us consider the case where some deterministic, continuous-time signal $s(t)$ is transmitted through a noisy channel. Let the channel impose both additive and multiplicative stochastic disturbances on the signal. Let the additive channel disturbances be modeled by a zero mean, white, variance σ^2, Gaussian process $N_2(t)$. Let the multiplicative disturbances be modeled by another zero mean, nonstationary, Gaussian process $N_1(t)$, whose autocovariance function $R_1(t_1, t_2)$ is as in Example 4.3.2—that is, $R_1(t_1, t_2) = \alpha \min(t_1, t_2)$. Let $N_1(t)$ and $N_2(t)$ be mutually independent. Let the observation interval be $[0, T]$, and let us denote by $V(t)$; $t \in [0, T]$ the observed stochastic waveform. Let us denote by $v(t)$; $t \in [0, T]$ a realization of this waveform. Let us denote by $n_1(t), n_2(t)$ realizations respectively from the stochastic waveforms $N_1(t), N_2(t)$, all in $t \in [0, T]$. Then, for $t \in [0, T]$

$$V(t) = \begin{cases} s(t)N_1(t) + N_2(t) & \text{if the signal is present} \\ N_2(t) & \text{if the signal is not present} \end{cases} \quad (4.5.31)$$

$$v(t) = \begin{cases} s(t)n_1(t) + n_2(t) & \text{if the signal is present} \\ n_2(t) & \text{if the signal is not present} \end{cases} \quad (4.5.32)$$

Let us consider a single hypothesis problem, where based on some observed realization $v(t)$ in $[0, T]$, it must be decided if the signal is present or not. We will denote the signal presence as H_1. We will denote the absence of the signal as H_0. We will also assume that the two hypotheses are equally probable, and that if the signal waveform is stochastic, then the process that generates it is well known and independent of the channel processes $N_1(t)$ and $N_2(t)$. As stated in Example 4.2.2, the white Gaussian process $N_2(t)$ has a Karhunen-Loéve expansion in $[0, T]$, for all sets $\{\phi_i(t)\}$, of deterministic time functions orthonormal in $[0, T]$. The Gaussian process, $N_1(t)$ has a Karhunen-Loéve expansion in $[0, T]$ via $\{\phi_i(t)\}$ which is such that [Papoulis (1965), Problem 13-3]

$$\phi_i(t) = 2^{1/2} \sin(\alpha \lambda_i^{-1})^{1/2} t \qquad t \in [0, T] \quad (4.5.33)$$

where

$$\lambda_i = \alpha T^2 \pi^{-2} \left(i + \frac{1}{\sqrt{2}}\right)^{-2} \quad (4.5.34)$$

and where if

$$\hat{N}_{1i} = \int_0^T \hat{N}_1(t) 2^{1/2} \sin(\alpha \lambda_i^{-1})^{1/2} t \, dt \quad (4.5.35)$$

then

$$E\{N_{1i}\}^2 = \lambda_i \quad (4.5.36)$$

4.5 Examples

and

$$\hat{N}_1(t) = \sum_i \hat{N}_{1i} \, 2^{1/2} \sin(\alpha \lambda_i^{-1})^{1/2} t \qquad t \in [0, T] \qquad (4.5.37)$$

where $\hat{N}_1(t)$ in (4.5.37) is a mean-squared approximation of the stochastic waveform $N_1(t)$ in $[0, T]$.

From now on we will denote,

$$U(t) = s(t)N_1(t) \qquad (4.5.38)$$

$$u(t) = s(t)n_1(t) \qquad (4.5.39)$$

where $U(t)$ denotes a stochastic waveform induced by the signal and the noise process $N_1(t)$.

If the stochastic waveform $U(t)$ in (4.5.38) has a Karhunen-Loéve expansion in $[0, T]$, we will denote by $\hat{U}(t)$ the mean-squared approximation of $U(t)$ so induced. We will denote by $\hat{N}_2(t)$ the Karhunen-Loéve approximation of the stochastic waveform $N_2(t)$ in $[0, T]$, via some set of time functions orthonormal in $[0, T]$.

Let the signal $S(t)$ be a deterministic, constant A. Then, denoting by $R_{s1}(t_1, t_2)$ the autocovariance function induced by the process $N_1(t)$ and the constant signal A, we have

$$R_{s1}(t_1, t_2) = \alpha A^2 \min(t_1, t_2) \qquad (4.5.40)$$

The orthonormal set $\{\phi_i(t)\}$ in (4.5.33), replacing α by αA^2, provides then Karhunen-Loéve expansions $\hat{U}(t)$ and $\hat{N}_2(t)$ for both the processes $U(t)$ in (4.5.38) and $N_2(t)$. In particular,

$$\hat{U}(t) = \sum_i \hat{U}_i \phi_i(t)$$

where

$$\hat{U}_i = \int_0^T U(t) \phi_i^*(t) \, dt$$

$$\hat{N}_2(t) = \sum_i \hat{N}_{2i} \phi_i(t) \qquad (4.5.41)$$

where

$$\hat{N}_{2i} = \int_0^T N_2(t) \phi_i^*(t) \, dt$$

In the expressions in (4.5.41), the random variables $\{\hat{N}_{2i}\}$ are mutually independent, zero mean, variance σ^2, and Gaussian. The random variables $\{\hat{U}_i\}$ are also mutually independent, zero mean, and Gaussian; the set $\{\hat{U}_i\}$ is independent from the set $\{\hat{N}_{2i}\}$; and the variance of the variable $\hat{U}_i$ is $\lambda_i A^2$, where λ_i is given by (4.5.34). If a realization $\{\hat{v}_i = \hat{u}_i + \hat{n}_{2i}\}$ of the set $\{\hat{V}_i = \hat{U}_i + \hat{N}_{2i}\}$ were observed, the single hypothesis Bayesian test would be

Given $\{\hat{v}_i\}$, decide in favor of the hypothesis H_1, iff

$$\frac{\prod_i \exp\left\{-\frac{\hat{v}_i^2}{2(\sigma^2 + \lambda_i A^2)}\right\}}{\prod_i \exp\left\{-\frac{\hat{v}_i^2}{2\sigma^2}\right\}} \geq \prod_i [1 + \sigma^{-2}A^2\lambda_i]^{1/2} \quad (4.5.42)$$

Indeed, under both the H_0 and H_1, $\{\hat{V}_i\}$ are independent, zero mean, and Gaussian. The variance of $\hat{V}_i$ is σ^2 under H_0, and $\sigma^2 + A^2\lambda_i$ under H_1. Reducing the test in (4.5.42) and taking the natural logarithm, we obtain

Given $\{\hat{v}_i\}$, decide in favor of hypothesis H_1 iff:

$$\sum_i \frac{\lambda_i}{\sigma^2 + \lambda_i A^2} \hat{v}_i^2 \geq \sigma^2 A^{-2} \sum_i \ln(1 + \sigma^{-2}A^2\lambda_i) = C \quad (4.5.43)$$

The constants $\{\hat{v}_i\}$ in (4.5.43) are given by

$$\hat{v}_i = \int_0^T v(t)\phi_i^*(t)\,dt \quad (4.5.44)$$

Let us denote by $R'(t, \tau)$, the autocovariance function whose Karhunen-Loéve expansion coefficients in $[0, T]$ are $\{\lambda_i/(\sigma^2 + \lambda_i A^2)\}$ via the set $\{\phi_i(t)\}$. That is,

$$R'(t, \tau): \int_0^T R'(t, \tau)\phi_i(\tau)\,d\tau = \frac{\lambda_i}{\sigma^2 + \lambda_i A^2} \phi_i(t) \quad t \in [0, T] \quad \forall\, i \quad (4.5.45)$$

Then, from (4.5.45), we easily obtain

$$\sum_i \lambda_i \phi_i(t_1)\phi_i(t_2) = \int_0^T d\tau\, R'(t_1, \tau) \sum_i (\sigma^2 + \lambda_i A^2)\phi_i(\tau)\phi_i(t_2) \quad (4.5.46)$$

where

$$A^2 \sum_i \lambda_i \phi_i(t_1)\phi_i(t_2) = R_{s1}(t_1, t_2) \quad (4.5.47)$$

$$\sum_i (\sigma^2 + \lambda_i A^2)\phi_i(\tau)\phi_i(t_2) = \sigma^2\delta(\tau - t_2) + R_{s1}(\tau, t_2) \quad (4.5.48)$$

where $\delta(x)$ is the delta function at the point zero.

Substituting (4.5.47) and (4.5.48) in (4.5.46), we find that the autocovariance function $R'(t, \tau)$ is the solution of the following integral equation.

$$A^{-2}R_{s1}(t_1, t_2) = \sigma^2 R'(t_1, t_2) + \int_0^T d\tau\, R'(t_1, \tau)R_{s1}(\tau, t_2) \quad t_1, t_2 \in [0, T]$$
$$(4.5.49)$$

where $R_{s1}(t_1, t_2)$ is given by (4.5.40).

4.5 Examples

Therefore, (4.5.49) takes the form

$$\alpha \cdot \min(t_1, t_2) = \sigma^2 R'(t_1, t_2) + A^2\alpha \int_0^{t_2} d\tau \cdot \tau \cdot R'(t_1, \tau)$$
$$+ A^2\alpha t_2 \int_{t_2}^T d\tau \cdot R'(t_1, \tau) \quad (4.5.50)$$

The solution of the integral equation in (4.5.50) results in the autocovariance function $R^*(t_1, t_2)$; $t_1, t_2 \in [0, T]$ given below.

$$R^*(t_1, t_2) = A^{-2}\delta(t_1 - t_2) - \sigma^2 A^{-2}[\sigma^2\delta(t_1 - t_2) + \alpha A^2 \min(t_1, t_2)]^{-1} \quad (4.5.51)$$

where $\delta(x)$ denotes the delta function at the point x.

On the other hand, by substituting (4.5.44) in (4.5.43), we find that in the continuous-time domain, the Bayesian test is

Given the observation waveform $v(t)$ in $[0, T]$, decide in favor of the hypothesis H_1, iff

$$\int\!\!\!\int_0^T d\tau\, dt\, R^*(t, \tau) v(t) v(\tau) \geq C \quad (4.5.52)$$

Substituting the autocovariance function in (4.5.51) into (4.5.52) we obtain,

Given the observation $v(t)$ in $[0, T]$, decide in favor of hypothesis H_1, iff

$$\int_0^T v^2(t)\, dt - \sigma^2 \int\!\!\!\int_0^T dt\, d\tau [\sigma^2\delta(t - \tau) + \alpha A^2 \min(t, \tau)]^{-1} v(t) v(\tau) \geq C \quad (4.5.53)$$

The threshold C in (4.5.53) is given in (4.5.43), and is clearly positive. It is also bounded, since, using the inequality $\ln x \leq x - 1$ and (4.4.6), we obtain

$$C = \sigma^2 A^2 \sum_i \ln(1 + \sigma^{-2} A^2 \lambda_i) \leq \sum_i \lambda_i = \int_0^T R_1(t, t)\, dt$$
$$= \alpha \int_0^T t\, dt = 2^{-1}\alpha T^2 < \infty \quad (4.5.54)$$

for finite α and T.

The statistics of the test in (4.5.53) are of the generalized χ^2 type, under both H_0 and H_1. The probability of error induced by the test cannot be found thus in closed form. We will search instead for an upper bound on the probability of error P_e using the Chernoff bound in Section 4.6. To gain insight, we will refer to the test in (4.5.43) rather than directly to the test in (4.5.53). From the former and Lemma 4.6.2, we obtain

$$P_e = 2^{-1} \Pr\left(-\sum_i \frac{\lambda_i}{\sigma^2 + \lambda_i A^2} \hat{V}_i^2 \geq -C \middle| H_1\right)$$

$$+ 2^{-1} \Pr\left(\sum_i \frac{\lambda_i}{\sigma^2 + \lambda_i A^2} \hat{V}_i^2 \geq C \middle| H_0\right)$$

$$\leq 2^{-1} e^{rC} E\left\{\exp\left[-r \sum_i \frac{\lambda_i \hat{V}_i^2}{\sigma^2 + \lambda_i A^2}\right] \middle| H_1\right\}$$

$$+ 2^{-1} e^{-rC} E\left\{\exp\left[r \sum_i \frac{\lambda_i \hat{V}_i^2}{\sigma^2 + \lambda_i A^2}\right] \middle| H_0\right\} \qquad (4.5.55)$$

for any real and positive number r.

At the same time, we have,

$$E\left\{\exp\left[-r \sum_i \frac{\lambda_i}{\sigma^2 + \lambda_i A^2} \hat{V}_i^2\right] \middle| H_1\right\}$$

$$= \int_{-\infty}^{\infty} \cdots \int_{-\infty}^{\infty} \prod_i dx_i \left\{\prod_i 2\pi[\sigma^2 + \lambda_i A^2]\right\}^{-1/2}$$

$$\times \exp\left\{-r \sum_i \frac{\lambda_i}{\sigma^2 + \lambda_i A^2} x_i^2 - \frac{1}{2} \sum_i \frac{x_i^2}{\sigma^2 + \lambda_i A^2}\right\}$$

$$= \int_{-\infty}^{\infty} \cdots \int_{-\infty}^{\infty} \prod_i dx_i \left\{\prod_i 2\pi[\sigma^2 + \lambda_i A^2]\right\}^{-1/2}$$

$$\times \exp\left\{-\frac{1}{2} \sum_i \frac{1 + 2r\lambda_i}{\sigma^2 + \lambda_i A^2} x_i^2\right\}$$

$$= \left\{\prod_i [1 + 2r\lambda_i]\right\}^{-1/2} \qquad \forall \, r > 0 \qquad (4.5.56)$$

$$E\left\{\exp\left[r \sum_i \frac{\lambda_i}{\sigma^2 + \lambda_i A^2} \hat{V}_i^2\right] \middle| H_0\right\}$$

$$= \int_{-\infty}^{\infty} \cdots \int_{-\infty}^{\infty} \prod_i dx_i \left\{\prod_i 2\pi\sigma^2\right\}^{-1/2}$$

$$\times \exp\left\{r \sum_i \frac{\lambda_i}{\sigma^2 + \lambda_i A^2} x_i^2 - \frac{1}{2} \sum_i \frac{x_i^2}{\sigma^2}\right\}$$

$$= \int_{-\infty}^{\infty} \cdots \int_{-\infty}^{\infty} \prod_i dx_i \left\{\prod_i 2\pi\sigma^2\right\}^{-1/2}$$

$$\times \exp\left\{-\frac{1}{2} \sum_i \frac{\sigma^2 + \lambda_i(A^2 - 2r\sigma^2)}{\sigma^2(\sigma^2 + \lambda_i A^2)} x_i^2\right\}$$

$$= \left\{\prod_i \frac{\sigma^2 + \lambda_i A^2}{\sigma^2 + \lambda_i(A^2 - 2r\sigma^2)}\right\}^{1/2} \qquad \forall \, r: 0 < r \leq 2^{-1}\sigma^{-2} A^2 \qquad (4.5.57)$$

4.5 Examples

Substituting (4.5.56) and (4.5.57) in (4.5.55) gives

$$P_e \leq 2^{-1} e^{rC} \left\{ \prod_i [1 + 2r\lambda_i] \right\}^{-1/2} + 2^{-1} e^{-rC} \left\{ \prod_i [\sigma^2 + \lambda_i A^2] \right\}^{1/2}$$

$$\times \left\{ \prod_i [\sigma^2 + \lambda_i(A^2 - 2r\sigma^2)] \right\}^{-1/2}$$

$$= 2^{-1} e^{rC} \left\{ \prod_i [1 + 2r\lambda_i] \right\}^{-1/2} + 2^{-1} e^{-rC} \left\{ \prod_i [1 + \sigma^{-2} A^2 \lambda_i] \right\}^{1/2}$$

$$\times \left\{ \prod_i [1 + \sigma^{-2} \lambda_i(A^2 - 2r\sigma^2)] \right\}^{-1/2}$$

$$\forall r : 0 < r \leq 2^{-1} \sigma^{-2} A^2 \qquad (4.5.58)$$

where from (4.5.43) we have,

$$C = \sigma^2 A^{-2} \ln\left(\prod_i [1 + \sigma^{-2} A^2 \lambda_i] \right) = \ln\left(\prod_i [1 + \sigma^{-2} A^2 \lambda_i] \right)^{\sigma^2 A^{-2}}$$

$$e^{rC} = \left\{ \prod_i [1 + \sigma^{-2} A^2 \lambda_i] \right\}^{r\sigma^2 A^{-2}} \qquad (4.5.59)$$

$$e^{-rC} = \left\{ \prod_i [1 + \sigma^{-2} A^2 \lambda_i] \right\}^{-r\sigma^2 A^{-2}} \qquad (4.5.60)$$

Substituting (4.5.59) and (4.5.60) in (4.5.58) gives

$$P_e \leq 2^{-1} \left\{ \prod_i [1 + 2r\lambda_i] \right\}^{-1/2} \left\{ \prod_i [1 + \sigma^{-2} A^2 \lambda_i] \right\}^{r\sigma^2 A^{-2}}$$

$$+ 2^{-1} \left\{ \prod_i [1 + \sigma^{-2} A^2 \lambda_i] \right\}^{1/2 - r\sigma^2 A^{-2}} \left\{ \prod_i [1 + \sigma^{-2} \lambda_i(A^2 - 2r\sigma^2)] \right\}^{-1/2}$$

$$\forall r : 0 < r \leq 2^{-1} \sigma^{-2} A^2 \qquad (4.5.61)$$

$$\ln P_e \leq -2 \ln 2 + 2^{-1} \sum_i \ln(1 + \sigma^{-2} A^2 \lambda_i) - 2^{-1} \sum_i \ln(1 + 2r\lambda_i)$$

$$- 2^{-1} \sum_i \ln(1 + \sigma^{-2} A^2 \lambda_i - 2r\lambda_i) = f(r)$$

$$\forall r : 0 < r \leq 2^{-1} \sigma^{-2} A^2 \qquad (4.5.62)$$

It can easily be found that the function $f(r)$ in (4.5.62) is convex and attains its minimum at $r = 2^{-2} \sigma^{-2} A^2$. This latter r value provides then the tightest upper bound P_e^u for the probability of error P_e within the class of the Chernoff upper bounds. By substitution we obtain

$$\ln P_e \le \ln P_e^u = -2 \ln 2 + 2^{-1} \sum_i \ln(1 + \sigma^{-2} A^2 \lambda_i) - \sum_i \ln(1 + 2^{-1} \sigma^{-2} A^2 \lambda_i)$$

$$= -2 \ln 2 + 2^{-1} \sum_i \ln(1 + \sigma^{-2} A^2 \lambda_i)(1 + 2^{-1} \sigma^{-2} A^2 \lambda_i)^{-2}$$

$$= -2 \ln 2 - 2^{-1} \sum_i \ln \left[\frac{1}{2} + \frac{1 + \sigma^{-2} A^2 \lambda_i}{4} + \frac{1}{4(1 + \sigma^{-2} A^2 \lambda_i)} \right] \quad (4.5.63)$$

Using the logarithmic inequality $\ln x \le x - 1$ in (4.5.63), we obtain

$$\ln P_e \le -2 \ln 2 - 2^{-1} \sum_i [1 + 2\sigma^2 A^{-2} \lambda_i^{-1}]^{-2} \quad (4.5.64)$$

We can now translate the bound in (4.5.64) to a continuous-time expression. Indeed, we can write

$$\sum_i [1 + 2\sigma^2 A^{-2} \lambda_i^{-1}]^{-2} = A^4 \sum_i \left[\frac{\lambda_i}{\lambda_i A^2 + 2\sigma^2} \right]^2 \quad (4.5.65)$$

where in parallel to previous derivations, we find that the set $\{\lambda_i/(\lambda_i A^2 + 2\sigma^2)\}$ corresponds to the autocovariance matrix $R(t_1, t_2) = A^{-2} \delta(t_1 - t_2) - 2\sigma^2 A^{-2} [2\sigma^2 \delta(t_1 - t_2) + \alpha A^2 \min(t_1, t_2)]^{-1}$ and the set $\{[\lambda_i/(\lambda_i A^2 + 2\sigma^2)]^2\}$ corresponds to the autocovariance matrix $R'(t_1, t_2) = \int_0^T R(t_1, \tau) R(\tau, t_2) \, d\tau$. In

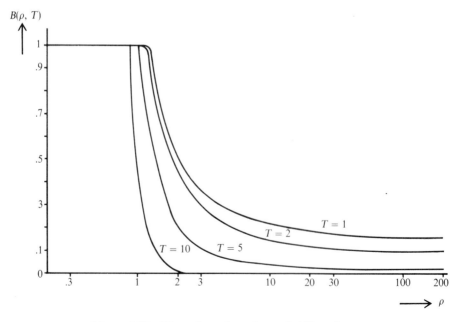

Figure 4.5.7 Upper bound on the probability of error

4.5 Examples

addition, due to (4.4.6) we also have

$$\sum_i \left[\frac{\lambda_i}{\lambda_i A^2 + 2\sigma^2} \right]^2 = \int_0^T R'(t,t)\,dt = \iint_0^T dt\,d\tau R(t,\tau)R(\tau,t)$$

Performing the operations, we find

$$A^4 \sum_i \left[\frac{\lambda_i}{\lambda_i A^2 + 2\sigma^2} \right]^2 = T - 4\sigma^4 \alpha^{-2} A^{-4} - 4\sigma^2 \alpha^{-1} A^{-2} \ln(1 + 2^{-1}\sigma^{-2}\alpha A^2 T)$$

(4.5.66)

Substituting (4.5.66) in (4.5.64), we finally get

$$\ln P_e \leq -2\ln 2 - 2^{-1}T + 2\rho^{-2} + 2\rho^{-1}\ln(1 + 2^{-1}\rho T) = B(\rho, T) \quad (4.5.67)$$

where $\rho = \alpha A^2 / \sigma^2$

The bound $B(\rho, T)$ in (4.5.67) has been expressed as a function of the observation time length T and the signal-to-noise ratio ρ. In Figure 4.5.7, we plot the bound $B(\rho, T)$ as a function of ρ for various T values.

Example 4.5.7

Let us now consider a problem that finds applications in quality control. Let n discrete observation data be available. Then let the n-dimensional observation vector be denoted x^n. Consider two well known and mutually independent discrete-time stochastic processes. Assuming that they exist, let the m-dimensional density functions of those processes at the vector point x^m be respectively denoted $f_0(x^m)$ and $f_1(x^m)$. Let $f_0(x_i | x_1^{i-1})$; $i \geq 2$ denote the corresponding conditional density functions conditioned on the vector point $x_1^{i-1} = [x_1 \ldots x_{i-1}]^T$. Given the observation vector x^n, let us consider $n + 1$ equally probable hypotheses, indexed by j; $j = 0, 1, \ldots, n$, where the jth hypotheses H_j is described by the density function

$$H_j : f_{cj}(x^n) = f_0(x_1^j) \cdot f_1(x_{j+1}^n) \qquad j = 0, 1, \ldots, n \quad (4.5.68)$$

where

$$f_0(x_1^0) = 1 \qquad f_1(x_{n+1}^n) = 1 \quad (4.5.69)$$

Given the observation vector x^n, let the objective be to decide in favor of one of the $n + 1$ hypotheses above via a Bayesian formalization with penalty coefficients $c_{ki} = 1 - \Delta_{ki} \;\forall\; k, i$, where Δ_{ki} is the Kronecker delta. Then, directly from Example 4.2.1, and for $p_i = p_j \;\forall\; i, j$ there, we obtain

$$\delta_j^*(x^n) = \begin{cases} 1 & x^n : \dfrac{f_{cj}(x^n)}{f_{ck}(x^n)} \geq 1 \quad \forall\; k = 0, 1, \ldots, n \\ 0 & \text{otherwise} \end{cases} \quad (4.5.70)$$

Using the identity,

$$f(x_1^m) = \prod_{i=1}^{m} f(x_i|x_1^{i-1}) \qquad f(x_1|x_1^0) = f(x_1)$$

and due to (4.5.68), we easily obtain the following form of the decision rule in (4.5.70)

Decide in favor of the hypothesis H_j iff

$$\sum_{i=j+1}^{n} \log \frac{f_1(x_i|x_{j+1}^{i-1})}{f_0(x_i|x_1^{i-1})} = \max_{0 \le k \le n} \left(\sum_{i=k+1}^{n} \log \frac{f_1(x_i|x_{k+1}^{i-1})}{f_0(x_i|x_1^{i-1})} \right) \qquad (4.5.71)$$

where

$$f_0(x_1|x_1^0) = f_0(x_1)$$
$$f_1(x_{m+1}|x_{m+1}^m) = f_1(x_{m+1})$$
$$\sum_{i=n+1}^{n} \log \frac{f_1(x_i|x_{n+1}^{i-1})}{f_0(x_i|x_1^{i-1})} = 0$$

If the stochastic processes that generate the distributions F_0 and F_1 are both memoryless, then the decision rule in (4.5.71) takes the form

Decide in favor of the hypothesis H_j iff

$$\sum_{i=j+1}^{n} \log \frac{f_1(x_i)}{f_0(x_i)} = \max_{0 \le k \le n} \left(\sum_{i=k+1}^{n} \log \frac{f_1(x_i)}{f_0(x_i)} \right) \qquad (4.5.72)$$

or equivalently, if

$$\min_{0 \le k \le n} \left(\sum_{i=1}^{k} \log \frac{f_1(x_i)}{f_0(x_i)} \right) = \sum_{i=1}^{j} \log \frac{f_1(x_i)}{f_0(x_i)} \qquad \sum_{i=1}^{0} \log \frac{f_1(x_i)}{f_0(x_i)} = 0 \quad (4.5.73)$$

It should be clear that the hypothesis testing problem in this example is actually testing a possible change from a distribution F_0 to a distribution F_1 within a preset data block of length n. The data points at which the change may occur are given equal weight, and the main crucial assumptions are that the data are initially controlled by the distribution F_0, and that if the control shifts to distribution F_1, it remains there throughout the remaining of the observation interval. Hypothesis H_0 reflects the case where the distribution F_0 is in control throughout the overall n-size data block. Hypothesis H_n reflects, instead, the case where the distribution F_1 is in control throughout the overall data block. Another crucial assumption is that the processes that generate the distributions F_0 and F_1 are mutually independent. A sequential modification of the problem has wide applications—in industrial quality control, in the monitoring of the performance of computer-communication networks, in automated image processing, and so on. Such a modification will be presented and analyzed in the chapter dedicated to sequential detection.

4.6 APPENDIX

We present two simple lemmas that are useful in computing upper bounds to the probability of errors induced by various statistical tests.

LEMMA 4.6.1: Let X be a real scalar random variable taking nonnegative values only. Let b be any real positive number, and let the density function $f(x)$ of the random variable X exist. Then, $P(X \geq b) \leq b^{-1}E\{X\}$.

Proof:

$$E\{X\} = \int_0^\infty xf(x)\,dx = \int_0^b xf(x)\,dx + \int_b^\infty xf(x)\,dx \geq \int_b^\infty xf(x)\,dx$$

$$\geq b\int_b^\infty f(x)\,dx = bP(X \geq b) \Rightarrow P(x \geq b) \leq b^{-1}E\{X\} \qquad \blacksquare$$

LEMMA 4.6.2: Let X be a real scalar random variable whose density function $f(x)$ exists $\forall\, x \in (-\infty, \infty)$. Let b be any real number, and let r be any positive and real number. Then

$$P(X \geq b) \leq e^{-rb}E\{e^{rX}\} \qquad (4.6.1)$$

The bound to the probability $P(X \geq b)$ in (4.6.1) is called the Chernoff bound (Chernoff is the statistician who devised it).

Proof: The random variable e^{rX} is real and scalar, it takes nonnegative values only, and its density function exists. Also, e^{rb} is a positive number. Thus, from Lemma 4.6.1, we obtain

$$P(e^{rX} \geq e^{rb}) \leq e^{-rb}E\{e^{rX}\} \qquad (4.6.2)$$

But, due to the monotonicity of the natural logarithm and since $r > 0$, we obtain $P(e^{rX} \geq e^{rb}) = P(\ln e^{rX} \geq \ln e^{rb}) = P(rX \geq rb) = P(X \geq b)$. Thus, (4.6.2) can be written as $P(X \geq b) \leq e^{-rb}E\{e^{rX}\}$.

4.7 PROBLEMS

4.7.1 Let $\{V_i;\, i = 1, \ldots, n\}$ be set of independent, identically distributed Gaussian random variables, such that

$$f(v_i) = \begin{cases} \dfrac{e^{-(1/2)(v_i - \theta_1)^2/\sigma^2}}{\sqrt{2\pi}\sigma} & \text{under } H_1 \\[2mm] \dfrac{e^{-(1/2)(v_i - \theta_0)^2/\sigma^2}}{\sqrt{2\pi}\sigma} & \text{under } H_0 \end{cases}$$

where $\theta_1 > \theta_0$. Let each of the two hypotheses occur with a priori probability 0.5. Find the ideal observer test.

4.7.2 Let $\{V_i; i = 1, \ldots, n\}$ be a set of independent, identically distributed random variables, such that

$$f(v_i) = \begin{cases} e^{-(v_i - \theta_1)} U(v_i - \theta_1) & \text{under } H_1 \\ e^{-(v_i - \theta_0)} U(v_i - \theta_0) & \text{under } H_0 \end{cases}$$

where

$$U(x) = \begin{cases} 1 & \text{if } x \geq 0 \\ 0 & \text{if } x < 0 \end{cases}$$

Let θ_1 and θ_0 be two distinct known parameter values, and let each of the two hypotheses occur with a priori probability 0.5. Find the ideal observer test for each of the two cases: (1) $\theta_1 > \theta_0$, (2) $\theta_0 > \theta_1$.

4.7.3 Let $\{V_i; i = 1, \ldots, n\}$ be a set of independent, identically distributed random variables

$$f(v_i | \theta) = e^{-(v_i - \theta)} U(v_i - \theta) \qquad i = 1, \ldots, n$$

where θ is a uniform random variable on $[-\pi, \pi]$. The penalty function is

$$L(\theta, H_0) = e^{\theta} \qquad L(\theta, H_1) = e^{-\theta}$$

where the decisions are

$$H_0: \theta \leq 0 \qquad H_1: \theta > 0$$

Find the optimum Bayesian test.

4.7.4 Suppose that the vector $V = [v_1, \ldots, v_n]^T$ is observed, where $V = \theta S + N$. S and N are jointly Gaussian, with moments

$$E\{S\} = S_0 \quad E\{N\} = 0 \quad E\{SS^T\} = R_s \quad E\{NN^T\} = I \quad E\{NS^T\} = R_{ns}$$

What is the appropriate test statistic for $H_0: \theta = \theta_0$ versus $H_1: \theta = \theta_1 > \theta_0$?

4.7.5 Let the vector V be observed with independent components of density p on $[0, 1]$. Suppose that it is desired to test $H_0: p(v_i) = 2(1 - v_i)$ versus $H_1: p(v_i) = 1$. What is the optimum Bayes test in terms of the costs c_{ij} and a priori probabilities π_i?

4.7.6 Let $\{v_i: i = 1, \ldots, N\}$ be a set of independent observations where

$$\Pr[v_i = 1] = \theta_i \qquad \Pr[v_i = 0] = 1 - \theta_i$$

Find an optimum test for

$$H_0: \theta_i = \frac{1}{2} \quad i = 1, \ldots, N \quad \text{vs} \quad H_1: \theta_i > \frac{1}{2} \quad i = 1, \ldots, N$$

4.7 Problems

4.7.7 For the decision problem

$$H_0: v(t) = n(t)$$
$$H_1: v(t) = s(t) + n(t) \quad 0 \le t \le T$$

it was found that the test which minimizes the probability of error computes the statistic

$$G = \int_0^T q(t)v(t)\,dt$$

and compares it to an appropriate threshold. The quantity $q(t)$ can be implemented as a matched filter and is the solution to the integral equation

$$s(t) = \int_0^T \phi(t, s)q(s)\,ds \quad 0 \le t \le T.$$

Show that $q(t)$ also maximizes the output signal-to-noise ratio

$$s_0^2(T)/E\{N_0^2(T)\}$$

where

$$s_0(T) = \int_0^T q(t)s(t)\,dt \qquad n_0(T) = \int_0^T q(t)n(t)\,dt.$$

4.7.8 Suppose V is a Gaussian vector with unknown component mean θ and variance σ^2, $E(v_i) = \theta$, and $E\{(v_i - \bar{v}_i)^T(v_j - \bar{v}_j)\} = \sigma^2 \Delta_{ij}$. The two decisions are, $H_0: \theta = 0$; $H_1: \theta > 0$. Assume the penalty functions $L(0, H_0) = L(\theta > 0, H_1) = 0$, $L(0, H_1) = L(\theta > 0, H_0) = 1$. Denote the sample mean $m = (1/k)\sum_{i=1}^k v_i$. Consider the two decision rules

$$d(v) = \begin{cases} H_0 & \text{if } m \ge 0 \\ H_1 & \text{if } m < 0 \end{cases} \qquad d'(v) = \begin{cases} H_0 & \text{if } m \le 0 \\ H_1 & \text{if } m > 0 \end{cases}$$

Show that $d(v)$ is not admissible.

4.7.9 Let $v = \theta s + n$ be the observation, where s is a known vector and n is zero mean Gaussian with independent components, $E(n_i n_j) = \sigma^2 \Delta_{ij}$. Consider testing $H_0: \theta = -1$ versus $H_1: \theta = +1$.
(a) Obtain the Bayes test for

$$\pi_0 = \tfrac{1}{4}, \qquad \pi_1 = \tfrac{3}{4}$$
$$c_{01} = 2 \qquad c_{10} = 1 \qquad c_{00} = c_{11} = 0.$$

(b) Calculate the test using the ideal observer.

4.7.10 A received waveform consists of one of the two signals

$$s_0(t) = A \sin \omega_1 t$$
$$s_1(t) = B \cos \omega_1 t \quad 0 \le t \le T_1 = 2\pi/\omega_1$$

plus zero mean white Gaussian noise. The a priori probability of each signal is 1/2, and the noise covariance is $R(\tau) = N_0 \delta(\tau)$. The waveform is passed through two filters matched to $s_0(t)$ and $s_1(t)$ respectively. The output of the filter is sampled at time T_1 and the signal corresponding to the largest output is said to be transmitted.

(a) What is the probability that an error is made in deciding which signal was transmitted?

(b) If the constraint $A^2 + B^2 = P$ is imposed, what values of A and B optimize the system performance? Prove your answer.

4.7.11 Let $V = \theta S + N$ where S is a known signal vector and N is Gaussian with mean 0 and covariance matrix R. θ has a priori density

$$p(\theta) = \frac{1}{4}\left\{ \frac{e^{-(\theta-\mu)^2/2\sigma_\theta^2}}{\sqrt{2\pi}\,\sigma_\theta} + \frac{e^{-(\theta+\mu)^2/2\sigma_\theta^2}}{\sqrt{2\pi}\,\sigma_\theta} \right\} + \frac{1}{2}\delta(\theta)$$

It is desired to test

$$H_0: \theta = 0 \quad \text{vs} \quad H_1: \theta \neq 0$$

Find the ideal observer solution.

4.7.12 You are given 3 signals defined on the interval $t \in [0, 1]$.

$$s_1(t) = a_1 + a_2\sqrt{2}\sin 2\pi t + a_3\sqrt{2}\sin 4\pi t$$
$$s_2(t) = b_1 + b_2\sqrt{2}\sin 2\pi t + b_3\sqrt{2}\sin 4\pi t$$
$$s_3(t) = c_1 + c_2\sqrt{2}\sin 2\pi t + c_3\sqrt{2}\sin 4\pi t$$

They are to be used for communication through a white Gaussian noise channel. We require them to have energy

$$\int_0^1 s_i^2(t)\,dt = 1 \quad i = 1, 2, 3$$

Write the set of equations that $\{a_i, b_i, c_i\}$; $i = 1, 2, 3$ must satisfy in order to achieve minimum probability of error. Show that $a_3 = b_3 = c_3 = 0$ are part of an optimal solution and find the solution.

Hint: Write

$$a_1 = \cos\theta \qquad b_1 = \cos\psi \qquad c_1 = \cos\omega$$
$$a_2 = \sin\theta \qquad b_2 = \sin\psi \qquad c_2 = \sin\omega$$

and solve for (θ, ψ, ω).

How is the fact that $a_3 = b_3 = c_3 = 0$ give an optimal solution explained by the theory you know?

REFERENCES

Arrow, K. J., D. Blackwell, and M. A. Girshick (1949), "Bayes and Minimax Solutions of Sequential Decision Problems," *Econometrica* 17, 213–244.

Bellman, R. (1970), *Introduction to Maxtrix Analysis*, 2d Edition, McGraw-Hill, New York.

Chernoff, H. and L. E. Moses (1959), *Elementary Decision Theory*, Wiley, New York.

Davenport, N. B., Jr. and W. L. Root (1958), *Random Signals and Noise*, McGraw-Hill, New York.

Doob, J. L. (1953), *Stochastic Processes*, Wiley, New York.

Feller, W. (1957), *An Introduction to Probability Theory and its Applications*, Vol. 1, 2d ed, Wiley, New York.

Ferguson, T. S. (1967), *Mathematical Statistics: A Decision Theoretic Approach*, Academic Press, New York.

Gray, R. M. (1972), "On the Asymptotic Eigenvalue Distribution of Toeplitz Matrices," *IEEE Trans. Inform. Th.* IT-18, 725–730.

Grenander, U. and G. Szego, (1958), *Toeplitz Forms and their Applications*, Univ. of Calif. Press, Berkeley.

Halmos, P. R. and L. J. Savage (1948), "Application of the Radon-Nikodym Theorem to the Theory of Sufficient Statistics," *Ann. Math. Statist.* 20, 225–241.

Lehmann, E. L. (1959), *Testing Statistical Hypotheses*, Wiley, New York.

Loéve, M. (1963), *Probability Theory*, Van Nostrand, Toronto, Ont.

Papoulis, A. (1965), *Probability, Random Variables, and Stochastic Processes*, McGraw-Hill, New York.

Van Trees, H. L. (1968), *Detection, Estimation, and Modulation Theory*, Wiley, New York.

Wozencraft, J. M., and I. M. Jacobs (1965), *Principles of Communication Engineering*, Wiley, New York.

CHAPTER 5

NEYMAN-PEARSON DETECTION RULE

5.1 INTRODUCTION

As summarized in Figure 3.2.1, among the parametric detection schemes, the Neyman-Pearson detection rule utilizes the minimum number of available assets; it is the result of a single hypothesis formalization. In fact, strictly speaking, the emphatic hypothesis H_1 in the Neyman-Pearson formalization is also well known or *noncomposite* while the nonemphatic hypothesis H_0 is, in general, parametric or *composite*. In particular, in the generalized, nonsequential Neyman-Pearson formalization, the following assets are available to the analyst (or system designer).

i. A parametrically known or composite, nonemphatic, hypothesis H_0 parametrized by a vector parameter $\theta^m \in \mathscr{E}^m$. A well known or noncomposite, emphatic hypothesis, H_1.
ii. A realization x^n or $x(t)$; $0 \le t \le T$ from the underlying hypothesis.
iii. A *false alarm rate* α, where $0 < \alpha < 1$.

As in Chapter 4, we will first consider discrete-time stochastic processes as hypotheses. Let $f_1(x^n)$ and $f_{\theta^m}(x^n)$ denote the n-dimensional density functions (assuming they exist) at the vector point x^n, induced respectively by H_1 and H_0 at the vector point θ^m in $\mathscr{E}^m$, where the parameter space $\mathscr{E}^m$ is assumed known. Let Γ^n denote the space where the data vector x^n takes its values. In the Neyman-Pearson single hypothesis setting, some decision rule δ is then fully described by the set, $\{\delta(x^n); x^n \in \Gamma^n\}$, of conditional probabilities that it induces, where

$$\delta(x^n) = \Pr\{H_1 \text{ decided} | x^n \text{ observed}\} \qquad x^n \in \Gamma^n \qquad (5.1.1)$$

$$1 - \delta(x^n) = \Pr\{H_0 \text{ decided} | x^n \text{ observed}\} \qquad x^n \in \Gamma^n \qquad (5.1.2)$$

5.1 Introduction

Given some decision rule δ let $P_{\theta^m}(\delta)$; $\theta^m \in \mathscr{E}^m$ denote the conditional probability that the hypothesis H_1 is decided, given that the acting hypothesis is H_0, at the parameter vector point θ^m. Let $P_1(\delta)$ denote the conditional probability that H_1 is decided, given that it is the acting hypothesis. This last probability is called the *power* induced by the decision rule δ. If the noncomposite hypothesis H_1 is determined via the parametrically known stochastic process of H_0 for some fixed θ_1^m such that $\theta_1^m \notin \mathscr{E}^m$, then given any δ, $P_1(\delta) = P_{\theta_1^m}(\delta)$ and $P_{\theta^m}(\delta)$ is then called the *power function* induced by δ, for θ^m values in $\theta_1^m \cup \mathscr{E}^m$. Given some δ the supremum $\sup_{\theta^m \in \mathscr{E}^m} P_{\theta^m}(\delta)$ is called the *false alarm* induced by δ. Among the assets available to the analyst, the false alarm rate α is used as an upper bound to the allowed false alarm value, narrowing, as a result, the choice of the analyst within a subclass $\mathscr{D}^*$ of decision rules, where

$$\delta \in \mathscr{D}^* \rightarrow \sup_{\theta^m \in \mathscr{E}^m} P_{\theta^m}(\delta) \leq \alpha \qquad (5.1.3)$$

$$P_{\theta^m}(\delta) = \int_{\Gamma^n} \delta(x^n) f_{\theta^m}(x^n) \, dx^n \qquad (5.1.4)$$

where (5.1.4) is as (3.1.13).

In the Neyman-Pearson formalization, the maximization of $P_1(\delta)$ is sought among all the decision rules in $\mathscr{D}^*$. We note that if no false alarm restrictions exist, then the realizations x^n become obsolete, and the power is maximized when H_1 is always and a priori decided. On the other hand, if the false alarm rate is zero and $f_{\theta^m}(x^n)$; $\theta^m \in \mathscr{E}^m$ and $f_1(x^n)$ exist for all x^n, then the maximum power is zero.

If the stochastic processes that determine the hypotheses in the Neyman-Pearson formalizations are continuous-time, then, as with the Bayesian formalizations, the available observations consist of continuous waveforms $x(t)$ in a fixed time interval $[0, T]$. A decision rule is then denoted $\delta = \delta(x(t); t \in [0, T])$, where $\delta(x(t); t \in [0, T])$ is defined as the conditional probability that H_1 is decided, given the observed waveform $x(t); t \in [0, T]$. As in the Bayesian case, the Neyman-Pearson rules for continuous-time hypotheses are derived from the parallel rules for discrete-time hypotheses, in conjunction with appropriate series expansions, such as the Karhunen-Loéve ones.

Throughout this chapter, we assume that the observation interval in the continuous-time case, and the number of observed data in the discrete-time case, are fixed. It is also assumed that throughout the observation interval, the underlying stochastic process remains unchanged. First, we will formulate and solve the optimization problem determined by the strict Neyman-Pearson formalization, as exhibited by the assets available to the analyst presented in this section. Then we will consider some special cases of composite H_1 hypotheses, which result in tests known as *uniformly most powerful*. Finally, we will present a variation of the Neyman-Pearson formalization that provides tests known as *locally most powerful*. The latter are useful when H_1 is composite and no uniformly most powerful tests exist. The concepts and approaches behind the uniformly most powerful and the locally most powerful tests will be clear, after the initial Neyman-Pearson problem has been fully explored.

5.2 THE OPTIMIZATION PROBLEM

Let us consider the generalized Neyman-Pearson formalization as stated in Section 5.1, and let us adopt the same notation. Let us consider discrete-time stochastic processes as hypotheses. Then the Neyman-Pearson decision rule is a solution (not necessarily unique) of the following constraint optimization problem:

Maximize
$$P_1(\delta) = \int_{\Gamma^n} \delta(x^n) f_1(x^n)\, dx^n$$

subject to (5.2.1)

$$0 \leq \delta(x^n) \leq 1 \quad \forall\, x^n \in \Gamma^n$$

$$\sup_{\theta^m \in \mathscr{E}^m} P_{\theta^m}(\delta) = \sup_{\theta^m \in \mathscr{E}^m} \int_{\Gamma^n} dx^n \delta(x^n) f_{\theta^m}(x^n) \leq \alpha$$

We will approach the solution of the constraint optimization problem in (5.2.1) progressively. We will first fix the parameter vector in hypothesis H_0 to some value θ^m in $\mathscr{E}^m$. The hypothesis H_0 is reduced then to a noncomposite hypothesis that is fully determined by the density function $f_{\theta^m}(x^n)$ at θ^m, and the constraint optimization problem in (5.2.1) reduces to the form:

Maximize
$$P_1(\delta) = \int_{\Gamma^n} \delta(x^n) f_1(x^n)\, dx^n$$

subject to (5.2.2)

$$0 \leq \delta(x^n) \leq 1 \quad \forall\, x^n \in \Gamma^n$$

$$P_{\theta^m}(\delta) = \int_{\Gamma^n} dx^n \delta(x^n) f_{\theta^m}(x^n) \leq \alpha$$

We state the general solution of the constraint optimization problem in (5.2.2) in a theorem known as the *Neyman-Pearson lemma*.

THEOREM 5.2.1: Given n, $f_1(x^n)$, $f_{\theta^m}(x^n)$, and $\alpha: 0 < \alpha < 1$, let $\lambda(\theta^m, \alpha)$ and $r(\theta^m, \alpha)$ be two scalars such that

$$\int f_{\theta^m}(x^n)\, dx^n \quad + r(\theta^m, \alpha) \int dx^n\, f_{\theta^m}(x^n) = \alpha$$

$$x^n: \frac{f_1(x^n)}{f_{\theta^m}(x^n)} > \lambda(\theta^m, \alpha) \qquad x^n: \frac{f_1(x^n)}{f_{\theta^m}(x^n)} = \lambda(\theta^m, \alpha) \qquad (5.2.3)$$

$$0 \leq r(\theta^m, \alpha) \leq 1 \qquad \lambda(\theta^m, \alpha) \geq 0$$

5.2 The Optimization Problem

Define the decision rule $\delta^*_{\theta^m,\alpha}(x^n)$ as

$$\delta^*_{\theta^m,\alpha}(x^n) = \begin{cases} 1 & x^n: \dfrac{f_1(x^n)}{f_{\theta^m}(x^n)} > \lambda(\theta^m, \alpha) \\ r(\theta^m, \alpha) & x^n: \dfrac{f_1(x^n)}{f_{\theta^m}(x^n)} = \lambda(\theta^m, \alpha) \\ 0 & x^n: \dfrac{f_1(x^n)}{f_{\theta^m}(x^n)} < \lambda(\theta^m, \alpha) \end{cases}$$

The decision rule $\delta^*_{\theta^m,\alpha}(x^n)$ is a solution of the constraint optimization problem in (5.2.2), called *most powerful* for the setting determined there.

Proof: It is clear that $\delta^*_{\theta^m,\alpha}(x^n)$ satisfies the constraints in (5.2.2). It remains to show that there is no other decision rule that satisfies the same constraints and also attains a power higher than the power attained by $\delta^*_{\theta^m,\alpha}(x^n)$. Let $\delta(x^n)$ be some arbitrary decision rule satisfying the constraints in (5.1.6), and let us observe that the sets $\Gamma^n_1 = \{x^n: f_1(x^n)/f_{\theta^m}(x^n) > \lambda(\theta^m, \alpha)\}$, $\Gamma^n_2 = \{x^n: f^n_1(x^n)/f_{\theta^m}(x^n) = \lambda(\theta^m, \alpha)\}$, $\Gamma^n_3 = \{x^n: f_1(x^n)/f_{\theta^m}(x^n) < \lambda(\theta^m, \alpha)\}$ are disjoint, and that their union covers the total space Γ^n. Then we obtain

$$\int_{\Gamma^n} \delta(x^n) f_{\theta^m}(x^n)\,dx^n = \int_{\Gamma^n_1} \delta(x^n) f_{\theta^m}(x^n)\,dx^n + \int_{\Gamma^n_2} \delta(x^n) f_{\theta^m}(x^n)\,dx^n$$

$$+ \int_{\Gamma^n_3} \delta(x^n) f_{\theta^m}(x^n)\,dx^n \leq \alpha \qquad (5.2.4)$$

$$P_1(\delta) = \int_{\Gamma^n} \delta(x^n) f_1(x^n)\,dx^n$$

$$= \int_{\Gamma^n_1} \delta(x^n) f_1(x^n)\,dx^n + \int_{\Gamma^n_2} \delta(x^n) f_1(x^n)\,dx^n + \int_{\Gamma^n_3} \delta(x^n) f_1(x^n)\,dx^n \qquad (5.2.5)$$

$$\int_{\Gamma^n_3} \delta(x^n) f_1(x^n)\,dx^n = \int \delta(x^n) f_1(x^n)\,dx^n < \lambda(\theta^m, \alpha) \int_{\Gamma^n_3} \delta(x^n) f_{\theta^m}(x^n)\,dx^n$$

$$x^n: \dfrac{f_1(x^n)}{f_{\theta^m}(x^n)} < \lambda(\theta^m, \alpha) \qquad (5.2.6)$$

$$\int_{\Gamma^n_2} \delta(x^n) f^n_1(x^n)\,dx^n = \int \delta(x^n) f_1(x^n)\,dx^n = \lambda(\theta^m, \alpha) \int_{\Gamma^n_2} \delta(x^n) f_{\theta^m}(x^n)\,dx^n$$

$$x^n: \dfrac{f_1(x^n)}{f_{\theta^m}(x^n)} = \lambda(\theta^m, \alpha) \qquad (5.2.7)$$

From (5.2.5), (5.2.6), and (5.2.7), we easily obtain

$$P_1(\delta) < \int_{\Gamma^n_1} \delta(x^n) f_1(x^n)\,dx^n + \lambda(\theta^m, \alpha)\left[\int_{\Gamma^n_2} \delta(x^n) f_{\theta^m}(x^n)\,dx^n + \int_{\Gamma^n_3} dx^n \delta(x^n) f_{\theta^m}(x^n)\right]$$

$$= \int_{\Gamma^n_1} dx^n \delta(x^n)[f_1(x^n) - \lambda(\theta^m, \alpha) f_{\theta^m}(x^n)] + \lambda(\theta^m, \alpha) \int_{\Gamma^n} \delta(x^n) f_{\theta^m}(x^n)\,dx^n$$

$$(5.2.8)$$

Now, since $\delta(x^n) \le 1 \ \forall \ x^n \in \Gamma^n$ and $f_1(x^n) - \lambda(\theta^m, \alpha) f_{\theta^m}(x^n) \ge 0 \ \forall \ x^n \in \Gamma_1^n$, we have

$$\int_{\Gamma_1^n} \delta(x^n)[f_1(x^n) - \lambda(\theta^m, \alpha) f_{\theta^m}(x^n)] \, dx^n \le \int_{\Gamma_1^n} f_1(x^n) \, dx^n - \lambda(\theta^m, \alpha) \int_{\Gamma_1^n} f_{\theta^m}(x^n) \, dx^n \quad (5.2.9)$$

Substituting (5.2.9) and (5.2.4) in (5.2.8), we obtain

$$P_1(\delta) < \int_{\Gamma_1^n} dx^n f_1(x^n) - \lambda(\theta^m, \alpha) \int_{\Gamma_1^n} dx^n f_{\theta^m}(x^n) + \lambda(\theta^m, \alpha) \cdot \alpha \quad (5.2.10)$$

where

$$\alpha = \int_{\Gamma_1^n} dx^n f_{\theta^m}(x^n) + r(\theta^m, \alpha) \int_{\Gamma_2^n} dx^n f_{\theta^m}(x^n) \quad (5.2.11)$$

From (5.2.10) and (5.2.11), we finally obtain

$$P_1(\delta) < \int_{\Gamma_1^n} dx^n f_1(x^n) + r(\theta^m, \alpha) \cdot \lambda(\theta^m, \alpha) \int_{\Gamma_2^n} dx^n f_{\theta^m}(x^n)$$

$$= \int_{\Gamma_1^n} dx^n f_1(x^n) + r(\theta^m, \alpha) \int_{\Gamma_2^n} dx^n f_1(x^n) = P_1(\delta_{\theta^m, \alpha}^*)$$

The proof is now complete. ∎

From Theorem 5.2.1, we observe that the most powerful decision rule is determined when the false alarm is set at the limit provided by the given false alarm rate α. This is evident by (5.2.3). The latter equation specifies the scalars $\lambda(\theta^m, \alpha)$ and $r(\theta^m, \alpha)$, which describe the decision rule $\delta_{\theta^m, \alpha}^*(x^n)$ completely. If the density function $f_{\theta^m}(x^n)$ is continuous everywhere in Γ^n then $\int_{\Gamma_2^n} dx^n f_{\theta^m}(x^n)$, where $\Gamma_2^n = \{x^n : f_1(x^n)/f_{\theta^m}(x^n) = \lambda(\theta^m, \alpha)\}$, is equal to zero and the scalar $r(\theta^m, \alpha)$ can take then any value in $[0, 1]$. The decision rule $\delta_{\theta^m, \alpha}^*(x^n)$ is then nonunique. If $r(\theta^m, \alpha) \ne 0, 1$ is selected, the decision rule becomes randomized; otherwise, it is nonrandomized. Thus, if $f_{\theta^m}(x^n)$ is continuous everywhere in Γ^n, we may select a nonrandomized decision rule without any loss in performance. We note that due to the monotonicity of the integral $\int_{\Gamma_1^n} dx^n f_{\theta^m}(x^n)$; $\Gamma_1^n = \{x^n : f_1(x^n)/f_{\theta^m}(x^n) > \lambda\}$ with respect to λ, the scalar $\lambda(\theta^m, \alpha)$ that satisfies (5.2.3) is then unique. In contrast to the Bayesian formalizations in Chapter 4, however, in the Neyman-Pearson formalization, randomized decision rules are not always obsolete. Indeed, if the density function $f_{\theta^m}(x^n)$ is not continuous at some vector points x^n then a value of the scalar $r(\theta^m, \alpha)$ that is unique in $(0, 1)$ may satisfy (5.2.3) in conjunction with a unique value of the scalar $\lambda(\theta^m, \alpha)$. In this latter case, a unique randomized decision rule may be the solution of the constraint optimization problem in (5.2.2). We note that in every case, the scalar $\lambda(\theta^m, \alpha)$ provides a threshold against which the likelihood ratio $f_1(x^n)/f_{\theta^m}(x^n)$ is tested. This pair together with the randomizing scalar $r(\theta^m, \alpha)$ comprise the test. As with the Bayesian tests, the maximum reduction of the test should be always sought. We will now present three examples to clarify some of the issues discussed above.

5.2 The Optimization Problem

Example 5.2.1

Consider two noncomposite, discrete-time hypotheses, and let n be the size of the observation vector. Let $f_1(x^n)$ be the n-dimensional density function of a Gaussian process, whose n-dimensional mean is θs^n, where θ is a given nonzero scalar and s^n is a given constant column vector, and whose $n \times n$ autocovariance matrix, denoted by R_n, is given. Let H_0 be specified by a zero mean Gaussian process, whose $n \times n$ autocovariance matrix is again R_n. Let $f_0(x^n)$ denote the n-dimensional density function that corresponds to H_0, and let the observation vector x^n be a column vector. Then we obtain

$$\text{LR}(\theta) = \frac{f_1(x^n)}{f_0(x^n)} = \frac{\exp\{-\frac{1}{2}[x^n - \theta s^n]^T R_n^{-1}[x^n - \theta s^n]\}}{\exp\{-\frac{1}{2}x^{n^T} R_n^{-1} x^n\}}$$

$$= \exp\left\{\theta s^{n^T} R_n^{-1} x^n - \frac{1}{2}\theta^2 s^{n^T} R_n^{-1} s^n\right\} \quad (5.2.12)$$

where T denotes transpose and -1 denotes inverse.

In the present example, $f_0(x^n)$ is continuous everywhere, thus randomized rules become obsolete. As a result, a most powerful decision rule δ_θ^* is given by the following expression

$$\delta_\theta^* = \begin{cases} 1 & \text{LR}(\theta) \geq \lambda \\ 0 & \text{LR}(\theta) < \lambda \end{cases} \quad (5.2.13)$$

where $\text{LR}(\theta)$ is the likelihood ratio in (5.2.12), and where the nonnegative threshold λ in (5.2.13) is uniquely determined by a given false alarm rate α, as in equation (5.2.3), for $r(\theta^m, \alpha) = 0$.

It is simpler to simplify the test in (5.2.13) for arbitrary λ first, and then determine the resulting modified threshold via the given false alarm restrictions. Specifically, from (5.2.13) we have $\delta_\theta^* = 1$, iff

$$\text{LR}(\theta) = \exp\left\{\theta s^{n^T} R_n^{-1} x^n - \frac{1}{2}\theta^2 s^{n^T} R_n^{-1} s^n\right\} \geq \lambda \quad (5.2.14)$$

or equivalently [by taking the natural logarithm of both parts in (5.2.14)] iff

$$\theta s^{n^T} R_n^{-1} x^n \geq \frac{1}{2}\theta^2 s^{n^T} R_n^{-1} s^n + \ln \lambda \quad (5.2.15)$$

At this point, two distinct cases evolve:

CASE 1: Let $\theta > 0$. Then, from (5.2.15), we obtain $\delta_\theta^* = 1$, iff

$$s^{n^T} R_n^{-1} x^n \geq \frac{1}{2}\theta s^{n^T} R_n^{-1} s^n + \theta^{-1} \ln \lambda = C(\theta) \quad (5.2.16)$$

In (5.2.16), the quadratic form, $s^{n^T}R_n^{-1}x^n$ is the test function; it is the only part of the test that is a function of the observation vector. The modified threshold is the function $C(\theta)$, which can be directly determined via the given false alarm rate α. Under H_0, the random variable $s^{n^T}R_n^{-1}X^n$ is zero mean Gaussian, with variance $E\{s^{n^T}R_n^{-1}X^n X^{n^T}R_n^{-1}s^n\} = s^{n^T}R_n^{-1}R_n R_n^{-1}s^n = s^{n^T}R_n^{-1}s^n$. Thus, the modified threshold $C(\theta)$ is uniquely determined by

$$\Pr\{\text{decide } H_1 | H_0\} = \Pr\{s^{n^T}R_n^{-1}X^n \geq C(\theta) | H_0\} = 1 - \Phi\left(\frac{C(\theta)}{[s^{n^T}R_n^{-1}s^n]^{1/2}}\right) = \alpha \quad (5.2.17)$$

where $\Phi(x)$ is the cumulative distribution of the zero mean, unit variance Gaussian random variable at the point x.

Due to the strict monotonicity of $\Phi(x)$ the equation in (5.2.17) has a unique solution with respect to $C(\theta)$. In particular, if $\Phi^{-1}(y)$ denotes the unique real scalar w such that $\Phi(w) = y$, the solution of (5.2.17) is

$$C(\theta) = [s^{n^T}R_n^{-1}s^n]^{1/2} \cdot \Phi^{-1}(1-\alpha) \quad (5.2.18)$$

From (5.2.18), we observe that the threshold $C(\theta)$ is independent of the particular value of θ as long as θ is positive. The nonunique most powerful decision rule δ_θ^* then takes the final form

$$\delta^* = \delta_\theta^* = \begin{cases} 1 & s^{n^T}R_n^{-1}x^n \geq [s^{n^T}R_n^{-1}s^n]^{1/2}\Phi^{-1}(1-\alpha) \\ 0 & s^{n^T}R_n^{-1}x^n < [s^{n^T}R_n^{-1}s^n]^{1/2}\Phi^{-1}(1-\alpha) \end{cases} \quad (5.2.19)$$

From (5.2.19) we observe that the test that corresponds to the most powerful decision rule is the same for all the positive θ values. As we will see in the next section, this is a significant observation, leading to the derivation of uniformly most powerful decision rules for composite H_1 hypotheses. The power induced by the decision rule in (5.2.19) is given by

$$P_1(\delta^*) = \Pr\{\text{decide } H_1 | H_1\} = \Pr\{s^{n^T}R_n^{-1}X^n \geq [s^{n^T}R_n^{-1}s^n]^{1/2}\Phi^{-1}(1-\alpha) | H_1\}$$

$$= 1 - \Phi\left(\frac{[s^{n^T}R_n^{-1}s^n]^{1/2}\Phi^{-1}(1-\alpha) - \theta s^{n^T}R_n^{-1}s^n}{[s^{n^T}R_n^{-1}s^n]^{1/2}}\right)$$

$$= 1 - \Phi(\Phi^{-1}(1-\alpha) - \theta[s^{n^T}R_n^{-1}s^n]^{1/2}) \quad (5.2.20)$$

where under H_1, the random variable $s^{n^T}R_n^{-1}X^n$ is Gaussian, with mean $\theta s^{n^T}R_n^{-1}s^n$, and with variance $s^{n^T}R_n^{-1}s^n$.

From (5.2.20), we observe that the power induced by the most powerful decision rule in (5.2.19) is a function of θ. In fact, as θ increases within the positive semiaxis, the power increases as well. This corresponds to increased power for increased signal-to-noise ratio. We also observe from (5.2.20) that the power increases monotonically as the false alarm rate α increases. ∎

5.2 The Optimization Problem

CASE 2: Let $\theta < 0$. Then from (5.2.15), we obtain $\delta_\theta^* = 1$, iff

$$s^{n^T} R_n^{-1} x^n \leq \frac{1}{2} \theta s^{n^T} R_n^{-1} s^n + \theta^{-1} \ln \lambda = C(\theta) \qquad (5.2.21)$$

Following procedures parallel to Case 1, we find

$$\Pr\{\text{decide } H_1 | H_0\} = \Pr\{s^{n^T} R_n^{-1} X^n \leq C(\theta) | H_0\} = \Phi\left(\frac{C(\theta)}{[s^{n^T} R_n^{-1} s^n]^{1/2}}\right) = \alpha$$

and

$$C(\theta) = [s^{n^T} R_n^{-1} s^n]^{1/2} \cdot \Phi^{-1}(\alpha) \qquad (5.2.22)$$

$$\delta^* = \delta_\theta^* = \begin{cases} 1 & s^{n^T} R_n^{-1} x^n \leq [s^{n^T} R_n^{-1} s^n]^{1/2} \cdot \Phi^{-1}(\alpha) \\ 0 & s^{n^T} R_n^{-1} x^n > [s^{n^T} R_n^{-1} s^n]^{1/2} \cdot \Phi^{-1}(\alpha) \end{cases} \qquad (5.2.23)$$

$$P_1(\delta^*) = \Pr\{s^{n^T} R_n^{-1} X^n \leq [s^{n^T} R_n^{-1} s^n]^{1/2} \cdot \Phi^{-1}(\alpha) | H_1\}$$

$$= \Phi\left(\frac{[s^{n^T} R_n^{-1} s^n]^{1/2} \Phi^{-1}(\alpha) - \theta s^{n^T} R_n^{-1} s^n}{[s^{n^T} R_n^{-1} s^n]^{1/2}}\right)$$

$$= \Phi(\Phi^{-1}(\alpha) - \theta [s^{n^T} R_n^{-1} s^n]^{1/2}) \qquad (5.2.24) \quad \blacksquare$$

We observe that the test in (5.2.23) is again independent of the particular negative value of θ and that the induced power again increases monotonically with increasing $|\theta|$ and increasing α. As compared to the test for positive θ values the test for negative θ values is reversed, and its threshold is different. In Figure 5.2.1, we plot the power $P_1(\delta^*)$ in (5.2.20) and (5.2.24) as a function of the parameter θ.

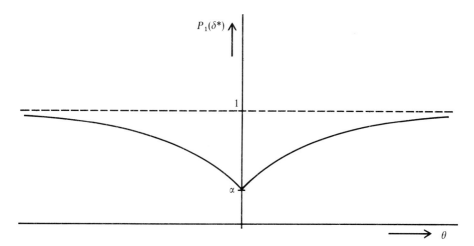

Figure 5.2.1

Example 5.2.2

Consider a slight variation in the hypotheses in Example 5.2.1. In particular, let the mean of the H_1 Gaussian process be $\theta_1 s^n$, where $\theta_1 > 0$. Let the mean of the H_0 Gaussian process be $\theta_0 s^n$, where $\theta_0 < 0$. Then we obtain

$$\text{LR}(\theta_1, \theta_0) = \frac{f_1(x^n)}{f_0(x^n)} = \frac{\exp\{-\frac{1}{2}[x^n - \theta_1 s^n]^T R_n^{-1}[x^n - \theta_1 s^n]\}}{\exp\{-\frac{1}{2}[x^n - \theta_0 s^n]^T R_n^{-1}[x^n - \theta_0 s^n]\}}$$

$$= \exp\left\{(\theta_1 - \theta_0)s^{n^T} R_n^{-1} x^n - \frac{1}{2}(\theta_1^2 - \theta_0^2)s^{n^T} R_n^{-1} s^n\right\} \quad (5.2.25)$$

As in Example 5.2.1, a most powerful rule $\delta^*_{\theta_1,\theta_0}$ has the form

$$\delta^*_{\theta_1,\theta_0} = \begin{cases} 1 & \text{LR}(\theta_1, \theta_0) \geq \lambda \\ 0 & \text{LR}(\theta_1, \theta_0) < \lambda \end{cases} \quad (5.2.26)$$

for some nonnegative constant λ that is uniquely determined by the given false alarm rate.

Proceeding as in Example 5.2.1, we derive the simplified decision rule as $\delta^*_{\theta_1,\theta_0} = 1$, iff

$$(\theta_1 - \theta_0)s^{n^T} R_n^{-1} x^n \geq \frac{1}{2}(\theta_1^2 - \theta_0^2)s^{n^T} R_n^{-1} s^n + \ln \lambda$$

or equivalently (since $\theta_1 - \theta_0 > 0$), $\delta^*_{\theta_1,\theta_0} = 1$, iff

$$s^{n^T} R_n^{-1} x^n \geq \frac{1}{2}(\theta_1 + \theta_0)s^{n^T} R_n^{-1} s^n + (\theta_1 - \theta_0)^{-1} \ln \lambda = C(\theta_1, \theta_0) \quad (5.2.27)$$

The random variable $s^{n^T} R_n^{-1} X^n$ is Gaussian with variance $s^{n^T} R_n^{-1} s^n$ under both H_0 and H_1. Its mean under H_0 is $\theta_0 s^{n^T} R_n^{-1} s^n$, and it is $\theta_1 s^{n^T} R_n^{-1} s^n$ under H_1. We thus obtain

$$\Pr\{\text{decide } H_1 | H_0\} = \Pr\{s^{n^T} R_n^{-1} X^n \geq C(\theta_1, \theta_0) | H_0\}$$

$$= 1 - \Phi\left(\frac{C(\theta_1, \theta_0) - \theta_0 s^{n^T} R_n^{-1} s^n}{[s^{n^T} R_n^{-1} s^n]^{1/2}}\right) = \alpha$$

and thus

$$C(\theta_1, \theta_0) = [s^{n^T} R_n^{-1} s^n]^{1/2} \Phi^{-1}(1 - \alpha) + \theta_0 s^{n^T} R_n^{-1} s^n \quad (5.2.28)$$

$$\delta^*_{\theta_0} = \delta^*_{\theta_1,\theta_0} = \begin{cases} 1 & s^{n^T} R_n^{-1} x^n \geq [s^{n^T} R_n^{-1} s^n]^{1/2} \Phi^{-1}(1 - \alpha) + \theta_0 s^{n^T} R_n^{-1} s^n \\ 0 & s^{n^T} R_n^{-1} x^n < [s^{n^T} R_n^{-1} s^n]^{1/2} \Phi^{-1}(1 - \alpha) + \theta_0 s^{n^T} R_n^{-1} s^n \end{cases}$$

$$\quad (5.2.29)$$

$$P_1(\delta^*_{\theta_0}) = \Pr\{s^{n^T} R_n^{-1} X^n \geq [s^{n^T} R_n^{-1} s^n]^{1/2} \Phi^{-1}(1 - \alpha) + \theta_0 s^{n^T} R_n^{-1} s^n | H_1\}$$

$$= 1 - \Phi\left(\frac{[s^{n^T} R_n^{-1} s^n]^{1/2} \Phi^{-1}(1 - \alpha) - (\theta_1 - \theta_0)s^{n^T} R_n^{-1} s^n}{[s^{n^T} R_n^{-1} s^n]^{1/2}}\right)$$

$$= 1 - \Phi(\Phi^{-1}(1 - \alpha) - (\theta_1 - \theta_0)[s^{n^T} R_n^{-1} s^n]^{1/2}) \quad (5.2.30)$$

5.2 The Optimization Problem

From (5.2.28) and (5.2.29) we observe that the test is independent of the specific θ_1 value, but it depends on the θ_0 value that determines the H_0 hypothesis. In particular, for a fixed α the threshold in (5.2.28) increases monotonically with increasing negative θ_0. Equivalently, if the threshold in (5.2.28) is set to a fixed value, then as $|\theta_0|$ increases, the induced false alarm rate increases as well. As we will see later in this section, this quality is significant for deriving solutions for the generalized Neyman-Pearson optimization problem in (5.2.1). From (5.2.30), we observe that the power induced by the decision rule in (5.2.29) increases monotonically, with increasing difference $\theta_1 - \theta_0$. Qualitatively speaking, this means that the more apart the two hypotheses are, the better they can be distinguished from each other.

Example 5.2.3

In the previous two examples of this section, the density functions evolving from the H_0 hypotheses were continuous everywhere in the vector observation space. Thus, randomized most powerful decision rules were obsolete. In this example, we will consider discrete-value hypotheses, and we will show how randomization becomes, in general, necessary in this case.

Let x^n denote a column n-dimensional observation vector, and let us consider two noncomposite hypotheses, specified as follows:

Under both hypotheses, the components $\{X_i; i = 1, \ldots, n\}$ of the stochastic vector X^n are independent and identically distributed Poisson variables. Under hypothesis H_1, the parameter of the Poisson distribution is θ_1, where $\theta_1 > 0$. Under hypothesis H_0, the parameter of the Poisson distribution is θ_0, where $\theta_0 > 0$ and $\theta_0 < \theta_1$.

Let us denote by $f_1(x^n)$ and $f_0(x^n)$ the density functions at the vector point x^n under respectively H_1 and H_0. For the model described above, we then obtain

$$f_1(x^n) = \prod_{i=1}^{n} f_1(x_i)$$

$$f_1(x_i) = \begin{cases} \dfrac{\theta_1^{x_i} e^{-\theta_1}}{x_i!} & x_i = 0, 1, 2, \ldots \\ 0 & \text{otherwise} \end{cases}$$

$$f_0(x^n) = \prod_{i=1}^{n} f_0(x_i)$$

$$f_0(x_i) = \begin{cases} \dfrac{\theta_0^{x_i} e^{-\theta_0}}{x_i!} & x_i = 0, 1, 2, \ldots \\ 0 & \text{otherwise} \end{cases}$$

where $\prod$ denotes product.

From the above expressions, we obtain, by straightforward substitution,

$$f_1(x^n) = e^{-n\theta_1} \frac{\theta_1^{\sum_{i=1}^{n} x_i}}{\prod_{i=1}^{n} x_i!}$$

$$f_0(x^n) = e^{-n\theta_0} \frac{\theta_0^{\sum_{i=1}^{n} x_i}}{\prod_{i=1}^{n} x_i!}$$

$$\text{LR}(\theta_0, \theta_1) = \frac{f_1(x^n)}{f_0(x^n)} = e^{-n(\theta_1 - \theta_0)} \left(\frac{\theta_1}{\theta_0}\right)^{\sum_{i=1}^{n} x_i} \tag{5.2.31}$$

Directly from Theorem 5.2.1, we conclude that a most powerful decision rule $\delta^*_{\theta_1, \theta_0}$ will have the general form

$$\delta^*_{\theta_1, \theta_0} = \begin{cases} 1 & \text{LR}(\theta_0, \theta_1) > \lambda \\ r & \text{LR}(\theta_0, \theta_1) = \lambda \\ 0 & \text{LR}(\theta_0, \theta_1) < \lambda \end{cases} \tag{5.2.32}$$

where the likelihood ratio $\text{LR}(\theta_0, \theta_1)$ is given by (5.2.31), where $\lambda \geq 0$, $0 \leq r \leq 1$, and λ and r are determined by a given α. As with Examples 5.2.1 and 5.2.2, we will first reduce the most powerful test to its simplest form for arbitrary r and λ. Then, we will determine the (possibly modified) scalars from the given α. From (5.2.31) and (5.2.32) we conclude that $\delta^*_{\theta_1, \theta_0}$ can be expressed as

$$\delta^*_{\theta_1, \theta_0} = \begin{cases} 1 & \text{iff } e^{-n(\theta_1 - \theta_0)} \cdot \left(\frac{\theta_1}{\theta_0}\right)^{\sum_{i=1}^{n} x_i} > \lambda \\ r & \text{iff } e^{-n(\theta_1 - \theta_0)} \cdot \left(\frac{\theta_1}{\theta_0}\right)^{\sum_{i=1}^{n} x_i} = \lambda \end{cases} \tag{5.2.33}$$

Taking the natural logarithm of all parts in (5.2.33), we obtain the equivalent expression

$$\delta^*_{\theta_1, \theta_0} = \begin{cases} 1 & \text{iff } \sum_{i=1}^{n} x_i \ln \frac{\theta_1}{\theta_0} > \ln \lambda + n(\theta_1 - \theta_0) \\ r & \text{iff } \sum_{i=1}^{n} x_i \ln \frac{\theta_1}{\theta_0} = \ln \lambda + n(\theta_1 - \theta_0) \end{cases} \tag{5.2.34}$$

Since we have assumed that $\theta_1 > \theta_0$, the logarithm $\ln(\theta_1/\theta_0)$ is positive. Thus, from (5.2.34) we obtain the equivalent expression

$$\delta^*_{\theta_1, \theta_0} = \begin{cases} 1 & \text{iff } \sum_{i=1}^{n} x_i > \dfrac{\ln \lambda}{\ln(\theta_1/\theta_0)} + \dfrac{n(\theta_1 - \theta_0)}{\ln(\theta_1/\theta_0)} = C(\theta_1, \theta_0) \\ r & \text{iff } \sum_{i=1}^{n} x_i = \dfrac{\ln \lambda}{\ln(\theta_1/\theta_0)} + \dfrac{n(\theta_1 - \theta_0)}{\ln(\theta_1/\theta_0)} = C(\theta_1, \theta_0) \end{cases} \tag{5.2.35}$$

5.2 The Optimization Problem

The form in (5.2.35) is the most reduced expression of the most powerful decision rule, the test function is then $\sum_{i=1}^{n} X_i$, and the modified threshold is $C(\theta_1, \theta_0)$. The parameters of the decision rule that remain unspecified are the randomizing scalar r and the modified threshold $C(\theta_1, \theta_0)$. Both those parameters will be determined by the false alarm rate α. We first note that the test variable $\sum_{i=1}^{n} X_i$ is under both hypotheses the sum of independent and identically distributed Poisson variables. Thus, $\sum_{i=1}^{n} X_i$ is Poisson under both hypotheses, with parameters $n\theta_0$ under H_0 and $n\theta_1$ under H_1. Due to the above and (5.2.35), we have

$$\Pr\{\text{decide } H_1 | H_0\} = \Pr\left\{\sum_{i=1}^{n} X_i > C(\theta_1, \theta_0) \bigg| H_0\right\}$$
$$+ r \cdot \Pr\left\{\sum_{i=1}^{n} X_i = C(\theta_1, \theta_0) \bigg| H_0\right\} = \alpha \quad (5.2.36)$$

Since $\sum_{i=1}^{n} X_i$ is Poisson under H_0, it can only take positive integer values. Thus, without lack in generality, we can restrict the $C(\theta_1, \theta_0)$ in (5.2.36) to positive integer values as well. Let us then denote this threshold by m. We can then write (5.2.36) as

$$r, m: \sum_{k=m+1}^{\infty} \frac{e^{-n\theta_0}(n\theta_0)^k}{k!} + r \frac{e^{-n\theta_0}(n\theta_0)^m}{m!} = \alpha$$

or

$$r, m: U(m-1) \sum_{k=0}^{m-1} \frac{e^{-n\theta_0}(n\theta_0)^k}{k!} + (1-r) \frac{e^{-n\theta_0}(n\theta_0)^m}{m!} = 1 - \alpha \quad (5.2.37)$$

where

$$U(x) = \begin{cases} 0 & x < 0 \\ 1 & x \geq 0 \end{cases}$$

Expression (5.2.37) determines a unique pair (r, m) for every fixed value α in $(0, 1)$. First, m is determined by

$$m = \max\left(l: U(l-1) \sum_{k=0}^{l-1} \frac{e^{-n\theta_0}(n\theta_0)^k}{k!} \leq 1 - \alpha\right) \quad (5.2.38)$$

If the integer m in (5.2.38) is such that $U(m-1)\sum_{k=0}^{m-1}[e^{-n\theta_0}(n\theta_0)^k]/k! = 1 - \alpha$, then the randomizing constant r equals one, and the most powerful decision rule is then nonrandomized. The decision rule is then equal to one if $\sum_{i=1}^{n} X_i \geq m$ and equal to zero otherwise. If, on the other hand, the integer m in (5.2.38) is such that $d(m, \alpha) = (1 - \alpha) - U(m-1)\sum_{k=0}^{m-1}[e^{-n\theta_0}(n\theta_0)^k]/k! > 0$ then r equals $1 - [m!d(m, \alpha)]/e^{-n\theta_0}(n\theta_0)^m$; it is strictly nonzero and not equal to one; and the resulting most powerful decision rule is unique and randomized. In Figure 5.2.2 we show graphically how the pair (r, m) is determined. From the above, we easily conclude that the unique pair (r, m) depends strictly on the parameter θ_0 of the H_0 Poisson process. It is independent of the parameter θ_1 of the H_1 Poisson process, under the constraint that $\theta_1 > \theta_0$. This property is the same as in Example 5.2.1, leading again to the derivation of uniformly most powerful decision rules for composite H_1 hypotheses. From (5.2.37), we also easily conclude that for fixed r and m the false alarm is monotonically nondecreasing, with θ_0 increasing. As in Example 5.2.2, this monotonicity is important for the derivation of solutions for the generalized Neyman-Pearson optimization problem in (5.2.1). Finally, if (r, m) is the pair that satisfies (5.2.37), the power $P_1(\delta^*_{\theta_1, \theta_0})$ induced by the most powerful decision rule in (5.2.35) is given by

$$P_1(\delta^*_{\theta_1,\theta_0}) = 1 - U(m-1) \sum_{k=0}^{m-1} \frac{e^{-n\theta_1}(n\theta_1)^k}{k!} - (1-r)\frac{e^{-n\theta_1}(n\theta_1)^m}{m!} \quad (5.2.39)$$

It is easily shown that the power in (5.2.39) is monotonically nondecreasing with increasing θ_1 values, where the fixed values r and m correspond to fixed values of θ_0 and α. Thus, the monotonicity of the power with respect to θ_1 means that H_1 and H_0 become more distinguishable as they are driven further apart. The power in (5.2.39) also increases monotonically with increasing n, converging asymptotically to one.

So far, we have solved and analyzed the constraint optimization problem that evolves from the Neyman-Pearson formalization when H_0 is well known or noncomposite. Now we will consider the generalized Neyman-Pearson formalization for composite or parametrically known H_0 hypotheses. Then, the constraint optimization problem in (5.2.1) evolves, where the parameter space $\mathscr{E}^m$ is assumed known. The latter problem does not always have a solution. A solution exists only when the density functions $f_1(x^n)$ and $f_{\theta m}(x^n)$ jointly satisfy certain conditions in $\mathscr{E}^m$. We state those conditions in a lemma.

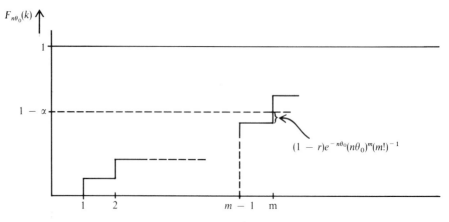

Figure 5.2.2 $F_{n\theta_0}(k) = \sum_{l=0}^{k} e^{-n\theta_0}(n\theta_0)^l (l!)^{-1}$

5.2 The Optimization Problem

LEMMA 5.2.1: Let $f_{\theta^m}(x^n)$; $\theta^m \in \mathscr{E}^m$ be the density functions in the constraint optimization problem in (5.2.1), under H_0. Let there exist some strictly monotonically increasing real and scalar function $f(\cdot)$, two scalar functions $g(\theta^m)$ and $c(\theta^m)$, and some scalar function $h(x^n)$, such that

$$f\left[\frac{f_1(x^n)}{f_{\theta^m}(x^n)}\right] = g(\theta^m)h(x^n) + c(\theta^m) \qquad \forall\, \theta^m \in \mathscr{E}^m \qquad (5.2.40)$$

where

$c(\theta^m)$ is not a function of x^n

$g(\theta^m)$ is not a function of x^n, and $g(\theta^m) > 0 \qquad \forall\, \theta^m \in \mathscr{E}^m$

$h(x^n)$ is not a function of θ^m

Let λ be some arbitrary nonnegative scalar. Let r be some arbitrary scalar in $[0, 1]$. Given λ and r as above, let us define

$$G(\lambda, r, \theta^m) = \int_{x^n:\, h(x^n) > \lambda} f_{\theta^m}(x^n)\, dx^n + r \int_{x^n:\, h(x^n) = \lambda} f_{\theta^m}(x^n)\, dx^n \qquad (5.2.41)$$

Let $\theta^m(\lambda, r)$ be the value θ^m that satisfies the supremum $\sup\limits_{\theta^m \in \mathscr{E}^m} G(\lambda, r, \theta^m)$. Then if $\theta^m(\lambda, r) = \theta^{m*}$—that is, if $\theta^m(\lambda, r)$ is independent of the specific values λ and r—the constraint optimization problem in (5.2.1) has a solution. The solution δ^* is given then by

$$\delta^*(x^n) = \begin{cases} 1 & x^n:\, h(x^n) > \lambda^* \\ r^* & x^n:\, h(x^n) = \lambda^* \\ 0 & x^n:\, h(x^n) < \lambda^* \end{cases} \qquad (5.2.42)$$

where λ^* and r^* satisfy

$$\sup_{\theta^m \in \mathscr{E}^m} G(\lambda^*, r^*, \theta^m) = G(\lambda^*, r^*, \theta^{m*}) = \alpha$$

Proof: Let us temporarily fix the value θ^m of the parameter vector. Then, for any given false alarm rate, a most powerful Neyman-Pearson decision rule has the form (directly from theorem 5.2.1)

$$\delta^*_{\theta^m}(x^n) = \begin{cases} 1 & x^n:\, \dfrac{f_1(x^n)}{f_{\theta^m}(x^n)} > \lambda \\ r & x^n:\, \dfrac{f_1(x^n)}{f_{\theta^m}(x^n)} = \lambda \\ 0 & x^n:\, \dfrac{f_1(x^n)}{f_{\theta^m}(x^n)} < \lambda \end{cases} \qquad (5.2.43)$$

for some appropriate λ and r.

Due to the strict monotonicity of the function $f(\cdot)$, the positiveness of the function $g(\cdot)$, and (5.2.40), the decision rule in (5.2.43) can be written

$$\delta^*_{\theta^m}(x^n) = \begin{cases} 1 & x^n: h(x^n) > g^{-1}(\theta^m)f(\lambda) - g^{-1}(\theta^m)c(\theta^m) = \lambda_1 \\ r & x^n: h(x^n) = g^{-1}(\theta^m)f(\lambda) - g^{-1}(\theta^m)c(\theta^m) = \lambda_1 \\ 0 & x^n: h(x^n) < g^{-1}(\theta^m)f(\lambda) - g^{-1}(\theta^m)c(\theta^m) = \lambda_1 \end{cases} \quad (5.2.44)$$

Therefore, for every θ^m in $\mathscr{E}^m$ the most powerful Neyman-Pearson decision rule has a form as in (5.2.44). Then, we obtain

$$P_{\theta^m}(\delta^*_{\theta^m}) = \int_{x^n: h(x^n) > \lambda_1} dx^n f_{\theta^m}(x^n) + r \int_{x^n: h(x^n) = \lambda_1} dx^n f_{\theta^m}(x^n) = G(\lambda_1, r, \theta^m)$$

But, due to the assumptions in the lemma, the supremum $\sup_{\theta^m \in \mathscr{E}^m} P_{\theta^m}(\delta^*_{\theta^m})$ is attained at θ^{m*} for every λ_1 and r. Thus, the false alarm constraint in Problem (5.2.1), reduces here to

$$\lambda^*, r^*: \int_{x^n: h(x^n) > \lambda^*} dx^n f_{\theta^{m*}}(x^n) + r^* \int_{x^n: h(x^n) = \lambda^*} dx^n f_{\theta^{m*}}(x^n) = \alpha$$

The proof of the lemma is now complete. ∎

We will now revisit Examples 5.2.2 and 5.2.3 in a modified form that is pertinent to the generalized Neyman-Pearson formalization, in (5.2.1).

Example 5.2.4

Consider the H_0 and H_1 hypotheses in Example 5.2.2, modified as follows. Let the parameter θ_1 in H_1 be some fixed scalar on the real axis R. Let H_0 be composite, where its parameter θ_0 takes values within some subset $\mathscr{E}$ of the real axis R that does not include the value θ_1. Let θ be some value in $\mathscr{E}$. Then, due to (5.2.25), we have

$$LR(\theta_1, \theta) = \exp\left\{ (\theta_1 - \theta)s^{n^T} R_n^{-1} x^n - \frac{1}{2}(\theta_1^2 - \theta^2)s^{n^T} R_n^{-1} s^n \right\} \quad (5.2.45)$$

Let us define,

$$g(\theta) = |\theta_1 - \theta|$$
$$c(\theta) = -\frac{1}{2}(\theta_1^2 - \theta^2)s^{n^T} R_n^{-1} s^n \quad (5.2.46)$$
$$f(x) = \ln x$$
$$h(x^n) = s^{n^T} R_n^{-1} x^n \cdot \operatorname{sgn}(\theta_1 - \theta)$$

From (5.2.45) and (5.2.46) we get

$$f[LR(\theta_1, \theta)] = \ln[LR(\theta_1, \theta)] = g(\theta)h(x^n) + c(\theta) \quad (5.2.47)$$

5.2 The Optimization Problem

Since the logarithm is a strictly monotone function, we observe from (5.2.47) that the conditions of Lemma 5.2.1 are satisfied only if the quantity $\text{sgn}(\theta_1 - \theta)$ in the definition of $h(x^n)$ remains unchanged for every θ in $\mathscr{E}$. Therefore, the generalized Neyman-Pearson problem may have a solution only if

$$\text{either} \quad E: \theta \in \mathscr{E} \to \theta < \theta_1$$

$$\text{or} \quad E: \theta \in \mathscr{E} \to \theta > \theta_1$$

Let us now examine $G(\lambda, r, \theta^m)$ in (5.2.41) as it evolves in the present example.

$$G(\lambda, r, \theta) = \Pr\{h(X^n) > \lambda | \theta\} + r \Pr\{h(X^n) = \lambda | \theta\}$$

$$= \Pr\{s^{n^T} R_n^{-1} X^n \cdot \text{sgn}(\theta_1 - \theta) > \lambda | \theta\}$$
$$+ r \Pr\{s^{n^T} R_n^{-1} X^n \cdot \text{sgn}(\theta_1 - \theta) = \lambda | \theta\}$$

$$= \begin{cases} 1 - \Phi\left(\dfrac{\lambda - \theta s^{n^T} R_n^{-1} s^n}{[s^{n^T} R_n^{-1} s^n]^{1/2}}\right) = \Phi\left(\dfrac{-\lambda + \theta s^{n^T} R_n^{-1} s^n}{[s^{n^T} R_n^{-1} s^n]^{1/2}}\right) \\ \qquad\qquad\qquad\qquad\qquad\qquad\qquad\qquad \mathscr{E}: \theta \in \mathscr{E} \to \theta < \theta_1 \\ \Phi\left(\dfrac{-\lambda - \theta s^{n^T} R_n^{-1} s^n}{[s^{n^T} R_n^{-1} s^n]^{1/2}}\right) \quad \mathscr{E}: \theta \in \mathscr{E} \to \theta > \theta_1 \end{cases} \quad (5.2.48)$$

where $\Phi(x)$ is the cumulative distribution function of the zero mean, unit variance, Gaussian random variable.

From (5.2.48), we observe that for any values λ and r the function $G(\lambda, r, \theta)$ is monotonically increasing with increasing θ for $\mathscr{E}: \theta \in \mathscr{E} \to \theta < \theta_1$, and it is monotonically decreasing with increasing θ for $\mathscr{E}: \theta \in \mathscr{E} \to \theta > \theta_1$. Therefore, in either case, the value θ^* that realizes the $\sup_{\theta \in \mathscr{E}} G(\lambda, r, \theta)$ is independent of λ and r. In fact, we obtain

a) If $\mathscr{E} = \mathscr{E}_1: \theta \in \mathscr{E}_1 \to \theta < \theta_1$
   Then $\theta^* = \theta_1^* = \sup\{\theta: \theta \in \mathscr{E}_1\}$ (5.2.49)

b) If $\mathscr{E} = \mathscr{E}_2: \theta \in \mathscr{E}_2 \to \theta > \theta_1$
   Then $\theta^* = \theta_1^* = \inf\{\theta: \theta \in \mathscr{E}_2\}$

Due to (5.2.48) and (5.2.49), we conclude that there exist solutions of the generalized Neyman-Pearson problem for $\mathscr{E} = \mathscr{E}_1$ and $\mathscr{E} = \mathscr{E}_2$, and we finally obtain:

a) For $\mathscr{E} = \mathscr{E}_1$, a most powerful generalized Neyman-Pearson test is given by

$$\delta_1^*(x^n) = \begin{cases} 1 & x^n: s^{n^T} R_n^{-1} x^n \geq \lambda_1^* \\ 0 & x^n: s^{n^T} R_n^{-1} x^n < \lambda_1^* \end{cases} \quad (5.2.50)$$

$$\lambda_1^*: \Phi\left(\dfrac{-\lambda_1^* + \theta_1^* s^{n^T} R_n^{-1} s^n}{[s^{n^T} R_n^{-1} s^n]^{1/2}}\right) = \alpha \to \lambda_1^*$$

$$= [s^{n^T} R_n^{-1} s^n]^{1/2} \Phi^{-1}(1 - \alpha) + \theta_1^* s^{n^T} R_n^{-1} s^n$$

b) For $\mathscr{E} = \mathscr{E}_2$, a most powerful generalized Neyman-Pearson test is given by

$$\delta_2^*(x^n) = \begin{cases} 1 & x^n: s^{n^T} R_n^{-1} x^n \leq \lambda_2^* \\ 0 & x^n: s^{n^T} R_n^{-1} x^n > \lambda_2^* \end{cases} \quad (5.2.51)$$

$$\lambda_2^*: \Phi\left(\frac{\lambda_2^* - \theta_2^* s^{n^T} R_n^{-1} s^n}{[s^{n^T} R_n^{-1} s^n]^{1/2}}\right) = \alpha \rightarrow \lambda_2^* = [s^{n^T} R_n^{-1} s^n]^{1/2} \Phi^{-1}(\alpha) + \theta_2^* s^{n^T} R_n^{-1} s^n$$

In the special cases where $\mathscr{E}_1 = (-\infty, \theta_1)$ and $\mathscr{E}_2 = (\theta_1, \infty)$, we easily conclude that $\theta_1^* = \theta_2^* = \theta_1$.

Example 5.2.5

Consider the hypotheses H_0 and H_1 in Example 5.2.3, modified as follows. Let the parameter θ_1 in H_1 be some fixed positive scalar. Let H_0 be composite where its parameter θ_0 takes values within some subset $\mathscr{E}$ of the real positive semiaxis. Let $\mathscr{E}$ not include the value θ_1. Let θ be some value in $\mathscr{E}$. Then, from (5.2.31) we obtain

$$\text{LR}(\theta, \theta_1) = e^{-n(\theta_1 - \theta)} \left(\frac{\theta_1}{\theta}\right)^{\sum_{i=1}^{n} x_i} \quad (5.2.52)$$

Let us define:

$$g(\theta) = \left|\ln \frac{\theta_1}{\theta}\right|$$

$$c(\theta) = -n(\theta_1 - \theta) \quad (5.2.53)$$

$$h(x^n) = \left[\sum_{i=1}^{n} x_i\right] \cdot \text{sgn}(\theta_1 - \theta)$$

Then, from (5.2.52) and (5.2.53), we obtain

$$\ln(\text{LR}(\theta, \theta_1)) = g(\theta)h(x^n) + c(\theta) \quad (5.2.54)$$

where in the place of the function, $f(\cdot)$ in Lemma 5.2.1, we have the natural logarithm.

As in Example 5.2.4, we conclude from (5.2.54), that a solution of the generalized Neyman-Pearson problem may exist if

either $\quad \mathscr{E} = \mathscr{E}_1 : \theta \in \mathscr{E} \rightarrow \theta < \theta_1$

or $\quad \mathscr{E} = \mathscr{E}_2 : \theta \in \mathscr{E} \rightarrow \theta > \theta_1$

Setting λ equal to some positive integer m as in Example 5.2.3, we also obtain, for $G(\lambda, r, \theta^m)$ in (5.2.41) as applied here,

5.2 The Optimization Problem

$$\mathscr{E} = \mathscr{E}_1 : G_1(m, r, \theta) = \sum_{k=m+1}^{\infty} \frac{e^{-n\theta}(n\theta)^k}{k!} + r \frac{e^{-n\theta}(n\theta)^m}{m!} \qquad (5.2.55)$$

$$\mathscr{E} = \mathscr{E}_2 : G_2(m, r, \theta) = U(m-1) \sum_{k=0}^{m-1} \frac{e^{-n\theta}(n\theta)^k}{k!} + r \frac{e^{-n\theta}(n\theta)^m}{m!} \qquad (5.2.56)$$

where

$$U(x) = \begin{cases} 1 & x \geq 0 \\ 0 & x < 0 \end{cases}$$

It can be easily shown that for arbitrary legitimate values m and r the function $G_1(m, r, \theta)$ is monotonically nondecreasing with θ increasing in value and $G_2(m, r, \theta)$ is monotonically nonincreasing with θ increasing in value. Thus, defining $\theta_1^* = \sup\{\theta : \theta \in \mathscr{E}_1\}$ and $\theta_2^* = \inf\{\theta : \theta \in \mathscr{E}_2\}$, we obtain

$$\begin{aligned} \sup_{\theta \in \mathscr{E}_1} G_1(m, r, \theta) &= G_1(m\, r, \theta_1^*) \qquad \forall\, m, \forall\, r : 0 \leq r \leq 1 \\ \sup_{\theta \in \mathscr{E}_1} G_2(m, r, \theta) &= G_2(m\, r, \theta_2^*) \qquad \forall\, m, \forall\, r : 0 \leq r \leq 1 \end{aligned} \qquad (5.2.57)$$

Due to (5.2.57) and Lemma 5.2.1, and as in Example 5.2.4, if δ_1^* and δ_2^* denote respectively the most powerful decision rules on $\mathscr{E}_1$ and $\mathscr{E}_2$, they are given by

$$\delta_1^*(x^n) = \begin{cases} 1 & x^n : \sum_{i=1}^n x_i > m_1^* \\ r_1^* & x^n : \sum_{i=1}^n x_i = m_1^* \\ 0 & x^n : \sum_{i=1}^n x_i < m_1^* \end{cases}$$

$$(m_1^*, r_1^*): \sum_{k=m_1^*+1}^{\infty} \frac{e^{-n\theta_1^*}(n\theta_1^*)^k}{k!} + r_1^* \frac{e^{-n\theta_1^*}(n\theta_1^*)^{m_1^*}}{m_1^*!} = \alpha$$

$$\delta_2^*(x^n) = \begin{cases} 1 & x^n : \sum_{i=1}^n x_i < m_2^* \\ r_2^* & x^n : \sum_{i=1}^n x_i = m_2^* \\ 0 & x^n : \sum_{i=1}^n x_i > m_2^* \end{cases}$$

$$(m_2^*, r_2^*): U(m_2^* - 1) \sum_{k=0}^{m_2^*-1} \frac{e^{-n\theta_2^*}(n\theta_2^*)^k}{k!} + r_2^* \frac{e^{-n\theta_2^*}(n\theta_2^*)^{m_2^*}}{m_2^*!} = \alpha$$

As in Example 5.2.4, we conclude again that if $\mathscr{E}_1 = (-\infty, \theta_1)$ and $\mathscr{E}_2 = (\theta_1, \infty)$, then $\theta_1^* = \theta_2^* = \theta_1$.

Example 5.2.6

Let us consider hypotheses generated by continuous-time stochastic processes. Let the observation time interval be $[0, T]$ and let the stochastic waveform be denoted by $V(t); t \in [0, T]$. Let $v(t); t \in [0, T]$ denote a realization of the stochastic waveforms $V(t)$. Let $s(t); t \in [0, T]$ be some known, finite energy deterministic waveform, let θ_1 be some known real scalar, and let $\mathscr{E} = (-\infty, \theta_0)$; $\theta_0 < \theta_1$. Let H_1 be generated by a Gaussian nonstationary process whose autocovariance function is $R(t_1, t_2)$ and whose mean is $\theta_1 s(t)$. Let H_0 be composite, and be generated by a nonstationary Gaussian process whose autocovariance function is again $R(t_1, t_2)$ and whose mean is $\theta s(t)$, where θ takes any value in $\mathscr{E} = (-\infty, \theta_0)$. Let $R(t_1, t_2)$ be known, and let the given false alarm rate be α. Let there exist a set $\{\phi_i(t)\}$ of deterministic time functions that is orthonormal and complete in $[0, T]$ for the class of finite energy deterministic waveforms, and that also provides a Karhunen-Loéve expansion in $[0, T]$ for stochastic processes whose autocovariance function is $R(t_1, t_2)$. Let $N(t)$ denote the zero mean Gaussian process whose autocovariance function is $R(t_1, t_2)$. Then, as in Section 4.4, we define,

$$s_i = \int_0^T s(t)\phi_i^*(t)\,dt \qquad \forall\, i$$

$$\hat{s}(t) = \sum_i s_i \phi_i(t) \qquad t \in [0, T]$$

$$\hat{N}_i = \int_0^T N(t)\phi_i^*(t)\,dt \qquad \forall\, i,\ \hat{n}_i \text{ a realization of } \hat{N}_i$$

$$\hat{N}(t) = \sum_i \hat{N}_i \phi_i(t) \qquad t \in [0, T] \quad \hat{n}(t) \text{ a realization of } \hat{N}(t)$$

$$\hat{V}(t) = \begin{cases} \theta_1 \hat{s}(t) + \hat{N}(t) & \text{under } H_1 \\ \theta \hat{s}(t) + \hat{N}(t) & \text{under } \theta \text{ in } H_0 \end{cases} \qquad \hat{v}(t) \text{ a realization of } \hat{V}(t)$$

$$V(t) = \begin{cases} \theta_1 s(t) + N(t) & \text{under } H_1 \\ \theta s(t) + N(t) & \text{under } \theta \text{ in } H_0 \end{cases} \qquad v(t) \text{ a realization of } V(t)$$

where the random variables $\{\hat{N}_i\}$ are Gaussian, zero mean, and mutually independent, and where the variance of $\hat{N}_i$ is λ_i, given as

$$\int_0^T R(t_1, t_2)\phi_i(t_2)\,dt_2 = \lambda_i \phi_i(t_1) \qquad \forall\, t_1 \in [0, T] \quad \text{for some } \lambda_i > 0$$

5.2 The Optimization Problem

For given θ in $\mathscr{E}$, and as in Example 4.4.1, we obtain

$$\ln\left[\frac{f_{\theta_1}(\{\hat{v}_i\})}{f_{\theta}(\{\hat{v}_i\})}\right] = -\frac{1}{2}\sum_i \frac{(\hat{v}_i - \theta_1 s_i)^2}{\lambda_i} + \frac{1}{2}\sum_i \frac{(\hat{v}_i - \theta s_i)^2}{\lambda_i}$$

$$= (\theta_1 - \theta)\sum_i \frac{s_i}{\lambda_i}\hat{v}_i + \frac{1}{2}(\theta^2 - \theta_1^2)\sum_i \frac{s_i}{\lambda_i} s_i \qquad (5.2.58)$$

But directly from the derivations in Example 4.4.1, we have

$$\sum_i \frac{s_i}{\lambda_i}\hat{v}_i = \int_0^T q(t)\hat{v}(t)\,dt \qquad (5.2.59)$$

$$\sum_i \frac{s_i}{\lambda_i} s_i = \int_0^T q(t)\hat{s}(t)\,dt \qquad (5.2.60)$$

where $q(t);\ t \in [0, T]$ is a solution of

$$\hat{s}(t) = \int_0^T q(\tau)R(t, \tau)\,d\tau \qquad t \in [0, T] \qquad (5.2.61)$$

Due to (5.2.58)–(5.2.60), and as in Example 4.4.1, we obtain

$$\ln\left[\frac{f_{\theta_1}(\hat{v}(t);\ t \in [0, T])}{f_{\theta}(\hat{v}(t);\ t \in [0, T])}\right] = (\theta_1 - \theta)\int_0^T q(t)\hat{v}(t)\,dt + \frac{1}{2}(\theta^2 - \theta_1^2)\int_0^T q(t)\hat{s}(t)\,dt \qquad (5.2.62)$$

Due to Lemma 5.2.1, and since $\theta_1 - \theta > 0\ \forall\ \theta \in \mathscr{E}$, if a most powerful rule $\hat{\delta}^*$ exists for the approximate generalized Neyman-Pearson problem, where the observations are described by the waveform $\hat{v}(t);\ t \in [0, T]$, it has the form

$$\hat{\delta}^*(\hat{v}(t);\ t \in [0, T]) = \begin{cases} 1 & \int_0^T q(t)\hat{v}(t)\,dt \geq \hat{\lambda}^* \\ 0 & \int_0^T q(t)\hat{v}(t)\,dt < \hat{\lambda}^* \end{cases} \qquad (5.2.63)$$

where $q(t)$ is given by (5.2.61), and where randomization is obsolete, since the random variable, $\int_0^T q(t)\hat{V}(t)\,dt$, is here Gaussian under both hypotheses, which implies that its density function is continuous everywhere.

Given θ in $\mathscr{E}$, the random variable $\int_0^T q(t)\hat{V}(t)\,dt$ is Gaussian, with mean $\int_0^T q(t)\mathscr{E}\{\hat{V}(t)|\theta\}\,dt = \theta \int_0^T q(t)\hat{s}(t)\,dt$ and variance

$$\iint_0^T q(t)q^*(\tau)E\{\hat{N}(t)\hat{N}^*(\tau)\}\,dt\,d\tau = \iint_0^T q(t)q^*(\tau)R(t, \tau)\,dt\,d\tau = \int_0^T q(t)\hat{s}(t)\,dt$$

as found in Example 4.4.1. Under H_1, the variable $\int_0^T q(t)\hat{V}(t)\,dt$, is still Gaussian with variance as above, and with mean $\theta_1 \int_0^T q(t)\hat{s}(t)\,dt$. We thus obtain

$$G(\lambda, \theta) = \Pr\left\{\int_0^T q(t)\hat{V}(t)\,dt \geq \lambda \bigg| \theta\right\} = 1 - \Phi\left(\frac{\lambda - \theta \int_0^T q(t)\hat{s}(t)\,dt}{[\int_0^T q(t)\hat{s}(t)\,dt]^{1/2}}\right) \qquad (5.2.64)$$

where $\Phi(x)$ is the cumulative distribution of the zero mean, unit variance, Gaussian random variable at the point x, and where due to (5.2.61) it is easily concluded that $\int_0^T q(t)\hat{s}(t)\,dt > 0$.

From (5.2.64), it is easily concluded that for any λ the supremum of the function $G(\lambda, \theta)$ in $\mathscr{E}$ is attained at $\theta = \theta_0$. Thus, the conclusions of Lemma 5.2.1 apply, a most powerful rule $\hat{\delta}^*$ for the approximate problem indeed exists, and it is described by (5.2.63), where

$$\hat{\lambda}^*: 1 - \Phi\left(\frac{\hat{\lambda}^* - \theta_0 \int_0^T q(t)\hat{s}(t)\,dt}{\left[\int_0^T q(t)\hat{s}(t)\,dt\right]^{1/2}}\right) = \alpha$$

or

$$\hat{\lambda}^* = \left[\int_0^T q(t)\hat{s}(t)\,dt\right]^{1/2} \Phi^{-1}(1-\alpha) + \theta_0 \int_0^T q(t)\hat{s}(t)\,dt \qquad (5.2.65)$$

Let us denote by δ^* the decision rule that evolves from the rule in (5.2.63) if $\hat{v}(t)$ is substituted by the truly observed waveform $v(t)$ in $[0, T]$, where $\hat{\lambda}^*$ is given by (5.2.65). Then, the rule δ^* is most powerful for the present example in a mean-squared sense. Finally, the power $P_1(\delta^*)$ induced by the rule δ^* is given by

$$P_1(\delta^*) = 1 - \Phi\left(\frac{\hat{\lambda}^* - \theta_1 \int_0^T q(t)s(t)\,dt}{\left[\int_0^T q(t)s(t)\,dt\right]^{1/2}}\right) = \Phi\left(\frac{\theta_1 \int_0^T q(t)s(t)\,dt - \hat{\lambda}^*}{\left[\int_0^T q(t)s(t)\,dt\right]^{1/2}}\right)$$

$$= \Phi\left([\theta_1 - \theta_0]\left[\int_0^T q(t)s(t)\,dt\right]^{1/2} - \Phi^{-1}(1-\alpha)\right) \qquad (5.2.66)$$

Clearly, $P_1(\delta^*)$ is monotonically decreasing with decreasing difference $\theta_1 - \theta_0$, approaching the value α as $\theta_1 - \theta_0$ approaches zero. This is so, because as $\theta_1 - \theta_0 \to 0$, the hypotheses H_1 and H_0 become indistinguishable. If H_1 and H_0 are both generated by white Gaussian processes, then $R(t, \tau) = \sigma^2 \delta(t - \tau)$, and from (5.2.61) we obtain $q(t) = \sigma^{-2}\hat{s}(t);\ t \in [0, T]$. Then

$\delta^*(v(t); t \in [0, T])$

$$= \begin{cases} 1 & \int_0^T s(t)v(t)\,dt \geq \sigma\left[\int_0^T s^2(t)\,dt\right]^{1/2} \Phi^{-1}(1-\alpha) + \theta_0 \int_0^T s^2(t)\,dt \\ 0 & \int_0^T s(t)v(t)\,dt < \sigma\left[\int_0^T s^2(t)\,dt\right]^{1/2} \Phi^{-1}(1-\alpha) + \theta_0 \int_0^T s^2(t)\,dt \end{cases}$$

(5.2.67)

The power $P_1(\delta^*)$ is given then by

$$P_1(\delta^*) = \Phi\left([\theta_1 - \theta_0]\left[\frac{\int_0^T s^2(t)\,dt}{\sigma^2}\right]^{1/2} - \Phi^{-1}(1-\alpha)\right) \qquad (5.2.68)$$

5.3 Uniformly Most Powerful Tests

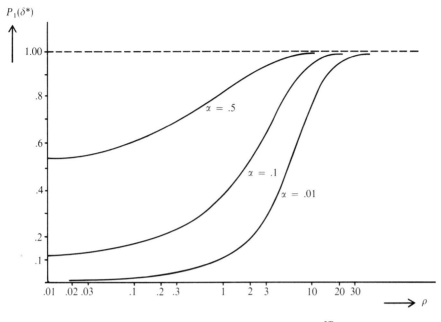

Figure 5.2.3 The power in (5.2.68) $\quad \rho = \sigma^{-2} \int_0^T s^2(t)\, dt$

From (5.2.68), it is clear that $P_1(\delta^*)$ increases monotonically with increasing signal-to-noise ratio $\sigma^{-2} \int_0^T s^2(t)\, dt$. In Figure 5.2.3, we plot the power in (5.2.68) against the signal-to-noise ratio.

With Example 5.2.6, we complete Section 5.2. In the next section, we will define and analyze uniformly most powerful tests.

5.3 UNIFORMLY MOST POWERFUL TESTS

As we saw in Section 5.2, in the Neyman-Pearson formalization, the emphatic hypothesis H_1 is always assumed well known or noncomposite. Then, the most powerful decision rule maximizes the power at this hypothesis, subject to a false alarm constraint. For unchanged H_0 and false alarm rate, the most powerful decision rule is, in general, different at different noncomposite hypotheses H_1. In mathematical terms, if H_1 is composite, a constraint optimization problem cannot be generally formalized. Under certain conditions, however, given some class $\mathscr{C}$ of stochastic processes, H_0, and the false alarm rate, there exists some decision rule δ^* that is most powerful at every member of $\mathscr{C}$. The decision rule δ^* is then called *uniformly most powerful in $\mathscr{C}$*. Without losing generality, we will consider hypotheses that are generated by discrete-time processes, and will adopt parametrically known hypotheses H_0 and H_1. We will thus define $\mathscr{C}$ via parameter spaces. We will then state conditions for the existence of uniformly most powerful decision rules or tests in a lemma.

Let the available observations be described by the n-dimensional column vector x^n. Let H_0 be composite, and described by the density functions $f_{\theta_0^m}(x^n)$;

$\theta_0^m \in \mathscr{E}_0^m$, where $\mathscr{E}_0^m$ is a given m-dimensional parameter space. Let H_1 also be composite, and described by the density functions $f_{\theta_1^m}(x^n)$; $\theta_1^m \in \mathscr{E}_1^m$, where $\mathscr{E}_1^m$ is a given n-dimensional parameter space, such that $\mathscr{E}_0^m \cap \mathscr{E}_1^m = \emptyset$. We then state the following lemma.

LEMMA 5.3.1: Let there exist some strictly monotonically increasing real and scalar function $f(\cdot)$, two scalar functions $g(\theta_1^m, \theta_0^m)$ and $c(\theta_1^m, \theta_0^m)$, and some scalar function $h(x^n)$ such that

$$f\left[\frac{f_{\theta_1^m}(x^n)}{f_{\theta_0^m}(x^n)}\right] = g(\theta_1^m, \theta_0^m)h(x^n) + c(\theta_1^m, \theta_0^m) \qquad \forall\, \theta_1^m \in \mathscr{E}_1^m, \forall\, \theta_0^m \in \mathscr{E}_0^m \quad (5.3.1)$$

where

$c(\theta_1^m, \theta_0^m)$ is not a function of x^n

$g(\theta_1^m, \theta_0^m)$ is not a function of x^n, and $g(\theta_1^m, \theta_0^m) > 0 \qquad \forall\, \theta_1^m \in \mathscr{E}_1^m, \forall\, \theta_0^m \in \mathscr{E}_0^m$

$h(x^n)$ is not a function of θ_1^m and θ_0^m

Let $G(\lambda, r, \theta_0^m)$, be as in (5.2.41), Lemma 5.2.1, where θ_0^m is substituted for θ^m. For arbitrary nonnegative scalars λ and r where $r \in [0, 1]$, let the supremum $\sup_{\theta_0^m \in \mathscr{E}_0^m} G(\lambda, r, \theta_0^m)$ be attained at some parameter value θ_0^{m*} that is independent of the λ and r values. Then, given a false alarm rate α there exists a decision rule δ^* that is uniformly most powerful in $\mathscr{E}_1^m$,

$$\delta^*(x^n) = \begin{cases} 1 & x^n: h(x^n) > \lambda^* \\ r^* & x^n: h(x^n) = \lambda^* \\ 0 & x^n: h(x^n) < \lambda^* \end{cases} \quad (5.3.2)$$

$$\lambda^*, r^*: G(\lambda^*, r^*, \theta_0^{m*}) = \alpha \quad (5.3.3)$$

Proof: Due to the conditions in the lemma, and due to Lemma 5.2.1, for any fixed parameter value θ_1^m in $\mathscr{E}_1^m$ the most powerful rule is given by (5.3.2) and (5.3.3). But the latter are independent of the particular parameter value θ_1^m. Thus, the decision rule in (5.3.2) and (5.3.3) is uniformly most powerful in $\mathscr{E}_1^m$. ∎

We note that the H_0 assumption in Lemma 5.3.1 is general. It includes both noncomposite and composite hypotheses H_0. We also note the similarities between the conditions in Lemma 5.3.1, and those in Lemma 5.2.1, and we point out that the uniformly most powerful decision rule in Lemma 5.3.1 induces, in general, a different power at different θ_1^m parameter values in $\mathscr{E}_1^m$. We now proceed with some examples.

Example 5.3.1

Let us revisit Example 5.2.4 in a modified form. Let the parameter space $\mathscr{E}_0$ of the hypothesis H_0 be $(-\infty, b_0)$ for some scalar b_0. Let the hypothesis H_1 be composite, and let its parameter space $\mathscr{E}_1$ be (b_1, ∞), where $b_1 > b_0$. Let us

5.3 Uniformly Most Powerful Tests

define,

$$f(x) = \ln(x)$$
$$g(\theta_1, \theta_0) = |\theta_1 - \theta_0| \qquad \theta_0 \in \mathscr{E}_0, \theta_1 \in \mathscr{E}_1$$
$$c(\theta_1, \theta_0) = -\frac{1}{2}(\theta_1^2 - \theta_0^2) s^{n^T} R_n^{-1} s^n \qquad \theta_0 \in \mathscr{E}_0, \theta_1 \in \mathscr{E}_1$$
$$h(x^n) = s^{n^T} R_n^{-1} x^n \operatorname{sgn}(\theta_1 - \theta_0) = s^{n^T} R_n^{-1} x^n \qquad \forall (\theta_1, \theta_0) \in \mathscr{E}_1 \times \mathscr{E}_0$$

Then, as in Example 5.2.4, we obtain

$$\ln[LR(\theta_1, \theta_0)] = g(\theta_1, \theta_0) h(x^n) + c(\theta_1, \theta_0)$$

where the functions $\ln(\cdot)$, $g(\cdot)$, $c(\cdot)$, and $h(\cdot)$ satisfy the conditions in Lemma 5.3.1.

It has already been found, in Example 5.2.4, that the function $G(\lambda, r, \theta_0)$ in (5.2.48), also satisfies the restrictions in Lemma 5.3.1. Thus, a decision rule δ^* that is uniformly most powerful in $\mathscr{E}_1$ exists, and it is given by (5.2.50). That is,

$$\delta^*(x^n) = \begin{cases} 1 & x^n: s^{n^T} R_n^{-1} x^n \geq \lambda^* \\ 0 & x^n: s^{n^T} R_n^{-1} x^n < \lambda^* \end{cases} \qquad (5.3.4)$$

where

$$\lambda^* = [s^{n^T} R_n^{-1} s^n]^{1/2} \Phi^{-1}(1 - \alpha) + b_0 s^{n^T} R_n^{-1} s^n \qquad (5.3.5)$$

Given θ_1 in $\mathscr{E}_1$, the power $P_{\theta_1}(\delta^*)$ induced by the rule in (5.3.4) and (5.3.5) at θ_1 is given by

$$P_{\theta_1}(\delta^*) = \Phi\left(\frac{-\lambda^* + \theta_1 s^{n^T} R_n^{-1} s^n}{[s^{n^T} R_n^{-1} s^n]^{1/2}}\right) = \Phi([\theta_1 - b_0][s^{n^T} R_n^{-1} s^n]^{1/2} - \Phi^{-1}(1 - \alpha)) \qquad (5.3.6)$$

From (5.3.6) we observe that $P_{\theta_1}(\delta^*)$ increases monotonically as θ_1 increases in $\mathscr{E}_1$.

Example 5.3.2

Let us revisit Example 5.2.5 in modified form. Let the parameter spaces $\mathscr{E}_0$ and $\mathscr{E}_1$ be as in Example 5.3.1, and let $\mathscr{E}_0$ and $\mathscr{E}_1$ be respectively the parameter spaces of hypotheses H_0 and H_1. Let us define

$$f(x) = \ln(x)$$
$$g(\theta_1, \theta_0) = \left|\ln \frac{\theta_1}{\theta_0}\right| \qquad \theta_1 \in \mathscr{E}_1, \theta_0 \in \mathscr{E}_0$$
$$c(\theta_1, \theta_0) = -n(\theta_1 - \theta_0) \qquad \theta_1 \in \mathscr{E}_1, \theta_0 \in \mathscr{E}_0$$
$$h(x^n) = \left[\sum_{i=1}^n x_i\right] \operatorname{sgn}(\theta_1 - \theta_0) = \sum_{i=1}^n x_i \qquad \forall (\theta_1, \theta_0) \in \mathscr{E}_1 \times \mathscr{E}_0.$$

where all the above functions satisfy the conditions in Lemma 5.3.1 and where

$$\ln[\mathrm{LR}(\theta_0, \theta_1)] = g(\theta_1, \theta_0)h(x^n) + c(\theta_1, \theta_0)$$

It has been found, in Example 5.2.5, that the function $G_1(m, r, \theta_0)$ in (5.2.55) satisfies the conditions in Lemma 5.3.1, and that its supremum is attained here at $\theta_0 = b_0$. Thus, a decision rule δ^* that is uniformly most powerful in $\mathscr{E}_1$ exists, and it is given by

$$\delta^*(x^n) = \begin{cases} 1 & x^n\colon \sum_{i=1}^{n} x_i > m^* \\ r^* & x^n\colon \sum_{i=1}^{n} x_i = m^* \\ 0 & x^n\colon \sum_{i=1}^{n} x_i < m^* \end{cases} \quad (5.3.7)$$

$$(m^*, r^*)\colon \sum_{k=m^*+1}^{\infty} \frac{e^{-nb_0}(nb_0)^k}{k!} + r^* \frac{e^{-nb_0}(nb_0)^{m^*}}{m^*!} = \alpha \quad (5.3.8)$$

Given θ_1 in $\mathscr{E}_1$, the power $P_{\theta_1}(\delta^*)$ induced by the rule in (5.3.7) and (5.3.8) at θ_1 is given by

$$P_{\theta_1}(\delta^*) = \sum_{k=m^*+1}^{\infty} \frac{e^{-n\theta_1}(n\theta_1)^k}{k!} + r^* \frac{e^{-n\theta_1}(n\theta_1)^{m^*}}{m^*!} \quad (5.3.9)$$

where the pair (m^*, r^*) is given by (5.3.8).

The power $P_{\theta_1}(\delta^*)$ in (5.3.9) is monotonically nondecreasing as the parameter θ_1 increases in $\mathscr{E}_1$.

Example 5.3.3

We revisit Example 5.2.6 with the modification $\mathscr{E}_0 = (-\infty, b_0)$ and $\mathscr{E}_1 = (b_1, \infty)$ with $b_1 > b_0$, where $\mathscr{E}_1$ is the parameter space of the now composite H_1. Defining

$$f(x) = \ln(x)$$

$$g(\theta_1, \theta_0) = |\theta_1 - \theta_0| \qquad \theta_1 \in \mathscr{E}_1, \theta_0 \in \mathscr{E}_0$$

$$c(\theta_1, \theta_0) = \frac{1}{2}(\theta_0^2 - \theta_1^2) \int_0^T q(t)s(t)\,dt \qquad \theta_1 \in \mathscr{E}_1, \theta_0 \in \mathscr{E}_0$$

$$h(\hat{v}(t); t \in [0, T]) = \mathrm{sgn}(\theta_1 - \theta_0) \int_0^T q(t)\hat{v}(t)\,dt$$

$$= \int_0^T q(t)\hat{v}(t)\,dt \qquad \forall\,(\theta_1, \theta_0) \in \mathscr{E}_1 \times \mathscr{E}_0 \qquad q(t)\colon s(t)$$

where, $\quad q(t)\colon s(t) = \int_0^T q(\tau)R(t, \tau)\,d\tau \qquad t \in [0, T]$

5.3 Uniformly Most Powerful Tests

$G(\lambda, \theta_0)$ as in (5.2.64), we have

$$\ln\left[\frac{f_{\theta_1}(\hat{v}(t); t \in [0, T])}{f_{\theta_0}(\hat{v}(t); t \in [0, T])}\right] = g(\theta_1, \theta_0)h(\hat{v}(t); t \in [0, T]) + c(\theta_1, \theta_0)$$

As found in Example 5.2.6, all the conditions of Lemma 5.3.1 are then satisfied for the waveform $\hat{v}(t); t \in [0, T]$. Thus, a decision rule $\hat{\delta}^*$ that is uniformly most powerful in $\mathscr{E}_1$ exists and it is given by (5.2.63), with $\hat{\lambda}^*$ as in (5.2.66), where in the latter equation, b_0 is substituted for θ_0. If in the test that determines the rule $\hat{\delta}^*$, the truly observed waveform $v(t)$ is substituted for $\hat{v}(t)$, the evolving rule δ^* is most powerful in a mean-squared sense. Given θ_1 in $\mathscr{E}_1$, the power $P_{\theta_1}(\delta^*)$ induced by δ^* at θ_1 is as in (5.2.66) if b_0 is substituted for θ_0. The power $P_{\theta_1}(\delta^*)$ clearly increases *monotonically* with increasing θ_1 in $\mathscr{E}_1$. Its minimum is attained at $\theta_1 = b_1$.

At this point, we will briefly discuss a particular modification of Lemma 5.3.1 that leads to the so called *uniformly most powerful two-sided* decision rules or tests. As we will see, the concept of two-sided tests evolves naturally from the concepts of the one-sided most powerful rules that we have already presented; thus, we will not discuss it extensively. Additional discussion can be found in Ferguson (1967). Consider the parameter space $\mathscr{E}_1^m$ in Lemma 5.3.1. Let $\mathscr{E}_{11}^m$ and $\mathscr{E}_{12}^m$ be two subspaces in $\mathscr{E}_1^m$ such that $\mathscr{E}_{11}^m \cap \mathscr{E}_{12}^m = \emptyset$ and $\mathscr{E}_{11}^m \cup \mathscr{E}_{12}^m = \mathscr{E}_1^m$. Let the function $g(\theta_1^m, \theta_2^m)$ in Lemma 5.3.1 be positive $\forall\, \theta_1^m \in \mathscr{E}_{11}^m$, $\forall\, \theta_0^m \in \mathscr{E}_0^m$, and let it be negative $\forall\, \theta_1^m \in \mathscr{E}_{12}^m$, $\forall\, \theta_0^m \in \mathscr{E}_0^m$. Then, the conditions in Lemma 5.3.1 are not strictly satisfied for the composite hypothesis H_1 that is determined via the parameter space $\mathscr{E}_1^m$. Let us now consider the two disjoint hypotheses H_{11} and H_{12} that are determined via the parameter spaces $\mathscr{E}_{11}^m$ and $\mathscr{E}_{12}^m$, respectively, and whose union is hypothesis H_1. Let us consider two separate hypothesis testing problems: one determined by H_{11} tested against H_0, and one determined by H_{12} tested against H_0. Let us then define

$$G_1(\lambda, r, \theta_0^m) = \int_{x^n:\, h(x^n) > \lambda} dx^n f_{\theta_0^m}(x^n) + r \int_{x^n:\, h(x^n) = \lambda} dx^n f_{\theta_0^m}(x^n) \qquad \theta_0^m \in \mathscr{E}_0^m \qquad (5.3.10)$$

$$G_2(\lambda, r, \theta_0^m) = \int_{x^n:\, -h(x^n) > \lambda} dx^n f_{\theta_0^m}(x^n) + r \int_{x^n:\, -h(x^n) = \lambda} dx^n f_{\theta_0^m}(x^n) \qquad \theta_0^m \in \mathscr{E}_0^m \qquad (5.3.11)$$

Let the functions $G_1(\lambda, r, \theta_0^m)$ and $G_2(\lambda, r, \theta_0^m)$ have suprema in $\mathscr{E}_0^m$ that are respectively attained at the vector points θ_{01}^{m*} and θ_{02}^{m*}, where θ_{01}^{m*} and θ_{02}^{m*} are both independent of the pair (λ, r) $\forall\, \lambda > 0$, $\forall\, r \in [0, 1]$. Then we easily conclude that Lemma 5.3.1 is satisfied for both the H_{11} against H_0 and H_{12} against H_0 decision problems. Thus, a decision rule δ_1^* that is uniformly most powerful in $\mathscr{E}_{11}^m$ and a decision rule δ_2^*, that is uniformly most powerful in $\mathscr{E}_{12}^m$ exist, and they are given by

$$\delta_1^*(x^n) = \begin{cases} 1 & x^n: h(x^n) > \lambda_1^* \\ r_1^* & x^n: h(x^n) = \lambda_1^* \\ 0 & x^n: h(x^n) < \lambda_1^* \end{cases} \qquad (5.3.12)$$

$$\lambda_1^*, r_1^*: G_1(\lambda_1^*, r_1^*, \theta_{01}^{m*}) = \alpha \qquad (5.3.13)$$

$$\delta_2^*(x^n) = \begin{cases} 1 & x^n: h(x^n) < \lambda_2^* \\ r_2^* & x^n: h(x^n) = \lambda_2^* \\ 0 & x^n: h(x^n) > \lambda_2^* \end{cases} \qquad (5.3.14)$$

$$\lambda_2^*, r_2^*: G_2(-\lambda_2^*, r_2^*, \theta_{02}^{m*}) = \alpha \qquad (5.3.15)$$

where the functions $G_1(\lambda, r, \theta_0^m)$ and $G_2(\lambda, r, \theta_0^m)$ are given respectively by (5.3.10) and (5.3.11).

Now let the characteristics of the overall problem be such that

$$\begin{aligned} r_1^* &= r_2^* = r^* \\ \lambda_2^* &< \lambda_1^* \end{aligned} \qquad (5.3.16)$$

Then we easily conclude that a most powerful decision rule δ^* exists, in testing the overall composite hypothesis H_1 against the composite hypothesis H_0. For λ_1^* as in (5.3.12) and (5.3.13), for λ_2^* as in (5.3.14) and (5.3.15), and for r^* as in (5.3.16), the rule δ^* is given by

$$\delta^*(x^n) = \begin{cases} 1 & x^n: \text{either} \quad h(x^n) < \lambda_2^* \quad \text{or} \quad h(x^n) > \lambda_1^* \\ r^* & x^n: \text{either} \quad h(x^n) = \lambda_1^* \quad \text{or} \quad h(x^n) = \lambda_2^* \\ 0 & x^n: \lambda_2^* < h(x^n) < \lambda_1^* \end{cases} \qquad (5.3.17)$$

The rule in (5.3.17) is called two-sided, because the test function $h(x^n)$ is tested against two thresholds λ_1^* and λ_2^*. We will now present an example of uniformly most powerful two-sided decision rules.

Example 5.3.4

Let us revisit Example 5.3.1 in a modified form. Let H_0 be specified by a Gaussian, discrete-time process whose autocovariance matrix is R_n, and whose mean is $\theta_0 s^n$, where $\theta_0 \in \mathscr{E}_0 = (b_0, b_1)$; $b_1 > b_0$. Let the hypothesis H_1 be specified by a Gaussian, discrete-time process whose autocovariance matrix is R_n and whose mean is $\theta_1 s^n$, where $\theta_1 \in \mathscr{E}_1 = (-\infty, c_0) \cup (c_1, \infty)$, where $c_0 \leq b_0$ and $c_1 \geq b_1$. Let us define the functions $f(\cdot)$, $g(\cdot)$, and $c(\cdot)$ as in Example 5.3.1. Let $h(x^n) = s^{n^T} R_n^{-1} x^n$, and let $\mathscr{E}_{11} = (c_1, \infty)$, $\mathscr{E}_{12} = (-\infty, c_0)$. Then we obtain

$$\ln[\text{LR}(\theta_1, \theta_0)] = \begin{cases} g(\theta_1, \theta_0)h(x^n) + c(\theta_1, \theta_0) & \theta_1 \in \mathscr{E}_{11}, \theta_0 \in \mathscr{E}_0 \\ -g(\theta_1, \theta_0)h(x^n) + c(\theta_1, \theta_0) & \theta_1 \in \mathscr{E}_{12}, \theta_0 \in \mathscr{E}_0 \end{cases} \qquad (5.3.18)$$

where $g(\theta_1, \theta_0) > 0 \; \forall \; \theta_1 \in \mathscr{E}_1, \; \forall \; \theta_0 \in \mathscr{E}_0$

5.3 Uniformly Most Powerful Tests

Due to (5.3.18), the problem clearly is initially subdivided into two decision rules δ_1^* and δ_2^* that are most powerful in $\mathscr{E}_1$ and $\mathscr{E}_0$, as in (5.3.12) and (5.3.14). The scalars $\lambda_1^*, r_1^*, \lambda_2^*$, and r_2^* that specify those two rules are given respectively by the functions $G_1(\lambda, r, \theta_0)$ and $G_2(-\lambda, r, \theta_1)$, where from (5.2.48) in Example 5.2.4, we have

$$G_1(\lambda, r, \theta_0) = \Phi\left(\frac{-\lambda + \theta_0 s^{n^T} R_n^{-1} s^n}{[s^{n^T} R_n^{-1} s^n]^{1/2}}\right) \qquad \theta_0 \in \mathscr{E}_0 = (b_0, b_1) \quad (5.3.19)$$

$$G_2(-\lambda, r, \theta_0) = \Phi\left(\frac{\lambda - \theta_0 s^{n^T} R_n^{-1} s^n}{[s^{n^T} R_n^{-1} s^n]^{1/2}}\right) \qquad \theta_0 \in \mathscr{E}_0 = (b_0, b_1) \quad (5.3.20)$$

For any $\lambda > 0$, the supremum $\sup_{\theta_0 \in \mathscr{E}_0} G_1(\lambda, r, \theta_0)$ is attained at $\theta_0 = b_1$, and the supremum $\sup_{\theta_0 \in \mathscr{E}_0} G_2(-\lambda, r, \theta_0)$ is attained at $\theta_0 = b_0$. Then, for a given false alarm rate α we get

$$\lambda_1^*: G_1(\lambda_1^*, r, b_1) = \alpha \rightarrow \lambda_1^* = b_1 s^{n^T} R_n^{-1} s^n - [s^{n^T} R_n^{-1} s^n]^{1/2} \Phi^{-1}(\alpha)$$
$$r_1^* = 0 \quad (5.3.21)$$

$$\lambda_2^*: G_2(-\lambda_2^*, r, b_0) = \alpha \rightarrow \lambda_2^* = b_0 s^{n^T} R_n^{-1} s^n + [s^{n^T} R_n^{-1} s^n]^{1/2} \Phi^{-1}(\alpha)$$
$$r_2^* = 0 \quad (5.3.22)$$

Let us now assume that

$$\alpha < \Phi\left([b_1 - b_0]\frac{[s^{n^T} R_n^{-1} s^n]^{1/2}}{2}\right) \quad (5.3.23)$$

where b_0 and b_1 are the limits of the parameter set $\mathscr{E}_0$.

If (5.3.23) is satisfied, then it is easily verified that $\lambda_1^* > \lambda_2^*$, where also $r_1^* = r_2^* = 0$. Then a uniformly most powerful, two-sided decision rule δ^* for the testing H_1 against H_0 is given by

$$\delta^*(x^n) = \begin{cases} 1 & x^n: \text{either } s^{n^T} R_n^{-1} x^n \leq \lambda_2^* \text{ or } s^{n^T} R_n^{-1} x^n \geq \lambda_1^* \\ 0 & x^n: \lambda_2^* < s^{n^T} R_n^{-1} x^n < \lambda_1^* \end{cases} \quad (5.3.24)$$

where λ_1^* and λ_2^* are given by (5.3.21) and (5.3.22).

Given θ_1 in $\mathscr{E}_1$, the power $P_{\theta_1}(\delta^*)$ induced by the rule in (5.3.24) at θ_1 is given by

$$P_{\theta_1}(\delta^*) = \Pr\{s^{n^T} R_n^{-1} x^n \leq \lambda_2^* | \theta_1\} + \Pr\{s^{n^T} R_n^{-1} x^n \geq \lambda_1^* | \theta_1\}$$

$$= \Phi\left(\frac{\lambda_2^* - \theta_1 s^{n^T} R_n^{-1} s^n}{[s^{n^T} R_n^{-1} s^n]^{1/2}}\right) + \Phi\left(\frac{-\lambda_1^* + \theta_1 s^{n^T} R_n^{-1} s^n}{[s^{n^T} R_n^{-1} s^n]^{1/2}}\right)$$

$$= \Phi[(b_0 - \theta_1)(s^{n^T} R_n^{-1} s^n)^{1/2} + \Phi^{-1}(\alpha)]$$
$$+ \Phi[(\theta_1 - b_1)(s^{n^T} R_n^{-1} s^n)^{1/2} + \Phi^{-1}(\alpha)]$$

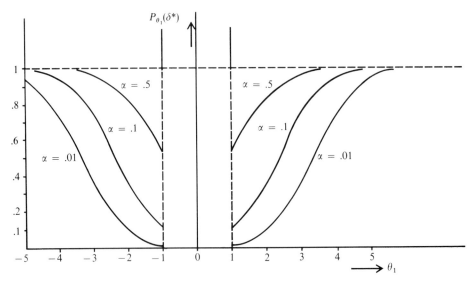

Figure 5.3.1 Power function

$$x = [s^{n^T} R_n^{-1} s^n]^{1/2} \qquad z_0: z_0(z_0 - w)^{-1} \exp[2^{-1} w(w - 2z_0)] = 1$$
$$w = -(b_1 - b_0)x + 2\Phi^{-1}(\alpha)$$

In Figure 5.3.1, we plot the power $P_{\theta_1}(\delta^*)$ against θ_1. If (5.3.23) is not satisfied, then $\lambda_2^* \geq \lambda_1^*$. Then, again, a uniformly most powerful, two-sided decision rule δ'^* for testing H_1 against H_0, exists; it is the image of the decision rule in (5.3.24); and it is given [for λ_1^* and λ_2^* as in (5.3.21) and (5.3.22)] by

$$\delta'^*(x^n) = \begin{cases} 1 & x^n: \lambda_1^* \leq s^{n^T} R_n^{-1} x^n \leq \lambda_2^* \\ 0 & x^n: \text{either } s^{n^T} R_n^{-1} x^n > \lambda_2^* \quad \text{or} \quad s^{n^T} R_n^{-1} x^n < \lambda_1^* \end{cases}$$

We conclude this section by pointing out that several variations of uniformly most powerful tests can be developed, if the constraints change. One such variation evolves when, instead of a false alarm rate, a lower bound α is imposed on the test power. Then, the *uniformly most powerful unbiased tests* evolve; they are discussed in Ferguson (1967). All the variations of the uniformly most powerful tests can be derived in a relatively straightforward fashion, via the formalizations and analyses presented in this section, and they do not present major conceptual differences. Each specific problem, however, has its own peculiarities that are induced by the nature of the hypotheses and the specifics of the imposed constraints.

5.4 LOCALLY MOST POWERFUL TESTS

As we saw in Section 5.3, when the H_1 hypothesis is composite, uniformly most powerful in H_1 decision rules may not exist. In such cases, a compromise is necessary, which translates to the adoption of a performance criterion weaker

5.4 Locally Most Powerful Tests

than the maximization of the power. Such a criterion may be the maximization of the slope of the power function at some selected point in H_1. To quantify our presentation in a clear fashion, let us assume that both H_0 and H_1 are generated by a discrete-time, parametrically known stochastic process whose n-dimensional density function at the vector point x^n and at the parameter point θ is $f_\theta(x^n)$, where θ is a scalar parameter. Let H_0 be determined by θ, taking values in some known space $\mathscr{E}_0$. Let H_1 be determined by $\theta \in \mathscr{E}_1$, where $\mathscr{E}_0 \cap \mathscr{E}_1 = \emptyset$. Given some decision rule δ let $P_\theta(\delta)$; $\theta \in \mathscr{E}_0 \cup \mathscr{E}_1$ denote the power function induced by the rule δ, as in Chapter 3. Given a false alarm rate α and some θ_1 in $\mathscr{E}_1$, let the constraint optimization problem in (5.2.1) have a solution $\delta_{\theta_1}^*$. However, let the most powerful decision rules $\delta_{\theta_1}^*$ be different for different values of θ_1 in $\mathscr{E}_1$. Then, there is no decision rule that is uniformly most powerful in $\mathscr{E}_1$. In this case, one choice would be to compromise our ambitions by being satisfied with the most powerful $\delta_{\theta_1}^*$ at some selected θ_1 in $\mathscr{E}_1$. A reduced compromise would be represented by the selection of a decision rule that is most powerful within a small parameter neighborhood around the selected value θ_1. The latter objective is closely represented by the maximization of the slope $(\partial/\partial\theta)P_\theta(\delta)|_{\theta=\theta_1}$ of $P_\theta(\delta)$ at θ_1. Then it is essentially required that $P_\theta(\delta)$ increase sharply for θ values close to θ_1 and such that $\theta > \theta_1$, and the following constraint optimization problem arises:

Maximize, at $\theta = \theta_1$, the slope,

$$\frac{\partial}{\partial \theta} P_\theta(\delta) = \frac{\partial}{\partial \theta} \int_{\Gamma^n} \delta(x^n) f_\theta(x^n) \, dx^n$$

subject to

$$0 \le \delta(x^n) \le 1 \qquad \forall \, x^n \in \Gamma^n$$

$$\sup_{\theta \in \mathscr{E}_0} P_\theta(\delta) = \sup_{\theta \in \mathscr{E}_0} \int_{\Gamma^n} dx^n \delta(x^n) f_\theta(x^n) \le \alpha$$

(5.4.1)

As with the optimization problem in (5.2.1), the optimization problem in (5.4.1) may not have a solution, unless some conditions parallel to those in Lemma 5.2.1 are satisfied. Since such conditions have been extensively discussed in Sections 5.2 and 5.3, here we will only consider the simpler problem that arises when H_0 is noncomposite and determined by a fixed θ_0. Then, (5.4.1) reduces to

Maximize, at $\theta = \theta_1$, the slope

$$\frac{\partial}{\partial \theta} \int_{\Gamma^n} \delta(x^n) f_\theta(x^n) \, dx^n$$

subject to

$$0 \le \delta(x^n) \le 1 \qquad \forall \, x^n \in \Gamma^n$$

$$\int_{\Gamma^n} dx^n \delta(x^n) f_{\theta_0}(x^n) \le \alpha$$

(5.4.2)

The optimization problem in (5.4.2) has a solution only if $(\partial/\partial\theta)\int_{A^n} f_\theta(x^n)\,dx^n = \int_{A^n}(\partial/\partial\theta)\,dx^n f_\theta(x^n)\ \forall\ A^n \subset \Gamma^n$. If this condition is not satisfied, the problem in (5.4.2) is compromised by substituting $\int_{\Gamma^n}\delta(x^n)(\partial/\partial\theta)f_\theta(x^n)\,dx^n$ for the objective function. Therefore, the final form of the optimization problem in (5.4.2) is

Maximize, at $\theta = \theta_1$, the objective function

$$\int_{\Gamma^n}\delta(x^n)\frac{\partial}{\partial\theta}f_\theta(x^n)\,dx^n$$

subject to

$$0 \leq \delta(x^n) \leq 1 \quad \forall\ x^n \in \Gamma^n$$

$$\int_{\Gamma^n} dx^n\delta(x^n)f_{\theta_0}(x^n) \leq \alpha$$

(5.4.3)

We now state the general solution of (5.4.3) in a theorem.

THEOREM 5.4.1: Let the derivative $(\partial/\partial\theta)f_\theta(x^n)$ exist everywhere in $\mathscr{E}_1$. Define δ^* as follows

$$\delta^*(x^n) = \begin{cases} 1 & x^n: \dfrac{\left.\frac{\partial}{\partial\theta}f_\theta(x^n)\right|_{\theta=\theta_1}}{f_{\theta_0}(x^n)} > \lambda^* \\[2ex] r^* & x^n: \dfrac{\left.\frac{\partial}{\partial\theta}f_\theta(x^n)\right|_{\theta=\theta_1}}{f_{\theta_0}(x^n)} = \lambda^* \\[2ex] 0 & x^n: \dfrac{\left.\frac{\partial}{\partial\theta}f_\theta(x^n)\right|_{\theta=\theta_1}}{f_{\theta_0}(x^n)} < \lambda^* \end{cases}$$

where λ^* and r^* are such that

$$\int f_{\theta_0}(x^n)\,dx^n \quad + \quad r^* \int f_{\theta_0}(x^n)\,dx^n = \alpha$$

$$x^n: \dfrac{\left.\frac{\partial}{\partial\theta}f_\theta(x^n)\right|_{\theta=\theta_1}}{f_{\theta_0}(x^n)} > \lambda^* \qquad x^n: \dfrac{\left.\frac{\partial}{\partial\theta}f_\theta(x^n)\right|_{\theta=\theta_1}}{f_{\theta_0}(x^n)} = \lambda^*$$

Then δ^* is a solution of (5.4.3), and is called *locally most powerful* at θ_1. ∎

The proof is exactly as for Theorem 5.2.1, if in the latter, $(\partial/\partial\theta)f_\theta(x^n)|_{\theta=\theta_1}$ is substituted for $f_1(x^n)$. The uniqueness or nonuniqueness of the rule δ^* in the theorem, and the necessity or obsoleteness of randomization are exactly as discussed in Section 5.2 for the decision rule in Theorem 5.2.1. If the false alarm constraint in (5.4.3) is replaced by a power constraint that is represented by a

5.4 Locally Most Powerful Tests

lower bound on the power function $P_\theta(\delta)$ at some parameter point θ_2 in $\mathscr{E}_1$, then the so-called *unbiased locally most powerful decision rules* evolve. The solution of the latter optimization problem is easily found; it has a form similar to that in Theorem 5.4.1 [for more on that see Ferguson (1967)]. We will conclude this section with an example.

Example 5.4.1

Let us consider the following two hypotheses. The hypothesis H_0 is described by n independent and identically distributed random variables whose distribution is Cauchy, with parameter $\theta_0 \in (-\infty, 0)$. The hypothesis H_1 is as above, where the Cauchy parameter θ_1 now takes values in (b, ∞); $b > 0$. Given the two composite hypotheses H_0 and H_1 as above, it can easily be seen that the conditions in Lemma 5.3.1 are *not* satisfied. In fact, this is true even when H_1 is noncomposite and specified by the Cauchy distribution at some point θ_1 in (b, ∞). Let us assume that a first compromise can be reached, by setting H_0 at $\theta_0 = 0$. The resulting H_0 hypothesis $\hat{H}_0$ is then noncomposite, and is described by $f_0(x^n) = \prod_{i=1}^{n} \pi^{-1}[1 + x_i^2]$. Considering H_1, let us assume that the maximization of the slope of the power function at $\theta_1 = b$ is satisfactory. Transforming then the optimization problem into a form as in (5.4.3) for $\theta_1 = b$, and due to Theorem 5.4.1, we find that a decision rule δ^* that is locally most powerful at $\theta_1 = b$ is given by

$$\delta^*(x^n) = \begin{cases} 1 & x^n: f(x^n, b) > \lambda^* \\ r^* & x^n: f(x^n, b) = \lambda^* \\ 0 & x^n: f(x^n, b) < \lambda^* \end{cases} \quad (5.4.4)$$

where

$$f(x^n, b) = \frac{\dfrac{\partial}{\partial \theta} \prod_{i=1}^{n} \pi^{-1}[1 + (x_i - \theta)^2]^{-1} \Big|_{\theta=b}}{\prod_{i=1}^{n} \pi^{-1}[1 + x_i^2]^{-1}} \quad (5.4.5)$$

$$(\lambda^*, r^*): \Pr\{f(X^n, b) > \lambda^* | \hat{H}_0\} + r^* \Pr\{f(X^n, b) = \lambda^* | \hat{H}_0\} = \alpha \quad (5.4.6)$$

It can easily be found that the following equation holds.

$$\frac{\partial}{\partial \theta} \prod_{i=1}^{n} \pi^{-1}[1 + (x_i - \theta)^2]^{-1} \Big|_{\theta=b}$$
$$= 2 \left\{ \prod_{i=1}^{n} \pi^{-1}[1 + (x_i - b)^2]^{-1} \right\} \left\{ \sum_{i=1}^{n} (x_i - b)[1 + (x_i - b)^2]^{-1} \right\} \quad (5.4.7)$$

From (5.4.5) and (5.4.7), we thus obtain

$$f(x^n, b) = 2 \left\{ \prod_{i=1}^{n} \frac{1 + x_i^2}{1 + (x_i - b)^2} \right\} \cdot \sum_{i=1}^{n} \frac{x_i - b}{1 + (x_i - b)^2} \quad (5.4.8)$$

Under $\hat{H}_0$ the density function of the random variable $f(X^n, b)$ clearly exists everywhere in R^n and is continuous. Thus, we can select $r^* = 0$. For the evaluation of the scalar λ^* from (5.4.6), however, numerical methods are necessary. Now let $b = 0$. Then, from (5.4.8) we have

$$f(x^n, 0) = 2 \sum_{i=1}^{n} \frac{x_i}{1 + x_i^2} \qquad (5.4.9)$$

Under $\hat{H}_0$, the random variables $Y_i = X_i/(1 + X_i^2)$; $1 \le i \le n$ are independent and identically distributed. Thus, for asymptotically large n values, the Gaussian approximation can be used for the random variable $f(X^n, 0)$. If we denote $m \triangleq E\{Y_i | \hat{H}_0\}$ and $\sigma^2 \triangleq E\{Y_i^2 | \hat{H}_0\} - m^2$, the mean m_n and the variance σ_n^2 of the Gaussian approximation are then given by

$$m_n = 2nm$$
$$\sigma_n^2 = 4n\sigma^2 \qquad (5.4.10)$$

We also easily obtain, via trigonometric transformations,

$$m = E\{Y_i | \hat{H}_0\} = \frac{1}{\pi} \int_{-\infty}^{\infty} \frac{u}{(1 + u^2)^2} \, du = 0$$

$$\sigma^2 = E\{Y_i^2 | \hat{H}_0\} - m^2 = E\{Y_i^2 | \hat{H}_0\} = \frac{1}{\pi} \int_{-\infty}^{\infty} \frac{u^2}{(1 + u^2)^3} \, du = \frac{1}{8} \qquad (5.4.11)$$

From (5.4.10) and (5.4.11) we finally obtain

$$m_n = 0$$
$$\sigma_n^2 = 2^{-1} n \qquad (5.4.12)$$

Due to (5.4.6) and (5.4.12) and the asymptotic Gaussian approximation, we finally conclude that for asymptotically large n the threshold λ^* is found by

$$1 - \Phi(\lambda^* 2^{1/2} \cdot n^{-1/2}) = \alpha \rightarrow \lambda^* = 2^{-1/2} \cdot n^{1/2} \cdot \Phi^{-1}(1 - \alpha) \qquad (5.4.13)$$

where $\Phi(x)$ is the cumulative distribution of the zero mean, unit variance, Gaussian random variable, and $w = \Phi^{-1}(x)$: $x = \Phi(w)$.

Now let θ_1 be some parameter value in $(0, \infty)$, let us assume asymptotically large n, and let us denote

$$m_{\theta_1} = E\{Y_i | \theta_1\} = \frac{1}{\pi} \int_{-\infty}^{\infty} \frac{u}{1 + u^2} \cdot \frac{1}{1 + (u - \theta_1)^2} \, du$$

$$m_{\theta_1}^2 + \sigma_{\theta_1}^2 = E\{Y_i^2 | \theta_1\} = \frac{1}{\pi} \int_{-\infty}^{\infty} \frac{u^2}{[1 + u^2]^2} \cdot \frac{1}{1 + (u - \theta_1)^2} \, du \qquad (5.4.14)$$

Then the power $P_{\theta_1}(\delta^*)$ induced by the asymptotic rule in (5.4.4) and (5.4.13) for $b = 0$ is given by

$$P_{\theta_1}(\delta^*) = 1 - \Phi\left(\frac{2^{-3/2}\Phi^{-1}(1-\alpha) - n^{1/2}m_{\theta_1}}{\sigma_{\theta_1}}\right) \qquad (5.4.15)$$

Now let $\theta_1 \to \infty$. Then, $2^{-3/2}\Phi^{-1}(1-\alpha) - n^{1/2}m_{\theta_1} \to 2^{-3/2}\Phi^{-1}(1-\alpha)$, and $\sigma_{\theta_1} \to 0$. At the same time, if $\alpha < .5$, then $\Phi^{-1}(1-\alpha) > 0$. Thus, for $\alpha < .5$ and $\theta_1 \to \infty$ we have $[2^{-3/2}\Phi^{-1}(1-\alpha) - n^{1/2}m_{\theta_1}]\sigma_{\theta_1}^{-1} \to \infty$, and therefore $P_{\theta_1}(\delta^*) \to 0$. The conclusion is that, for a false alarm rate less than .5, the asymptotic rule δ^* induces a power function $P_{\theta_1}(\delta^*)$ whose value tends to zero as θ_1 tends to infinity. That is, the asymptotic rule is useless when H_0 and H_1 are far apart. This is a very serious drawback to the rule. We are reminded, however, that no uniformly most powerful in $(0, \infty)$ decision rule exists for the problem at hand, and that the decision rule δ^* is locally most powerful at $\theta_1 = 0$, thus providing good protection against small positive deviations from the parameter value zero.

5.5 EXAMPLES AND APPLICATIONS

The applications of the Neyman-Pearson tests are numerous. Such tests are especially important in the design of high-performance communication systems. To discuss all possible applications is an impossible task. We are hoping that after studying this chapter, the reader when confronted with appropriate real problems, will recognize the applicability of the Neyman-Pearson tests. In this section, we will present just two additional examples, which represent radar applications.

Example 5.5.1

Let $s(t)$ be a deterministic, real, scalar, and continuous-time signal. Within a preset time interval $[0, T]$, let it be possible that a particular transformation of the signal may be transmitted through the atmosphere. Let the atmospheric disturbances be modeled as additive, Gaussian, zero mean, white, at level σ^2 noise, whose generated stochastic waveforms are denoted $N(t)$. Let the pretransmission transformation of the signal consist of amplitude modulation and random time delay. That is, if transmitted in $[0, T]$, the signal $s(t)$ is first transformed to the form $\theta s(t - \tau)$, where θ is the modulation amplitude and τ is the time delay. Let the delay τ be random, and uniformly distributed within some interval $[0, b]$. Let the objective be the design of a radar, for the detection of the signal presence. Let the statistics of the time delay τ and of the atmospheric disturbances be included in the design of the operation performed by the radar. In the design of the same operation, let no a priori probability of the signal presence be included, let the modulation amplitude θ be initially included, and let a false alarm rate α be required. Then, a Neyman-Pearson formalization evolves. In this formalization, H_0 is represented by the zero mean, white, and Gaussian waveform $N(t)$; $t \in [0, T]$ whose realizations are denoted $n(t)$; $t \in [0, T]$. Under fixed delay value τ H_1 is represented by a Gaussian and white waveform $\theta s(t - \tau) + N(t)$; $t \in [0, T]$ whose mean is $\theta s(t - \tau)$; $t \in [0, T]$. Let $s(t)$ be square integrable on $(-\infty, \infty)$, and let $\{\phi_i(t)\}$ be an orthonormal set complete in $[0, T]$, for the class of square integrable deterministic functions. Then, as established in Chapter 4, $\{\phi_i(t)\}$ also provides a Karhunen-Loéve expansion in $[0, T]$ for the white atmospheric noise. As in Section 4.4, we now write

$$\hat{N}(t) = \sum_i \hat{N}_i \phi_i(t) \qquad t \in [0, T] \qquad \hat{N}_i = \int_0^T N(t) \phi_i^*(t)\, dt$$

Given

$$\tau: \hat{s}(t - \tau) = \sum_i s_i(\tau) \phi_i(t) \qquad s_i(\tau) = \int_0^T s(t - \tau) \phi_i^*(t)\, dt$$

where the random variables $\{\hat{N}_i\}$ are Gaussian, zero mean, variance σ^2 and independent.

Let us denote by $\hat{X}(t)$; $t \in [0, T]$ the mean-squared approximations of the stochastic waveforms actually observed, let us denote by $\hat{x}(t)$; $t \in [0, T]$ the realizations of the stochastic waveform, $\hat{X}(t)$, and let us denote by $\hat{n}_i$ a realization of the random variable $\hat{N}_i$ in (5.5.1). Let $f_{\theta,\tau}(\cdot)$ denote the density function under H_1, and under the condition that the signal delay equals τ. Let $f_0(\cdot)$ denote the density function under H_0. Then we obtain

$$f_{\theta,\tau}(\{\hat{x}_i\}) = \prod_i (2\pi)^{-1/2} \sigma^{-1} \exp\left\{-\frac{[\hat{x}_i - \theta s_i(\tau)]^2}{2\sigma^2}\right\} \qquad (5.5.2)$$

where $\hat{X}_i = \int_0^T X(t) \phi_i^*(t)\, dt$ and $\hat{x}_i$ is a realization of $\hat{X}_i$, and

$$f_0(\{\hat{x}_i\}) = \prod_i (2\pi)^{-1/2} \sigma^{-1} \exp\left\{-\frac{\hat{x}_i^2}{2\sigma^2}\right\} \qquad (5.5.3)$$

where $\prod$ signifies product.

Now let $f_\theta(\cdot)$ denote a density function under H_1 via the incorporation of all possible delay values. Then it is clear that

$$f_\theta(\{\hat{x}_i\}) = \frac{1}{b} \int_0^b f_{\theta,\tau}(\{\hat{x}_i\})\, d\tau \qquad (5.5.4)$$

Substituting (5.5.2) in (5.5.4) gives

$$f_\theta(\{\hat{x}_i\}) = \frac{1}{b} \prod_i \{(2\pi)^{-1/2} \sigma^{-1}\} \int_0^b \exp\left\{-\frac{1}{2\sigma^2} \sum_i [\hat{x}_i - \theta s_i(\tau)]^2\right\} d\tau \qquad (5.5.5)$$

From (5.5.3) and (5.5.5) we obtain the likelihood ratio

$$\frac{f_\theta(\{\hat{x}_i\})}{f_0(\{\hat{x}_i\})} = b^{-1} \int_0^b \exp\left\{\frac{\theta}{\sigma^2} \sum_i s_i(\tau) \hat{x}_i - \frac{\theta^2}{2\sigma^2} \sum_i s_i^2(\tau)\right\} d\tau \qquad (5.5.6)$$

5.5 Examples and Applications

Replacing the quantities $s_i(\tau)$ and $\hat{x}_i$ in the integrand of (5.5.6) by their integral definitions, we finally obtain

$$\frac{f_\theta(\hat{x}(t); t \in [0, T])}{f_0(\hat{x}(t); t \in [0, T])} = b^{-1} \int_0^b \exp\left\{\frac{\theta}{\sigma^2} \int_0^T \hat{s}(t - \tau)\hat{x}(t)\,dt - \frac{\theta^2}{2\sigma^2} \int_0^T \hat{s}^2(t - \tau)\,dt\right\} d\tau \tag{5.5.7}$$

Replacing $\hat{x}(t)$ and $\hat{s}(t - \tau)$ in (5.5.7) by, respectively, $x(t)$ and $s(t - \tau)$, we obtain a mean-squared approximation $\hat{\delta}_\theta^*$ of a most powerful at θ decision rule.

$$\hat{\delta}_\theta^*(x(t); t \in [0, T]) = \begin{cases} 1 & x(t); t \in [0, T]: f_\theta(x(t); t \in [0, T]) \geq \lambda^* \\ 0 & x(t); t \in [0, T]: f_\theta(x(t); t \in [0, T]) < \lambda^* \end{cases} \tag{5.5.8}$$

where

$$f_\theta(x(t); t \in [0, T]) = \int_0^b \exp\left\{\frac{\theta}{\sigma^2} \int_0^T s(t - \tau)x(t)\,dt - \frac{\theta^2}{2\sigma^2} \int_0^T s^2(t - \tau)\,dt\right\} d\tau \tag{5.5.9}$$

$$\lambda^*: \Pr\{f_\theta(X(t); t \in [0, T]) \geq \lambda^* | H_0\} = \Pr\{f_\theta(N(t); t \in [0, T]) \geq \lambda^*\} = \alpha \tag{5.5.10}$$

For arbitrary signal waveform $s(t)$ the analytical study of the rule in (5.5.8) is clearly intractable. In addition, it is then clear that the conditions in Lemma 5.3.1 are not, in general, satisfied. Thus, if the amplitude θ takes values within a whole interval, a uniformly most powerful decision rule does not, in general, exist. We may thus search for the locally most powerful decision rule $\hat{\delta}_0^*$ at $\theta = 0$. We easily obtain

$$\frac{\frac{\partial}{\partial \theta} f_\theta(x(t); t \in [0, T])\big|_{\theta=0}}{f_0(x(t); t \in [0, T])} = b^{-1} \cdot \frac{\partial}{\partial \theta} f_\theta(x(t); t \in [0, T])\big|_{\theta=0}$$

$$= \frac{1}{b\sigma^2} \int_0^b d\tau \int_0^T s(t - \tau)x(t)\,dt$$

$$\stackrel{\Delta}{=} \frac{1}{b\sigma^2} g(x(t); t \in [0, T]) \tag{5.5.11}$$

The rule $\hat{\delta}_0^*$ is thus given by

$$\hat{\delta}_0^*(x(t); t \in [0, T]) = \begin{cases} 1 & g(x(t); t \in [0, T]) \geq \lambda_0^* \\ 0 & g(x(t); t \in [0, T]) < \lambda_0^* \end{cases}$$

$$\lambda_0^*: \Pr\{g(X(t); t \in [0, T]) \geq \lambda^* | H_0\} = \Pr\left\{\int_0^b d\tau \int_0^T s(t - \tau)N(t)\,dt \geq \lambda_0^*\right\} = \alpha$$

Example 5.5.2

Let the signal in Example 5.5.1 be instead sinusoidal with random phase. In particular, let $s(t) = \cos(\omega t + \phi)$, where the frequency ω is deterministic and known, and where the phase ϕ is a random variable, uniformly distributed in $[0, 2\pi]$. Then the analysis in Example 5.5.1 is valid for the Fourier set $\{\phi_i(t)\}$ in $[0, T]$ and the function $f_\theta(x(t); t \in [0, T])$ in (5.5.9) now takes the form

$$f_\theta(x(t); t \in [0, T]) = \int_0^{2\pi} \exp\left\{\frac{\theta}{\sigma^2} \int_0^T \cos(\omega t + \tau) x(t)\, dt \right. $$
$$\left. - \frac{\theta^2}{2\sigma^2} \int_0^T \cos^2(\omega t + \tau)\, dt\right\} d\tau \quad (5.5.12)$$

Let us now assume that the observation interval $[0, T]$ corresponds to a period of the sinusoidal signal. That is, let $T: \omega T = 2\pi$. Then, using the trigonometric identity $\cos^2 x = 2^{-1} + 2^{-1} \cos 2x$, we find

$$\omega T = 2\pi \to \int_0^T \cos^2(\omega t + \tau)\, dt = \frac{T}{2} + \frac{1}{2}\int_0^T \cos(2\omega t + 2\tau)\, dt$$

$$= \frac{T}{2} + \frac{1}{4\omega}[\sin(2\omega T + 2\tau) - \sin(2\tau)]$$

$$= \frac{T}{2} + \frac{1}{4\omega}[\sin(4\pi + 2\tau) - \sin(2\tau)] = \frac{T}{2}$$

$$(5.5.13)$$

Expanding now $\cos(\omega t + \tau)$, we also find

$$\int_0^T \cos(\omega t + \tau) x(t)\, dt = a \cos \tau - b \sin \tau \quad (5.5.14)$$

where

$$a = \int_0^T x(t) \cos \omega t\, dt \qquad b = \int_0^T x(t) \sin \omega t\, dt \quad (5.5.15)$$

Substituting (5.5.13) and (5.5.14) in (5.5.12), we find

$$\omega T = 2\pi \to f_\theta(x(t); t \in [0, T])$$
$$= \exp\left\{-\frac{\theta^2 T}{4\sigma^2}\right\} \int_0^{2\pi} \exp\left\{\frac{\theta \cdot a}{\sigma^2} \cos \tau - \frac{\theta \cdot b}{\sigma^2} \sin \tau\right\} d\tau \quad (5.5.16)$$

where a and b are given by (5.5.15).

Let us define

$$g_\theta(a, b) = \int_0^{2\pi} \exp\left\{\frac{\theta \cdot a}{\sigma^2} \cos \tau - \frac{\theta \cdot b}{\sigma^2} \sin \tau\right\} d\tau \quad (5.5.17)$$

where a and b are functions of the observed waveform $x(t); t \in [0, T]$.

5.5 Examples and Applications

Then the decision rule $\hat{\delta}_\theta^*$ in (5.5.8) takes the form

$$\hat{\delta}_\theta^*(x(t); t \in [0, T]) = \begin{cases} 1 & g_\theta(a, b) \geq \lambda^* \\ 0 & g_\theta(a, b) < \lambda^* \end{cases} \quad (5.5.18)$$

But the function $g_\theta(a, b)$ is the familiar Bessel function $I_0([\theta/\sigma^2][a^2 + b^2]^{1/2})$, which is known to be a monotone function of its argument. Let us assume that θ is positive. Then, squaring the argument of the Bessel function, we clearly obtain the following equivalent form of the test in (5.5.18):

$$\hat{\delta}_\theta^*(x(t); t \in [0, T]) = \begin{cases} 1 & a^2 + b^2 \geq \lambda_0^* \\ 0 & a^2 + b^2 < \lambda_0^* \end{cases} \quad (5.5.19)$$

Let us now denote

$$A = \int_0^T X(t) \cos \omega t \, dt \qquad B = \int_0^T X(t) \sin \omega t \, dt \quad (5.5.20)$$

The parameters a and b in (5.5.15) are clearly realizations of the random variables A and B in (5.5.20). The random variables A and B are uncorrelated and Gaussian, under both H_0 and H_1, with identical variances equal to $2^{-1}\sigma^2 T$. The mean values of A and B under H_0 are both zero. Let $m_A(\theta, \tau)$ and $m_B(\theta, \tau)$ denote respectively the mean values of the variables A and B under hypothesis H_1, conditioned on signal amplitude equal to θ and phase fixed at τ. Then

$$\omega T = 2\pi \rightarrow m_A(\theta, \tau) = \theta \int_0^T \cos(\omega t + \tau) \cos \omega t \, dt = \frac{\theta \cdot T}{2} \cos \tau$$

$$\omega T = 2\pi \rightarrow m_B(\theta, \tau) = \int_0^T \cos(\omega t + \tau) \sin \omega t \, dt = -\frac{\theta \cdot T}{2} \sin \tau \quad (5.5.21)$$

The threshold λ_0^* in (5.5.19) is found from

$$\Pr\{A^2 + B^2 \geq \lambda_0^* | H_0\} = \alpha \quad (5.5.22)$$

The power of the test is then given by the conditional probability

$$P(\delta_\theta^*) = (2\pi)^{-1} \int_0^{2\pi} d\tau \cdot \Pr\{A^2 + B^2 \geq \lambda_0^* | H_1, \theta, \tau\} \quad (5.5.23)$$

It is left to the reader to verify as an exercise that

$$\Pr\{A^2 + B^2 \geq \lambda_0^* | H_0\} = \exp\{-\lambda_0^*/\sigma^2 T\} \quad (5.5.24)$$

$$P(\delta_\theta^*) = 1 - (2\pi)^{-1} \exp\left\{-\frac{\theta^2 T}{4\sigma^2}\right\} \int_0^{\sqrt{2\lambda_0^*/\sigma\sqrt{T}}} x e^{-x^2/2} I_0\left(\frac{\theta}{\sigma}\sqrt{\frac{T}{2}} x\right) dx \quad (5.5.25)$$

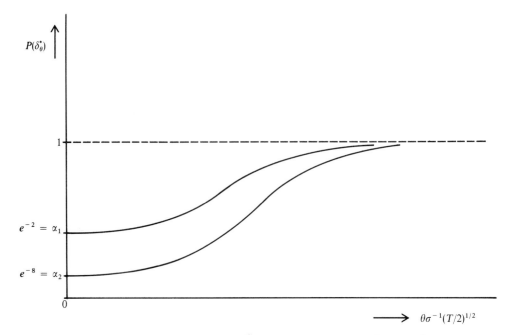

Figure 5.5.1 Power function for false alarm rates $\alpha = e^{-2}$, $\alpha = e^{-8}$

where $I_0(\cdot)$ is the previously mentioned Bessel function. From (5.5.22) and (5.5.24) we find

$$\lambda_0^* = -\sigma^2 T \ln \alpha \tag{5.5.26}$$

and by substitution in (5.5.25) we obtain

$$P(\delta_\theta^*) = 1 - (2\pi)^{-1} \exp\left\{-\frac{T}{4}\frac{\theta^2}{\sigma^2}\right\} \int_0^{\sqrt{-2\ln\alpha}} xe^{-x^2/2} I_0\left(\frac{\theta}{\sigma}\sqrt{\frac{T}{2}}x\right) dx \tag{5.5.27}$$

In Figure 5.5.1, we plot the power in (5.5.27), against the quantity $(\theta/\sigma)[T/2]^{1/2}$ for $\alpha = e^{-2}$ (where $\sqrt{(-2\ln\alpha)} = 2$), and $\alpha = e^{-8}$ [where $\sqrt{(-2\ln\alpha)} = 4$]. The ratio $\theta/\sigma\sqrt{2}$ signifies signal-to-noise ratio, while T represents the observation interval. For fixed T, the power change in Figure 5.5.1 represents change with respect to $\theta/\sigma\sqrt{2}$. For a fixed signal-to-noise ratio $\theta/\sigma\sqrt{2}$ the power change in Figure 5.5.1 represents change with respect to the observation interval T.

5.6 PROBLEMS

5.6.1 Consider the hypotheses as defined in Problem 4.7.1. Given a false alarm rate α, find the optimum Neyman-Pearson test and the probability of detection.

5.6.2 Consider the hypotheses as defined in Problem 4.7.2, for $\theta_1 > \theta_0$. Is the optimum Neyman-Pearson test different for different false alarm values?

5.6 Problems

5.6.3 Let $V = \{v_1, v_2, \ldots, v_n\}$ be an observed vector of independent uniform random variables on $[0, \theta]$. That is

$$f(v_i|\theta) = \begin{cases} \frac{1}{\theta} & 0 \le v_i \le \theta \\ 0 & \text{otherwise} \end{cases}$$

θ takes on one of two values, θ_0 or θ_1, with respective a priori probabilities σ_0 and σ_1. The loss function (L_{ij} = loss in deciding θ_j when $\theta = \theta_i$) is given. What is the Bayes rule? Find the average risk. What is the optimum Neyman-Pearson test at level α. Evaluate the probability of detection.

5.6.4 Find the optimum Neyman-Pearson test for

$$H_0: a = 0 \quad \text{vs} \quad H_1: a > 0$$

where the observable process is $v(t) = a + n(t);\ 0 \le t \le T;\ n(t)$ Gaussian with mean μ and covariance, $E[n(t)n(\tau)] = \sigma^2[1 + \min(t^2, \tau^2)]$.

5.6.5 Let $v(t) = \theta s(t - \tau) + n(t);\ 0 \le t \le T$, be observed, where $s(t)$ is a signal of known waveform, $s(t) = 0$ for $t < 0;\ t > T_1 \ll T,\ \tau$ is a uniform random variable on $[0, T - T_1]$, $n(t)$ is white Gaussian noise with zero mean.
(a) Find the optimum Neyman-Pearson test for $H_0: \theta = 0$ and $H_1: \theta = \theta_1 > 0$.
(b) Find the locally most powerful test.

5.6.6 Let $V = \{v(t);\ 0 \le t \le T\}$, where $v(t) = \theta + n(t)$, $n(t)$ is Gaussian with zero mean and covariance $R(\tau) = e^{-|\tau|}$.
(a) Determine the optimum Neyman-Pearson test for $H_0: \theta = 0$ and $H_1: \theta > 0$.
(b) What is the probability of detection?

5.6.7 Suppose: $v(t) = \theta s_\omega(t) + n(t)$ is observed for $0 \le t \le T$; $n(t)$ is Gaussian with mean zero and covariance $R(t, \tau)$; $s_\omega(t) = \cos \omega t$, where ω is an *unknown* frequency with probability density $p(\omega)$. What is the optimum detector for $H_0: \theta = 0$ versus $H_1: \theta = \theta_1$? Is it uniformly most powerful for $\theta_1 \in [0, \infty]$? Now suppose $p(\omega)$ is not known. What detection procedure might be used?

5.6.8 Let $v(t);\ 0 \le t \le T$ be observed, where $v(t) = \theta \cos(\omega t + \phi) + n(t)$. It is desired to make one of the two decisions

$$d_0: \theta = 0 \qquad d_1: \theta = \theta_1 > 0.$$

The a priori probabilities and loss functions are

$$\Pr[\theta = 0] = q \qquad \Pr[\theta = \theta_1] = p$$
$$f(\phi) = \text{density of } \phi \text{ on } [0, 2\pi]$$
$$L_{00} = L_{11} = 0 \qquad L_{10} = L_1 > 0 \qquad L_{01} = L_0 > 0$$

$n(t)$ is white Gaussian noise; $E[n(t)] = 0$; $E[n(t)n(\tau)] = \sigma^2 \delta(t - \tau)$. Find the Bayes rule, conditional risk, and Bayes risk. What is the Neyman-Pearson rule? For what $f(\phi)$, if any, is the test UMP?

5.6.9 Let $V = [v(0), v(\Delta t), \ldots, v(n\Delta t)]$ be the sample values of a continuous time waveform $v(t)$; $0 \leq t \leq T$ where $\Delta t = T/n$; and $v(t)$ is a Gaussian random process with mean zero and covariance $R(t, \tau)$. Let $R_{\Delta t}$ be the sampled covariance matrix. Find the optimum Neyman-Pearson test for

$$H_0: R_{\Delta t} = Q_{\Delta t} \quad \text{vs} \quad H_1: R_{\Delta t} = \mu Q_{\Delta t} \quad \mu > 1$$

Calculate the power of the test. What happens as $\Delta t \to 0$? What does this imply in terms of detection for non-Gaussian random processes with these hypotheses? [Hint: recall that $R^{-1} = W^T W$]

5.6.10 Let $\chi_1, \ldots, \chi_n$ be a sample of size n from the uniform distribution $U(0, \theta)$. Sufficiency reduces the problem to $T = \max_i \chi_i$.

(a) Find the class of all Neyman-Pearson best tests of $H_0: \theta = \theta_0$ against $H_1: \theta = \theta_1$, where $\theta_1 > \theta_0$.
(b) Find the subclass of the tests that are independent of θ_1. These are UMP tests of H_0 against $H_1: \theta > \theta_0$.
(c) Show that the test

$$\phi(t) = \begin{cases} 1 & \text{if } t > \theta_0 \\ a & \text{if } t \leq \theta_0 \end{cases}$$

is UMP of size a for testing $H'_0: \theta \leq \theta_0$ against $H'_1: \theta > \theta_0$ but that ϕ is not admissible.

(d) Show that

$$\phi(t) = \begin{cases} 1 & \text{if } t > \theta_0 \text{ or } t \leq b \\ 0 & \text{if } b < t \leq \theta_0 \end{cases}$$

where $b = \theta_0 \sqrt[n]{a}$ is a UMP test of size a for testing $H_0: \theta = \theta_0$ against $H: \theta \neq \theta_0$.

REFERENCES

Bahadur, R. R. (1954), "Sufficiency and Statistical Decision Functions," *Ann. Math. Statist.* 25, 423–462.

Chernoff, H. and L. E. Moses (1959). *Elementary Decision Theory*, Wiley, New York.

Dvoretzky, A., A. Wald, and Wolfowitz, J. (1951), "Elimination of Randomization in Certain Statistical Decision Procedures and Zero-Sum Two-Person Games," *Ann. Math. Statist.* 22, 1–21.

Ferguson, T. S. (1967), *Mathematical Statistics: A Decision Theoretic Approach*, Academic Press, New York.

References

Fox, M. (1956), "Charts of the Power of the F-Test," *Ann. Math. Statist.* 27, 484–497.

Karlin, S. and H. Rubin (1956), "The Theory of Decision Procedures for Distributions with Monotone Likelihood Ratio," *Ann. Math. Statist.* 27, 272–299.

Lehmann, E. L. (1959). *Testing Statistical Hypotheses*, Wiley, New York.

Neyman, J. (1935), "Su un Teorema Concernente le Cosiddete Statistiche Sufficienti," *Giornale dell' Instituto degli Attuari* 6, 320–334.

——— (1952), *Lectures and Conferences on Mathematical Statistics and Probability*, 2d ed., Graduate School, U.S. Dep. of Agriculture.

Neyman, J. and E. S. Pearson (1933), "On the Problem of the Most Efficient Tests of Statistical Hypotheses," *Pilos. Trans. Roy. Soc. London Series A* 231, 289–337.

——— (1936, 1938), "Contributions to the Theory of Testing Statistical Hypotheses. I. Unbiased Critical Regions of Type A and Type A_1. II. Certain Theorems on Unbiased Critical Regions of Type A. III. Unbiased Tests of Simple Statistical Hypotheses Specifying the Values of More than One Unknown Parameter," *Statist. Res. Mem.* 1, 1–37, and 2, 25–57.

Pearson, E. S. and H. O. Hartley (1951), "Charts of the Power Function for Analysis of Variance Tests derived from the Noncentral F-Distribution," *Biometrika* 38, 112–130.

Savage, L. J. (1954), *The Foundations of Statistics*, Wiley, New York.

Wald, A. (1950), *Statistical Decision Functions*, Wiley, New York.

Van Trees, H. L. (1968), *Detection, Estimation, and Modulation Theory*, Wiley, New York.

CHAPTER 6

MINIMAX AND ROBUST DETECTION

The Bayesian and the Neyman-Pearson detection schemes, in Chapters 4 and 5, are the basis for the development of all the remaining detection procedures. In this chapter we study the minimax and the robust detection formalizations. The former evolves from the Bayesian schemes when no a priori distributions are available. The latter results from the Neyman-Pearson format, where the two hypotheses are nonparametrically described and are represented by two distinct classes of stochastic processes. The common characteristic between the minimax and the robust detection schemes is that they both evolve as the solutions of saddle-point games.

6.1 THE GAME IN MINIMAX DETECTION

Let us assume that a parameter vector Θ^m and the space $\mathscr{E}^m$ where this vector takes its values are given. Let also a discrete-time, parametrically known stochastic process be given, such that for each value θ^m in $\mathscr{E}^m$ and for given n the process is described by the well known n-dimensional density function $f_{\theta^m}(x^n)$; $x^n \in \Gamma^n$. Let the parameter space $\mathscr{E}^m$ be broken into M disjoint subspaces $\mathscr{E}_i^m$; $1 \leq i \leq M$ such that $\bigcup_{i=1}^{m} \mathscr{E}_i^m = \mathscr{E}^m$ and let $\mathscr{E}_i^m$ determine hypothesis H_i. Let $\{c_k(\theta^m); 1 \leq k \leq M, \theta^m \in \mathscr{E}^m\}$ be a set of given scalar penalty functions; that is, $c_k(\theta^m) \geq 0; 1 \leq k \leq M, \theta^m \in \mathscr{E}^m$, where $c_k(\theta^m)$ denotes the penalty assigned when the value of the parameter vector is θ^m and hypothesis H_k is decided. Given all the above, let us consider some arbitrary density function $p(\theta^m)$ defined on $\mathscr{E}^m$ and some decision rule $\delta = \{\delta_k(x^n); 1 \leq k \leq M, x^n \in \Gamma^n\}$. Then the expected penalty $C(\delta, p)$ induced by $p(\theta^m)$ and δ is given by (see Chapter 4)

$$C(\delta, p) = \int_{\mathscr{E}^m} d\theta^m p(\theta^m) C(\delta, \theta^m) \qquad (6.1.1)$$

6.1 The Game in Minimax Detection

where, as in (3.1.17),

$$C(\delta, \theta^m) = \int_{\Gamma^n} dx^n \sum_{k=1}^{M} \delta_k(x^n) c_k(\theta^m) f_{\theta^m}(x^n) \qquad (6.1.2)$$

If $p(\theta^m)$ is available, then a Bayesian detection formalization evolves, and the Bayesian decision rule minimizes $C(\delta, p)$ in (6.1.1) as in Chapter 4. In the minimax setup, no $p(\theta^m)$ in $\mathscr{E}^m$ is available, however. The available assets are then: the subspaces $\mathscr{E}_i^m$; $1 \leq i \leq M$, the density function $f_{\theta^m}(x^n)$; $\theta^m \in \mathscr{E}^m$, $x^n \in \Gamma^n$, the penalty functions $\{c_k(\theta^m); 1 \leq k \leq M, \theta^m \in \mathscr{E}^m\}$, and an observation vector x^n. Those assets give rise to a game-theory formalization in a natural and conceptually pleasing fashion. The players in the game are Nature and the analyst (or system designer). Nature selects an a priori parameter density function $p(\theta^m)$ in $\mathscr{E}^m$ which, in conjunction with the conditional density function $f_{\theta^m}(x^n)$; $\theta^m \in \mathscr{E}^m$, $x^n \in \Gamma^n$, generates the data vector x^n that the analyst observes. Based on x^n and the assets available to him, the analyst selects a decision rule δ. Nature and the analyst make their selections, independently of each other. For any given, independently selected pair (δ, p) the analyst pays a penalty $C(\delta, p)$, as in (6.1.1). To protect his interest, and trying to outsmart Nature, the analyst imagines that Nature is trying to maximize his minimum, with respect to the δ selection penalty. On the other hand, the analyst's objective is to minimize the maximum, with respect to the $p(\theta^m)$ selection penalty. In conclusion, the analyst is then searching for a pair (δ^*, p^*) such that

$$\forall\, p(\theta^m) \in \mathscr{E}^m \quad C(\delta^*, p) \leq C(\delta^*, p^*) \leq C(\delta, p^*) \quad \forall\, \delta \in \mathscr{D} \qquad (6.1.3)$$

where $C(\delta, p)$ is given by (6.1.1), and where the class $\mathscr{D}$ includes all the valid decision rules $\delta = \{\delta_k(x^n); 1 \leq k \leq M, x^n \in \Gamma^n\}$, that is

$$\begin{aligned} \delta_k(x^n) &\geq 0 \quad 1 \leq k \leq M, x^n \in \Gamma^n \\ \sum_{k=1}^{M} \delta_k(x^n) &= 1 \quad \forall\, x^n \in \Gamma^n \end{aligned} \qquad (6.1.4)$$

Expression (6.1.3) respresents a saddle-point game, with objective function $C(\delta, p)$. Any pair (δ^*, p^*) that satisfies (6.1.3) is a solution of the game, and is called the saddle point. The decision rule δ^* is then called a *minimax rule*. We point out that, in general, such a solution may not exist, and that even if it does, it is generally nonunique. From the theory of saddle-point games, we conclude that the game in (6.1.3) has a solution (δ^*, p^*) if

$$C(\delta^*, p^*) = \inf_{\delta \in \mathscr{D}} \sup_{p \in \mathscr{P}} C(\delta, p) = \sup_{p \in \mathscr{P}} \inf_{\delta \in \mathscr{D}} C(\delta, p) \qquad (6.1.5)$$

where $\mathscr{P}$ denotes the class of all the valid density functions $p(\theta^m)$ in $\mathscr{E}^m$.

The expression in (6.1.5) is significant and very useful. It states that, if the game in (6.1.3) has a solution (δ^*, p^*) then this solution can be found as $C(\delta^*, p^*) = \sup_{p \in \mathscr{P}} \inf_{\delta \in \mathscr{D}} C(\delta, p)$. But, if $p(\theta^m)$ is first fixed, the $\inf_{\delta \in \mathscr{D}} C(\delta, p)$ is then optimal Bayesian expected penalty, at $p(\theta^m)$; the decision rule δ_p^* that attains this minimum is the optimal Bayesian decision rule, as in Chapter 4. Therefore, given $p(\theta^m)$, the infimum $C(\delta_p^*, p) = \inf_{\delta \in \mathcal{D}} C(\delta, p)$ exists and is known. Next, the supremum $\sup_{p \in \mathcal{P}} C(\delta_p^*, p)$ is sought. If this supremum exists, it is satisfied by some density function $p^*(\theta^m)$ in $\mathcal{E}^m$ and its value is $C(\delta_{p^*}, p^*)$. Finally, the satisfaction of the double inequality in (6.1.3) by the pair (δ_{p^*}, p^*) can be studied. If (δ_{p^*}, p^*) satisfies this double inequality, it clearly provides a solution for the saddle-point game. Otherwise, it is concluded that the game has no solution. Let us now assume that, following the procedure explained above, we find that the game has a solution, given by (δ^*, p^*). Then, the equalities in (6.1.5) hold and thus, $C(\delta^*, p^*) = \inf_{\delta \in \mathcal{D}} \sup_{p \in \mathcal{P}} C(\delta, p)$. Let us first examine $\sup_{p \in \mathcal{P}} C(\delta, p)$ in this latter expression. Given a decision rule δ and due to (6.1.1) and the nonnegativeness of $C(\delta, \theta^m)$ in it, we easily conclude that $\sup_{p \in \mathcal{P}} C(\delta, p)$ equals $\sup_{\theta^m \in \mathcal{E}^m} C(\delta, \theta^m)$ and is attained by a $p_\delta(\theta^m)$ that assigns all its mass at any vector parameter value (there may be more than one such value) that satisfies $\sup_{\theta^m \in \mathcal{E}^m} C(\delta, \theta^m)$. As a result, $C(\delta^*, p^*) = \inf_{\delta \in \mathcal{D}} \sup_{\theta^m \in \mathcal{E}^m} C(\delta, \theta^m)$, as discussed in Chapter 3 and shown there by Figures 3.1.2 and 3.1.3.

The methodology of the search for a saddle point (δ^*, p^*) in (6.1.3), explained in the above paragraph, is clearly Bayesian. That is, at each $p(\theta^m)$ in $\mathcal{E}^m$ an optimal Bayesian decision rule δ_p is found. The search is then limited among the latter rules for the final identification of the (δ_{p^*}, p^*) such that, $C(\delta_{p^*}, p^*) = \sup_{p \in \mathcal{P}} C(\delta_p, p)$. Therefore, if a pair (δ_{p^*}, p^*) exists, the decision rule $\delta^* = \delta_{p^*}$ is Bayesian; thus, it is not in general unique. As discussed in Chapter 4, we then select a nonrandomized such rule for simplicity in implementation, and at no cost in performance. We now present a definition pertinent to the above discussion.

DEFINITION 6.1.1: Given $\mathcal{E}^m$, given $f_{\theta^m}(x^n)$; $\theta^m \in \mathcal{E}^m$, $x^n \in \Gamma^n$, and given $\{c_k(\theta^m); 1 \leq k \leq M, \theta^m \in \mathcal{E}^m\}$, a decision rule δ is called *admissible*, if it is an optimal Bayesian rule at some density function $p(\theta^m)$ in $\mathcal{E}^m$. A density function $p^*(\theta^m)$ in $\mathcal{E}^m$ is called *least favorable* if the minimum Bayesian expected penalty at $p^*(\theta^m)$ is not exceeded by the minimum Bayesian expected penalty at any other density function $p(\theta^m)$ in $\mathcal{E}^m$. ∎

In view of Definition 6.1.1, the search for a δ^* that satisfies (6.1.3) is limited among all the admissible decision rules, and if a $p^*(\theta^m)$ in $\mathcal{E}^m$ that satisfies (6.1.3) exists, then it is least favorable. At this point, we will proceed with two examples.

Example 6.1.1

Let us consider the case where a constant scalar signal is transmitted through discrete-time, additive, Gaussian, and white noise. Let the power of the Gaussian noise be known and equal to σ^2. Let the constant signal be a random variable Θ whose value θ is selected before transmission, via some density function, $p(\theta)$,

6.1 The Game in Minimax Detection

defined on $(-\infty, -\alpha] \cup [\alpha, \infty)$, where α is a known positive constant. Let the density function $p(\theta)$ be unknown, and let us consider the two hypotheses

$$H_0: -\infty < \theta \leq -\alpha \qquad H_1: \alpha \leq \theta < \infty$$

Let the penalty functions $c_0(\theta)$ and $c_1(\theta)$ be assigned as

$$c_0(\theta) = \begin{cases} 1 & \theta \geq \alpha \\ 0 & \theta \leq -\alpha \end{cases} \qquad c_1(\theta) = \begin{cases} 1 & \theta \leq -\alpha \\ 0 & \theta \geq \alpha \end{cases} \qquad (6.1.6)$$

Given a parameter value θ and an observation vector $x^n = [x_1, \ldots, x_n]^T$, the conditional density function $f_\theta(x^n)$ induced by the model is given as

$$f_\theta(x^n) = (2\pi)^{-n/2} \sigma^{-n} \exp\left\{-\frac{1}{2\sigma^2} \sum_{i=1}^{n} (x_i - \theta)^2\right\} \qquad (6.1.7)$$

Given some $p(\theta)$ on $(-\infty, -\alpha] \cup [\alpha, \infty)$ that is also symmetric around zero, an optimal Bayesian decision rule δ_p that minimizes the expected penalty $C(\delta, p)$ is (see Chapter 4), for $\delta_p(x^n) = P\{H_1 \text{ decided} | x^n\}$,

$$\delta_p(x^n) = \begin{cases} 1 & x^n: g_0(x^n) \geq g_1(x^n) \\ 0 & \text{otherwise} \end{cases} \qquad (6.1.8)$$

where

$$g_0(x^n) = \int_\alpha^\infty d\theta\, p(\theta) f_\theta(x^n) = \int_{-\infty}^{-\alpha} d\theta\, p(\theta) f_{-\theta}(x^n) \qquad (6.1.9)$$

$$g_1(x^n) = \int_{-\infty}^{-\alpha} d\theta\, p(\theta) f_\theta(x^n) \qquad (6.1.10)$$

Substituting (6.1.7) in (6.1.9) and (6.1.10), and applying some straightforward simplifications, we obtain the following form of (6.1.8):

$$\delta_p(x^n) = \begin{cases} 1 & x^n: \int_{-\infty}^{-\alpha} d\theta\, p(\theta) e^{-n\theta^2/2\sigma^2} \left[e^{-(\theta/\sigma^2) \sum_{i=1}^{n} x_i} - e^{(\theta/\sigma^2) \sum_{i=1}^{n} x_i}\right] \geq 0 \\ 0 & \text{otherwise} \end{cases} \qquad (6.1.11)$$

But, it can be easily seen that the integral in (6.1.11) is nonnegative if and only if $\sum_{i=1}^{n} x_i \geq 0$. Thus, the decision rule in (6.1.11) is simplified as

$$\delta_p(x^n) = \begin{cases} 1 & x^n: \sum_{i=1}^{n} x_i \geq 0 \\ 0 & \text{otherwise} \end{cases} \qquad (6.1.12)$$

We observe that the Bayesian decision rule in (6.1.12) is identical at all symmetric around zero density functions $p(\theta)$ in $(-\infty, -\alpha] \cup [\alpha, \infty)$; we thus drop the subscript p in $\delta_p(x^n)$, denoting it $\delta^*(x^n)$, instead. Clearly $\delta^*(x^n)$ in (6.1.12) is admissible. The conditional expected penalty $C(\delta^*, \theta)$ induced by the admissible rule δ^* at the parameter value θ is given by

$$C(\delta^*, \theta) = \int_{\sum_{i=1}^n x_i \geq 0} dx^n c_1(\theta) f_\theta(x^n) + \int_{\sum_{i=1}^n x_i < 0} dx^n c_0(\theta) f_\theta(x^n)$$

$$= \begin{cases} \int_{\sum_{i=1}^n x_i \geq 0} dx^n f_\theta(x^n) & \text{if } \theta \leq -\alpha \\ \int_{\sum_{i=1}^n x_i < 0} dx^n f_\theta(x^n) & \text{if } \theta \geq \alpha \end{cases}$$

$$= \begin{cases} \Phi\left(\dfrac{-n\theta}{\sigma\sqrt{n}}\right) & \text{if } \theta \geq \alpha \\ \Phi\left(\dfrac{n\theta}{\sigma\sqrt{n}}\right) & \text{if } \theta \leq -\alpha \end{cases}$$

$$= \Phi\left(-\dfrac{\sqrt{n}}{\sigma}|\theta|\right) \qquad \forall\, \theta \in (-\infty, -\alpha] \cup [\alpha, \infty) \quad (6.1.13)$$

where $\Phi(x)$ is the cumulative distribution of the zero mean, unit variance Gaussian random variable, at the point x.

Now let $p(\theta)$ be some density function symmetric around zero, defined on $\theta \in (-\infty, -\alpha] \cup [\alpha, \infty)$. Then, the Bayesian decision rule at $p(\theta)$ is as in (6.1.12), the induced conditional expected penalty $C(\delta^*, \theta)$ is as in (6.1.13), and the Bayesian expected penalty $C(\delta^*, p)$ is given by

$$C(\delta^*, p) = \int_{-\infty}^{-\alpha} d\theta\, p(\theta) \Phi\left(-\dfrac{\sqrt{n}}{\sigma}|\theta|\right) + \int_\alpha^\infty d\theta\, p(\theta) \Phi\left(-\dfrac{\sqrt{n}}{\sigma}|\theta|\right)$$

$$= 2\int_{-\infty}^{-\alpha} d\theta\, p(\theta) \Phi\left(-\dfrac{\sqrt{n}}{\sigma}|\theta|\right) = 2\int_{-\infty}^{-\alpha} d\theta\, p(\theta) \Phi\left(\dfrac{\sqrt{n}}{\sigma}\theta\right) \quad (6.1.14)$$

Let us define

$$p^*(\theta) = \dfrac{1}{2}[\delta(\theta - \alpha) + \delta(\theta + \alpha)] \qquad (6.1.15)$$

where $\delta(x)$ is the delta function at the point x.

Clearly, we then conclude from (6.1.14) and (6.1.15) that

$$C(\delta^*, p) \leq C(\delta^*, p^*) = \Phi\left(-\dfrac{\sqrt{n}}{\sigma}\alpha\right) \forall\, p(\theta) \text{ in } (-\infty, -\alpha] \cup [\alpha, \infty) \quad (6.1.16)$$

6.1 The Game in Minimax Detection

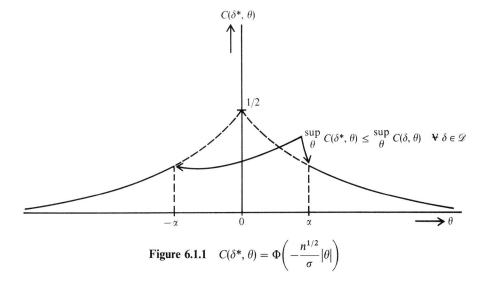

Figure 6.1.1 $C(\delta^*, \theta) = \Phi\left(-\dfrac{n^{1/2}}{\sigma}|\theta|\right)$

Since $p^*(\theta)$ in (6.1.15) is symmetric around zero, the Bayesian rule at $p^*(\theta)$ is the rule δ^* in (6.1.12). Thus, we also conclude

$$C(\delta^*, p^*) \leq C(\delta, p^*) \ \forall \ \delta \in \mathscr{D} \tag{6.1.17}$$

Due to (6.1.16) and (6.1.17), we conclude that the pair (δ^*, p^*) in (6.1.12) and (6.1.15) satisfies the game in (6.1.3). Thus δ^* is minimax. It can be easily verified that the Bayesian expected penalties at $p(\theta)$ that are nonsymmetric around zero are no larger than the same expected penalties at $p(\theta)$ that are symmetric around zero. So, the $p^*(\theta)$ in (6.1.15) is least favorable for the present problem; it is not a unique least favorable density function, however. Since (δ^*, p^*) is a saddle point for the game, then $C(\delta^*, p^*) = \inf_{\delta \in \mathscr{D}} \sup_{p \in \mathscr{P}} C(\delta, p) = \inf_{\delta \in \mathscr{D}} \sup_{\theta} C(\delta, \theta)$. Thus $\sup_{\theta} C(\delta^*, \theta)$ is the smallest among the suprema $\sup_{\theta} C(\delta, \theta)$ induced by all possible decision rules. In Figure 6.1.1, we plot the conditional expected penalty $C(\delta^*, \theta)$ given by (6.1.13). In Figure 6.1.2, we plot the saddle value $C(\delta^*, p^*)$ of the game, given by (6.1.16) as a function of the observation length n.

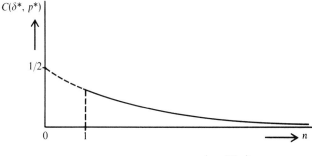

Figure 6.1.2 $C(\delta^*, p^*) = \Phi\left(-\dfrac{n^{1/2}}{\sigma}\alpha\right)$

Example 6.1.2

Let us consider a radar problem. Let the signal be a positive known constant scalar θ, and let the transmission medium be represented by discrete-time, zero mean, power σ^2, Gaussian additive noise. Let hypothesis H_0 signify the absence of the signal, and let hypothesis H_1 signify its presence. Let the a priori probability q of H_1 be unknown, and let the penalty coefficients $c_{10} = c_{01} = 1$, $c_{00} = c_{11} = 0$ be given. The objective is the search for a minimax decision rule. Given n and an observation sequence x^n, let $f_0(x^n)$ and $f_1(x^n)$ denote the density functions at x^n, under H_0 and H_1 respectively. Then

$$f_0(x^n) = (2\pi)^{-n/2}\sigma^{-n}\exp\left\{-\frac{1}{2\sigma^2}\sum_{i=1}^{n}x_i^2\right\}$$

$$f_1(x^n) = (2\pi)^{-n/2}\sigma^{-n}\exp\left\{-\frac{1}{2\sigma^2}\sum_{i=1}^{n}(x_i-\theta)^2\right\}$$
(6.1.18)

Given some a priori probability q the optimal Bayesian decision rule δ_q at q is

$$\delta_q(x^n) = \begin{cases} 1 & x^n: \sum_{i=1}^{n} x_i \geq 2^{-1}n\theta + \sigma^2\theta^{-1}\ln\frac{1-q}{q} \\ 0 & \text{otherwise} \end{cases}$$
(6.1.19)

The conditional expected penalty $C(\delta_q, q)$ at q is then expressed as

$$C(\delta_q, q) = q\Phi\left(-\frac{\theta\sqrt{n}}{2\sigma} + \frac{\sigma}{\theta\sqrt{n}}\ln\frac{1-q}{q}\right) + (1-q)\Phi\left(-\frac{\theta\sqrt{n}}{2\sigma} - \frac{\sigma}{\theta\sqrt{n}}\ln\frac{1-q}{q}\right)$$
(6.1.20)

where $\Phi(x)$ is the cumulative distribution of the zero mean, unit variance Gaussian random variable, at the point x.

From (6.1.20) we easily conclude

$$C(\delta_q, q) = C(\delta_{1-q}, 1-q) \qquad \forall\, q: 0 \leq q \leq 1$$

$$C(\delta_0, 0) = C(\delta_1, 1) = 0 \qquad (6.1.21)$$

$$C(\delta_{.5}, .5) = \Phi\left(-\frac{\theta\sqrt{n}}{2\sigma}\right)$$

We are now searching for the supremum $\sup_q C(\delta_q, q)$. Due to the symmetry property $C(\delta_q, q) = C(\delta_{1-q}, 1-q)$ it is sufficient to search for the supremum of the expected penalty $C(\delta_q, q)$ in the q region $(0, .5]$, where the following condition holds.

$$\ln\frac{1-q}{q} \geq 0 \qquad q \in (0, .5] \qquad (6.1.22)$$

6.1 The Game in Minimax Detection

From (6.1.20), it is easily found that the derivative, $(\partial/\partial q)C(\delta_q, q)$ is nonnegative everywhere in $q \in (0, .5]$. Therefore, the conditional expected penalty $C(\delta_q, q)$ assumes its maximum at $q^* = .5$. The selection $q^* = .5$ is thus least favorable, the pair $(\delta_{.5}, .5)$ is a saddle-point solution of the game, and the decision rule $\delta_{.5}$ is minimax, where

$$\delta_{.5}(x^n) = \begin{cases} 1 & x^n: \sum_{i=1}^{n} x_i \geq 2^{-1}n\theta \\ 0 & \text{otherwise} \end{cases} \quad (6.1.23)$$

By substitution, we find

$$C(\delta_{.5}, .5) = C(\delta_{.5}, q) = \Phi\left(-\frac{\theta\sqrt{n}}{2\sigma}\right) \quad \forall q: 0 \leq q \leq 1 \quad (6.1.24)$$

As a function of the observation length n the saddle value $C(\delta_{.5}, .5)$ of the game behaves as in Figure 6.1.2, Example 6.1.1. The expected conditional penalty $C(\delta_{.5}, q)$ is constant as a function of q and is plotted in Figure 6.1.3.

We will now discuss a useful special case. Consider the general description of the minimax problem presented in this section. Let there exist some density function $p^*(\theta^m)$ in $\mathcal{E}^m$ such that the Bayesian decision rule δ^* at $p^*(\theta^m)$ induces a conditional expected penalty $C(\delta^*, \theta^m)$ that is a constant as a function of the vector parameter values θ^m in $\mathcal{E}^m$. That is, $C(\delta^*, \theta^m) = C(\delta^*)$. Then from (6.1.1) we obtain

$$C(\delta^*, p) = \int_{\mathcal{E}^m} d\theta^m\, p(\theta^m) C(\delta^*) = C(\delta^*) \quad \forall\, p(\theta^m) \text{ in } \mathcal{E}^m \quad (6.1.25)$$

Therefore, we conclude that

$$\forall\, p(\theta^m) \text{ in } \mathcal{E}^m \quad C(\delta^*, p) = C(\delta^*, p^*) \leq C(\delta, p^*) \quad \forall\, \delta \in \mathcal{D} \quad (6.1.26)$$

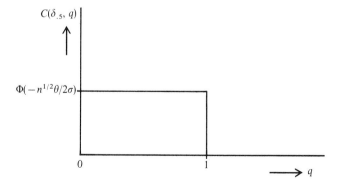

Figure 6.1.3 $C(\delta_{.05}, q)$ given n

Due to (6.1.26), we conclude that the pair (δ^*, p^*) satisfies the game in (6.1.3). The Bayesian rule δ^* is then a minimax rule, and the density function $p^*(\theta^m)$ in $\mathscr{E}^m$ is least favorable. Due to the usefulness of the above conclusions, we summarize them in the form of a lemma.

LEMMA 6.1.1: Given $\mathscr{E}^m$; given $f_{\theta^m}(x^n)$; $\theta^m \in \mathscr{E}^m$, $x^n \in \Gamma^n$; given penalty functions $\{c_k(\theta^m); 1 \leq k \leq M, \theta^m \in \mathscr{E}^m\}$ let there exist an admissible decision rule $\delta^* = \{\delta_i^*(x^n); 1 \leq i \leq M, x^n \in \Gamma^n\}$ such that the conditional expected penalty $C(\delta^*, \theta^m)$ is a constant as a function of the vector parameter value θ^m. Then the rule δ^* is a minimax rule for the problem. If the rule δ^* is an optimal Bayesian rule at the density function $p^*(\theta^m)$ in $\mathscr{E}^m$ then $p^*(\theta^m)$ is a least favorable density function for the problem. ∎

Let us revisit Example 6.1.2. As found there, the admissible rule $\delta_{.5}$ in (6.1.23) induces a conditional expected penalty $C(\delta_{.5}, q)$ that is a constant as a function of q. Applying Lemma 6.1.1, we thus immediately conclude that the decision rule $\delta_{.5}$ is minimax for the problem in the example. We will complete this section with three additional examples.

Example 6.1.3

Let us consider the case where, in a time interval $[0, T]$, either pure noise or noise plus some signal is transmitted. Let the noise be modeled by a zero mean, Gaussian, white, level σ^2 process. Let the signal be represented by a sinusoidal waveform $\theta \cos(\omega t + \phi)$, where $\omega = 2\pi T^{-1}$; θ is fixed and known; and ϕ is a random variable, taking values in $[0, 2\pi]$. Let hypothesis H_0 correspond to the presence of pure noise. Let hypothesis H_1 correspond to the presence of signal plus noise. Let the a priori probabilities of the two hypotheses, and the density function $f(\phi)$ of the random phase ϕ be unknown. Let the penalty coefficients $c_{11} = c_1(\phi) = 0 \, \forall \, \phi \in [0, 2\pi]$; $c_{00} = 0$; $c_{10} = c_{01} = 1$ be given. The objective is to derive a minimax detection scheme for testing H_1 versus H_0.

Let us denote by q; $0 < q < 1$ the a priori probability of the H_0 hypothesis. Then, the a priori probability of the H_1 hypothesis is $1 - q$. Let us denote by $x(t)$; $t \in [0, T]$ the observed waveform. Given q and $f(\phi)$, a Bayesian decision rule $\delta_{q,f}^*$ that is optimal in the mean-squared sense can be derived via the approach presented in Examples 5.5.1 and 5.5.2. Directly from Example 5.5.2, we conclude

$$\delta_{q,f}^*(x(t); t \in [0, T])$$

$$= \begin{cases} 1 & \int_0^{2\pi} d\tau \cdot f(\tau) \exp\left\{\frac{\theta \cdot a}{\sigma^2} \cos \tau - \frac{\theta \cdot b}{\sigma^2} \sin \tau\right\} \geq \frac{q}{1-q} \exp\left\{\frac{\theta^2 T}{4\sigma^2}\right\} \\ 0 & \text{otherwise} \end{cases}$$

(6.1.27)

6.1 The Game in Minimax Detection

where

$$a = \int_0^T x(t) \cos \omega t \, dt \qquad b = \int_0^T x(t) \sin \omega t \, dt \qquad (6.1.28)$$

Let us now select $f^*(\phi) = (2\pi)^{-1}$; $\phi \in [0, 2\pi]$. Then, from the results in Example 5.5.2, we conclude that the Bayesian decision rule δ^*_{q,f^*} at f^* can be expressed in the simplified form

$$\delta^*_{q,f^*}(x(t); t \in [0, T]) = \begin{cases} 1 & a^2 + b^2 \geq \dfrac{\sigma^4}{\theta^2} h\left(\dfrac{2\pi q}{1-q} \exp\left\{\dfrac{\theta^2 T}{4\sigma^2}\right\}\right) = \lambda_q \\ 0 & \text{otherwise} \end{cases} \qquad (6.1.29)$$

where a and b are in (6.1.28), and where $h(\cdot)$ is a monotone function determined by the inverse of the Bessel function $I_0(\cdot)$.

Let us denote by $C(\delta^*_{q,f^*}, H_0)$ and $C(\delta^*_{q,f^*}, H_1, \phi)$ the conditional penalties induced by the decision rule in (6.1.29), given respectively H_0 and H_1 at the phase point ϕ. Then we obtain

$$\begin{aligned} C(\delta^*_{q,f^*}, H_0) &= \Pr\{A^2 + B^2 \geq \lambda_q | H_0\} \\ C(\delta^*_{q,f^*}, H_1, \phi) &= \Pr\{A^2 + B^2 < \lambda_q | H_1, \phi\} \end{aligned} \qquad (6.1.30)$$

where A and B are the random variables

$$A = \int_0^T X(t) \cos \omega t \, dt \qquad B = \int_0^T X(t) \sin \omega t \, dt \qquad (6.1.31)$$

As found in Example 5.5.2, the random variables A and B in (6.1.31) are Gaussian, uncorrelated, and with common variance $2^{-1}\sigma^2 T$ under both H_0 and H_1. In addition, under H_0 both variables have mean zero and

$$E\{A|H_1, \phi\} = \frac{\theta T}{2} \cos \phi \qquad E\{B|H_1, \phi\} = -\frac{\theta T}{2} \sin \phi \qquad (6.1.32)$$

Directly from (5.5.24) in Example 5.5.2, we find

$$C(\delta^*_{q,f^*}, H_0) = \Pr\{A^2 + B^2 \geq \lambda_q | H_0\} = \exp\left\{-\frac{\lambda_q}{\sigma^2 T}\right\} \qquad (6.1.33)$$

In parallel, following the same approach, and as in (5.5.25), Example 5.5.2, we find

$$\begin{aligned} C(\delta^*_{q,f^*}, H_1, \phi) &= \Pr\{A^2 + B^2 < \lambda_q | H_1, \phi\} \\ &= (2\pi)^{-1} \exp\left\{-\frac{\theta^2 T}{4\sigma^2}\right\} \int_0^{\sqrt{2\lambda_q}/\sigma\sqrt{T}} x e^{-x^2/2} I_0\left(\frac{\theta}{\sigma}\sqrt{\frac{T}{2}} x\right) dx \end{aligned} \qquad (6.1.34)$$

where $I_0(\cdot)$ is the previously mentioned Bessel function.

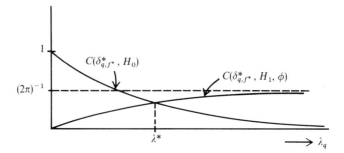

Figure 6.1.4

From (6.1.34), we observe that the conditional expected penalty $C(\delta^*_{q,f^*}, H_1, \phi)$ is a constant as a function of the phase value ϕ; it is only a function of the a priori probability q via the threshold λ_q in (6.1.29), and so is $C(\delta^*_{q,f^*}, H_0)$ in (6.1.33). In addition, λ_q in (6.1.29) increases monotonically to infinity, with q increasing from zero to one. At the same time, the conditional penalty in (6.1.33) decreases monotonically with increasing λ_q while the conditional penalty in (6.1.34) increases monotonically with increasing λ_q, reaching the asymptotic value $(2\pi)^{-1}$. Thus, as exhibited in Figure 6.1.4, there exists some value λ^* such that the conditional penalties in (6.1.33) and (6.1.34) are equal. Via (6.1.29), there exists a unique value q^* in $(0, 1)$ such that $\lambda^* = \lambda_{q^*}$. Therefore, q^* in conjunction with $f^*(\phi) = (2\pi)^{-1}$; $\forall \phi \in [0, 2\pi]$ induces a conditional penalty that is a constant, as a function of H_0 and H_1 and the signal phase ϕ. Due to Lemma 6.1.1, we then conclude that the decision rule δ^* given below is minimax, and that the pair (q^*, f^*) is least favorable.

$$\delta^*(x(t); t \in [0, T])$$
$$= \begin{cases} 1 & \left[\int_0^T x(t) \cos \omega t \, dt\right]^2 + \left[\int_0^T x(t) \sin \omega t \, dt\right]^2 \geq h\left(\frac{2\pi q^*}{1-q^*} \exp\left\{\frac{\theta^2 T}{4\sigma^2}\right\}\right) \\ 0 & \text{otherwise} \end{cases}$$

(6.1.35)

The expected penalty $C(\delta^*, q^*)$ induced by the minimax rule in (6.1.35) is the saddle value of the game for the present problem. For λ^* as in Figure 6.1.4 and $q^*: \lambda_{q^*} = \lambda^*$ it is given by

$$C(\delta^*, q^*) = \exp\left\{-\frac{\lambda^*}{\sigma^2 T}\right\} = \exp\left\{-\frac{\lambda_{q^*}}{\sigma^2 T}\right\} \qquad (6.1.36)$$

Example 6.1.4

Consider the same problem as in Example 6.1.3, with one difference. In particular, let it be known here that H_0 and H_1 are equally probable. The density

6.1 The Game in Minimax Detection

function $f(\phi)$ in $[0, 2\pi]$ of the phase ϕ is still unknown. As in Example 6.1.3, let $f^*(\phi)$ be the uniform density function in $[0, 2\pi]$. Then the optimal Bayesian decision rule at $f^*(\phi)$ is as in (6.1.29), with $q = 1 - q = .5$. That is

$$\delta^*_{f*}(x(t); t \in [0, T]) = \begin{cases} 1 & a^2 + b^2 \geq h\left(2\pi \exp\left\{\dfrac{\theta^2 T}{4\sigma^2}\right\}\right) = \lambda \\ 0 & \text{otherwise} \end{cases} \quad (6.1.37)$$

where a and b are as in (6.1.28).

The conditional penalty function $C(\delta^*_{f*}, \phi)$ for given ϕ and for the decision rule in (6.1.37) is given as

$$C(\delta^*_{f*}, \phi) = \frac{1}{2}\Pr\{A^2 + B^2 \geq \lambda | H_0\} + \frac{1}{2}\Pr\{A^2 + B^2 < \lambda | H_1, \phi\} \quad (6.1.38)$$

The conditional probabilities in (6.1.38) are respectively as in (6.1.33) and (6.1.34), with λ_q replaced by the threshold λ in (6.1.37). Performing the substitutions in (6.1.38), we obtain,

$$C(\delta^*_{f*,q}, \phi) = \frac{1}{2}\exp\left\{-\frac{\lambda}{\sigma^2 T}\right\}$$
$$+ \frac{1}{2}(2\pi)^{-1}\exp\left\{-\frac{\theta^2 T}{4\sigma^2}\right\} \int_0^{\sqrt{2\lambda}/\sigma\sqrt{T}} xe^{-x^2/2} I_0\left(\frac{\theta}{\sigma}\sqrt{\frac{T}{2}}x\right) dx \quad (6.1.39)$$

The expression in (6.1.39) is clearly not a function of the phase ϕ. Thus, due to Lemma 6.1.1, we conclude that the uniform density function $f^*(\phi)$ in $[0, 2\pi]$ is least favorable, and that the decision rule δ^*_{f*} in (6.1.37) is minimax for the problem. The expected penalty $C(\delta^*_{f*}, f^*)$ induced by (δ^*_{f*}, f^*) is as in (6.1.39). In Figure 6.1.5, we plot this expected penalty as it varies with varying $(\theta/\sigma)\sqrt{(T/2)}$.

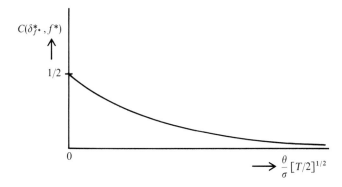

Figure 6.1.5

For fixed observation length T the expected penalty decreases monotonically, with increasing signal-to-noise ratio $\theta^2/2\sigma^2$. The expected penalty also decreases monotonically with increasing observation interval T when the signal-to-noise ratio remains unchanged.

Example 6.1.5

Let us revisit Example 4.5.1, where we will now assume that the a priori probabilities of the M hypotheses are unknown. Let us consider the decision rule in (4.5.1), which evolves from the assumption of equiprobable hypotheses. From (4.5.6) in Example 4.5.1 we conclude that the conditional expected penalty (which here coincides with the probability of error) induced by the rule in (4.5.1) is not a function of the set $\{q_k; 1 \leq k \leq M\}$ of a priori probabilities on the M hypotheses. Thus, Lemma 6.1.1 applies, and it is concluded that the admissible decision rule in (4.5.1) is minimax. The set $\{q_k = M^{-1}; \forall k\}$ is then the least favorable set of a priori probabilities. The minimax expected penalty is given by (4.5.7) in Example 4.5.1 and its behavior for $M = 2$ is exhibited in Figure 4.5.2.

6.2 THE GAME IN ROBUST DETECTION

To this point, we have studied optimal detection schemes in the case where the hypotheses considered are described by parametrically known stochastic processes. As we have seen, among the parametric detection schemes, the Neyman-Pearson formalization, in Chapter 5, assumes the minimum number of available assets. In this section, we consider further reduction of the available assets. In particular, starting with noncomposite hypotheses H_0 and H_1 in the Neyman-Pearson formalization, we assume that the stochastic processes that describe the two hypotheses are both nonparametrically described. Each hypothesis is then represented by a class of parametrically known stochastic processes, and the robust models arise. Physically speaking, the robust models reflect the analyst's lack of confidence in his own conclusions regarding the statistical description of the hypotheses. Indeed, the analyst draws such conclusions, applying data analysis methods, on data generated by each hypothesis. Data analysis methods are never precise, however, and so the statistical conclusions they induce cannot be taken with blind faith. This, in conjunction with the established fact that small deviations from the assumed stochastic models may result in significant performance deterioration, gave rise to the robust formalizations.

Let us consider discrete-time stochastic processes, and let n be the number of observed data. Let $\mathscr{F}_1$ and $\mathscr{F}_0$ be two disjoint classes of n-dimensional density functions, and let $\mathscr{F}_1$ and $\mathscr{F}_0$ represent respectively hypothesis H_1 and hypothesis H_0. Given some observation vector x^n let the objective be to decide in favor of one of the hypotheses H_0 or H_1, where the crucial assumption is that x^n is generated by a single density function in $\mathscr{F}_1 \cup \mathscr{F}_0$. In addition to x^n let $\mathscr{F}_1$ and $\mathscr{F}_0$ and some false alarm rate α be available to the analyst (or system designer). The above assets naturally induce a game, the players in which are Nature and the analyst. Nature selects the density function

6.2 The Game in Robust Detection

in $\mathscr{F}_1 \cup \mathscr{F}_0$ that generates x^n while the analyst selects a decision rule. Those two selections are done independently, and the payoff function is induced by the scheme power, subject to a false alarm constraint. Given some decision rule $\delta(x^n) = \Pr\{H_1 \text{ decided} | x^n\}$, given some density function f in $\mathscr{F}_1 \cup \mathscr{F}_0$, let $\mathscr{D}$ be the class of all valid decision rules, and let us denote

$$P(\delta, f) = \int_{\Gamma^n} \delta(x^n) f(x^n) \, dx^n \qquad \Gamma^n \text{ the space where } x^n \text{ takes its values} \quad (6.2.1)$$

The game is then formally expressed as

Find a pair (δ^*, f_1^*) such that $\delta^* \in \mathscr{D}$, $f_1^* \in \mathscr{F}_1$, and

$$\forall \delta \in \mathscr{D} \qquad P(\delta, f_1^*) \leq P(\delta^*, f_1^*) \leq P(\delta^*, f) \qquad \forall f \in \mathscr{F}_1 \quad (6.2.2)$$

$$P(\delta^*, f) \leq \alpha \qquad \forall f \in \mathscr{F}_0 \quad (6.2.3)$$

Expression (6.2.3) represents the false alarm constraint. If a decision rule δ^* that satisfies both (6.2.2) and (6.2.3) exists, it is called the *robust rule* for the problem. Let some density function f_0^* in $\mathscr{F}_0$ exist such that $P(\delta^*, f_0^*) = \sup\{P(\delta^*, f): f \in \mathscr{F}_0\}$ and let the rule δ^* and the density function f_1^* in $\mathscr{F}_1$ satisfy (6.2.2). The pair (f_0^*, f_1^*) is then called *least favorable* in $\mathscr{F}_0 \cup \mathscr{F}_1$ and δ^* is then the Neyman-Pearson rule at (f_0^*, f_1^*), given the false alarm rate α. The latter statement is clear due to the left part of the double inequality in (6.2.2), and the robust decision rule δ^* is then expressed as

$$\delta^*(x^n) = \begin{cases} 1 & x^n: \log \dfrac{f_1^*(x^n)}{f_0^*(x^n)} > \lambda^* \\ r^* & x^n: \log \dfrac{f_1^*(x^n)}{f_0^*(x^n)} = \lambda^* \\ 0 & x^n: \log \dfrac{f_1^*(x^n)}{f_0^*(x^n)} < \lambda^* \end{cases} \quad (6.2.4)$$

where

$$r^*: 0 \leq r^* \leq 1$$

$$r^*, \lambda^*: \quad r^* \int f_0^*(x^n) \, dx^n \quad + \quad \int f_0^*(x^n) \, dx^n \quad = \alpha \quad (6.2.5)$$

$$x^n: \log \dfrac{f_1^*(x^n)}{f_0^*(x^n)} = \lambda^* \qquad x^n: \log \dfrac{f_1^*(x^n)}{f_0^*(x^n)} > \lambda^*$$

As we discussed in Chapter 5, the randomizing constant r^* is zero if the density functions f_1^* and f_0^* are both continuous everywhere. The constant r^* may be nonzero otherwise, in which case the robust rule in (6.2.4) is randomized.

The concept of the least favorable density functions, introduced above, gives rise to the following version of the game in (6.2.2) and (6.2.3).

Given a false alarm rate α and the classes $\mathscr{F}_1$ and $\mathscr{F}_0$, search for a pair (λ^*, r^*) of constants and a pair (f_0^*, f_1^*) of density functions, such that

$$f_0^* \in \mathscr{F}_0 \quad f_1^* \in \mathscr{F}_1 \quad 0 \le r^* \le 1$$

$$\alpha = \Pr\left\{\log\frac{f_1^*(X^n)}{f_0^*(X^n)} > \lambda^* \,\Big|\, f_0^*\right\} + r^* \Pr\left\{\log\frac{f_1^*(X^n)}{f_0^*(X^n)} = \lambda^* \,\Big|\, f_0^*\right\}$$

$$\ge \Pr\left\{\log\frac{f_1^*(X^n)}{f_0^*(X^n)} > \lambda^* \,\Big|\, f\right\} + r^* \Pr\left\{\log\frac{f_1^*(X^n)}{f_0^*(X^n)} = \lambda^* \,\Big|\, f\right\} \quad \forall f \in \mathscr{F}_0$$

(6.2.6)

$$\Pr\left\{\log\frac{f_1^*(X^n)}{f_0^*(X^n)} > \lambda^* \,\Big|\, f_1^*\right\} + r^* \Pr\left\{\log\frac{f_1^*(X^n)}{f_0^*(X^n)} = \lambda^* \,\Big|\, f_1^*\right\}$$

$$\le \Pr\left\{\log\frac{f_1^*(X^n)}{f_0^*(X^n)} > \lambda^* \,\Big|\, f\right\} + r^* \Pr\left\{\log\frac{f_1^*(X^n)}{f_0^*(X^n)} = \lambda^* \,\Big|\, f\right\} \quad \forall f \in \mathscr{F}_1$$

(6.2.7)

If pairs (λ^*, r^*) and (f_0^*, f_1^*) exist as above, then the decision rule δ^* in (6.2.4) is robust for the problem. The rule δ^* is then also the Neyman-Pearson rule at the pair (f_0^*, f_1^*) and at the false alarm rate α.

The objective of robust detection rules is protection against unfavorable conditions. The least favorable such conditions are represented by the pair (f_0^*, f_1^*) of density functions in (6.2.6) and (6.2.7). The existence and nature of this pair and the performance characteristics of the resulting robust rule δ^* depend on the specific structure of the classes $\mathscr{F}_1$ and $\mathscr{F}_0$. All the presently known robust detection schemes are limited to the case where both $\mathscr{F}_1$ and $\mathscr{F}_0$ contain only stationary and memoryless stochastic processes. We study this interesting special case in the next section.

6.3 ROBUST DETECTION FOR CLASSES OF STATIONARY AND MEMORYLESS PROCESSES

Let us consider the game in (6.2.2) and (6.2.3), where both classes $\mathscr{F}_0$ and $\mathscr{F}_1$ contain only stationary and memoryless processes. Then, given n, an observation vector $x^n = [x_1, x_2, \ldots, x_n]^T$, and some density function f in $\mathscr{F}_0 \cup \mathscr{F}_1$ we have

$$f(x^n) = \prod_{i=1}^{n} f(x_i) \quad (6.3.1)$$

where $\prod$ denotes product.

Given some f_1 in $\mathscr{F}_1$ and some f_0 in $\mathscr{F}_0$, we then obtain

$$\log\frac{f_1(x^n)}{f_0(x^n)} = \sum_{i=1}^{n} \log\frac{f_1(x_i)}{f_0(x_i)} \quad (6.3.2)$$

Therefore, the game in (6.2.6) and (6.2.7) is here expressed as

6.3 Robust Detection for Classes of Stationary and Memoryless Processes

Given a false alarm rate, α, given the classes, $\mathscr{F}_1$ and $\mathscr{F}_0$, of stationary and memoryless processes, search for a pair (λ^*, r^*) of constants, and a pair, (f_0^*, f_1^*), of density functions, such that,

$$f_0^* \in \mathscr{F}_0, f_1^* \in \mathscr{F}_1 \qquad 0 \leq r^* \leq 1$$

and

$$\text{for} \quad T^*(X^n) = n^{-1} \sum_{i=1}^{n} \log \frac{f_1^*(X_i)}{f_0^*(X_i)} \tag{6.3.3}$$

$$\alpha = \Pr\{T^*(X^n) > \lambda^* | f_0^*\} + r^* \Pr\{T^*(X^n) = \lambda^* | f_0^*\}$$
$$\geq \Pr\{T^*(X^n) > \lambda^* | f\} + r^* \Pr\{T^*(X^n) = \lambda^* | f\} \qquad \forall f \in \mathscr{F}_0 \tag{6.3.4}$$

$$\Pr\{T^*(X^n) > \lambda^* | f_1^*\} + r^* \Pr\{T^*(X^n) = \lambda^* | f_1^*\}$$
$$\leq \Pr\{T^*(X^n) > \lambda^* | f\} + r^* \Pr\{(T^*(X^n) = \lambda^* | f\} \qquad \forall f \in \mathscr{F}_1 \tag{6.3.5}$$

If the pairs (λ^*, r^*) and (f_0^*, f_1^*) above exist, then the robust decision rule δ^* is as follows:

$$\delta^*(x^n) = \begin{cases} 1 & x^n: T^*(x^n) > \lambda^* \\ r^* & x^n: T^*(x^n) = \lambda^* \\ 0 & x^n: T^*(x^n) < \lambda^* \end{cases} \tag{6.3.6}$$

Let us now study the random variable $T^*(X^n)$ in (6.3.3), assuming that the random vector X^n is generated by some density function f in $\mathscr{F}_0 \cup \mathscr{F}_1$. Since f is the density function of a stationary and memoryless process, the random variables $Y_i = \log[f_1^*(X_i)/f_0^*(X_i)]$; $1 \leq i \leq n$ are independent and identically distributed. Their identical density functions are controlled by f. At this point, let us perform asymptotic analysis. In particular, let the length n of the observation vector be asymptotically large. Then the law of large numbers applies, and we conclude that the random variable $W = n^{-1} \sum_{i=1}^{n} Y_i$ is asymptotically ($n \to \infty$) Gaussian, with mean m_f and variance σ_f^2 given by

$$m_f = E\{W | f\} = n^{-1} \sum_{i=1}^{n} E\{Y_i | f\} = E\{Y | f\} = \int_\Gamma \log \frac{f_1^*(x)}{f_0^*(x)} f(x) \, dx \tag{6.3.7}$$

$$\sigma_f^2 = E\{(W - m_f)^2 | f\} = E\left\{\left[n^{-1} \sum_{i=1}^{n} (Y_i - m_f)\right]^2 \bigg| f\right\}$$

$$= n^{-2} \sum_{i=1}^{n} E\{(Y_i - m_f)^2 | f\} = n^{-1} E\{(Y - m_f)^2 | f\}$$

$$= n^{-1} E\left\{\left[\log \frac{f_1^*(X)}{f_0^*(X)} - m_f\right]^2 \bigg| f\right\}$$

$$= n^{-1} \left\{\int_\Gamma \left[\log \frac{f_1^*(x)}{f_0^*(x)}\right]^2 f(x) \, dx - m_f^2\right\} \tag{6.3.8}$$

where Γ denotes the space where f is defined.

Therefore, given that the random vector X^n is generated by the density function f in $\mathscr{F}_0 \cup \mathscr{F}_1$, the random variable $T^*(X^n)$ in (6.3.3) is asymptotically ($n \to \infty$) Gaussian, with mean m_f as in (6.3.7) and with variance σ_f^2 as in (6.3.8). Denoting by $\Phi(x)$ the cumulative distribution function of the zero mean, unit variance, Gaussian random variable at the point x, we can thus write the asymptotic ($n \to \infty$) versions of (6.3.4) and (6.3.5) as (where $r^* = 0$, due to the continuity of the Gaussian density function).

$$\alpha = \Phi\left(\frac{-\lambda^* + m_{f_0^*}}{\sigma_{f_0^*}}\right) \geq \Phi\left(\frac{-\lambda^* + m_f}{\sigma_f}\right) \quad \forall f \in \mathscr{F}_0 \quad (6.3.9)$$

$$\Phi\left(\frac{-\lambda^* + m_{f_1^*}}{\sigma_{f_1^*}}\right) \leq \Phi\left(\frac{-\lambda^* + m_f}{\sigma_f}\right) \quad \forall f \in \mathscr{F}_1 \quad (6.3.10)$$

Due to the strict monotonicity of $\Phi(x)$ we can simplify (6.3.9) and (6.3.10) as

$$\Phi^{-1}(\alpha) = \frac{-\lambda^* + m_{f_0^*}}{\sigma_{f_0^*}} \geq \frac{-\lambda^* + m_f}{\sigma_f} \quad \forall f \in \mathscr{F}_0 \quad (6.3.11)$$

$$\frac{-\lambda^* + m_{f_1^*}}{\sigma_{f_1^*}} \leq \frac{-\lambda^* + m_f}{\sigma_f} \quad \forall f \in \mathscr{F}_1 \quad (6.3.12)$$

where $w = \Phi^{-1}(x)$ implies $x = \Phi(w)$.

Solving for λ^* from the equality in (6.3.11) and substituting in (6.3.11) and (6.3.12), we obtain

$$\sigma_f \Phi^{-1}(\alpha) - m_f \geq \sigma_{f_0^*}\Phi^{-1}(\alpha) - m_{f_0^*} \quad \forall f \in \mathscr{F}_0 \quad (6.3.13)$$

$$\frac{\sigma_{f_0^*}\Phi^{-1}(\alpha) - m_{f_0^*} + m_{f_1^*}}{\sigma_{f_1^*}} \leq \frac{\sigma_{f_0^*}\Phi^{-1}(\alpha) - m_{f_0^*} + m_f}{\sigma_f} \quad \forall f \in \mathscr{F}_1 \quad (6.3.14)$$

$$\lambda^* = m_{f_0^*} - \sigma_{f_0^*}\Phi^{-1}(\alpha) \quad (6.3.15)$$

where m_f and σ_f are given respectively by (6.3.7) and (6.3.8).

Asymptotically ($n \to \infty$), the search for a robust decision rule reduces to the following problem:

Search for a pair (f_0^*, f_1^*) of density functions such that

$$\sigma_f \Phi^{-1}(\alpha) - m_f \geq \sigma_{f_0^*}\Phi^{-1}(\alpha) - m_{f_0^*} \quad \forall f \in \mathscr{F}_0 \quad (6.3.16)$$

$$\frac{\sigma_{f_0^*}\Phi^{-1}(\alpha) - m_{f_0^*} + m_{f_1^*}}{\sigma_{f_1^*}} \leq \frac{\sigma_{f_0^*}\Phi^{-1}(\alpha) - m_{f_0^*} + m_f}{\sigma_f} \quad \forall f \in \mathscr{F}_1 \quad (6.3.17)$$

where

$$m_f = \int_\Gamma \log \frac{f_1^*(x)}{f_0^*(x)} f(x)\, dx \quad (6.3.18)$$

$$\sigma_f^2 = n^{-1}\left\{\int_\Gamma \left[\log \frac{f_1^*(x)}{f_0^*(x)}\right]^2 f(x)\, dx - m_f^2\right\} \quad (6.3.19)$$

6.3 Robust Detection for Classes of Stationary and Memoryless Processes

If such a pair (f_0^*, f_1^*) exists, then the asymptotic $(n \to \infty)$ robust decision rule δ^* is

$$\delta^*(x^n) = \begin{cases} 1 & x^n: n^{-1} \sum_{i=1}^{n} \log \frac{f_1^*(x_i)}{f_0^*(x_i)} \geq m_{f_0^*} - \sigma_{f_0^*} \Phi^{-1}(\alpha) \\ 0 & \text{otherwise} \end{cases} \quad (6.3.20)$$

Given two disjoint classes $\mathscr{F}_0$ and $\mathscr{F}_1$ of stationary and memoryless processes, let us assume the existence of a pair of density functions (f_0^*, f_1^*); $f_0^* \in \mathscr{F}_0$, $f_1^* \in \mathscr{F}_1$ that satisfy the above asymptotic game. Let us further assume that the pair (f_0^*, f_1^*) is such that the quantity $n\sigma_f^2$ in (6.3.19) equals a bounded constant c_1 for every f in F_1, and equals another bounded constant c_0 for every f in $\mathscr{F}_0$. Then, for asymptotically large values of n, the terms in expressions (6.3.16) and (6.3.17) that involve variances σ_f become infinitesimal, and those expressions take respectively the forms

$$m_{f_0^*} = \int_\Gamma \log \frac{f_1^*(x)}{f_0^*(x)} f_0^*(x)\, dx \geq m_f = \int_\Gamma \log \frac{f_1^*(x)}{f_0^*(x)} f(x)\, dx \quad \forall f \in \mathscr{F}_0$$

(6.3.21)

$$m_{f_1^*} = \int_\Gamma \log \frac{f_1^*(x)}{f_0^*(x)} f_1^*(x)\, dx \leq m_f = \int_\Gamma \log \frac{f_1^*(x)}{f_0^*(x)} f(x)\, dx \quad \forall f \in \mathscr{F}_1$$

(6.3.22)

where, due to the logarithmic inequality, $\log x \leq x - 1$,

$$\int_\Gamma \log \frac{f_1^*(x)}{f_0^*(x)} f(x)\, dx \leq \int_\Gamma \log \frac{f(x)}{f_0^*(x)} f(x)\, dx \quad (6.3.23)$$

$$\int_\Gamma \log \frac{f_1^*(x)}{f_0^*(x)} f(x)\, dx \geq \int_\Gamma \log \frac{f_1^*(x)}{f(x)} f(x)\, dx \quad (6.3.24)$$

From (6.3.21) through (6.3.24), we conclude that if the pair (f_0^*, f_1^*) that satisfies the asymptotic game in (6.3.16) through (6.3.20) is such that $n\sigma_f^2 = c_1$; $\forall f \in \mathscr{F}_1$ and $n\sigma_f^2 = c_0$; $\forall f \in \mathscr{F}_0$, then this pair necessarily satisfies

$$I(f_0^*, f_1^*) = \int_\Gamma \log \frac{f_0^*(x)}{f_1^*(x)} f_0^*(x)\, dx \leq \int_\Gamma \log \frac{f(x)}{f_1^*(x)} f(x)\, dx = I(f, f_1^*) \quad \forall f \in \mathscr{F}_0$$

(6.3.25)

$$I(f_1^*, f_0^*) = \int_\Gamma \log \frac{f_1^*(x)}{f_0^*(x)} f_1^*(x)\, dx \leq \int_\Gamma \log \frac{f(x)}{f_0^*(x)} f(x)\, dx = I(f, f_0^*) \quad \forall f \in \mathscr{F}_1$$

(6.3.26)

That is, among all density functions in $\mathscr{F}_0$, f_0^* is then the closest to f_1^*, in Kullback-Leibler distance (6.3.25). Similarly, among all density functions in $\mathscr{F}_1$,

f_1^* is then the closest to f_0^*, also in Kullback-Leibler distance (6.2.26). Given two density functions f_1 and f_2 the Kullback-Leibler distance $I(f_1, f_2)$ equals $\int \log[f_1(x)/f_2(x)] f_1(x) dx$ and is a pseudo-distance. The Kullback-Leibler closeness between f_0^* and f_1^*, explained above, is intuitively pleasing. It quantifies the qualitative statement that the least favorable density functions in the robust game represent the shortest distance points between the two disjoint classes $\mathscr{F}_0$ and $\mathscr{F}_1$. Those two points are the hardest to distinguish from each other; they thus correspond to the Neyman-Pearson test with the least power.

Given two disjoint classes $\mathscr{F}_0$ and $\mathscr{F}_1$ of stationary and memoryless processes, let us now assume that there exist some density functions f_1^* in $\mathscr{F}_1$ and f_0^* in $\mathscr{F}_0$, such that

$$\Pr\{\gamma^*(X) < y | f_0\} \geq \Pr\{\gamma^*(X) < y | f_0^*\} \geq \Pr\{\gamma^*(X) < y | f_1^*\}$$
$$\geq \Pr\{\gamma^*(X) < y | f_1\} \quad \forall f_0 \in \mathscr{F}_0, \forall f_1 \in \mathscr{F}_1, \forall y \text{ in } \Gamma$$
(6.3.27)

where

$$\gamma^*(x) = \log \frac{f_1^*(x)}{f_0^*(x)} \tag{6.3.28}$$

We can then express the following lemma.

LEMMA 6.3.1: If density functions f_1^* in $\mathscr{F}_1$ and f_0^* in $\mathscr{F}_0$ exist that satisfy (6.3.27), then they comprise the least favorable pair of density functions in $\mathscr{F}_0 \times \mathscr{F}_1$ for every dimensionality n of the observation vector. Thus, given n and a false alarm rate α the robust decision rule $\delta^*(x^n)$ is

$$\delta^*(x^n) = \begin{cases} 1 & x^n: T^*(x^n) = n^{-1} \sum_{i=1}^{n} \gamma^*(x_i) > \lambda^* \\ r^* & x^n: T^*(x^n) = n^{-1} \sum_{i=1}^{n} \gamma^*(x_i) = \lambda^* \\ 0 & x^n: T^*(x^n) = n^{-1} \sum_{i=1}^{n} \gamma^*(x_i) < \lambda^* \end{cases} \tag{6.3.29}$$

where

$$\lambda^*, r^*: \Pr\left\{n^{-1} \sum_{i=1}^{n} \gamma^*(X_i) > \lambda^* \Big| f_0^*\right\} + r^* \Pr\left\{n^{-1} \sum_{i=1}^{n} \gamma^*(X_i) = \lambda^* \Big| f_0^*\right\} = \alpha$$
(6.3.30)

Proof: The condition in (6.3.27) implies that, given $f_0 \in \mathscr{F}_0$ and $f_1 \in \mathscr{F}_1$, a random variable Y and nondecreasing functions $g_0(y)$, $g_0^*(y)$, $g_1^*(y)$ and $g_1(y)$ exist, such that $g_0(y) \leq g_0^*(y) \leq g_1^*(y) \leq g_1(y) \; \forall \, y \in \Gamma$ and $g_0(y)$, $g_0^*(y)$, $g_1^*(y)$, and $g_1(y)$ coincide respectively with the probabilities $\Pr\{\gamma^*(X) < y | f_0\}$, $\Pr\{\gamma^*(X) < y | f_0^*\}$, $\Pr\{\gamma^*(X) < y | f_1^*\}$, and $\Pr\{\gamma^*(X) < y | f_1\}$. Therefore, given n, we can

6.3 Robust Detection for Classes of Stationary and Memoryless Processes

write

$P(\delta^*, f_0)$

$$= \Pr\left\{n^{-1}\sum_{i=1}^{n}\gamma^*(X_i) > \lambda^* \mid f_0\right\} + r^*\Pr\left\{n^{-1}\sum_{i=1}^{n}\gamma^*(X_i) = \lambda^* \mid f_0\right\}$$

$$= r^*\Pr\left\{n^{-1}\sum_{i=1}^{n}\gamma^*(X_i) \geq \lambda^* \mid f_0\right\} + (1-r^*)\Pr\left\{n^{-1}\sum_{i=1}^{n}\gamma^*(X_i) > \lambda^* \mid f_0\right\}$$

$$= r^*\Pr\left\{n^{-1}\sum_{i=1}^{n}g_0(Y_i) \geq \lambda^*\right\} + (1-r^*)\Pr\left\{n^{-1}\sum_{i=1}^{n}g_0(Y_i) > \lambda^*\right\}$$

$$\leq r^*\Pr\left\{n^{-1}\sum_{i=1}^{n}g_0^*(Y_i) \geq \lambda^*\right\} + (1-r^*)\Pr\left\{n^{-1}\sum_{i=1}^{n}g_0^*(Y_i) > \lambda^*\right\}$$

$$= r^*\Pr\left\{n^{-1}\sum_{i=1}^{n}\gamma^*(X_i) \geq \lambda^* \mid f_0^*\right\} + (1-r^*)\Pr\left\{n^{-1}\sum_{i=1}^{n}\gamma^*(X_i) > \lambda^* \mid f_0^*\right\}$$

$$= \Pr\left\{n^{-1}\sum_{i=1}^{n}\gamma^*(X_i) > \lambda^* \mid f_0^*\right\} + r^*\Pr\left\{n^{-1}\sum_{i=1}^{n}\gamma^*(X_i) = \lambda^* \mid f_0^*\right\}$$

$$= P(\delta^*, f_0^*)$$

So

$$P(\delta^*, f_0) \leq P(\delta^*, f_0^*) = \alpha \qquad \forall n, \forall f_0 \in \mathscr{F}_0$$

Similarly, we find

$$\forall \delta \qquad P(\delta, f_1^*) \leq P(\delta^*, f_1^*) \leq P(\delta^*, f_1) \qquad \forall n, \forall f_1 \in \mathscr{F}_1$$

The pair (f_0^*, f_1^*) thus satisfies the game in (6.2.2) and (6.2.3) for all n and the proof is complete. ∎

An interesting application of Lemma 6.3.1, was first presented by Huber (1965). Huber considered classes $\mathscr{F}_0$ and $\mathscr{F}_1$ such that each represents a "linear" contamination of some well known nominal, stationary, and memoryless process. In particular, if $f_0(x)$ and $f_1(x)$ denote respectively the one-dimensional density functions induced by the nominal stochastic processes in $\mathscr{F}_0$ and $\mathscr{F}_1$ at the scalar point x, Huber's classes are defined as

$$\begin{aligned}\mathscr{F}_0 &= \{f(x) = (1-\varepsilon_0)f_0(x) + \varepsilon_0 h(x) \qquad \forall x \in R, h \in \mathscr{H}\} \\ \mathscr{F}_1 &= \{f(x) = (1-\varepsilon_1)f_1(x) + \varepsilon_1 h(x) \qquad \forall x \in R, h \in \mathscr{H}\}\end{aligned} \qquad (6.3.31)$$

where R is the real line; $\mathscr{H}$ is the class of all the legitimate, one-dimensional density functions designed on R; ε_0 and ε_1 are two given real constants whose values lie in $(0, 1)$; and $f(x)$ denotes a one-dimensional density function defined on R.

Each one of the classes in (6.3.31) models the case where, in a sequence of data generated by the corresponding nominal process, contaminations may appear independently per datum with probabilities ε_0 for class $\mathscr{F}_0$ and ε_1 for class $\mathscr{F}_1$. Thus, the classes in (6.3.31) represent contaminations of the nominal data due to occasional, independent data outliers. Let us now consider the density functions f_0^* and f_1^* that belong respectively to $\mathscr{F}_0$ and $\mathscr{F}_1$ in (6.3.31) and are defined as

$$f_0^*(x) = \begin{cases} (1-\varepsilon_0)f_0(x) & x: f_1(x)/f_0(x) < c_0 \\ c_0^{-1}(1-\varepsilon_0)f_1(x) & x: f_1(x)/f_0(x) \geq c_0 \end{cases} \quad (6.3.32)$$

$$f_1^*(x) = \begin{cases} (1-\varepsilon_1)f_1(x) & x: f_1(x)/f_0(x) > c_1 \\ c_1(1-\varepsilon_1)f_0(x) & x: f_1(x)/f_0(x) \leq c_1 \end{cases} \quad (6.3.33)$$

where $0 \leq c_1 < c_0 \leq \infty$, and c_0, c_1 are unique and such that

$$\Pr\left\{\frac{f_1(X)}{f_0(X)} < c_0 \Big| f_0\right\} + c_0^{-1} \Pr\left\{\frac{f_1(X)}{f_0(X)} \geq c_0 \Big| f_1\right\} = (1-\varepsilon_0)^{-1}$$

$$\Pr\left\{\frac{f_1(X)}{f_0(X)} > c_1 \Big| f_1\right\} + c_1 \Pr\left\{\frac{f_1(X)}{f_0(X)} \leq c_1 \Big| f_0\right\} = (1-\varepsilon_1)^{-1} \quad (6.3.34)$$

It can easily be verified that the density functions f_0^* and f_1^* in (6.3.32) and (6.3.33) satisfy the condition in (6.3.27) for $\Gamma = R$; thus, Lemma 6.3.1 applies. Therefore, f_0^* and f_1^* induce the robust decision rule for the classes in (6.3.31) and for any dimensionality n of the observation vector. We state the robust rule in the corollary below.

COROLLARY 6.3.1: Given the classes $\mathscr{F}_0$ and $\mathscr{F}_1$ in (6.3.31), n, and the false alarm rate α the robust decision rule $\delta^*(x^n)$ is

$$\delta^*(x^n) = \begin{cases} 1 & x^n: T^*(x^n) > \lambda^* \\ r^* & x^n: T^*(x^n) = \lambda^* \\ 0 & x^n: T^*(x^n) < \lambda^* \end{cases}$$

where

$$\lambda^*, r^*: \Pr\{T^*(X^n) > \lambda^* | f_0^*\} + r^* \Pr\{T^*(X^n) = \lambda^* | f_0^*\} = \alpha$$

where f_0^* is as in (6.3.32), the constants c_0 and c_1 are as in (6.3.34), and

$$T^*(x^n) = n^{-1} \sum_{i=1}^{n} \gamma^*(x_i)$$

$$\gamma^*(x) = \log \frac{f_1^*(x)}{f_0^*(x)}$$

$$= \begin{cases} \log[c_1(1-\varepsilon_1)(1-\varepsilon_0)^{-1}] & x: f_1(x)/f_0(x) \leq c_1 \\ \log\left[(1-\varepsilon_1)(1-\varepsilon_0)^{-1}\frac{f_1(x)}{f_0(x)}\right] & x: c_1 < f_1(x)/f_0(x) < c_0 \\ \log[c_0(1-\varepsilon_1)(1-\varepsilon_0)^{-1}] & x: f_1(x)/f_0(x) \geq c_0 \end{cases} \quad (6.3.35)$$

∎

6.3 Robust Detection for Classes of Stationary and Memoryless Processes

We note that if the contamination constants ε_0 and ε_1 in (6.3.31) are not small enough, then $\mathscr{F}_0$ and $\mathscr{F}_1$ may overlap, and c_0 and c_1 in (6.3.35) may then be such that $c_0 < c_1$. Lemma 6.3.1 does not hold then. If $\varepsilon_0 = \varepsilon_1$, then $c_0 = c_1 = 1$ results. Therefore, as $c_0 \searrow 1$ and $c_1 \nearrow 1$, the normalized log likelihood ratio sum $\sum_{i=1}^n [\log[f_1^*(x_i)/f_0^*(x_i)] - \log c_1][\log c_0 - \log c_1]^{-1}$ tends to the number of times the inequality $f_1(x_1)|f_0(x_1) > 1$ holds, where f_1 and f_0 are the nominal density functions in $\mathscr{F}_1$ and $\mathscr{F}_0$. The latter limit test is called the *sign test*. We will see more on the sign test in the next chapter.

From (6.3.35), we observe that the test function $T^*(x^n)$ of the robust decision rule for the classes in (6.3.31) is the normalized sum of bounded data transformations. Such transformations protect the system from extreme outliers, which would otherwise cause highly erroneous decisions. We also note that the pair (f_0^*, f_1^*) in (6.3.32) and (6.3.33) satisfies the conditions in (6.3.25) and (6.3.26); it is thus the pair of density functions closest in Kullback-Leibler distance across $\mathscr{F}_0$ and $\mathscr{F}_1$. We will complete this section with an example that represents a case with wide applicability.

Example 6.3.1

Consider the case where a deterministic scalar constant signal may be transmitted in additive, white noise. Let the noise process be discrete-time, and let θ denote the scalar signal value. Let θ be positive and known to the receiver, and let the objective be to detect the absence or presence of the signal from an n-dimensional observation vector x^n. Let us suppose that it has been deduced, via initial data analysis that the noise process is Gaussian, zero mean, stationary and memoryless, with power σ^2. Extreme noise data, in the form of occasional outliers, do occur, however, due to unusual atmospheric disturbances. In the absence of the signal, let the observed frequency of those outliers be ε_0. Let the presence of the signal involve a frequency ε_1 of observed outliers. Thus, the absence of the signal is then represented by a class $\mathscr{F}_0$ of discrete-time, stationary and memoryless processes, while the presence of the signal is represented by another such class $\mathscr{F}_1$. The two classes are as in (6.3.31), with nominal density functions f_0 and f_1 given respectively as

$$f_0(x) = (2\pi)^{-1/2} \sigma^{-1} \exp\left\{-\frac{x^2}{2\sigma^2}\right\}$$
$$f_1(x) = (2\pi)^{-1/2} \sigma^{-1} \exp\left\{-\frac{(x-\theta)^2}{2\sigma^2}\right\} \quad (6.3.36)$$

The robust receiver is designed for protection against outliers, and for a given false alarm rate α and observation length n. This receiver operates as in Corollary 6.3.1, where here

$$\frac{f_1(x)}{f_0(x)} = \exp\left\{\frac{\theta}{2\sigma^2}(2x-\theta)\right\} \quad \ln\frac{f_1(x)}{f_0(x)} = \frac{\theta}{2\sigma^2}(2x-\theta) \quad (6.3.37)$$

$$\gamma^*(x) = \ln \frac{f_1^*(x)}{f_0^*(x)}$$

$$= \begin{cases} \ln c_1 + \ln[(1-\varepsilon_1)(1-\varepsilon_0)^{-1}] & x: \dfrac{\theta}{2\sigma^2}(2x-\theta) \leq \ln c_1 \\ \ln[(1-\varepsilon_1)(1-\varepsilon_0)^{-1}] + \dfrac{\theta}{2\sigma^2}(2x-\theta) & x: \ln c_1 < \dfrac{\theta}{2\sigma^2}(2x-\theta) < \ln c_0 \\ \ln[(1-\varepsilon_1)(1-\varepsilon_0)^{-1}] + \ln c_0 & x: \dfrac{\theta}{2\sigma^2}(2x-\theta) \geq \ln c_0 \end{cases}$$

(6.3.38)

Let us define

$$z(x) = \frac{\sigma^2}{\theta}\gamma^*(x) - \frac{\sigma^2}{\theta}\ln[(1-\varepsilon_1)(1-\varepsilon_0)^{-1}] + \frac{\theta}{2} \quad (6.3.39)$$

$$d_1 = \frac{\sigma^2}{\theta}\ln c_1 + \frac{\theta}{2}$$

$$d_0 = \frac{\sigma^2}{\theta}\ln c_0 + \frac{\theta}{2}$$

(6.3.40)

Then, via some simple transformations in (6.3.38), we find

$$z(x) = \begin{cases} d_1 & x \leq d_1 \\ x & d_1 < x < d_0 \\ d_0 & x \geq d_0 \end{cases} \quad (6.3.41)$$

$$T^*(x^n) = n^{-1} \sum_{i=1}^n \gamma^*(x_i)$$

$$= n^{-1} \frac{\theta}{\sigma^2} \sum_{i=1}^n \left(z(x_i) + \frac{\sigma^2}{\theta}\ln[(1-\varepsilon_1)(1-\varepsilon_0)^{-1}] - \frac{\theta}{2} \right)$$

$$= \frac{\theta}{\sigma^2} n^{-1} \sum_{i=1}^n z(x_i) + \ln[(1-\varepsilon_1)(1-\varepsilon_0)^{-1}] - \frac{\theta^2}{2\sigma^2}$$

The robust decision rule $\delta^*(x^n)$ in Corollary 6.3.1, now takes the equivalent form

$$\delta^*(x^n) = \begin{cases} 1 & x^n: n^{-1} \sum_{i=1}^n z(x_i) > \lambda^* \\ r^* & x^n: n^{-1} \sum_{i=1}^n z(x_i) = \lambda^* \\ 0 & x^n: n^{-1} \sum_{i=1}^n z(x_i) < \lambda^* \end{cases} \quad (6.3.42)$$

6.3 Robust Detection for Classes of Stationary and Memoryless Processes

where $z(x)$ is as in (6.3.41), and where λ^* and r^* are such that

$$\Pr\left\{n^{-1}\sum_{i=1}^{n} z(X_i) > \lambda^* \mid f_0^*\right\} + r^* \Pr\left\{n^{-1}\sum_{i=1}^{n} z(X_i) = \lambda^* \mid f_0^*\right\} = \alpha \quad (6.3.43)$$

Directly from (6.3.32), (6.3.33), and (6.3.34) we conclude that there the least favorable density functions f_0^* and f_1^* and the constants d_0 and d_1 in (6.3.41) are

$$f_0^*(x) = \begin{cases} (1-\varepsilon_0)(2\pi)^{-1/2}\sigma^{-1}\exp\left\{-\dfrac{x^2}{2\sigma^2}\right\} & x: x < d_0 \\[2mm] (1-\varepsilon_0)(2\pi)^{-1/2}\sigma^{-1}\exp\left\{-\dfrac{(x-\theta)^2}{2\sigma^2} + \dfrac{\theta^2}{2\sigma^2} - \dfrac{\theta}{\sigma^2}d_0\right\} & x: x \geq d_0 \end{cases}$$

(6.3.44)

$$f_1^*(x) = \begin{cases} (1-\varepsilon_1)(2\pi)^{-1/2}\sigma^{-1}\exp\left\{-\dfrac{(x-\theta)^2}{2\sigma^2}\right\} & x: x > d_1 \\[2mm] (1-\varepsilon_1)(2\pi)^{-1/2}\sigma^{-1}\exp\left\{-\dfrac{x^2}{2\sigma^2} - \dfrac{\theta^2}{2\sigma^2} + \dfrac{\theta}{\sigma^2}d_1\right\} & x \leq d_1 \end{cases}$$

(6.3.45)

$$(1-\varepsilon_0)\left[\Pr\{X < d_0 \mid f_0\} + \exp\left\{\dfrac{\theta^2}{2\sigma^2} - \dfrac{\theta}{\sigma^2}d_0\right\}\Pr\{X \geq d_0 \mid f_1\}\right] = 1$$

$$(1-\varepsilon_1)\left[\Pr\{X > d_1 \mid f_1\} + \exp\left\{\dfrac{\theta}{\sigma^2}d_1 - \dfrac{\theta^2}{2\sigma^2}\right\}\Pr\{X \leq d_1 \mid f_0\}\right] = 1$$

(6.3.46)

where f_0 and f_1 are as in (6.3.36). Denoting by $\Phi(x)$ the cumulative distribution function of the zero mean, unit variance, Gaussian random variable at the point x, we can provide the following versions of (6.3.46).

$$(1-\varepsilon_0)\left[\Phi\left(\dfrac{d_0}{\sigma}\right) + \exp\left\{-\dfrac{\theta}{\sigma^2}d_0 + \dfrac{\theta^2}{2\sigma^2}\right\}\Phi\left(\dfrac{-d_0+\theta}{\sigma}\right)\right] = 1$$

$$(1-\varepsilon_1)\left[\Phi\left(\dfrac{-d_1+\theta}{\sigma}\right) + \exp\left\{\dfrac{\theta}{\sigma^2}d_1 - \dfrac{\theta^2}{2\sigma^2}\right\}\Phi\left(\dfrac{d_1}{\sigma}\right)\right] = 1$$

(6.3.47)

It can be easily verified that the system in (6.3.47) has no (d_0, d_1) solution if either one of the contaminating frequencies ε_0 and ε_1 is larger than .5. Indeed, the classes $\mathscr{F}_0$ and $\mathscr{F}_1$ overlap then, and no robust decision rule exists. Qualitatively, this means that when either $\varepsilon_1 > .5$ or $\varepsilon_0 > .5$ the outliers take over, and the presence versus the absence of the signal hypotheses cannot be distinguished. If $\varepsilon_0 = \varepsilon_1 < .5$, we easily conclude from (6.3.47) that $d_0 = -d_1 + \theta$. But due to (6.3.41), we also need $d_1 \leq d_0$. Thus, if $\varepsilon_0 = \varepsilon_1 < .5$ we necessarily have $2^{-1}\theta \geq d_1 = -d_0 + \theta$. We will study this case in detail. It represents the occurrence of occasional outliers that are not influenced by the presence or absence of the signal but represent external atmospheric disturbances.

CASE: $\varepsilon_0 = \varepsilon_1 = \varepsilon < .5$

As established above, we then have

$$2^{-1}\theta \geq d = d_1 = -d_0 + \theta \tag{6.3.48}$$

where the constant d is the unique solution of [from (6.3.47)]

$$G(\varepsilon, d) = (1 - \varepsilon)\left[\Phi\left(\frac{-d + \theta}{\sigma}\right) + \exp\left\{\frac{\theta}{\sigma^2}d - \frac{\theta^2}{2\sigma^2}\right\}\Phi\left(\frac{d}{\sigma}\right)\right] = 1 \tag{6.3.49}$$

For given $\varepsilon < .5$, the function $G(\varepsilon, d)$ in (6,3.49) increases monotonically with increasing d. For given d, it decreases monotonically with increasing ε. Thus, if we denote by d_ε^* the solution of the equation in (6.3.49) for given ε, then $d_{\varepsilon_1}^* > d_{\varepsilon_2}^*$ for $\varepsilon_1 > \varepsilon_2$. But in (6.3.48) we established that $d \leq 2^{-1}\theta$. Therefore, the largest value ε^* of ε for which the equation in (6.3.49) has a solution, is such that $\varepsilon^* = \min(.5, \varepsilon_0)$ where $G(\varepsilon_0, 2^{-1}\theta) = 1$. Substituting in (6.3.49), we obtain

$$\varepsilon_0: 2(1 - \varepsilon_0)\Phi\left(\frac{\theta}{2\sigma}\right) = 1 \quad \text{or} \quad \varepsilon_0 = 1 - \left[2\Phi\left(\frac{\theta}{2\sigma}\right)\right]^{-1}$$

and

$$\varepsilon^* = \min\left(.5, 1 - \left[2\Phi\left(\frac{\theta}{2\sigma}\right)\right]^{-1}\right) = 1 - \left[2\Phi\left(\frac{\theta}{2\sigma}\right)\right]^{-1} \tag{6.3.50}$$

The value ε^* in (6.3.50) is called the *breakdown point* of the problem. For every $\varepsilon > \varepsilon^*$, no robust decision rule exists. In Figure 6.3.1, we plot the breakdown point ε^* as a function of the ratio $\theta/2\sigma$. ∎

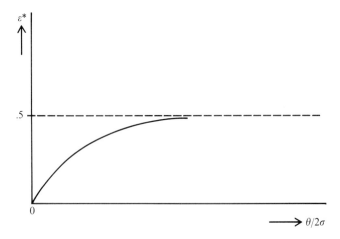

Figure 6.3.1 Breakdown point.

6.3 Robust Detection for Classes of Stationary and Memoryless Processes

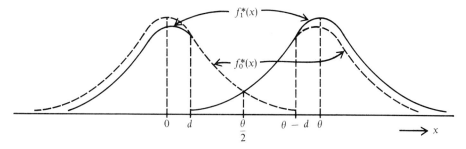

Figure 6.3.2

Given θ and $\varepsilon \leq \varepsilon^*$, the least favorable density functions f_0^* and f_1^* in (6.3.44) and (6.3.45) are

$$f_0^*(x) = \begin{cases} (1-\varepsilon)(2\pi)^{-1/2}\sigma^{-1} \exp\left\{-\dfrac{x^2}{2\sigma^2}\right\} & x: x < -d + \theta \\[2mm] (1-\varepsilon)(2\pi)^{-1/2}\sigma^{-1} \exp\left\{-\dfrac{(x-\theta)^2}{2\sigma^2} + \dfrac{\theta}{\sigma^2}d - \dfrac{\theta^2}{2\sigma^2}\right\} & x: x \geq -d + \theta \end{cases}$$

(6.3.51)

$$f_1^*(x) = \begin{cases} (1-\varepsilon)(2\pi)^{-1/2}\sigma^{-1} \exp\left\{-\dfrac{(x-\theta)^2}{2\sigma^2}\right\} & x: x > d \\[2mm] (1-\varepsilon)(2\pi)^{-1/2}\sigma^{-1} \exp\left\{-\dfrac{x^2}{2\sigma^2} + \dfrac{\theta}{\sigma^2}d - \dfrac{\theta^2}{2\sigma^2}\right\} & x: x \leq d \end{cases}$$

(6.3.52)

The above density functions are both exponential; they are such that $f_1^*[x + (\theta/2)] = f_0^*[-x + (\theta/2)]$; and they are sketched in Figure 6.3.2. From (6.3.41), (6.3.42), and (6.3.43) the robust decision rule $\delta^*(x^n)$ is here

$$\delta^*(x^n) = \begin{cases} 1 & x^n: n^{-1} \sum_{i=1}^{n} z(x_i) > \lambda^* \\[2mm] r^* & x^n: n^{-1} \sum_{i=1}^{n} z(x_i) = \lambda^* \\[2mm] 0 & x^n: n^{-1} \sum_{i=1}^{n} z(x_i) < \lambda^* \end{cases}$$

(6.3.53)

where

$$z(x) = \begin{cases} d & x \leq d \\ x & d < x < -d + \theta \\ -d + \theta & x \geq -d + \theta \end{cases}$$

(6.3.54)

$$(\lambda^*, r^*): \Pr\left\{n^{-1} \sum_{i=1}^{n} z(X_i) > \lambda^* \mid f_0^*\right\} + r^* \Pr\left\{n^{-1} \sum_{i=1}^{n} z(X_i) = \lambda^* \mid f_0^*\right\} = \alpha$$

(6.3.55)

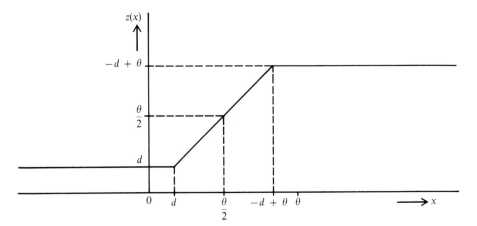

Figure 6.3.3

From (6.3.54), we observe that the robust receiver performs a truncation per datum, shown in Figure 6.3.3; this operation cuts off the extreme outliers. For given $\theta > 0$ and $\varepsilon < \varepsilon^*$ let us now study the robust receiver for asymptotically large values of n. As discussed earlier in this section, since the random variables $\{X_i\}$ are independent and identically distributed, when conditioned on any f in $\mathscr{F}_0 \cup \mathscr{F}_1$, the random variables $\{Y_i = z(X_i)\}$ are too, and the random variable $n^{-1} \sum_{i=1}^{n} z(X_i)$ is then asymptotically Gaussian. Let us denote

$$\rho = \frac{\theta}{\sigma} \cdot \quad c(\varepsilon) = \frac{d}{\sigma} \tag{6.3.56}$$

The quantity ρ in (6.3.56) is the signal-to-noise ratio. Given $\varepsilon < \varepsilon^*$ $c(\varepsilon)$ in (6.3.56) is the solution of the equation in (6.3.49).

From (6.3.51), (6.3.52), (6.3.54), and (6.3.56) we then find

$$m_{f_0^*} = E\left\{ n^{-1} \sum_{i=1}^{n} z(X_i) \bigg| f_0^* \right\} = E\{z(X) | f_0^*\}$$

$$= (1 - \varepsilon) d\Phi\left(\frac{d}{\sigma}\right) + (1 - \varepsilon)(-d + \theta) \exp\left\{\frac{\theta}{\sigma^2} d - \frac{\theta^2}{2\sigma^2}\right\} \Phi\left(\frac{d}{\sigma}\right)$$

$$+ (1 - \varepsilon)\sigma \int_{d/\sigma}^{(-d+\theta)/\sigma} x\phi(x) \, dx$$

$$= (1 - \varepsilon)\sigma\Phi\left(\frac{d}{\sigma}\right)\left[\frac{d}{\sigma} + \frac{\theta - d}{\sigma} \exp\left\{\frac{\theta}{\sigma^2} d - \frac{\theta^2}{2\sigma^2}\right\}\right]$$

$$+ (1 - \varepsilon)\sigma\left[\phi\left(\frac{d}{\sigma}\right) - \phi\left(\frac{\theta - d}{\sigma}\right)\right]$$

or

$$m_{f_0^*} = (1 - \varepsilon)\sigma\{\Phi[c(\varepsilon)][c(\varepsilon) + [\rho - c(\varepsilon)] \exp\{\rho c(\varepsilon) - \tfrac{1}{2}\rho^2\}]$$
$$+ \phi[c(\varepsilon)] - \phi[c(\varepsilon) - \rho]\} \tag{6.3.57}$$

6.3 Robust Detection for Classes of Stationary and Memoryless Processes

$$v_{f_0^*}^2 = \mathrm{Var}\left\{n^{-1} \sum_{i=1}^{n} z(X_i) \Big| f_0^*\right\} = n^{-1}\,\mathrm{Var}\{z(X)|f_0^*\}$$

$$= n^{-1}\left((1-\varepsilon)\sigma^2 \Phi\left(\frac{d}{\sigma}\right)\left[\left(\frac{d}{\sigma}\right)^2 + \left(\frac{d-\theta}{\sigma}\right)^2 \exp\left\{\frac{\theta}{\sigma^2}d - \frac{\theta^2}{2\sigma^2}\right\} - 1\right]\right.$$

$$\left. + (1-\varepsilon)\sigma^2 \Phi\left(\frac{-d+\theta}{\sigma}\right) + (1-\varepsilon)\sigma^2\left[\frac{d}{\sigma}\phi\left(\frac{d}{\sigma}\right) + \frac{d-\theta}{\sigma}\phi\left(\frac{\theta-d}{\sigma}\right)\right] - m_{f_0^*}^2\right)$$

or

$$v_{f_0^*}^2 = (1-\varepsilon)\sigma^2 n^{-1}(\Phi[c(\varepsilon)][c^2(\varepsilon) + [\rho - c(\varepsilon)]^2 \exp\{\rho c(\varepsilon) - \tfrac{1}{2}\rho^2\} - 1]$$
$$+ \Phi[\rho - c(\varepsilon)] + c(\varepsilon)\phi[c(\varepsilon)] + [c(\varepsilon) - \rho]\phi[\rho - c(\varepsilon)]$$
$$- (1-\varepsilon)^{-1}\sigma^{-2}m_{f_0^*}^2) \tag{6.3.58}$$

where $\phi(x)$ denotes the zero mean, unit variance Gaussian density function at the point x and Var means variance.

For asymptotically large values of n the constant r^* in (6.3.53) and (6.3.55) can be taken equal to zero, and the threshold λ^* in (6.3.55) is then found as follows: for $n \to \infty$, $1 - \alpha = \Phi[(\lambda^* - m_{f_0^*})/v_{f_0^*}]$, thus

$$\lambda^* = m_{f_0^*} + v_{f_0^*}\Phi^{-1}(1-\alpha) \tag{6.3.59}$$

where $m_{f_0^*}$ and $v_{f_0^*}$ are given respectively by (6.3.57) and (6.3.58).

Given some density function $f = (1-\varepsilon)f_1 + \varepsilon h$ in $\mathscr{F}_1$, where f_1 is as in (6.3.36), we obtain

$$m_f = E\left\{n^{-1}\sum_{i=1}^{n} z(X_i)\Big| f\right\} = E\{z(X)|f\}$$

$$= d\,\mathrm{Pr}\{X \le d|f\} + (-d+\theta)\,\mathrm{Pr}\{X \ge -d+\theta|f\} + \int_{d}^{-d+\theta} xf(x)\,dx$$

$$= (1-\varepsilon)m_{f_1} + \varepsilon m_h \tag{6.3.60}$$

where

$$m_{f_1} = E\left\{n^{-1}\sum_{i=1}^{n} z(X_i)\Big| f_1\right\}$$

$$= \sigma\left[\frac{\theta}{\sigma} - \phi\left(\frac{d}{\sigma}\right) + \phi\left(\frac{d-\theta}{\sigma}\right) + \frac{d-\theta}{\sigma}\Phi\left(\frac{d-\theta}{\sigma}\right) - \frac{d}{\sigma}\Phi\left(\frac{d}{\sigma}\right)\right]$$

$$= \sigma[\rho - \phi(c(\varepsilon)) + \phi(c(\varepsilon) - \rho) + (c(\varepsilon) - \rho)\Phi(c(\varepsilon) - \rho)$$
$$- c(\varepsilon)\Phi(c(\varepsilon))] \tag{6.3.61}$$

$$m_h = E\left\{n^{-1}\sum_{i=1}^{n} z(X_i)\Big| h\right\} = d\,\mathrm{Pr}\{X \le d|h\} + (-d+\theta)\,\mathrm{Pr}\{X \ge -d+\theta|h\}$$

$$+ \int_{d}^{-d+\theta} xh(x)\,dx \tag{6.3.62}$$

Also

$$v_f^2 = \text{Var}\left\{n^{-1} \sum_{i=1}^n z(X_i) \middle| f\right\} = n^{-1} \text{Var}\{z(X)|f\}$$
$$= (1-\varepsilon)v_{f_1}^2 + \varepsilon v_h^2 + n^{-1}\varepsilon(1-\varepsilon)[m_{f_1} - m_h]^2 \quad (6.3.63)$$

where

$$v_{f_1}^2 = n^{-1}\sigma^2\{c^2(\varepsilon) + [1 + \rho^2 - c^2(\varepsilon)]\Phi[\rho - c(\varepsilon)] + (c^2(\varepsilon) - 2\rho c(\varepsilon) - 1)\Phi[c(\varepsilon)]$$
$$- [-c(\varepsilon) + 2\rho]\phi[c(\varepsilon)] + [\rho + c(\varepsilon)]\phi[\rho - c(\varepsilon)] - \sigma^{-2}m_{f_1}^2\} \quad (6.3.64)$$

$$v_h^2 = n^{-1}\left(d^2 \Pr\{X \le d|h\} + (d-\theta)^2 \Pr\{X \ge -d+\theta|h\}\right.$$
$$\left.+ \int_d^{-d+\theta} x^2 h(x)\,dx - m_h^2\right) \quad (6.3.65)$$

If the density function h in $f = (1-\varepsilon)f_1 + \varepsilon h$ is Gaussian with variance σ^2 and mean m then we denote it g_m. We define

$$\mu = m/\sigma \quad (6.3.66)$$

and obtain, via (6.3.60) through (6.3.66),

$$m_{g_m} = \sigma\{\mu + [\rho - \mu - c(\varepsilon)]\Phi(c(\varepsilon) - \rho + \mu) + [c(\varepsilon) - \mu]\Phi(c(\varepsilon) - \mu)$$
$$+ \phi(c(\varepsilon) - \mu) - \phi[\rho - \mu - c(\varepsilon)]\} \quad (6.3.67)$$

$$v_{g_m}^2 = n^{-1}\sigma^2\{1 + \mu^2 + [c^2(\varepsilon) - \mu^2 - 1]\Phi(c(\varepsilon) - \mu)$$
$$+ [(c(\varepsilon) - \rho)^2 - \mu^2 - 1]\Phi(c(\varepsilon) - \rho + \mu) + [c(\varepsilon) + \mu]\phi(c(\varepsilon) - \mu)$$
$$+ [c(\varepsilon) - \rho - \mu]\phi[\rho - \mu - c(\varepsilon)] - \sigma^{-2}m_{g_m}^2\} \quad (6.3.68)$$

For

$$f = (1-\varepsilon)f_1 + \varepsilon g_m \quad (6.3.69)$$

we also have

$$m_f = (1-\varepsilon)m_{f_1} + \varepsilon m_{g_m} \quad (6.3.70)$$
$$v_f^2 = (1-\varepsilon)v_{f_1}^2 + \varepsilon v_{g_m}^2 + n^{-1}\varepsilon(1-\varepsilon)[m_{f_1} - m_{g_m}]^2 \quad (6.3.71)$$

Let us denote by $P_n(f)$ the power induced by the robust decision rule in (6.3.53) at the density function f in (6.3.69). Due to (6.3.59), we conclude: for $n \to \infty$,

$$P_n(f) = \Pr\left\{n^{-1}\sum_{i=1}^n z(X_i) > \lambda^* \middle| f\right\}$$
$$= \Phi\left(\frac{-\lambda^* + m_f}{v_f}\right) = \Phi\left(\frac{m_f - m_{f_0^*} - v_{f_0^*}\Phi^{-1}(1-\alpha)}{v_f}\right) \quad (6.3.72)$$

6.3 Robust Detection for Classes of Stationary and Memoryless Processes

where $m_{f_0^*}$, $v_{f_0^*}$, m_f, and v_f are given respectively by (6.3.57), (6.3.58), (6.3.70), and (6.3.71).

Given a false alarm rate α, let us now consider the optimal Neyman-Pearson test at the Gaussian pair of density functions f_1 and f_0 in (6.3.36). As we saw in Chapter 5, given n, the Neyman-Pearson decision rule $\delta_{10}(x^n)$ is

$$\delta_{10}(x^n) = \begin{cases} 1 & x^n : n^{-1} \sum_{i=1}^{n} x_i \geq (\sigma/\sqrt{n})\Phi^{-1}(1-\alpha) \\ 0 & x^n : n^{-1} \sum_{i=1}^{n} x_i < (\sigma/\sqrt{n})\Phi^{-1}(1-\alpha) \end{cases} \quad (6.3.73)$$

Given n, the power $P_n^0(f_1)$, that the rule $\delta_{10}(x^n)$ induces at the Gaussian density f_1 in (6.3.36) is given by

$$P_n^0(f_1) = \Pr\left\{ n^{-1} \sum_{i=1}^{n} X_i \geq (\sigma/\sqrt{n})\Phi^{-1}(1-\alpha) \bigg| f_1 \right\} = \Phi\left[\sqrt{n}\frac{\theta}{\sigma} - \Phi^{-1}(1-\alpha) \right] \quad (6.3.74)$$

Given some arbitrary density function $f = (1-\varepsilon)f_1 + \varepsilon h$ in $\mathscr{F}_1$, and for asymptotically large values of n, the power $P_n^0(f)$ induced by the rule in (6.3.73) at f is given, for $n \to \infty$, by

$$P_n^0(f) = \Pr\left\{ n^{-1} \sum_{i=1}^{n} X_i \geq (\sigma/\sqrt{n})\Phi^{-1}(1-\alpha) \bigg| f \right\}$$
$$= \Phi\left(\frac{m(f)\sqrt{n} - \sigma\Phi^{-1}(1-\alpha)}{v(f)} \right) \quad (6.3.75)$$

where

$$m(f) = \int_{-\infty}^{\infty} xf(x)\,dx = (1-\varepsilon)\theta + \varepsilon \int_{-\infty}^{\infty} xh(x)\,dx = (1-\varepsilon)\theta + \varepsilon m(h) \quad (6.3.76)$$

$$\begin{aligned} v^2(f) &= \int_{-\infty}^{\infty} x^2 f(x)\,dx - m^2(f) \\ &= (1-\varepsilon)(\sigma^2 + \theta^2) + \varepsilon \int_{-\infty}^{\infty} x^2 h(x)\,dx - m^2(f) \\ &= (1-\varepsilon)\sigma^2 + \varepsilon(1-\varepsilon)\theta^2 + \varepsilon(1-\varepsilon)m^2(h) + \varepsilon v^2(h) - 2\varepsilon(1-\varepsilon)\theta m(h) \\ &= (1-\varepsilon)\sigma^2 + \varepsilon v^2(h) + \varepsilon(1-\varepsilon)[\theta - m(h)]^2 \end{aligned}$$
$$(6.3.77)$$

In (6.3.76) and (6.3.77), $m(h)$ and $v^2(h)$ denote respectively the mean and variance induced by the density function $h(x)$. Now consider the density function f in (6.3.69). Then, $h = g_m$, $m(h) = m$, and $v^2(h) = \sigma^2$. Substituting those values in (6.3.76), (6.3.77), and (6.3.75), and using the quantities in (6.3.56) and (6.3.66), we

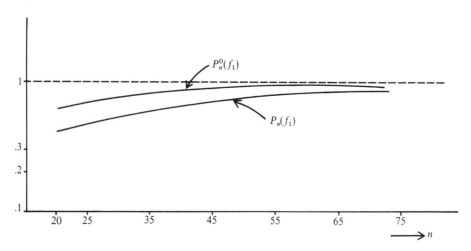

Figure 6.3.4 Power curves for $\rho = 0.5$, $\alpha = .05$, $\varepsilon = .1$

obtain

$$m(f) = (1 - \varepsilon)\theta + \varepsilon m = \sigma\{(1 - \varepsilon)\rho + \varepsilon\mu\}$$
$$v^2(f) = \sigma^2 + \varepsilon(1 - \varepsilon)[\theta - m]^2 = \sigma^2\{1 + \varepsilon(1 - \varepsilon)[\rho - \mu]^2\} \quad (6.3.78)$$

For $n \to \infty$

$$P_n^0(f) = \Phi\left(\frac{\sqrt{n}[(1 - \varepsilon)\rho + \varepsilon\mu] - \Phi^{-1}(1 - \alpha)}{[1 + \varepsilon(1 - \varepsilon)(\rho - \mu)^2]^{1/2}}\right) \quad (6.3.79)$$

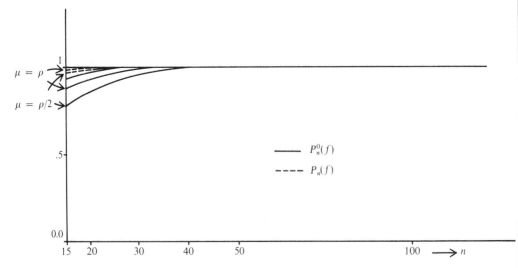

Figure 6.3.5 Power curves for $f = (1 - \varepsilon)f_1 + \varepsilon g_m$, $\sigma = \rho = 1$, $\alpha = .05$, $\varepsilon = .2$

When the two hypotheses are described strictly by the nominal density functions f_0 and f_1 in (6.3.36), the decision rule in (6.3.73) is the optimal; that is, the power $P_n^0(f_1)$ in (6.3.79) is uniformly superior to the power $P_n(f_1)$ in (6.3.72). In Figure 6.3.4, we plot the powers $P_n^0(f_1)$ and $P_n(f_1)$ as functions of n for $\sigma = 1$, $\alpha = .05$, $\rho = 0.5$, $\varepsilon = .1$, and $n \geq 25$.

Let us now suppose that outliers do occur, with probability ε, and that each outlier is represented by a Gaussian random variable, with variance σ^2 and mean $m - \theta$. Then, in the presence of the signal θ the data are generated by the density function $f = (1 - \varepsilon)f_1 + \varepsilon g_m$, which is Gaussian with mean $m(f)$ and variance $v^2(f)$, both given by (6.3.78). For such data, the robust decision rule performs better than the optimal Neyman-Pearson rule at the pair (f_1, f_0). This is shown quantitatively in Figures 6.3.5 and 6.3.6. In Figure 6.3.5, the powers $P_n(f)$ in (6.3.72) and $P_n^0(f)$ in (6.3.79) are plotted as functions of n for f as above, $\sigma = 1$, $\rho = 1$, $\alpha = .05$, $\varepsilon = .2$, $n \geq 50$, and for various values of the parameter $\mu = m\sigma^{-1} = m$. In Figure 6.3.6, we plot the powers $P_n(f)$ and $P_n^0(f)$ as functions of ε for $\sigma = 1$, $\rho = 1$, $\alpha = .05$, $n = 100$, and for various values of the parameter $\mu = m\sigma^{-1} = m$.

6.4 EXAMPLES

In this section, we will study two problems, first presented in Chapter 4, in the presence of noise outliers—the matched filter problem, and the detection of one of two first order autoregressive models.

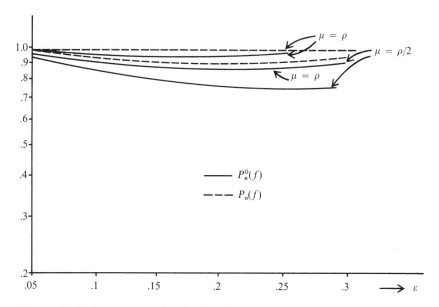

Figure 6.3.6 Power curves for $f = (1 - \varepsilon)f_1 + \varepsilon g_m$, $\sigma = \rho = 1$, $\alpha = .05$, $n = 100$

Example 6.4.1

Let us consider the M periodic and orthogonal signals $\{s_l^n; 1 \leq l \leq M\}$ in Example 4.5.1. In addition, let the m nonzero components of each signal, within a period, be identical and equal to θ. The power c^2 of each signal then equals $M^{-1}\theta^2$. Let the M signals be equally probable, let one of the M signals be transmitted in $n = KmM$ discrete-time observations, and let the transmission media be represented by additive, white, and stationary noise. Let the nominal noise process be zero mean, unit variance Gaussian, and let outliers occur with probability ε, due to occasional unusual disturbances. The first order density function f of the noise process is then given by

$$f(x) = (1 - \varepsilon)\phi(x) + \varepsilon h(x) \qquad \forall\, x \in R,\, h \in \mathcal{H} \qquad (6.4.1)$$

where ϕ denotes the zero mean, unit variance Gaussian density function, R is the real line, and $\mathcal{H}$ is the class of all first order density functions, defined on R.

Given some density function f as in (6.4.1), let us denote by f_l the density induced by f and the signal s_l^n. Let us denote by ϕ_l the density function induced by the Gaussian density ϕ and the signal s_l^n. Given some density function f in (6.4.1), the matched filter decides as follows:

Given the observation vector x^n; $n = KmM$, decide in favor of the signal s_k^n, iff

$$\sum_{i=1}^{n} \ln \frac{f_k(x_i)}{f_l(x_i)} \geq 0 \qquad \forall\, l;\, 1 \leq l \leq M$$

Since the actual density function f is unknown to the system designer, a game is formalized. Given k and l, the least favorable pair (f_k^*, f_l^*) of density functions is as in (6.3.32), (6.3.33), and (6.3.34), and takes here the form

$$f_l^*(x) = \begin{cases} (1 - \varepsilon)\phi_l(x) & x: \dfrac{\phi_k(x)}{\phi_l(x)} < c_l \\ c_l^{-1}(1 - \varepsilon)\phi_k(x) & x: \dfrac{\phi_k(x)}{\phi_l(x)} \geq c_l \end{cases} \qquad (6.4.2)$$

$$c_l: (1 - \varepsilon)\Pr\left\{\frac{\phi_k(X)}{\phi_l(X)} < c_l \Big| \phi_l\right\} + c_l^{-1}\Pr\left\{\frac{\phi_k(X)}{\phi_l(X)} \geq c_l \Big| \phi_k\right\} = 1$$

$$f_k^*(x) = \begin{cases} (1 - \varepsilon)\phi_k(x) & x: \dfrac{\phi_k(x)}{\phi_l(x)} > c_k \\ c_k(1 - \varepsilon)\phi_l(x) & x: \dfrac{\phi_k(x)}{\phi_l(x)} \leq c_k \end{cases} \qquad (6.4.3)$$

$$c_k: (1 - \varepsilon)\left[\Pr\left\{\frac{\phi_k(X)}{\phi_l(X)} > c_k \Big| \phi_k\right\} + c_k \Pr\left\{\frac{\phi_k(X)}{\phi_l(X)} \leq c_k \Big| \phi_l\right\}\right] = 1$$

6.4 Examples

Let us denote by s_{il} and s_{ik}, and ith component of respectively the s_l^n and the s_k^n signal. Then, due to the orthogonality of the signals, we have

$$s_{il} = \begin{cases} \theta & p[(l-1)m+1] \leq i \leq plm \quad p = 1, \ldots, K \\ 0 & \text{otherwise} \end{cases} \qquad (6.4.4)$$

Expressions (6.4.2) and (6.4.3) can be then easily simplified as follows for $l \neq k$, where Φ denotes the cumulative distribution of the zero mean, unit variance, Gaussian random variable.

a) If $i: p[(k-1)m+1] \leq i \leq pkm$, for some $p = 1, \ldots, K$

$$f_l^*(x_i) = \begin{cases} (1-\varepsilon)\phi_l(x_i) & x_i < \dfrac{d}{\theta} \\ (1-\varepsilon)e^{-d+\theta^2/2}\phi_k(x_i) & x_i \geq \dfrac{d}{\theta} \end{cases}$$

$$f_k^*(x_i) = \begin{cases} (1-\varepsilon)\phi_k(x_i) & x_i > -\dfrac{d}{\theta} + \theta \\ (1-\varepsilon)e^{-d+\theta^2/2}\phi_l(x_i) & x_i \leq -\dfrac{d}{\theta} + \theta \end{cases}$$

$$\ln \frac{f_k^*(x_i)}{f_l^*(x_i)} = \begin{cases} d - \dfrac{\theta^2}{2} & x_i \geq \dfrac{d}{\theta} \\ \theta x_i - \dfrac{\theta^2}{2} & -\dfrac{d}{\theta} + \theta < x_i < \dfrac{d}{\theta} \\ -d + \dfrac{\theta^2}{2} & x_i \leq -\dfrac{d}{\theta} + \theta \end{cases} \qquad (6.4.5)$$

$$d: (1-\varepsilon)\left[\Phi\left(-\dfrac{d}{\theta}\right) + e^{-d+\theta^2/2}\Phi\left(-\dfrac{d}{\theta}+\theta\right)\right] = 1 \qquad (6.4.6)$$

b) If $i: p[(l-1)m+1] \leq i \leq plm$, for some $p = 1, \ldots, K$

$$f_l^*(x_i) = \begin{cases} (1-\varepsilon)\phi_l(x_i) & x_i > -\dfrac{c}{\theta} \\ (1-\varepsilon)e^{-c-\theta^2/2}\phi_k(x_i) & x_i \leq -\dfrac{c}{\theta} \end{cases} \qquad (6.4.7)$$

$$f_k^*(x_i) = \begin{cases} (1-\varepsilon)\phi_k(x_i) & x_i < \dfrac{c}{\theta} + \theta \\ (1-\varepsilon)e^{-c+\theta^2/2}\phi_l(x_i) & x_i \geq \dfrac{c}{\theta} + \theta \end{cases} \qquad (6.4.8)$$

$$c: (1-\varepsilon)\left[\Phi\left(\dfrac{c}{\theta}+\theta\right) + e^{-c-\theta^2/2}\Phi\left(-\dfrac{c}{\theta}\right)\right] = 1 \qquad (6.4.9)$$

c) For i different than both cases a) and b), $\ln[f_k^*(x_i)/f_l^*(x_i)] = 0$. From (6.4.6) and (6.4.9), we observe that

$$\frac{c}{\theta} = \frac{d}{\theta} - \theta \qquad (6.4.10)$$

Then we can combine cases a), b), and c) as follows:

Find d from

$$(1-\varepsilon)\left[\Phi\left(\frac{d}{\theta}\right) + e^{-d+\theta^2/2}\Phi\left(-\frac{d}{\theta}+\theta\right)\right] = 1 \qquad (6.4.11)$$

Then

i. For $i: p[(k-1)m+1] \leq i \leq pkm$, for some set $p = 1, 2, \ldots, K$

$$\gamma^*(x_i) = \ln\frac{f_k^*(x_i)}{f_l^*(x_i)} = \begin{cases} d - \dfrac{\theta^2}{2} & x_i \geq \dfrac{d}{\theta} \\ \theta x_i - \dfrac{\theta^2}{2} & -\dfrac{d}{\theta}+\theta < x_i < \dfrac{d}{\theta} \\ -d + \dfrac{\theta^2}{2} & x_i \leq -\dfrac{d}{\theta}+\theta \end{cases} \qquad (6.4.12)$$

ii. For $i: p[(l-1)m+1] \leq i \leq plm$, for some set $p = 1, 2, \ldots, K$

$$\gamma^*(x_i) = \ln\frac{f_k^*(x_i)}{f_l^*(x_i)} = \begin{cases} -d + \dfrac{\theta^2}{2} & x_i \geq \dfrac{d}{\theta} \\ -\theta x_i + \dfrac{\theta^2}{2} & -\dfrac{d}{\theta}+\theta < x_i < \dfrac{d}{\theta} \\ d - \dfrac{\theta^2}{2} & x_i \leq -\dfrac{d}{\theta}+\theta \end{cases} \qquad (6.4.13)$$

Let us define $C_l = \{i: p[(l-1)m+1] \leq i \leq plm, \text{ for some } p = 1, \ldots, K\}$. For $\gamma^*(x_i)$ as in (6.4.12), the robust matched filter can be expressed as

Given $x^n = [x_1, \ldots, x_n]^T$, decide in favor of signal s_l^n iff

$$Mn^{-1}\sum_{i \in C_k} \gamma^*(x_i) \geq Mn^{-1}\sum_{i \in C_l} \gamma^*(x_i) \qquad \forall\, l = 1, \ldots, M \quad (6.4.14)$$

For $l \neq k$, the random variables $Y_k = Mn^{-1}\sum_{i \in C_k}\gamma^*(X_i)$ and $Y_l = Mn^{-1}\sum_{i \in C_l}\gamma^*(X_i)$ are mutually independent for every f as in (6.4.1).

6.4 Examples

Let us select

$$f(x) = (1 - \varepsilon)\phi(x) + \varepsilon \frac{e^{-(1/2)(x-\rho)^2}}{(2\pi)^{1/2}} \quad \forall\, x \in R \qquad (6.4.15)$$

where ρ is some real scalar.

Let us denote,

$$\begin{aligned} m_0 &= E\{Y_k | f_k\} & \sigma_0^2 &= E\{(Y_k - m_0)^2 | f_k\} \\ m_1 &= E\{Y_l | f_k\} & \sigma_1^2 &= E\{(Y_l - m_1)^2 | f_k\} \quad k \neq l \end{aligned} \qquad (6.4.16)$$

Then, from (6.4.12) and (6.4.15), we find

$$\begin{aligned} m_0 &= \frac{\theta^2}{2} + \varepsilon\theta\rho - (1-\varepsilon)\,d\Phi\!\left(-\frac{d}{\theta}\right) - \varepsilon(d+\theta\rho)\Phi\!\left(-\frac{d}{\theta} - \rho\right) \\ &\quad + (1-\varepsilon)[d - \theta^2]\Phi\!\left(-\frac{d}{\theta} + \theta\right) + \varepsilon[d - \theta^2 - \theta\rho]\Phi\!\left(-\frac{d}{\theta} + \theta + \rho\right) \\ &\quad - (1-\varepsilon)\theta\!\left[\phi\!\left(\frac{d}{\theta} - \theta\right) - \phi\!\left(\frac{d}{\theta}\right)\right] - \varepsilon\theta\!\left[\phi\!\left(\frac{d}{\theta} - \theta - \rho\right) - \phi\!\left(\frac{d}{\theta} + \rho\right)\right] \end{aligned}$$

$$(6.4.17)$$

$$\begin{aligned} m_1 &= -\frac{\theta^2}{2} + \varepsilon\theta\rho + (1-\varepsilon)\,d\Phi\!\left(-\frac{d}{\theta}\right) + \varepsilon(d - \theta\rho)\Phi\!\left(-\frac{d}{\theta} + \rho\right) \\ &\quad + (1-\varepsilon)(\theta^2 - d)\Phi\!\left(-\frac{d}{\theta} + \theta\right) + \varepsilon(\theta^2 - d - \theta\rho)\Phi\!\left(-\frac{d}{\theta} + \theta - \rho\right) \\ &\quad - (1-\varepsilon)\theta\!\left[\phi\!\left(\frac{d}{\theta}\right) - \phi\!\left(\frac{d}{\theta} - \theta\right)\right] - \varepsilon\theta\!\left[\phi\!\left(\frac{d}{\theta} - \rho\right) - \phi\!\left(\frac{d}{\theta} + \rho - \theta\right)\right] \end{aligned}$$

$$(6.4.18)$$

$$\begin{aligned} \sigma_0 &= Mn^{-1}\bigg\{\left(d - \frac{\theta^2}{2}\right)^2 + \varepsilon(\theta^2 + \theta^2\rho^2 + \theta^3\rho - d^2 + d\theta^2)\Phi\!\left(\frac{d}{\theta} - \theta - \rho\right) \\ &\quad + (1-\varepsilon)(\theta^2 - d^2 + d\theta^2)\Phi\!\left(\frac{d}{\theta} - \theta\right) + (1-\varepsilon)(d^2 - d\theta^2 - \theta^2)\Phi\!\left(-\frac{d}{\theta}\right) \\ &\quad + \varepsilon(d^2 - d\theta^2 - \theta^2 - \theta^2\rho^2 - \theta^3\rho)\Phi\!\left(-\frac{d}{\theta} - \rho\right) \\ &\quad + (1-\varepsilon)\!\left[-d\theta\,\phi\!\left(\frac{d}{\theta} - \theta\right) + (\theta^3 - \theta d)\phi\!\left(\frac{d}{\theta}\right)\right] \\ &\quad + \varepsilon\!\left[(\theta^3 - d\theta)\phi\!\left(\frac{d}{\theta} + \rho\right) - (\theta^2\rho + d\theta)\phi\!\left(\frac{d}{\theta} - \theta - \rho\right)\right] - m_0^2\bigg\} \end{aligned}$$

$$(6.4.19)$$

$$\sigma_1 = Mn^{-1}\left\{\left(d - \frac{\theta^2}{2}\right)^2 + (1-\varepsilon)(\theta^2 - d^2 + d\theta^2)\Phi\left(\frac{d}{\theta}\right)\right.$$

$$+ (1-\varepsilon)(d^2 - \theta^2 - d\theta^2)\Phi\left(-\frac{d}{\theta} + \theta\right)$$

$$+ \varepsilon(\theta^2 + \theta^2\rho^2 - \theta^3\rho - d^2 + d\theta^2)\Phi\left(\frac{d}{\theta} - \rho\right)$$

$$+ \varepsilon(-\theta^2 - \theta^2\rho^2 + \theta^3\rho + d^2 - d\theta^2)\Phi\left(-\frac{d}{\theta} + \theta - \rho\right)$$

$$- (1-\varepsilon)\left[(d\theta - \theta^3)\phi\left(\frac{d}{\theta}\right) + d\theta\,\phi\left(\frac{d}{\theta} - \theta\right)\right]$$

$$\left. + \varepsilon\left[(\rho\theta^2 - d\theta)\phi\left(-\frac{d}{\theta} + \theta - \rho\right) - (d\theta - \theta^3 - \rho\theta^2)\phi\left(\frac{d}{\theta} - \rho\right) - m_1^2\right]\right\}$$

(6.4.20)

For asymptotically large values of n the random variables $\{Y_l;\, 1 \le l \le M\}$ are Gaussian and mutually independent. As in Example 4.5.1, we then find that the probability of detection $P_d^*(n, f)$ induced by the robust matched filter at the density function f in (6.4.15) is given asymptotically ($n \to \infty$) by

$$P_d^*(n, f) = \int_{-\infty}^{\infty} dx\, q(x)\Phi^{M-1}\left(\frac{\sigma_0}{\sigma_1}x + \frac{m_0 - m_1}{\sigma_1}\right) \qquad (6.4.21)$$

where m_0, m_1, σ_0, and σ_1 are given by (6.4.17) through (6.4.20).

Let us denote by $P_d(n, f)$ the probability of detection induced by the matched filter in Example 4.5.1 at the density function f in (6.4.15) for observation size n. Then we easily find

$$P_d(n, f) = \int_{-\infty}^{\infty} dx\, \phi(x)\Phi^{M-1}\left(x + \frac{\theta}{[1 + \varepsilon(1-\varepsilon)\rho^2]^{1/2}}\sqrt{\frac{n}{M}}\right) \qquad (6.4.22)$$

We note that the matched filter in Example 4.5.1 is optimal under the assumption that the white noise is exactly Gaussian. If the Gaussian assumption is violated, however, this filter may induce poor performance. The robust matched filter would then be superior. We justify this claim quantitatively in Figure 6.4.1, where we plot the probabilities of detection $P_d^*(n, f)$ in (6.4.21) and $P_d(n, f)$ in (6.4.22) as functions of θ for $n = 100$, $\rho = \theta/2$, $M = 4$, and $\varepsilon = .1$.

Example 6.4.2

As in Example 4.5.4, let us consider two first order autoregressive models,

(1) $Y_n = \alpha Y_{n-1} + W_n$ (2) $Y_n = -\alpha Y_{n-1} + W_n$ (6.4.23)

where $\{W_n\}$ is a sequence of independent, zero mean, unit variance random variables, and where the autoregressive coefficient α is such that $0 < \alpha < 1$.

6.4 Examples

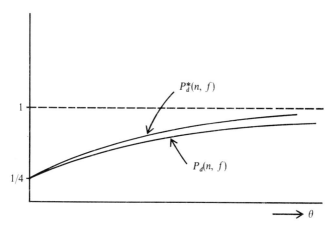

Figure 6.4.1

Let the two models be equally probable, let one of the two models be active in n discrete-time observations, and, given an observation sequence x^n, let it be desirable to decide which model is active. Let the active model be transmitted through additive white noise. Let the nominal first order density function of the noise be zero mean and variance Gaussian, and let independent outliers occur with probability ε. Then the first order density function f of the noise is represented as in (6.4.1), Example 6.4.1. This noise can be then incorporated into the sequence $\{W_n\}$ in (6.4.23), where the independent and identically distributed random variables $\{W_n\}$ now have a density function f given as

$$f(x) = (1 - \varepsilon)\frac{e^{-x^2/4}}{(4\pi)^{1/2}} + \varepsilon h(x) \qquad \forall\, x \in R,\, h \in \mathcal{H} \qquad (6.4.24)$$

Given f as in (6.4.24) and an observation vector $x^n = [x_1, \ldots, x_n]^T$, let $f_1(x^n)$ and $f_0(x^n)$ denote the density functions induced by f and the autoregressive models (1) and (2) respectively in (6.4.23) at the vector point x^n. Then

Model (1) is decided iff

$$\ln \frac{f_1(x^n)}{f_0(x^n)} \geq 0 \quad \text{or} \quad \ln \frac{f_1(x_1)}{f_0(x_1)} + \sum_{l=1}^{n-1} \ln \frac{f_1(x_l - \alpha x_{l-1})}{f_0(x_l + \alpha x_{l-1})} \geq 0 \qquad (6.4.25)$$

Directly from (6.3.38) we conclude that the least favorable pair (f_1^*, f_0^*) of density functions are here such that

$$\gamma^*(x_l, x_{l-1}) = \ln \frac{f_1^*(x_l - \alpha x_{l-1})}{f_0^*(x_l + \alpha x_{l-1})} = \begin{cases} d_1 & x_l x_{l-1} \geq \dfrac{d_1}{\alpha} \\ \alpha x_l x_{l-1} & \dfrac{d_0}{\alpha} < x_l x_{l-1} < \dfrac{d_1}{\alpha} \\ d_0 & x_l x_{l-1} \leq \dfrac{d_0}{\alpha} \end{cases} \quad l \geq 2$$

$$(6.4.26)$$

$$\ln \frac{f_1^*(x_1)}{f_0^*(x_1)} = 0 \tag{6.4.27}$$

$$d_0: (1-\varepsilon)\left[\Pr\left\{ X_l X_{l-1} < \frac{d}{\alpha}\bigg|(2), g \right\} + e^{-d_0} \Pr\left\{ X_l X_{l-1} \geq \frac{d_0}{\alpha}\bigg|(1), g \right\} \right] = 1 \tag{6.4.28}$$

$$d_1: (1-\varepsilon)\left[\Pr\left\{ X_l X_{l-1} > \frac{d_1}{\alpha}\bigg|(1), g \right\} + e^{d_1} \Pr\left\{ X_l X_{l-1} \leq \frac{d_1}{\alpha}\bigg|(2), g \right\} \right] = 1 \tag{6.4.29}$$

where $g(x) = e^{-x^2/4}/\sqrt{4\pi}$ and (1) or (2) denotes the corresponding autoregressive models in (6.4.23).

From (6.4.28) and (6.4.29) we easily conclude that $d_1 = -d_0$. We can thus summarize the robust test as

Compute d such that

$$\int_{-\infty}^{0} du\, \phi(u) \left\{ (1-\varepsilon)\Phi\left(\frac{d\sqrt{1-\alpha^2}}{2}\frac{1}{u} + \frac{\alpha}{(1-\alpha^2)^{1/2}} u \right) \right.$$
$$\left. + e^{-\alpha d}\Phi\left(-\frac{d\sqrt{1-\alpha^2}}{2}\frac{1}{u} + \frac{\alpha}{(1-\alpha^2)^{1/2}} u \right) \right\} = 0.5 \tag{6.4.30}$$

where $\phi(x)$ and $\Phi(x)$ denote respectively the density function and the cumulative distribution function of the zero mean, unit variance Gaussian random variable at the point x.

For

$$\gamma^*(x_l, x_{l-1}) = \begin{cases} d & x_l x_{l-1} \geq d \\ x_l x_{l-1} & -d < x_l x_{l-1} < d \\ -d & x_l x_{l-1} \leq -d \end{cases} \tag{6.4.31}$$

decide in favor of the model (1), iff

$$T^*(x^n) = \sum_{l=2}^{n} \gamma^*(x_l, x_{l-1}) \geq 0 \tag{6.4.32}$$

From (6.4.31), we observe that the robust test involves truncation. On the other hand, the test in (4.5.20) does not; it decides in favor of model (1) iff

$$T^0(x^n) = \sum_{l=2}^{n} x_l x_{l-1} \geq 0 \tag{6.4.33}$$

Let us consider the following density function f

$$f(x) = (1-\varepsilon)\frac{e^{-x^2/4}}{(4\pi)^{1/2}} + \varepsilon \frac{e^{-(x-\rho)^2/4}}{(4\pi)^{1/2}} \quad \forall\, x \in R \tag{6.4.34}$$

where ρ is some positive scalar.

6.4 Examples

Let $P_e^*(n, f)$ and $P_e^0(n, f)$ denote respectively the probability of error induced by the robust test and the probability of error incuded by the parametric test in (6.4.33), both for noise represented by f in (6.4.34) and for n observations. Then, applying the Chernoff bound in Chapter 4 for some $r > 0$ and some $\mu > 0$, we obtain

$$P_e^*(n, f) = \frac{1}{2} \Pr\left\{ \sum_{l=2}^{n} \gamma^*(X_l, X_{l-1}) \geq 0 \,\Big|\, (2), f \right\}$$
$$+ \frac{1}{2} \Pr\left\{ -\sum_{l=2}^{n} \gamma^*(X_l, X_{l-1}) > 0 \,\Big|\, (1), f \right\}$$
$$\leq \frac{1}{2} E\left\{ \exp\left[r \sum_{l=2}^{n} \gamma^*(X_l, X_{l-1}) \right] \Big|\, (2), f \right\}$$
$$+ \frac{1}{2} E\left\{ \exp\left[-\mu \sum_{l=2}^{n} \gamma^*(X_l, X_{l-1}) \right] \Big|\, (1), f \right\}$$

(6.4.35)

where, for $l = 2$ actually representing asymptotic time, we have

$$E\left\{ \exp\left[r \sum_{l=2}^{n} \gamma^*(X_l, X_{l-1}) \right] \Big|\, (2), f \right\}$$

$$= \int_{R^n} dx_1^n \exp\left\{ r \sum_{l=2}^{n} \gamma^*(x_l, x_{l-1}) \right\} \frac{\exp\left\{ \left(x_1 - \frac{\varepsilon\rho}{1+\alpha}\right)^2 \sqrt{1-\alpha^2}\, / 4 \right\}}{(4\pi)^{1/2}} \sqrt{1-\alpha^2}$$
$$\cdot \prod_{l=2}^{n} \left[(1-\varepsilon) \frac{\exp\{-\tfrac{1}{4}(x_l + \alpha x_{l-1})^2\}}{(4\pi)^{1/2}} + \varepsilon \frac{\exp\{-\tfrac{1}{4}(x_l + \alpha x_{l-1} - \rho)^2\}}{(4\pi)^{1/2}} \right]$$

$$= (1-\varepsilon)^{n-1} \frac{\sqrt{1-\alpha^2}}{2^n \pi^{n/2}} \int_{R^n} dx_1^n \exp\left\{ r \sum_{l=2}^{n} \gamma^*(x_l, x_{l-1}) - \frac{1-\alpha^2}{4}\left(x_1 - \frac{\varepsilon\rho}{1+\alpha}\right)^2 \right.$$
$$\left. - \frac{1}{4} \sum_{l=2}^{n} (x_l + \alpha x_{l-1})^2 \right\} \prod_{l=2}^{n} \left[1 + \frac{\varepsilon}{1-\varepsilon} \exp\left\{ \frac{\rho}{2}(x_l + \alpha x_{l-1}) - \frac{\rho^2}{4} \right\} \right]$$

$$\underset{\rho \to \infty}{\simeq} \frac{1-\alpha^2}{2^n \pi^{n/2}} (1-\varepsilon)^{n-1} \left[1 + (n-1)\frac{\varepsilon}{1-\varepsilon} e^{-\rho^2/4} \right] \int_{R^n} dx_1^n \exp\left\{ r \sum_{l=2}^{n} \gamma^*(x_l, x_{l-1}) \right.$$
$$\left. - \frac{1}{4}[x_1 - (1-\alpha)\varepsilon\rho]^2 - \frac{x_n^2}{4} - \frac{1+\alpha^2}{2} \sum_{l=2}^{n-1} x_l^2 - \frac{\alpha}{2} \sum_{l=2}^{n} x_l x_{l-1} \right\}$$

$$\underset{\substack{\rho \to \infty \\ r < 1}}{\simeq} \frac{\sqrt{1-\alpha^2}(1-\varepsilon)^{n-1}}{[2(1+\alpha^2)]^{(n/2)-1}} \left[1 + (n-1)\frac{\varepsilon}{1-\varepsilon} e^{-\rho^2/4} \right]$$

(6.4.36)

Similarly, for any $\mu > 0$, we have

$$E\left\{\exp\left[-\mu \sum_{l=2}^{n} \gamma^*(X_l, X_{l-1})\right]\Big|(1), f\right\}$$

$$\underset{\rho \to \infty}{\simeq} \frac{1-\alpha^2}{2^n \pi^{n/2}}(1-\varepsilon)^{n-1}\left[1+(n-1)\frac{\varepsilon}{1-\varepsilon}e^{-\rho^2/4}\right]$$

$$\cdot \int_{R^n} dx_1^n \exp\left\{-\mu \sum_{l=2}^{n} \gamma^*(x_l, x_{l-1}) - \frac{1}{4}[x_1 - (1+\alpha)\varepsilon\rho]^2\right.$$

$$\left. -\frac{x_n^2}{4} - \frac{1+\alpha^2}{2}\sum_{l=2}^{n-1} x_l^2 + \frac{\alpha}{2}\sum_{l=2}^{n} x_l x_{l-1}\right\}$$

$$\underset{\substack{\rho \to \infty \\ \mu < 1}}{\simeq} \frac{\sqrt{1-\alpha^2}(1-\varepsilon)^{n-1}}{[2(1+\alpha^2)]^{(n/2)-1}}\left[1+(n-1)\frac{\varepsilon}{1-\varepsilon}e^{-\rho^2/4}\right] \quad (6.4.37)$$

From (6.4.35), (6.4.36), and (6.4.37), we conclude

$$P_e^*(n, f) \underset{\rho \to \infty}{\leq} \frac{\sqrt{1-\alpha^2}(1-\varepsilon)^{n-1}}{[2(1+\alpha^2)]^{(n/2)-1}}\left[1+(n-1)\frac{\varepsilon}{1-\varepsilon}e^{-\rho^2/4}\right] = B_u^*(n, f) \quad (6.4.38)$$

Similarly, we find

$$P_e^0(n, f) = 1 - P_d^0(n, f)$$

$$P_d^0(n, f) = \frac{1}{2}\Pr\left\{\sum_{l=2}^{n} X_l X_{l-1} \geq 0 \Big|(1), f\right\} + \frac{1}{2}\Pr\left\{-\sum_{l=2}^{n} X_l X_{l-1} > 0 \Big|(2), f\right\}$$

$$\leq \frac{1}{2}E\left\{\exp\left[\frac{\alpha}{2}\sum_{l=2}^{n} X_l X_{l-1}\right]\Big|(1), f\right\} + \frac{1}{2}E\left\{\exp\left[-\frac{\alpha}{2}\sum_{l=2}^{n} X_l X_{l-1}\right]\Big|(2), f\right\}$$

$$\underset{\rho \to \infty}{\simeq} \frac{\sqrt{1-\alpha^2}(1-\varepsilon)^{n-1}}{[2(1-\alpha^2)]^{(n/2)-1}}\left[1+(n-1)\frac{\varepsilon}{1-\varepsilon}e^{-\rho^2/4}\right]$$

and

$$P_e^0(n, f) \underset{\rho \to \infty}{\geq} 1 - \frac{\sqrt{1-\alpha^2}(1-\varepsilon)^{n-1}}{[2(1+\alpha^2)]^{(n/2)-1}}\left[1+(n-1)\frac{\varepsilon}{1-\varepsilon}e^{-\rho^2/4}\right] = B_l^0(n, f)$$

$$(6.4.39)$$

In Figure 6.4.2, we plot the bounds $B_u^*(n, f)$ in (6.4.38) and $B_l^0(n, f)$ in (6.4.39) as functions of n for $\alpha = 0.5$, $\varepsilon = .1$, $\rho \to \infty$, and $n < 1000$. We observe that as n increases, the two bounds diverge from each other. Therefore, for extreme ($\rho \to \infty$) outliers, occurring with probability .1, the robust test offers excellent statistical protection, while the parametric at the Gaussian noise test does not. In fact, the probability of error induced by the parametric test can then be arbitrarily close to one.

6.5 Problems

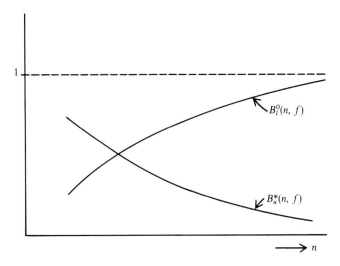

Figure 6.4.2

6.5 PROBLEMS

6.5.1 Let $V = \theta S + N$, where $V = \{v_i\}$ $i = 1, \ldots, n$, S is a known signal, and $N = G(0, R)$. Prior probabilities:

$$P(\theta = 0) = \frac{1}{2} \qquad P(\theta = \theta_1) = P(\theta = -\theta_1) = \frac{1}{4}$$

Decisions: $H_0: \theta = 0 \qquad H_1: |\theta| = \theta_1$

Penalty functions: $L(H_i, H_j) = 1 - \Delta_{ij}$

(a) Find the Bayes decision rule and Bayes risk.
(b) Find the minimax rule.
(c) Find an admissible decision for this problem.

6.5.2 Let $\theta = \{\theta_1, \theta_2\}$, $a = [0, \pi/2]$ = the closed interval on the real line from 0 to $\pi/2$, $C(\theta_1, \alpha) = -\cos \alpha$, and $C(\theta_2, \alpha) = -\sin \alpha$. A coin is tossed once with probability P_θ of heads, where $P_{\theta_1} = \frac{1}{3}$ and $P_{\theta_2} = \frac{2}{3}$. Find $R(\theta_1, \delta)$ and $R(\theta_2, \delta)$ for all $\delta \in \mathcal{D}$. Find the Bayes rule δ_0 with respect to the prior distribution τ_0 which gives probability $\frac{1}{2}$ to each state of nature. Show that δ_0 is minimax and that τ_0 is least favorable.

6.5.3 Let Θ consist of two points $\{\theta_1, \theta_2\}$, let α be the closed unit interval $[0, 1]$, and let the loss function be

$$L(\theta_1, \alpha) = \alpha^2 \qquad L(\theta_2, \alpha) = 1 - \alpha$$

Note that this penalty is convex in α for each $\theta \in \Theta$. A coin is tossed once. The probability of heads is $\frac{1}{3}$ if θ_1 is the true state of nature and $\frac{2}{3}$ if θ_2 is the true state of nature.

(a) Represent the class $\mathscr{D}$ of decision rules as a subset of the plane.
(b) Find $R[\theta_1,(x,y)]$ and $R[\theta_2,(x,y)]$ for $(x,y) \in \mathscr{D}$.
(c) Find the class of all nonrandomized Bayes rules. Plot this class as a subset of $\mathscr{D}$.
(d) Find a minimax rule among the class of all Bayes rules.

6.5.4 Consider the robust decision rule in Corollary 6.3.1, and let the nominal density functions be Poisson. That is,

$$f_0(x) = e^{-\lambda_0}\frac{\lambda_0^x}{x!} \quad x \text{ a positive integer}$$

$$f_1(x) = e^{-\lambda_1}\frac{\lambda_1^x}{x!} \quad x \text{ a positive integer}$$

$$\lambda_0 \neq \lambda_1$$

Find the complete robust test, and study its asymptotic behavior for changing ε_0 and ε_1 values. Compare this behavior with that of the parametric at f_0 and f_1 test.

6.5.5 Consider single hypothesis testing between two memoryless stationary processes, respectively represented by two disjoint classes, $\mathscr{F}_0$ and $\mathscr{F}_1$, of first order density functions. Let $\mathscr{F}_0$ contain all density functions defined on $[a,b] \subset R$, with fixed masses c_i; $i = 1,\ldots,m$ on respectively $[a,e_1)$, $[e_1,e_2),\ldots,[e_{m-1},b]$. Let $\mathscr{F}_1$ contain all density functions defined on $[a,b]$, with fixed masses d_i; $i=1,\ldots,m$ on respectively $[a,e_1)$, $[e_1,e_2),\ldots,[e_{m-1},b]$. Find the robust decision rule for testing $\mathscr{F}_0$ against $\mathscr{F}_1$, and study its asymptotic probability of error as a function of $\{c_i\}$ and $\{d_i\}$. What can you conclude about the least favorable pair of density functions?

6.5.6 Let the densities in the classes of Problem 6.5.5 be defined on R. Let their masses on $[a,b]$ be as in the latter problem, and let the mass on $(-\infty,a)$ (b,∞) be c_0 for class $\mathscr{F}_0$ and d_0 for class $\mathscr{F}_1$. Find the robust decision rule, and the least favorable pair of density functions. Study the asymptotic probability of error induced by the robust rule, as it varies with the active density functions in $\mathscr{F}_0$ and $\mathscr{F}_1$.

REFERENCES

Agarwal, O. P. (1955), "Some Minimax Invariant Procedures for Estimating a Cumulative Distribution Function," *Ann. Math. Statist.* 26, 450–463.

Anscombe, F. J. (1960), "Rejection of Outliers," *Technometrics* 2, 123–147.

Arrow, K. J., D. Blackwell, and M. A. Girshick (1949), "Bayes and Minimax Solutions of Sequential Decision Problems," *Econometrica*, 17, 213–244.

Banos, A. (1967), On Pseudo-Games, Ph.D. Thesis, UCLA.

Blackwell, D., and M. A. Girshick (1954), *Theory of Games and Statistical Decisions*, Wiley, New York.

References

Blyth, C. R. (1951), "On Minimax Statistical Decision Procedures and their Admissibility," *Ann. Math. Statist.* 52, 22–42.

Ferguson, T. S. (1961), "On the Rejection of Outliers," *Proc. 4th Berkeley Symp. on Math. Statist. and Prob.* 1, Univ. of Calif. Press, Berkeley, CA. 253–287.

———— (1967), *Mathematical Statistics: A Decision Theoretic Approach*, Academic Press, New York.

Huber, P. (1965), "A Robust Version of the Probability Ratio Test," *Ann. Math. Statist.* 36, 1753–1758.

———— (1966), "Strict Efficiency Excludes Superefficiency," (abstract) *Ann. Math. Statist.* 37, 1425.

———— (1967), "The Behavior of Maximum Likelihood Estimates under Nonstandard Conditions," *Proc. 5th Berkeley Symp. on Math. Statist. and Prob.* 1, Univ. of Calif. Press, Berkeley.

———— (1968), "Robust Confidence Limits," *Z. Wahrscheinlichkeitstheorie Verw. Gebiete* 10, 269–278.

———— (1981), *Robust Statistics*, Wiley, New York.

Huber, P. J. and V. Strassen (1973), "Minimax Tests and the Neyman-Pearson Lemma for Capacities," *Ann. Statist.* 1, 251–263, and 2, 223–224.

Karlin, S. (1959), *Mathematical Methods and Theory in Games, Programming, and Economics*, Vols. 1 and 2, Addison-Wesley, Reading, MA.

Kiefer, J. (1957), "Invariance, Minimax Sequential Estimation and Continuous Time Processes," *Ann. Math. Statist.* 28, 573–601.

Kudo, H. (1955), "On Minimax Invariant Estimates of the Transformation Parameter," *Natural Science Report* 6, Ochanomizu Univ., 31–73.

Rieder, H. (1978), "A Robust Asymptotic Testing Model," *Ann. Statist.* 6, 1080–1094.

Tukey, J. W. (1977), *Exploratory Data Analysis*, Addison-Wesley, Reading, Mass.

Wald, A. and J. Wolfowitz (1951), "Two Methods of Randomization in Statistics and the Theory of Games," *Ann. Math. Statist.* 23, 581–586.

CHAPTER 7

NONPARAMETRIC DETECTION

7.1 INTRODUCTION

As already stated in Chapter 6, the nonparametric detection schemes generally arise from the Neyman-Pearson formalization when the two hypotheses are nonparametrically described. While the robust detection schemes in Chapter 6 address the case where each hypothesis is described by a "relatively small" family of stochastic processes, the nonparametric schemes in this chapter consider "large" such families. Alternatively, while the robust schemes model the analyst's uncertainty about possible small deviations from some nominal stochastic model, the objective of the classical nonparametric schemes in this chapter is the design of a multipurpose detector. In particular, the existing nonparametric detection schemes provide *location parameter* detectors, for a class of nonparametrically described, memoryless, discrete-time, and stationary stochastic processes with very general characteristics. Such a widely used class $\mathscr{F}_{-\theta}$ of processes is, for example, the one described by all the one-dimensional density functions that are symmetric around the scalar point $-\theta$ and whose variance is finite. The objective of the location parameter nonparametric detector is, then, to identify the existence of a constant, known, scalar, additive signal parameter, when the observed data may be generated by either one of the processes in class $\mathscr{F}_{-\theta}$.

As we will see, some of the existing nonparametric tests can be derived by taking the classes of stochastic processes in some robust detection tests to their limit. In general, given some class $\mathscr{F}_{-\theta}$ of memoryless, discrete-time, and stationary processes, whose common mean is $-\theta$ we will denote each member of the class by its first order density function $g_{-\theta}$. Let the H_1 and H_0 nonpara-

7.1 Introduction

metric hypotheses be respectively described by $\mathscr{F}_\theta$ and $\mathscr{F}_{-\theta}$. The assets available to the analyst are, then, $\mathscr{F}_\theta$ and $\mathscr{F}_{-\theta}$, an observation vector x^n of fixed length n and a false alarm rate α. The classical nonparametric detection schemes are not necessarily solutions of some optimization formalizations. Rather, they are the results of some ad hoc empirical observations. Therefore, the comparison among different nonparametric schemes is impossible unless some performance criteria are determined. One such criterion should provide the means for classifying detection schemes as nonparametric or not. The other two well-established performance criteria for nonparametric detection procedures are the *asymptotic relative efficiency* and the *efficacy*. The latter two criteria are both asymptotic, and they will be defined precisely in this section.

Let the H_1 and H_0 hypotheses be described respectively by $\mathscr{F}_\theta$ and $\mathscr{F}_{-\theta}$ as above. Let a decision rule be described by a test function, $T_n(x^n)$, a randomizing constant r, and a threshold λ, where x^n denotes the observation vector, and H_1 is decided with probability one when $T_n(x^n) > \lambda$, and with probability r when $T_n(x^n) = \lambda$. We then denote the decision rule by the triple $(T_n(x^n), \lambda, r)$ and we call, $(T_n(x^n), \lambda, r)$, *the test*. We now present a definition.

DEFINITION 7.1.1: Given the classes, $\mathscr{F}_\theta$ and $\mathscr{F}_{-\theta}$, given that the H_0 hypothesis is described by the class $\mathscr{F}_{-\theta}$, given the false alarm rate α, given n, given the test $(T_n(x^n), \lambda, r)$, the test, $(T_n(x^n), \lambda, r)$, is *nonparametric* in $(\mathscr{F}_\theta, \mathscr{F}_{-\theta})$ if and only if it induces the same false alarm at each member $g_{-\theta}$ in $\mathscr{F}_{-\theta}$—that is, iff

$$\Pr\{T_n(X^n) > \lambda | g_{-\theta}\} + r \Pr\{T_n(X^n) = \lambda | g_{-\theta}\} = \alpha \qquad \forall\, g_{-\theta} \in \mathscr{F}_{-\theta} \qquad \blacksquare$$

Given $\mathscr{F}_\theta$ and $\mathscr{F}_{-\theta}$, let $\mathscr{F}$ denote the class induced by either $\mathscr{F}_\theta$ or $\mathscr{F}_{-\theta}$ when the mean of their members is set equal to zero. Let g^* be some density function in $\mathscr{F}$. Let g^*_θ denote the density function induced by g^* when its mean is changed from the value zero to the value θ. Given n, and the nonparametric in $(\mathscr{F}_\theta, \mathscr{F}_{-\theta})$ test $(T_n(x^n), \lambda, r)$, let $E_{g^*_\theta}\{T_n(X^n)\}$ and $\text{Var}_{g^*_\theta}\{T_n(X^n)\}$ denote respectively the mean and the variance of the random variable $T_n(X^n)$ when the random vector X^n is generated by the density function g^*_θ. Given n, let $(T_{n,g^*_\theta}(x^n), \lambda_{g^*_\theta}, r_{g^*_\theta})$ be the Neyman-Pearson test for testing g^*_θ against the $g^*_{-\theta}$ (H_1 versus H_0 respectively) noncomposite hypotheses via the observation vector x^n. Given a false alarm rate α and a power level β let $n(\alpha, \beta, T_{g^*_\theta})$ denote the number of data needed by the Neyman-Pearson test $(T_{n,g^*_\theta}(x^n), \lambda_{g^*_\theta}, r_{g^*_\theta})$ to attain power equal to β with false alarm rate equal to α at the hypothesis pair $(g^*_\theta, g^*_{-\theta})$. Let $n(\alpha, \beta, T, g^*_\theta, g^*_{-\theta})$ denote the number of data needed by the nonparametric in $(\mathscr{F}_\theta, \mathscr{F}_{-\theta})$ test $(T_n(x^n), \lambda, r)$ to attain power equal to β with false alarm rate equal to α when used to test the H_1 hypothesis g^*_θ against the H_0 hypothesis $g^*_{-\theta}$. We now present a definition.

DEFINITION 7.1.2: Given the classes, $\mathscr{F}_\theta$ and $\mathscr{F}_{-\theta}$, given the nonparametric in $(\mathscr{F}_\theta, \mathscr{F}_{-\theta})$ test, $(T_n(x^n), \lambda, r)$; for any given n, given the distribution G^*, in class $\mathscr{F}$, the *efficacy*, $E_f(G^*)$, and the *asymptotic relative efficiency*, $\text{ARE}(G^*)$, of the test $(T_n(x^n), \lambda, r)$, at G^*, are respectively defined as

$$E_f(G^*) = \lim_{n \to \infty} \frac{\left[\dfrac{\partial}{\partial \theta} E_{g_\theta^*}\{T_n(X^n)\}\bigg|_{\theta=0}\right]^2}{n \operatorname{Var}_{g^*}\{T_n(X^n)\}}$$

$$\text{ARE}(G^*) = \lim_{\theta \to 0} \frac{n(\alpha, \beta, T_{g_\theta^*})}{n(\alpha, \beta, T, g_\theta^*, g_{-\theta}^*)} \quad\blacksquare$$

From Definition 7.1.2, we observe that both the efficacy and the asymptotic relative efficiency are defined at particular members of the nonparametric classes, and they generally induce different values at different such members. The efficacy measures how well a nonparametric test performs asymptotically ($n \to \infty$) in terms of discriminating among hypotheses that are asymptotically close to each other. The asymptotic relative efficiency is a relative performance criterion. It measures the performance loss induced by the nonparametric test, at some given distribution, as compared to the performance induced by the optimal Neyman-Pearson test at the same distribution, when the two hypotheses are asymptotically close to each other ($\theta \to 0$). We will discuss the efficacy and the asymptotic relative efficiency further when we analyze specific nonparametric tests. We point out that the reason why both the efficacy and the asymptotic relative efficiency are defined only asymptotically is that the nonasymptotic analysis of nonparametric tests is, in general, nonfeasible via analytic methods. In addition, as we will discuss in Chapter 12, the existing nonparametric tests are notoriously nonrobust, when the size of the observed data block is relatively small.

In the remainder of this chapter, we will present and analyze some well-established nonparametric tests. Our analysis will be mostly asymptotic. We will not cover topics such as adaptive and two-sample nonparametric tests. For information on such topics, the interested reader may refer to P. Papantoni-Kazakos and D. Kazakos (1977) and to the additional references included at the end of this chapter.

7.2 THE SIGN TEST

Let $\mathscr{F}$ be the class of all scalar, real, memoryless, and stationary processes whose first order density function exists, is symmetric around zero, and induces finite variance. The sign test is the oldest and most widely used nonparametric in $(\mathscr{F}_\theta, \mathscr{F}_{-\theta})$ test, for any real and scalar parameter θ. Although initially derived in an ad hoc fashion, the sign test evolves when Huber's nonparametric classes in Chapter 6, are taken to their limit. Thus, the sign test can be considered as the solution of a saddle-point game formalization, as in Chapter 6. To justify the above statements, let us consider the classes $\mathscr{F}_0$ and $\mathscr{F}_1$ in (6.3.31). Let the nominal density functions $f_0(x)$ and $f_1(x)$ in those classes be both Gaussian and unit variance, with mean values that are respectively equal to $-\theta$ and θ, where

7.2 The Sign Test

$\theta > 0$. Let the class $\mathscr{H}$ in $\mathscr{F}_0$ be identical to the class $\mathscr{F}_{-\theta}$ above. Let the class $\mathscr{H}$ in $\mathscr{F}_1$ be identical to the class $\mathscr{F}_\theta$ above. Let $\varepsilon_0 = \varepsilon_1 = \varepsilon$ and let ε approach the value one. Then $\mathscr{F}_0$ and $\mathscr{F}_1$ in (6.3.31) approach respectively the classes $\mathscr{F}_{-\theta}$ and $\mathscr{F}_\theta$. In addition, from the comments included after Corollary 6.3.1, we conclude that Huber's robust, normalized log likelihood test tends then to the number of times the inequality $f_1(x_i)|f_0(x_i) > 1$ holds. But in the present model we have

$$f_1(x_i)|f_0(x_i) = \frac{\exp\{-\frac{1}{2}(x_i - \theta)^2\}}{\exp\{-\frac{1}{2}(x_i + \theta)^2\}} = \exp\{2\theta x_i\} \quad (7.2.1)$$

and

$$f_1(x_i)|f_0(x_i) > 1 \rightarrow x_i > 0 \quad (7.2.2)$$

Noting that when $\varepsilon_0 \nearrow 1$, $\varepsilon_1 \nearrow 1$, the choice of the nominal distributions in classes $\mathscr{F}_{-\theta}$ and $\mathscr{F}_\theta$ can be ad hoc, and due to the above, we conclude that Huber's robust in $(\mathscr{F}_{-\theta}, \mathscr{F}_\theta)$ test has the test function

$$T_n(x^n) = n^{-1} \sum_{i=1}^n \operatorname{sgn}(x_i) \qquad \operatorname{sgn}(x) = \begin{cases} 1 & x > 0 \\ 0 & x \leq 0 \end{cases} \quad (7.2.3)$$

The test function in (7.2.3), together with some threshold λ_n^* and some randomizing constant r_n^* comprise the sign test. The threshold λ_n^* and the constant r_n^* are determined via a false alarm constraint. Let us now observe that instead of testing the class $\mathscr{F}_\theta$ against the class $\mathscr{F}_{-\theta}$ we can equivalently test class $\mathscr{F}_{2\theta}$ against class $\mathscr{F}$. Let θ be positive. We now state a proposition.

PROPOSITION 7.2.1: The sign test is nonparametric in $(\mathscr{F}_{2\theta}, \mathscr{F})$, for any given n.

Proof: Given n, given some density function g in $\mathscr{F}$, given r in $[0, 1]$, given some λ such that λn is a positive integer ($0 \leq \lambda n \leq n$), we obtain

$$\Pr\left\{n^{-1} \sum_{i=1}^n \operatorname{sgn}(X_i) > \lambda \big| g\right\} + r \Pr\left\{n^{-1} \sum_{i=1}^n \operatorname{sgn}(X_i) = \lambda \big| g\right\}$$

$$= U(n - \lambda n - 1) 2^{-n} \sum_{k=\lambda n+1}^n \binom{n}{k} + r 2^{-n} \binom{n}{\lambda n} \qquad \forall g \in \mathscr{F} \quad (7.2.4)$$

where

$$U(x) = \begin{cases} 1 & x \geq 0 \\ 0 & x < 0 \end{cases}$$

Due to (7.2.4) and Definition 7.1.1, the proof of the proposition is complete. ∎

From the proof of Proposition 7.2.1, and drawing from the discussions and analysis in Chapter 5, we conclude that given n and a false alarm rate α, the threshold λ_n^* and the randomizing parameter r_n^* in the sign test are uniquely determined by

$$\lambda_n^*, r_n^*: U(n - \lambda_n^* n - 1) 2^{-n} \sum_{k=\lambda_n^* n + 1}^{n} \binom{n}{k} + r_n^* 2^{-n} \binom{n}{\lambda_n^* n} = \alpha \qquad (7.2.5)$$

Given n and α, let us now compute the power induced by the sign test at some given density function $g_{2\theta}$ in class $\mathscr{F}_{2\theta}$. Denoting this power $P_n(g_{2\theta})$ and denoting $G(x) = \int_{-\infty}^{x} g(u)\,du$, we obtain

$$P_n(g_{2\theta}) = \Pr\left\{ n^{-1} \sum_{i=1}^{n} \text{sgn}(X_i) > \lambda_n^* \Big| g_{2\theta} \right\} + r_n^* \Pr\left\{ n^{-1} \sum_{i=1}^{n} \text{sgn}(X_i) = \lambda_n^* \Big| g_{2\theta} \right\}$$

$$= U(n - \lambda_n^* n - 1) \sum_{k=\lambda_n^* n + 1}^{n} \binom{n}{k} G^k(2\theta) G^{n-k}(-2\theta)$$

$$+ r_n^* \binom{n}{\lambda_n^* n} G^{\lambda_n^* n}(2\theta) G^{n - \lambda_n^* n}(-2\theta)$$

$$= U(n - \lambda_n^* n - 1) G^n(-2\theta) \sum_{k=\lambda_n^* n + 1}^{n} \binom{n}{k} [G(2\theta) G^{-1}(-2\theta)]^k$$

$$+ r_n^* G^n(-2\theta) \binom{n}{\lambda_n^* n} [G(2\theta) G^{-1}(-2\theta)]^{\lambda_n^* n} \qquad (7.2.6)$$

where the constants λ_n^* and r_n^* are given by (7.2.5) and where $g \in \mathscr{F}$ is the density function induced by $g_{2\theta}$ when the mean value changes from 2θ to zero.

From (7.2.6), we observe that the power $P_n(g_{2\theta})$ depends on the value of the density function g at the scalar point -2θ. This value depends explicitly on the exact form of g when $\theta \neq 0$, and so does then the power $P_n(g_{2\theta})$. Furthermore, for finite n no closed analytic form of the expression in (7.2.6) exists. Let us now consider asymptotically large n. Then, given $g_{2\theta}$ in $\mathscr{F}_{2\theta}$, the central limit theorem applies to the random variable $n^{-1} \sum_{i=1}^{n} \text{sgn}(X_i)$. Indeed, given $g_{2\theta}$, the random variable $T_n(X^n) = n^{-1} \sum_{i=1}^{n} \text{sgn}(X_i)$ is asymptotically ($n \to \infty$) Gaussian, with mean $E_{g_{2\theta}}\{T_n(X^n)\}$ and variance $\text{Var}_{g_{2\theta}}\{T_n(X^n)\}$, where

$$E_{g_{2\theta}}\{T_n(X^n)\} = G(2\theta) \qquad (7.2.7)$$

$$\text{Var}_{g_{2\theta}}\{T_n(X^n)\} = n^{-1} G(2\theta) G(-2\theta) \qquad (7.2.8)$$

where $G(x) = \int_{-\infty}^{x} g(u)\,du$.

Therefore, asymptotically ($n \to \infty$), the randomizing constant r_n^* in the sign test tends to zero. The threshold λ_n^* of the sign test and the power it induces

7.2 The Sign Test

at $g_{2\theta}$ in $\mathscr{F}_{2\theta}$ are then given, for $n \to \infty$, by

$$\lambda_n^*: \Phi\left(\frac{2^{-1} - \lambda_n^*}{2^{-1}n^{-1/2}}\right) = \alpha \to \lambda_n^* = 2^{-1} + 2^{-1}n^{-1/2}\Phi^{-1}(1-\alpha) \quad (7.2.9)$$

$$P_n(g_{2\theta}) = \Phi\left(\frac{G(2\theta) - \lambda_n^*}{[G(2\theta)G(-2\theta)]^{1/2}n^{-1/2}}\right)$$

$$= \Phi\left(\frac{n^{1/2}[G(2\theta) - 2^{-1}] - 2^{-1}\Phi^{-1}(1-\alpha)}{[G(2\theta)G(-2\theta)]^{1/2}}\right) \quad (7.2.10)$$

where $\Phi(x)$ is the cumulative distribution of the zero mean, unit variance, Gaussian random variable at the point x.

From (7.2.10), we observe that the power induced by the sign test at some density function in $\mathscr{F}_{2\theta}$ can be found in a closed form for asymptotically large sample sizes n. For such values of n, we observe that the power $P_n(g_{2\theta})$ increases monotonically with increasing sample size. For false alarm rate α strictly positive, the limit $\lim_{n\to\infty} P_n(g_{2\theta})$ equals one at every density function g in $\mathscr{F}$.

Now, let g^* be the zero mean, variance σ^2, Gaussian density function, where $\sigma^2 < \infty$. Then for each point x on the real axis, we obtain $G^*(x) = \Phi(\sigma^{-1}x)$. Given θ positive, let us consider the parametric Neyman-Pearson test when the H_0 and H_1 hypotheses are respectively described by the distributions $G^*(x) = \Phi(\sigma^{-1}x)$ and $G_{2\theta}^*(x) = \Phi(\sigma^{-1}[x - 2\theta])$ and the size of the observation vector is n. Then, as found in case 1, Example 5.2.1, when R_n is the identity matrix the above test is described by the pair $(T_{g^*}(x^n) = n^{-1}\sum_{i=1}^n x_i, \lambda_n^0)$, where

Given the false alarm rate α, given n,

$$\lambda_n^0 = n^{-1/2}\sigma \Phi^{-1}(1-\alpha)$$

$$\text{Decide } H_1 \text{ iff } n^{-1}\sum_{i=1}^n x_i \geq \lambda_n^0 \quad (7.2.11)$$

Given n, given α, the power $P_n^0(g_{2\theta}^*)$ induced by the Neyman-Pearson test, in (7.2.11) at the density function $g_{2\theta}^*$ is given by

$$P_n^0(g_{2\theta}^*) = \Phi[2\theta\sigma^{-1}n^{1/2} - \Phi^{-1}(1-\alpha)] \quad (7.2.12)$$

Let us now study the efficacy and the asymptotic relative efficiency induced by the sign test at the Gaussian distribution $G^*(x) = \Phi(\sigma^{-1}x)$. From (7.2.7) and (7.2.8), we find

$$\frac{\partial}{\partial \theta}E_{g_{2\theta}^*}\{T_n(X^n)\}\bigg|_{\theta=0} = \frac{\partial}{\partial \theta}G^*(2\theta)\bigg|_{\theta=0} = \frac{\partial}{\partial \theta}\Phi(\sigma^{-1}2\theta)\bigg|_{\theta=0}$$

$$= 2\sigma^{-1}\cdot\phi(0) = \sigma^{-2}\left(\frac{2}{\pi}\right)^{1/2} \quad (7.2.13)$$

$$\text{Var}_{g^*}\{T_n(X^n)\} = n^{-1}[G^*(0)]^2 = 4^{-1}n^{-1} \quad (7.2.14)$$

where $\phi(x)$ denotes the density function of the zero mean, unit variance, Gaussian random variable at the point x.

Referring to Definition 7.1.2 and to (7.2.13) and (7.2.14), we compute the efficacy $E_f(G^*)$ induced by the sign test at the Gaussian distribution $G^*(x) = \Phi(\sigma^{-1}x)$ as follows,

$$E_f(G^*) = 8\pi^{-1}\sigma^{-4} \tag{7.2.15}$$

The number in (7.2.15) represents the per datum squared difference between the values of the sign test asymptotically ($n \to \infty$) converges to under the two hypotheses when those hypotheses are asymptotically close to each other ($\theta \to 0$). This squared difference is normalized by the variance of the test. Let now the false alarm rate α and some power value β be given. Given the Gaussian distribution $G^*(x) = \Phi(\sigma^{-1}x)$, let us then consider the optimal in $(g_{2\theta}^*, g^*)$ Neyman-Pearson test in (7.2.11). From (7.2.12), we then conclude that the number $n(\alpha, \beta, T_{g_{2\theta}^*})$ of data needed by the Neyman-Pearson test above to attain false alarm rate and power that are respectively equal to α and β is found by

$$n(\alpha, \beta, T_{g_{2\theta}^*}): \beta = \Phi[2\theta\sigma^{-1}n^{1/2} - \Phi^{-1}(1-\alpha)]$$

and thus,

$$n(\alpha, \beta, T_{g_{2\theta}^*}) = 4^{-1}\theta^{-2}\sigma^2[\Phi^{-1}(\beta) + \Phi^{-1}(1-\alpha)]^2 \tag{7.2.16}$$

For asymptotically large n and from (7.2.10), we conclude that the number $n(\alpha, \beta, T, g_{2\theta}^*, g^*)$ of data needed by the sign test to attain false alarm rate α and power β at the Gaussian density function g^* is found as

$$n(\alpha, \beta, T, g_{2\theta}^*, g^*): \beta = \Phi\left(\frac{n^{1/2}[G^*(2\theta) - 2^{-1}] - 2^{-1}\Phi^{-1}(1-\alpha)}{[G^*(2\theta)G^*(-2\theta)]^{1/2}}\right)$$

where $G^*(x) = \Phi(\sigma^{-1}x)$ and, thus,

$$n(\alpha, \beta, T, g_{2\theta}^*, g^*) = [2\Phi(2\theta\sigma^{-1}) - 1]^{-2}\{\Phi^{-1}(1-\alpha) + 2[\Phi(2\theta\sigma^{-1}) \\ \times \Phi(-2\theta\sigma^{-1})]^{1/2}\Phi^{-1}(\beta)\}^2 \tag{7.2.17}$$

We now observe that when $\theta \to 0$, the g^* and $g_{2\theta}^*$ hypotheses approach infinitesimally close to each other. When then nonzero power β is demanded, it can only be attained if the sample size n is asymptotically large. Thus, in the definition of the asymptotic relative efficiency (Definition 7.1.2) the collection of asymptotically large sample sizes $n(\alpha, \beta, T_{g_{2\theta}^*})$ and $n(\alpha, \beta, T, g_\theta^*, g_{-\theta}^*)$ is implied. Therefore, (7.2.17) is applicable when seeking the asymptotic relative efficiency of the sign test at the Gaussian distribution $G^*(x) = \Phi(\sigma^{-1}x)$. Via Taylor expansion, let us now consider the first order approximations

$$\lim_{x \to 0} \Phi(x) = 2^{-1} + (2\pi)^{-1/2}x \tag{7.2.18}$$

$$\lim_{x \to 0} \Phi(x)\Phi(-x) = 4^{-1}$$

7.3 Rank Tests

Due to the approximations in (7.2.18) and via direct substitution in (7.2.17), we easily obtain

$$\lim_{\theta \to 0} n(\alpha, \beta, T, g^*_{2\theta}, g^*) = 8^{-1}\theta^{-2}\pi\sigma^2[\Phi^{-1}(1-\alpha) + \Phi^{-1}(\beta)]^2 \quad (7.2.19)$$

From (7.2.16) and (7.2.19), we then conclude

$$\mathrm{ARE}(G^*) = \lim_{\theta \to 0} \frac{n(\alpha, \beta, T_{g^*_{2\theta}})}{n(\alpha, \beta, T, g^*_{2\theta}, g^*)} = 2\pi^{-1} \quad (7.2.20)$$

Summarizing the results in (7.2.15) and (7.2.20), we express a corollary.

COROLLARY 7.2.1: The efficacy $E_f(G^*)$ and the asymptotic relative efficiency $\mathrm{ARE}(G^*)$ induced by the sign test at the zero mean, variance σ^2, Gaussian distribution, are respectively given by

$$E_f(G^*) = 8\pi^{-1}\sigma^{-4}$$

$$\mathrm{ARE}(G^*) = 2\pi^{-1}$$
∎

From the number $\mathrm{ARE}(G^*)$ in the corollary, we conclude that when the sign test is applied on Gaussian, noncomposite, location parameter hypotheses, it asymptotically ($\theta \to 0$) requires $\pi/2$ times the data optimal Neyman-Pearson test requires for the same false alarm rate and power. We note that, if there existed strong suspicions indicating that the hypotheses are close to Gaussian, we would be much better off by considering robust tests (as in Chapter 6) rather than the sign test. In general, the larger the nonparametric class $\mathscr{F}$ considered, the lower the efficiency induced by the corresponding nonparametric tests at any given distribution in $\mathscr{F}$. In that spirit, we compare in Section 7.5 the asymptotic relative efficiency induced by Huber's robust test in Section 6.3, at the Gaussian distribution, with the efficiency $2\pi^{-1}$ induced by the sign test at the same distribution, when the nominal density functions in Huber's classes are respectively $f_0(x) = \phi(x)$ and $f_1(x) = \phi(x - \theta)$ and the contaminating parameters ε_0 and ε_1 are relatively small.

7.3 RANK TESTS

Given the class $\mathscr{F}$ of all scalar, real, memoryless, and stationary processes whose first order density function is symmetric around zero and induces finite variance, and given a real, scalar, positive parameter θ, the objective of the *rank tests* is to test the hypothesis $\mathscr{F}_\theta$ against the hypothesis $\mathscr{F}_{-\theta}$, where the emphatic H_1 hypothesis is represented by the class $\mathscr{F}_\theta$. The idea preceding the formalization of the rank tests is that, if large and positive data values are heavily weighted, then the hypothesis H_1 will be highly visible. Thus, given some observation vector $x^n = \{x_i; 1 \le i \le n\}$, the rank tests first order the data $\{x_i\}$

from absolutely smaller to absolutely larger and then take the sign of the data so ordered. The vector $z^n = \{z_j; 1 \leq j \leq n\}$ of the signs is called the *rank vector*; it is such that

$$z_j = \begin{cases} 1 & \text{if the } j\text{th absolutely smallest datum in the sequence } \{x_i\} \\ & \text{is nonnegative} \\ -1 & \text{if the } j\text{th absolutely smallest datum in the sequence } \{x_i\} \\ & \text{is negative} \end{cases} \quad (7.3.1)$$

Given n, we will denote by $Z^n = \{Z_j; 1 \leq j \leq n\}$ the random rank vector that corresponds to some stochastic observation vector X^n. Given some density function g in $\mathscr{F}$, we will denote by g_θ and $g_{-\theta}$ the mappings of g in classes $\mathscr{F}_\theta$ and $\mathscr{F}_{-\theta}$ respectively. We will then denote by $P_g(z^n)$, $P_{g_\theta}(z^n)$, and $P_{g_{-\theta}}(z^n)$ the distributions of the random rank vector Z^n at the vector point z^n as they are induced respectively by the density functions g, g_θ, and $g_{-\theta}$. Denoting $z^n = \{z_j; 1 \leq j \leq n\}$, the following expressions can then be derived, in a relatively straightforward fashion, for $g(x) = g(-x)$.

$$P_g(z^n) = n! \int \cdots \int \prod_{j=1}^n g(z_j x_j) \, dx^n$$
$$0 \leq x_1 \leq \cdots \leq x_n \leq \infty$$

$$P_{g_\theta}(z^n) = n! \int \cdots \int \prod_{j=1}^n g(x_j - \theta z_j) \, dx^n \qquad (7.3.2)$$
$$0 \leq x_1 \leq \cdots \leq x_n \leq \infty$$

$$P_{g_{-\theta}}(z^n) = n! \int \cdots \int dx^n \prod_{j=1}^n g(x_j + \theta z_j) = P_{g_\theta}(-z^n)$$
$$0 \leq x_1 \leq \cdots \leq x_n \leq \infty$$

In the evaluation of rank tests, computing the distributions in (7.3.2) is necessary. This computation, however, can generally be performed only via computational methods. The interested reader will find such methods in Chapter 5 of P. Papantoni-Kazakos and D. Kazakos (1977). Analytical techniques for the computation of the distributions in (7.3.2) can be devised only in two asymptotic cases: either when the length n of the observation vector is asymptotically large and the parameter θ is infinitesimally small, or when the length n is finite and the parameter θ is asymptotically large. The former case will be considered in this section. Some results on the latter case will be included in the next section.

Let us now study the distribution of the rank vector $P_{g_\theta}(z^n)$ in (7.3.2) when the parameter θ is positive and very close to the zero value. Given some density function g in $\mathscr{F}$, let us assume that $g(x)$ has a Taylor expansion everywhere in R. Let us then denote by $g'(x)$ the first order derivative of $g(x)$ at the point x. Then, for θ values small enough, we use first order approximations in the

7.3 Rank Tests

Taylor expansion of $g(x_j - \theta z_j)$; $1 \leq j \leq n$ around the point x_j, obtaining

$$0 < \theta \ll 1; \qquad g(x_j - \theta z_j) \simeq [g(x_j) - \theta z_j g'(x_j)] \qquad (7.3.3)$$

$$0 < \theta \ll 1; \qquad \prod_{j=1}^{n} g(x_j - \theta z_j) \simeq \left[1 - \theta \sum_{j=1}^{n} z_j \frac{g'(x_j)}{g(x_j)}\right] \prod_{j=1}^{n} g(x_j) \qquad (7.3.4)$$

Let us now denote

$$C_n(\mathscr{F}) = n! \int \cdots \int \prod_{j=1}^{n} g(x_j) \, dx_j \qquad (7.3.5)$$
$$0 \leq x_1 \leq \cdots \leq x_n \leq \infty$$

$$F_n(z^n, g) = n! \int \cdots \int \left[\sum_{j=1}^{n} z_j \frac{g'(x_j)}{g(x_j)}\right] \prod_{k=1}^{n} g(x_k) \, dx_k \qquad (7.3.6)$$
$$0 \leq x_1 \leq \cdots \leq x_n \leq \infty$$

It can be easily shown, via Klotz's (1964) recursive method, that for the class $\mathscr{F}$ of density functions g symmetric around zero the quantity $C_n(\mathscr{F})$ in (7.3.5) is independent of g, and equal to 2^{-n}. Klotz's method is explained in Chapter 5 of P. Papantoni-Kazakos and D. Kazakos (1977). For a parameter θ such that $0 < \theta \ll 1$, and due to (7.3.2), (7.3.4), (7.3.5), and (7.3.6), we obtain, by substitution,

$$0 < \theta \ll 1 \qquad P_{g_\theta}(z^n) \simeq C_n(\mathscr{F}) - \theta F_n(z^n, g) \qquad (7.3.7)$$

We note that for given rank vector z^n the quantity $F_n(z^n, g)$ in (7.3.6) is basically the expected value of the random variable $T(z^n, X_R^n) = \sum_{i=1}^{n} z_i g'(X_{Ri})/g(X_{Ri})$ at the density function g, conditioned on the event that the rank vector of the ranked random vector X_R^n is given by z^n. Consider the random variable $F_n(Z^n, g)$ that evolves from $F_n(z^n, g)$, by replacing the vector point z^n by the random rank vector Z^n. Let $\Pr\{n^{-1} F_n(Z^n, g) < \lambda | g\}$ denote the probability that, given the density function g in $\mathscr{F}$, the rank vector Z^n induces less than λn values of the quantity in (7.3.6). Then we conclude

$$\Pr\{n^{-1} F_n(Z^n, g) < \lambda | g\} = \sum_{z^n:\, F_n(z^n, g) < \lambda n} P_g(z^n) n^{-1} F_n(z^n, g)$$

$$= \Pr\left\{n^{-1} \sum_{i=1}^{n} Z_i \frac{g'(X_{Ri})}{g(X_{Ri})} < \lambda \Big| g\right\}$$

$$= \Pr\left\{n^{-1} \sum_{i=1}^{n} \frac{g'(X_i)}{g(X_i)} < \lambda \Big| g\right\} \qquad (7.3.8)$$

where $X^n = \{X_i; 1 \leq i \leq n\}$ denotes the random observation vector generated by the density function g.

We note that given g, the terms in the sum $n^{-1}\sum_{i=1}^{n} g'(X_i)/g(X_i)$ are independent and identically distributed random variables. Thus, for asymptotically large n, the random variable $T(X^n) = n^{-1}\sum_{i=1}^{n} g'(X_i)/g(X_i)$ is Gaussian with mean $m(T, g) = E\{g'(X)/g(X)|g\} = 0$ and variance $\sigma^2(T, g) = n^{-1}E\{[g'(X)|g(X) - m(T, g)]^2|g\} = n^{-1}\int_R [g'(x)]^2 g^{-1}(x)\,dx$, which is called the Fischer information measure.

We will now present and analyze two well-established rank tests: the *optimal rank test*, and the *Wilcoxon rank test*. We first define the two tests.

DEFINITION 7.3.1: Given $\theta > 0$, given the classes $\mathscr{F}$, $\mathscr{F}_\theta$, $\mathscr{F}_{-\theta}$ as in this section, given n, and given a false alarm rate α,

i. Given g in $\mathscr{F}$, the optimal-at-g, rank test is defined as

$$\delta_{g_\theta}(x^n) = \begin{cases} 1 & x^n\colon P_{g_\theta}(z^n)|P_{g_{-\theta}}(z^n) > \lambda \\ r & x^n\colon P_{g_\theta}(z^n)|P_{g_{-\theta}}(z^n) = \lambda \\ 0 & \text{otherwise} \end{cases}$$

$$\lambda, r\colon \Pr\{P_{g_\theta}(Z^n)|P_{g_{-\theta}}(Z^n) > \lambda|g_{-\theta}\} + r\Pr\{P_{g_\theta}(Z^n)|P_{g_{-\theta}}(Z^n) = \lambda|g_{-\theta}\} = \alpha$$

where $\delta_{g_\theta}(x^n)$ denotes the probability with which the density function g is decided, given the observation vector x^n.

ii. The Wilcoxon rank test, is defined as

$$\delta(x^n) = \begin{cases} 1 & x^n\colon \sum_{i=1}^{n} iz_i > \lambda \\ r & x^n\colon \sum_{i=1}^{n} iz_i = \lambda \\ 0 & \text{otherwise} \end{cases}$$

$$\lambda, r\colon \sup_{g_{-\theta} \in \mathscr{F}_{-\theta}} \left(\Pr\left\{\sum_{i=1}^{n} iZ_i > \lambda|g_{-\theta}\right\} + r\Pr\left\{\sum_{i=1}^{n} iZ_i = \lambda|g_{-\theta}\right\}\right) = \alpha$$

where $\delta(x^n)$ denotes the probability with which the H_1 hypothesis $\mathscr{F}_\theta$ is decided, given the observation vector x^n. ∎

We note that the optimal rank test is designed for a fixed density function g in $\mathscr{F}$, and it is the Neyman-Pearson test for the rank vector z^n at g. On the other hand, the Wilcoxon rank test is designed for the testing of the whole class $\mathscr{F}_\theta$ against the class $\mathscr{F}_{-\theta}$ or equivalently, for the testing of the class $\mathscr{F}_{2\theta}$ against the class $\mathscr{F}$. We now proceed with two lemmas which include some asymptotic properties of the optimal and the Wilcoxon rank tests.

LEMMA 7.3.1: Let g be any density function in $\mathscr{F}$ that possesses a Taylor expansion. Then the asymptotic relative efficiency of the optimal-at-g rank test, equals one. The efficacy of the same test at g equals the Fischer information measure $\int_R [g'(x)]^2 g^{-1}(x)\,dx$.

7.3 Rank Tests

Proof:

i. Given g, let us consider the likelihood ratios $\text{LR}_0(x^n) = g_\theta(x^n)|g_{-\theta}(x^n) = [\prod_{i=1}^n g(x_i - \theta)]|[\prod_{i=1}^n g(x_i + \theta)]$ and $\text{LR}(z^n) = P_{g_\theta}(z^n)|P_{g_{-\theta}}(z^n)$. As we have established, and via (7.3.3), (7.3.4), and (7.3.7), for θ values such that $0 < \theta \ll 1$, we obtain

$$\text{LR}_0(x^n) \simeq \frac{1 - \theta \sum_{j=1}^n g'(x_j)|g(x_j)}{1 + \theta \sum_{j=1}^n g'(x_j)|g(x_j)} \qquad (7.3.9)$$

$$\text{LR}(z^n) \simeq \frac{C_n(\mathscr{F}) - \theta F_n(z^n, g)}{C_n(\mathscr{F}) + \theta F_n(z^n, g)} \qquad (7.3.10)$$

The likelihood ratio in (7.3.9) is utilized by the optimal-at-g Neyman-Pearson test. The likelihood ratio in (7.3.10) is utilized by the optimal-at-g rank test. We easily conclude that the conditions $\text{LR}_0(x^n) > \lambda_1$ and $\text{LR}(z^n) > \lambda_2$ are respectively equivalent to

$$n^{-1} \sum_{j=1}^n g'(x_j)|g(x_j) < n^{-1} \frac{1 - \lambda_1}{1 + \lambda_1} \theta^{-1} = \lambda_0^* \qquad (7.3.11)$$

$$n^{-1} F_n(z^n, g) < n^{-1} \frac{1 - \lambda_2}{1 + \lambda_2} \theta^{-1} C_n(\mathscr{F}) = \lambda^* \qquad (7.3.12)$$

The threshold λ_0^* in (7.3.11), together with a randomizing parameter r_0^*, is given by the Neyman-Pearson constraint for a given false alarm rate α. That is

$$\lambda_0^*, r_0^*: \Pr\left\{n^{-1} \sum_{j=1}^n g'(X_j)|g(X_j) > \lambda_0^* \Big| g_{-\theta}\right\}$$
$$+ r_0^* \Pr\left\{n^{-1} \sum_{j=1}^n g'(X_j)|g(X_j) = \lambda_0^* \Big| g_{-\theta}\right\} = \alpha \qquad (7.3.13)$$

For the same false alarm rate α the threshold λ^* in (7.3.12), together with a randomizing parameter r^*, is given by the following constraint, as established in part i, Definition 7.3.1.

$$\lambda^*, r^*: \Pr\{n^{-1} F_n(Z^n, g) > \lambda^* | g_{-\theta}\}$$
$$+ r^* \Pr\{n^{-1} F_n(Z^n, g) = \lambda^* | g_{-\theta}\} = \alpha \qquad (7.3.14)$$

But, as established by (7.3.8), (7.3.14) and (7.3.13) are identical. Thus, $\lambda^* = \lambda_0^*$, $r^* = r_0^*$, and the optimal-at-g, rank test is asymptotically ($\theta \to 0$) identical to the optimal Neyman-Pearson test at g. Thus, the asymptotic relative efficiency of the optimal-at-g rank test is trivially equal to one.

ii. As already established, for asymptotically small θ values, the test function induced by the optimal-at-g rank test is $T(X^n) = n^{-1} \sum_{i=1}^{n} g'(X_i)|g(X_i)$. For asymptotically large values of n we have established that the random variable $T(X^n)$ is Gaussian for each g_θ in $\mathscr{F}_\theta$, and for some g in $\mathscr{F}$ its variance equals $n^{-1} \int_R [g'(x)]^2 g^{-1}(x)\,dx$. Let us now consider the expected value $E\{T(X^n)|g_\theta\}$ for θ such that $0 < \theta \ll 1$. Then, we obtain,

$$E\{T(X^n)|g_\theta\} = E\left\{\frac{g'(X)}{g(X)}\bigg|g_\theta\right\} = \int_R \frac{g'(x)}{g(x)} g(x-\theta)\,dx$$

$$\simeq \int_R \frac{g'(x)}{g(x)} [g(x) - \theta g'(x)]\,dx$$

$$= \int_R g'(x)\,dx - \theta \int_R [g'(x)]^2 g^{-1}(x)\,dx$$

$$= -\theta \int_R [g'(x)]^2 g^{-1}(x)\,dx$$

and thus,

$$\frac{\partial}{\partial \theta} E\{T(X^n)|g_\theta\} = -\int_R [g'(x)]^2 g^{-1}(x)\,dx \qquad (7.3.15)$$

From (7.3.15), and via Definition 7.1.2, we now obtain the efficacy $E_f(G)$ of the optimal-at-g rank test as

$$E_f(G) = \lim_{n\to\infty} \frac{(\int_R [g'(x)]^2 g^{-1}(x)\,dx)^2}{\int_R [g'(x)]^2 g^{-1}(x)\,dx} = \int_R [g'(x)]^2 g^{-1}(x)\,dx$$

The proof of the lemma is now complete. ∎

We note that at the Gaussian density function, the Fischer information measure is equal to the variance of this density function, and so is the efficacy induced by the optimal rank test. We note that asymptotically ($\theta \to 0$, $n \to \infty$), the optimal rank test performs as well as the optimal Neyman-Pearson test does. This means that the rank vector is then a sufficient statistical measure. Observe that this vector essentially counts zero crossings of the absolutely ranked data. We now proceed with a lemma that states some of the important properties induced by the Wilcoxon rank test.

LEMMA 7.3.2: The Wilcoxon rank test is nonparametric in $(\mathscr{F}_{2\theta}, \mathscr{F})$. Its asymptotic relative efficiency, at the variance σ^2 Gaussian density function in $\mathscr{F}$ equals $3\pi^{-1}$. Its efficacy at the same density function is equal to $6\sigma^{-2}\pi^{-1}$.

Proof:

i. Consider the class $\mathscr{F}$ of density functions symmetric around zero whose variance is finite. Consider, then, the distribution $P_g(z^n)$ in (7.3.2) at some g

7.3 Rank Tests

in $\mathscr{F}$. Then, due to the symmetry of g, and for $z_j = \pm 1$, we obtain

$$P_g(z^n) = n! \int \cdots \int_{0 \leq x_1 \leq \cdots \leq x_n < \infty} \prod_{j=1}^{n} dx_j \, g(z_j x_j)$$

$$= n! \int \cdots \int_{0 \leq x_1 \leq \cdots \leq x_n < \infty} \prod_{j} dx_j \, g(x_j) = C_n(\mathscr{F}) = 2^{-n}$$

where $C_n(\mathscr{F})$, is given by (7.3.5), and where the equality $C_n(\mathscr{F}) = 2^{-n}$ has been previously established.

The statistics of the test function $T(Z^n) = n^{-1} \sum_{i=1}^{n} i Z_i$ in the Wilcoxon test at g are fully determined by the distribution $P_g(z^n)$ above, which is independent of g in $\mathscr{F}$. Thus, given λ and r, the false alarm induced by the Wilcoxon test at g is constant for every g in $\mathscr{F}$. So, the test is nonparametric in $(\mathscr{F}_{2\theta}, \mathscr{F})$.

ii. Given n, let x^n be the observation vector, and let it be such that, $|x_i| \leq |x_j| \; \forall \; i < j$. Let $z_i = \text{sgn } x_i = \begin{cases} 1; & x_i \geq 0 \\ -1; & x_i < 0 \end{cases}$. Then we can write

$$\sum_{i=1}^{n} i z_i = \sum_{i=1}^{n} i[U(x_i) - U(-x_i)] = T(x^n) \qquad (7.3.16)$$

where $U(x) = \begin{cases} 1; & x \geq 0 \\ 0; & x < 0 \end{cases}$

Let g^* be the Gaussian, zero mean, and variance $\sigma^2 < \infty$ density function in $\mathscr{F}$. Let the observation vector x^n above be generated by g^*. Then, due to the ergodic theorem, we conclude that asymptotically $(n \to \infty)$, the random variable $T(X^n)$ is Gaussian. In addition, due to the symmetry around zero of the zero mean Gaussian density, the expected value $E\{T(X^n)|g^*\}$ equals zero. The variance $\text{Var}\{T(X^n)|g^*\}$ is then given by

$$\text{Var}\{T(X^n)|g^*\} = \sum_{i=1}^{n} \sum_{j=1}^{n} ij \, E\{[U(X_i) - U(-X_i)][U(X_j) - U(-X_j)]|g^*\}$$

(7.3.17)

Remembering that the random variables $\{X_i\}$ are absolutely ordered, we find

$$E\{[U(X_i) - U(-X_i)][U(X_j) - U(-X_j)]|g^*\} = \begin{cases} 1 & i = j \\ 0 & i \neq j \end{cases} \qquad (7.3.18)$$

From (7.3.17) and (7.3.18), we conclude

$$\text{Var}\{T(X^n)|g^*\} = \sum_{i=1}^{n} i^2 = \frac{1}{6} n(n+1)(2n+1) \qquad (7.3.19)$$

Let us now consider, the Gaussian distribution $g_{2\theta}^*$ whose mean value is 2θ. Then we find

$$E\{T(X^n)|g_{2\theta}^*\} = [1 - 2\Phi(-2\theta\sigma^{-1})]2^{-1}n(n + 1) \qquad (7.3.20)$$

$$E\{[U(X_i) - U(-X_i)][U(X_j) - U(-X_j)]|g_{2\theta}^*\} = \begin{cases} 1 & i = j \\ 0 & i \neq j \end{cases} \qquad (7.3.21)$$

$$\text{Var}\{T(X^n)|g_{2\theta}^*\} = \frac{1}{6}n(n + 1)(2n + 1) \qquad (7.3.22)$$

where $\Phi(x)$ is the cumulative distribution of the zero mean, unit variance, Gaussian random variable at the point x.

From (7.3.19) and (7.3.20), we find easily the efficacy $E_f(G^*)$ of the Wilcoxon rank test.

$$E_f(G^*) = \lim_{n \to \infty} \frac{[(\partial/\partial\theta)E\{T(X^n)|g_{2\theta}^*|_{\theta=0}]^2}{n^{-1}\text{Var}\{T(X^n)|g^*\}}$$

$$= \lim_{n \to \infty} \frac{12}{\sigma^2\pi} \frac{n+1}{2n+1} = 6\sigma^{-2}\pi^{-1} \qquad (7.3.23)$$

Now let a false alarm rate α and a power level β be given. Let us consider testing the hypothesis $g_{2\theta}^*$ against the hypothesis g^* for asymptotically large n. In Section 7.2, we found the number of data then needed by the Neyman-Pearson test to attain α and β. This number is given by (7.2.16) as

$$n(\alpha, \beta, T_{g_{2\theta}^*}) = 4^{-1}\theta^{-2}\sigma^2[\Phi^{-1}(\beta) + \Phi^{-1}(1 - \alpha)]^2 \qquad (7.3.24)$$

The number $n(\alpha, \beta, T, g_{2\theta}^*, g^*)$ needed, then, by the Wilcoxon rank test is found by

$$\lambda: \alpha = 1 - \Phi\left(\frac{\lambda\sqrt{6}}{[n(n+1)(2n+1)]^{1/2}}\right) \to \lambda$$

$$= \left[\frac{1}{6}n(n+1)(2n+1)\right]^{1/2}\Phi^{-1}(1 - \alpha) \qquad (7.3.25)$$

$$\beta = 1 - \Phi\left(\left[\frac{1}{6}n(n+1)(2n+1)\right]^{-1/2}\{\lambda - 2^{-1}n(n+1)[1 - 2\Phi(-2\theta\sigma^{-1})]\}\right) \qquad (7.3.26)$$

Substituting (7.3.25) in (7.3.26), we obtain

$$n(\alpha, \beta, T, g_{2\theta}^*, g^*): \frac{n(n+1)}{2n+1}$$

$$= \frac{2}{3}[\Phi^{-1}(\beta) + \Phi^{-1}(1 - \alpha)]^2[1 - 2\Phi(-2\theta\sigma^{-1})]^{-2} \qquad (7.3.27)$$

7.3 Rank Tests

For asymptotically large n we use the approximation $n(n+1)(2n+1)^{-1} \simeq 2^{-1}n$. Then, we obtain from (7.3.27)

$$n(\alpha, \beta, T, g_{2\theta}^*, g^*) = \frac{4}{3}[\Phi^{-1}(\beta) + \Phi^{-1}(1-\alpha)]^2 [1 - 2\Phi(-2\theta\sigma^{-1})]^{-2}$$
(7.3.28)

For very small θ values, we also use the approximation

$$\Phi(-2\theta\sigma^{-1}) \simeq 2^{-1} - \frac{\theta}{\sigma}\left[\frac{2}{\pi}\right]^{1/2}$$
(7.3.29)

From (7.3.24), (7.3.28), and (7.3.29), we then finally obtain the asymptotic relative efficiency at g^* of the Wilcoxon rank test as

$$\text{ARE}(G^*) = \lim_{\theta \to 0} \frac{n(\alpha, \beta, T_{g_{2\theta}^*})}{n(\alpha, \beta, T, g_{2\theta}^*, g^*)} = 3\pi^{-1}$$
(7.3.30)
∎

We observe that at the Gaussian distribution, the asymptotic relative efficiency of the Wilcoxon rank test is 3/2 times better, than that of the sign test. The efficacy of the Wilcoxon rank test at the variance σ^2 Gaussian distribution is $4/3\,\sigma^2$ times better, than that of the sign test for σ^2 values less than $4/3$. The optimal rank test is the most superior, in terms of its induced asymptotic relative efficiency. Its disadvantage, however, is that generally it is not nonparametric in $(\mathscr{F}_{2\theta}, \mathscr{F})$. We note that good asymptotic performance does not imply or guarantee good small-sample performance. In fact, as we will see in the next section, the performance of the nonparametric tests deteriorates drastically when the length of the observation vector is relatively small. In addition, as we mentioned in Section 7.2, the existing nonparametric tests are grossly nonrobust on infinitesimally small families of stochastic processes (see Chapter 12) and in the presence of relatively small data sample sizes.

We conclude this section by briefly mentioning the *t*-test, which is used for evaluating rank tests. Given an observation vector $x^n = \{x_i; 1 \le i \le n\}$ the *t*-test consists of a test function $T_t(x^n)$ tested against some threshold λ, where

$$T_t(x^n) \triangleq \left[n^{-1/2} \sum_{i=1}^{n} x_i\right]\left[(n-1)^{-1} \sum_{i=1}^{n} (x_i - n^{-1/2} \sum_{j=1}^{n} x_j)^2\right]^{-1}$$
(7.3.31)

If the stochastic process that generates the observation vector x^n is stationary and memoryless, the numerator in the $T_t(x^n)$ ratio is asymptotically ($n \to \infty$) Gaussian via the central limit theorem. The denominator in the $T_t(x^n)$ ratio converges asymptotically ($n \to \infty$) to the variance of the underlying process. The *t*-test is thus asymptotically ($n \to \infty$) nonparametric in $(\mathscr{F}_{2\theta}, \mathscr{F})$. It is, however, asymptotically ($n \to \infty$) grossly nonrobust, for infinitesimally small families of stochastic processes (see Chapter 12). That is, the asymptotic (for $n \to \infty$) power induced by the t-test deteriorates drastically for small deviations from some nominal stochastic process.

7.4 SMALL-SAMPLE RESULTS

As we mentioned earlier, the performance induced by the existing nonparametric tests deteriorates drastically, when the size n of the observation vector x^n is relatively small. In addition, for the evaluation of the small-sample performance of such tests, only numerical methods can be generally applied. Some analytical results can be found, however, when then the location parameter θ in the classes $\mathscr{F}_\theta$ and $\mathscr{F}_{-\theta}$ is asymptotically large and $\mathscr{F}_\theta$ and $\mathscr{F}_{-\theta}$ are as in Sections 7.2 and 7.3. In this section, we will consider such asymptotic ($\theta \to \infty$) small-sample results for rank tests. We will not get into extensive analytial derivations. Such derivations can be found in Chapter 5 of P. Papantoni-Kazakos and D. Kazakos (1977).

Let us consider the distributions $P_{g_\theta}(z^n)$ and $p_{g_{-\theta}}(z^n)$ of the rank vector in (7.3.2). As we have already mentioned, for finite n the analytical computation of those distributions is generally impossible. Let us now consider the case where the positive parameter θ is asymptotically large, and the density function g in $\mathscr{F}$ is Gaussian with unit variance. Then for arbitrary sample size n the distributions $P_{g_\theta}(z^n)$ and $P_{g_{-\theta}}(z^n)$ can be evaluated analytically. This evaluation is quite complicated and involves both dynamic programming and asymptotic approximation methods. The interested reader may look into Chapter 5 of P. Papantoni-Kazakos and D. Kazakos (1977). Here we will only present the final result. Let us consider the distribution $P_{g_\theta}(z^n)$ in (7.3.2), and let us denote by $x_0^n = \{x_{0i}; 1 \leq i \leq n\}$ the vector x^n that maximizes the distribution $P_{g_\theta}(z^n)$. Let us consider the clusters of equally valued consecutive x_{0i} values in the vector x_0^n. Let there be l such clusters, and let the size of the ith such cluster be m_i. Consider now the rank vector z^n whose components are as in (7.3.1). Going from z_1 to z_n and given z^n, there exist, in general, clusters of $(+1)$-valued and (-1)-valued consecutive components. Let us order those clusters from left to right, and let us then denote by k_i and μ_i the size of the ith $(+1)$-versus (-1)-valued cluster. Let us then define the following parameters.

$$l_i = \sum_{j=1}^{i} m_j \qquad v_i = m_i^{-1} \sum_{k=1}^{m_i} z_{l_{i-1}+k}$$

$$\rho_1 = \begin{cases} 1 & \mu_1 < k_1 \\ 1 - v_1^2 & \mu_1 \geq k_1 \end{cases} \qquad s = \begin{cases} m_1 & \mu_1 \geq k_1 \\ 0 & \mu_1 < k_1 \end{cases} \qquad (7.4.1)$$

$$\rho_i = 1 - v_i^2 \quad i \geq 2 \qquad \alpha = \begin{cases} 1 & k_1 > \mu_1 \\ v_1^{-m_1} & k_1 \leq \mu_1 \end{cases}$$

Via the parameters defined above, the limiting value of the distribution $P_{g_\theta}(z^n)$, is expressed as

$$\lim_{\theta \to \infty} P_{g_\theta}(z^n) = \frac{\alpha}{\prod_{i=1}^{l} m_i!} \theta^{-s} \exp\left\{-\frac{\theta^2}{2} \sum_{i=1}^{l} m_i \rho_i\right\} \qquad (7.4.2)$$

7.4 Small-Sample Results

We note that the parameters l, α, and $\{m_i; 1 \le i \le l\}$ in (7.4.2) turn out to depend on the rank vector z^n only. Also, due to the relationship $P_{g-\theta}(z^n) = P_{g\theta}(-z^n)$ in (7.3.2), the limit, $\lim_{\theta \to \infty} P_{g-\theta}(z^n)$, has a form similar to that in (7.4.2). Considering now rank tests, whose evaluation is controlled by the distribution $P_{g\theta}(z^n)$ we first present a definition.

DEFINITION 7.4.1: Given the classes $\mathscr{F}_\theta$ and $\mathscr{F}_{-\theta}$ and a test $(T_n(x^n), \lambda, r)$ that is nonparametric in $(\mathscr{F}_\theta, \mathscr{F}_{-\theta})$ for any given n, given the density function g^* in $\mathscr{F}$, the *small sample efficiency*, SSE(g^*) of the test $(T_n(x^n), \lambda, r)$, at g^* is defined as

$$\text{SSE}(g^*) = \lim_{\theta \to \infty} \frac{n(\alpha, \beta, T_{g_\theta^*})}{n(\alpha, \beta, T, g_\theta^*, g_{-\theta}^*)}$$

where $n(\alpha, \beta, T_{g_\theta^*})$ and $n(\alpha, \beta, T, g_\theta^*, g_{-\theta}^*)$ are as in Section 7.1, and Definition 7.1.2. ∎

We note that, while the asymptotic relative efficiency is a performance measure for the cases where the $\mathscr{F}_\theta$ and $\mathscr{F}_{-\theta}$ hypotheses are infinitesimally close to each other, the small-sample efficiency measures performance when the $\mathscr{F}_\theta$ and $\mathscr{F}_{-\theta}$ hypotheses are, instead, asymptotically far from each other. Also, while the model corresponding to the asymptotic relative efficiency implies asymptotically long data sequences, the model for the small-sample efficiency requires relatively small data samples. In Chapter 5 of P. Papantoni-Kazakos and D. Kazakos (1977), the small-sample efficiencies of the optimal rank test and the Wilcoxon rank test have been found at the unit variance Gaussian distribution. Here we only present the results in a lemma.

LEMMA 7.4.1: The small sample efficiencies, at the Gaussian, unit variance, distribution of the optimal rank test, and the Wilcoxon rank test, are respectively equal to $3/4$ and $2^{-1/2}$. ∎

Comparing the results in the above lemma with those in Lemmas 7.3.1 and 7.3.2, we observe that the small-sample efficiencies of the optimal rank and the Wilcoxon rank tests are inferior to their corresponding asymptotic relative efficiencies. Indeed, the optimal rank test suffers then a 25% performance reduction; it requires $4/3$ the number of data needed by the Neyman-Pearson test to attain the same false alarm and power, when two asymptotically distant Gaussian hypotheses are tested. It also requires as many data as the Neyman-Pearson test does when the two Gaussian hypotheses are infinitesimally close to each other and the data sequences are asymptotically long. As compared to its asymptotic $(n \to \infty, \theta \to 0)$ performance, the small-sample performance (for $\theta \to \infty$) of the Wilcoxon rank test is reduced by more than 25%. From the above results, it is not hard to conclude that in the presence of hypotheses that are not asymptotically far from each other, and when the data sequences are not asymptotically long, the rank tests are practically useless. On the other hand, tests that are robust in relatively small nonparametric families of processes, as in Chapter 6, perform very well in this case.

7.5 AN ASYMPTOTIC COMPARISON

In Section 7.2, we studied the sign test. We found that the asymptotic relative efficiency of this test at the Gaussian density function equals $2\pi^{-1}$. In this section, we will compute the asymptotic relative efficiency of Huber's robust test (Section 6.3) at the Gaussian density function, for comparison with the sign test. Let the nominal density functions $f_1(x)$ and $f_0(x)$ in Huber's contaminated families be unit variance Gaussian, with respective means $\theta > 0$ and zero. Let $\varepsilon_1 = \varepsilon_2 = \varepsilon < .5$. Then, as we found in Section 6.3, given an observation vector x^n, the robust test is

$$\delta^*(x^n) = \begin{cases} 1 & x^n : n^{-1} \sum_{i=1}^n z(x_i) > \lambda^* \\ r^* & x^n : n^{-1} \sum_{i=1}^n z(x_i) = \lambda^* \\ 0 & x^n : n^{-1} \sum_{i=1}^n z(x_i) < \lambda^* \end{cases} \quad (7.5.1)$$

where,

$$z(x) = \begin{cases} d & x \leq d \\ x & d < x < -d + \theta \\ -d + \theta & x \geq -d + \theta \end{cases} \quad (7.5.2)$$

$$d : (1 - \varepsilon)\left[\Phi(-d + \theta) + \exp\left\{\theta d - \frac{\theta^2}{2}\right\}\Phi(d)\right] = 1 \quad (7.5.3)$$

Let us consider the asymptotic ($n \to \infty$, $\theta \to 0$) use of the robust test at the Gaussian density function. Then, given some false alarm rate α, given n, the pair (λ^*, r^*) of constants in (7.5.1) are found from the equation below, where $\phi(x)$ denotes the unit variance, zero mean Gaussian density function at the point x and $f_0(x) = \phi(x)$.

$$\Pr\left\{n^{-1} \sum_{i=1}^n z(X_i) > \lambda^* \Big| \phi\right\} + r^* \Pr\left\{n^{-1} \sum_{i=1}^n z(X_i) = \lambda^* \Big| \phi\right\} = \alpha \quad (7.5.4)$$

When the random vector X^n is generated by the density function ϕ and for $n \to \infty$, the random variable $n^{-1} \sum_{i=1}^n z(X_i)$ is Gaussian with mean m_0 and variance σ_{0n}^2 given below.

$$m_0 = E\left\{n^{-1} \sum_{i=1}^n z(X_i) \Big| \phi\right\} = E\{z(X) | \phi\}$$

$$= d \Pr\{X \leq d | \phi\} + (\theta - d) \Pr\{X \geq -d + \theta | \phi\} + \int_d^{-d+\theta} x\phi(x)\,dx$$

$$= d\Phi(d) + (\theta - d)\Phi(d - \theta) + \phi(d) - \phi(d - \theta) \quad (7.5.5)$$

7.5 An Asymptotic Comparison

where $\Phi(x)$ is the cumulative distribution of the zero mean, unit variance Gaussian random variable at the point x.

$$\sigma_{0n}^2 = E\left\{\left[n^{-1}\sum_{i=1}^{n} z(X_i) - m_0\right]^2 \Big| \phi\right\} = n^{-1}E\{[z(X) - m_0]^2 | \phi\}$$

$$= n^{-1}\bigg(d^2\Pr\{X \leq d | \phi\} + (\theta - d)^2 \Pr\{X \geq -d + \theta | \phi\}$$

$$+ \int_d^{-d+\theta} x^2 \phi(x)\,dx - m_0^2\bigg)$$

$$= n^{-1}(d^2\Phi(d) + (\theta - d)^2\Phi(d - \theta) + d\,\phi(d) + (d - \theta)\phi(d - \theta)$$
$$+ \Phi(-d + \theta) - \Phi(d) - m_0^2)$$
$$= n^{-1}([d^2 - 1]\Phi(d) + [(\theta - d)^2 - 1]\Phi(d - \theta) + d\,\phi(d)$$
$$+ (d - \theta)\phi(d - \theta) + 1 - m_0^2) \tag{7.5.6}$$

Let now $\theta \to 0$. We first observe that for $\theta \to 0$, the value of the constant d in (7.5.3) is asymptotically large. Then we use the following first order approximations, via Taylor series expansions.

$$\exp\left\{\theta d - \frac{\theta^2}{2}\right\} \simeq 1 + \theta(d - \theta) \simeq 1 + \theta d$$

$$\Phi(d - \theta) \simeq \Phi(d) - \theta\,\phi(d) \simeq \Phi(d) \tag{7.5.7}$$

$$\Phi(-d + \theta) \simeq 1 - \Phi(d) + \theta\,\phi(d) \simeq 1 - \Phi(d)$$

$$\phi(d - \theta) \simeq \phi(d)[1 + d\theta]$$

Applying the approximations in (7.5.7) to (7.5.3), (7.5.5), and (7.5.6), we obtain

$$d:(1 - \varepsilon)[1 + \theta d\,\Phi(d)] = 1 \tag{7.5.8}$$

$$m_0 \simeq \theta\,\Phi(d) - \theta^2\phi(d) \simeq \theta\,\Phi(d) \tag{7.5.9}$$

$$\sigma_{0n}^2 \simeq n^{-1}([2d^2 - 2 - 2\theta d + \theta^2]\Phi(d) + [2d + \theta d^2 - \theta^3]\phi(d) + 1 - m_0^2)$$
$$= n^{-1}(1 + 2[d^2 - \theta d - 1]\Phi(d) + 2d\,\phi(d))$$
$$\simeq n^{-1}(1 + 2d^2\Phi(d) + 2d\,\phi(d)) \simeq n^{-1}[1 + 2d^2\Phi(d)] \tag{7.5.10}$$

Since asymptotically $(n \to \infty)$ the random variable $n^{-1}\sum_{i=1}^{n} z(X_i)$ is Gaussian, we may set $r^* = 0$ in (7.5.4). From this latter equation and from (7.5.9) and (7.5.10), we then obtain, for $n \to \infty$, $\theta \to 0$,

$$\Pr\left\{n^{-1}\sum_{i=1}^{n} z(X_i) > \lambda^* \Big| \phi\right\} = 1 - \Phi\left(\frac{\lambda^* - m_0}{\sigma_{0n}}\right) = \alpha$$

or

$$\Phi\left(\frac{\lambda^* - m_0}{\sigma_{0n}}\right) = 1 - \alpha$$

and

$$\lambda^* = m_0 + \sigma_{0n}\Phi^{-1}(1-\alpha) \underset{\theta \to 0}{\simeq} n^{-1/2}[1 + 2d^2\Phi(d)]^{1/2}\phi^{-1}(1-\alpha) \quad (7.5.11)$$

Let ϕ_θ denote the unit variance, mean θ Gaussian density function. Then, the asymptotic ($n \to \infty$, $\theta \to 0$) power $P_n^*(\phi_\theta)$ of the robust test in (7.5.1) and (7.5.2) at ϕ_θ with d as in (7.5.8), with $r^* = 0$, and with λ^* as in (7.5.11) is

$$P_n^*(\phi_\theta) = \Pr\left\{n^{-1}\sum_{i=1}^n z(X_i) > \lambda^* \bigg| \phi_\theta\right\} = \Phi\left(\frac{m_1 - \lambda^*}{\sigma_{1n}}\right) \quad (7.5.12)$$

where

$$m_1 = E\left\{n^{-1}\sum_{i=1}^n z(X_i)\bigg|\phi_\theta\right\} = E\{z(X)|\phi_\theta\}$$
$$= d\Pr\{X \le d|\phi_\theta\} + (\theta - d)\Pr\{X \ge -d + \theta|\phi_\theta\}$$
$$+ \int_d^{-d+\theta} x\phi_\theta(x)\,dx = \theta - d\,\Phi(d) + (d-\theta)\Phi(d-\theta)$$
$$+ (d-\theta)\phi(d-\theta) + d\,\phi(d) \quad (7.5.13)$$

$$\sigma_{1n}^2 = E\left\{\left[n^{-1}\sum_{i=1}^n z(X_i) - m_1\right]^2\bigg|\phi_\theta\right\} = n^{-1}E\{[z(X) - m_1]^2|\phi_\theta\}$$
$$= n^{-1}\bigg(d^2\Pr\{X \le d|\phi_\theta\} + (\theta - d)^2\Pr\{X \ge -d + \theta|\phi_\theta\}$$
$$+ \int_d^{-d+\theta} x^2\phi_\theta(x)\,dx - m_1^2\bigg)$$
$$= n^{-1}[1 + \theta^2 - m_1^2 + [d^2 - 1 - 2\theta d]\Phi(d) + [d^2 - \theta^2 - 1]\Phi(d-\theta)$$
$$+ [d - 2\theta]\phi(d) - (d-\theta)\phi(d-\theta)] \quad (7.5.14)$$

Due to the approximations in (7.5.7), in conjunction with the fact that if $\theta \to 0$ then $d \to \infty$ we conclude from (7.5.13) and (7.5.14) that

$$m_1 \underset{\theta \to 0}{\simeq} \theta - \theta\,\Phi(d) + [2d - \theta - \theta d - d^2\theta]\phi(d) \simeq \theta - \theta\,\Phi(d) + d(2 - \theta d)\phi(d)$$
$$\simeq \theta + d(2 - \theta d)\phi(d) \quad (7.5.15)$$

$$\sigma_{1n}^2 \underset{\theta \to 0}{\simeq} n^{-1}(1 - m_1^2 + 2(d^2 - 1 - \theta d)\Phi(d) - 2\theta d^2\phi(d))$$
$$= n^{-1}(1 - \theta^2 - d^2(2 - \theta d)^2\phi^2(d) + 2\theta d(2 - d\theta)\phi(d)$$
$$+ 2(d^2 - 1 - d\theta)\Phi(d) - 2\theta d^2\phi(d))$$
$$\simeq n^{-1}(1 + 2d^2\Phi(d)) \quad (7.5.16)$$

Substituting (7.5.11), (7.5.15), and (7.5.16) in (7.5.12), we conclude

$$P_n^*(\phi_\theta) \underset{\substack{n \to \infty \\ \theta \to 0}}{\simeq} \Phi[\sqrt{n}[\theta + d(2 - \theta d)\phi(d)][1 + 2d^2\Phi(d)]^{-1/2} - \Phi^{-1}(1 - \alpha)] \quad (7.5.17)$$

Let us now consider the optimal parametric Neyman-Pearson test, at the Gaussian pair (ϕ, ϕ_θ) and at false alarm rate α. For given n, let us denote by $P_n^0(\phi_\theta)$ the power of this test at ϕ_θ. Then, from (7.2.12) in Section 7.2, and replacing 2θ by θ, we obtain

$$P_n^0(\phi_\theta) = \Phi[\theta\sqrt{n} - \Phi^{-1}(1 - \alpha)] \quad (7.5.18)$$

Let n_1 and n_2 be the sample sizes that, for $n_2 \to \infty$ and $\theta \to 0$, induce powers $P_{n_1}^0(\phi_\theta)$ and $P_{n_2}^*(\phi_\theta)$ that are equal to each other. Then by substitution in (7.5.17) and (7.5.18), we conclude that those sample sizes are such that

$$\sqrt{n_2}\,\frac{\theta + d(2 - \theta d)\phi(d)}{[1 + 2d^2\Phi(d)]^{1/2}} = \sqrt{n_1}\,\theta \quad \text{or} \quad \frac{n_1}{n_2} = \frac{[1 + d(2\theta^{-1} - d)\phi(d)]^2}{1 + 2d^2\Phi(d)} \quad (7.5.19)$$

The ratio n_1/n_2 in (7.5.19) is the asymptotic relative efficiency of Huber's robust test at the Gaussian density function. We selected $\varepsilon = 0.1$, and $\theta \leq 0.01$. For those values, we computed the d values from (7.5.8), and we subsequently computed the ratio n_1/n_2 in (7.5.19) as a function of $\theta \leq 0.01$. We found

$$n_1/n_2 \geq 0.9 > 2\pi^{-1} \quad (7.5.20)$$

From (7.5.20) we observe that at the Gaussian density function the robust test is asymptotically significantly superior to the sign test.

7.6 CONCLUDING REMARKS

In this chapter, we presented the basic concepts in nonparametric detection, and the basic existing nonparametric detection tests. A variety of additional nonparametric tests can be devised via appropriate use of the concept represented by the rank vector, such as the one-sample normal scores test, and a variety of two-sample nonparametric tests. The objective of the latter is to test if the distributions that generate two different sample populations are identical. Information on all the above tests, as well as on some adaptive nonparametric tests, can be found in P. Papantoni-Kazakos and D. Kazakos (1977).

In his efforts to model and subsequently design detection schemes, the analyst usually possesses a stochastic model more accurate than the nonparametric models in this chapter. The analyst forms such a model via histograms, whose accuracy is certainly imprecise, but not completely unreliable. Then, the analyst's model is closer to that represented by the robust detection schemes in Chapter 6 than it is to the nonparametric models in this chapter. In addition, the robust models induce performance, that is generally far superior to the performance induced by the nonparametric ones. We thus believe that the emphasis should generally be on robust tests, and that the classical nonparametric schemes should be used only with caution.

7.7 PROBLEMS

7.7.1 Let $\{X_1, X_2, \ldots, X_n\}$ be a sequence of independent and identically distributed random variables each with density function $f(x)$ and c.d.f. $F(x)$.
   (a) Find the Neyman-Pearson detector operating on this sequence to test the hypothesis that the data has zero median versus the alternative that the median is greater than zero.
   (b) Find a nonparametric detector for the same test and calculate its asymptotic relative efficiency with respect to the detector in (a).

7.7.2 Let $\{v_i\}$ be a sequence of independent Cauchy random variables

$$f(v|\theta) = \frac{1}{\pi} \frac{1}{1 + (v - \theta)^2}$$

Find the asymptotic relative efficiency of the sign test relative to the locally most powerful test.

7.7.3 Let $\{v_i\}$ be a sequence of independent identically distributed random variables. For testing $H_0: \theta = 0$ vs. $H_1: \theta > 0$, the following nonparametric test statistic is proposed.

$$\frac{1}{N} \sum_{i=0}^{N-1} \text{sgn}(v_i + v_{i+1})$$

What is the asymptotic relative efficiency of this detector for Gaussian observations with respect to the optimum?

7.7.4 A laser communication system transmits a burst of photons every second for N seconds. Assume the number of photons emitted at the ith second is k_i and is Poisson distributed. The receiver observes the independent sequence $\{k_1, k_2, \ldots, k_N\}$ and is to decide which of two possible lasers is being used. The only distinguishing feature between the lasers is the average number of emissions—that is, $E\{k_i\} = \lambda_0$ or λ_1.

Determine the receiver structure that minimizes the probability of incorrectly deciding which laser is transmitting. Obtain an expression for the probability of error. *Do not evaluate it.* Is the receiver UMP for testing λ_0 versus $\lambda_1 > \lambda_0$?

A locally optimum detector may be defined as one that maximizes the slope of the power curve as $\lambda_1 \to \lambda_0$. Determine the locally optimum detector and compare it to the Neyman-Pearson detector in terms of ARE.

REFERENCES

Ahrens, E. H., Sr. (1969), "Mass Field Trials of the Diet-Heart Question (Their Significance, Timeliness, Feasibility and Applicability)," *American Heart Association Monograph No. 28.*

Berk, R. H. and I. Savage (1968), "The Information in a Rank-Order and the Stopping Time of Some Associated SPRT's," *Ann. Math. Statist.* 39, 1661–1674.

Carlyle, J. W. and J. B. Thomas (1964), "On Nonparametric Signal Detectors," *IEEE Trans. Inf. Th.* IT-10, 146–152.

Chernoff, H. and I. R. Savage (1958), "Asymptotic Normality and Efficiency of Certain Nonparametric Test Statistics," *Ann. Math. Statist.* 29, 972–994.

Davisson, L. D., E. A. Feustel, and J. W. Modestino (1970), "The Effects of Dependence on Nonparametric Detection," *IEEE Trans. Inf. Th.* IT-16, 32–41.

Davisson, L. D. and P. Papantoni-Kazakos (1972), "Small Sample Efficiencies of Rank Tests," *Proc. 1972 International Symp. on Inf. Th.*, 57.

Feustel, E. A. and Davisson, D. L. (1967) "The Asymptotic Relative Efficiency of Mixed Statistical Tests," *IEEE Trans. Inf. Th.* IT-13, 247–255.

——— (1968), "On the Efficacy of Mixed Locally-Most-Powerful One- and Two-Sample Rank Tests," *IEEE Trans. Inf. Th.* IT-14, 776–778.

Fraser, D. A. S. (1957), *Nonparametric Methods in Statistics*, Wiley, New York.

Gaswirth, J. L. and H. Rubin (1971), "Effect of Dependence on the Level of Some One-Sample Tests," *J. Amer. Statist. Assoc.* 66, 816–820.

Hájek, J. and Z. Šidák (1967), *Theory of Rank Tests.* Academic Press, New York.

Hodges, J. L. and E. L. Lehmann (1962), "Probabilities of Rankings for Two Widely Separated Normal Distributions," in *Studies in Mathematical Analysis and Related Topics* (Gilbarg, Solomon, et al, eds.), Stanford Univ. Press, Stanford, CA, 146–151.

Hoeffding, W. (1951), "Optimum Nonparametric Tests," *Proc. 2nd Berkeley Symp. Math. Statist. and Prob.* (J. Neymann, ed.), Univ. of Calif. Press, Berkeley, 83–93.

Kazakos, D. and P. Papantoni-Kazakos (1977), *Nonparametric Methods in Communications*, Marcel Dekker, New York.

Klotz, J. H. (1963), "Small Sample Power and Efficiency for the One-Sample Wilcoxon and Normal Scores Test," *Ann. Math. Statist.* 34, 335–337.

Klotz, J. H. (1964), "On the Normal Scores Two-Sample Tests", *J. Amer. Stat. Assoc.* 59, 652–664.

Lehmann, E. L. (1953), "The Power of Rank Tests," *Ann. Math. Statist.* 24, 23–42.

Mahamunulu, D. M. (1967), "Some Fixed-Sample Ranking and Selection Problems," *Ann. Math. Statist.* 38, 1079–1091.

Millard, J. B. and L. Kurz (1962), "Nonparametric signal detection—An Application to the Kolmogorov-Smirnov, Cramer-von Mises Tests," (Abstract only) *IEEE Trans. Inf. Th.* IT-12, 275.

Miller, J. H. and J. B. Thomas (1972), "Detectors for Discrete-Time Signal in Non-Gaussian Noise," *IEEE Trans. Inf. Th.* IT-18(2), 241–250.

Milton, R. C. (1970), *Rank Order Probabilities: Two-Sample Normal Shift Alternatives.* Wiley, New York.

Modestino, J. W. (1969), Nonparametric and Adaptive Detection of Dependent Data, Ph. D. Dissertation, Dept. of Elec. Engrg., Princeton Univ., Princeton, NJ.

Papantoni-Kazakos, P. (1975), "Small Sample Efficiencies of Rank Tests," *IEEE Trans. Inf. Th.* IT-21, 150–157.

Puri, M. L. (ed.), (1970), *Nonparametric Techniques in Statistical Inference.* Cambridge Univ. Press, London, England.

Rao, U. V. R., I. R. Savage, and M. Sobel (1960), "Contributions to the Theory of Rank-Order Statistics: The Two-Sample Censored Case," *Ann. Math. Statist.* 31, 415–426.

Savage, I. R. (1956), "Contributions to the Theory of Rank-Order Statistics—the Two-sample Case," *Ann. Math. Statist.* 27, 590–615.

―――― (1957), "Contributions to the Theory of Rank-Order Statistics—the 'Trend' Case," *Ann. Math. Statist.* 28, 968–977.

―――― (1960), "Contributions to the Theory of Rank-Order Statistics: Computation Rules for Probabilites of Rank Orders," *Ann. Math. Statist.* 31, 519–520.

―――― (1962), *Bibliography of Nonparametric Statistics.* Harvard Univ. Press, Cambridge, Mass.

Smirnov, N. V. (1948), "Table for Estimating the Goodness of Fit of Empirical Distributions," *Ann. Math. Statist.* 19, 279–281.

Teichroew, D. (1955), "Empirical Power Functions for Nonparametric Two-Sample Tests for Small Samples," *Ann. Math. Statist.* 26, 340–344.

―――― (1955), Probabilities Associated with Order Statistics in Samples from Two Normal Populations with Equal Variance, Engineering Agency Statistical Tables. Chemical Corps Engineering Agency, Army Chemical Center, MD.

References

———— (1956), "Tables of Expected Values of Order Statistics and Products of Order Statistics for Sample of Size Twenty and Less from the Normal Distribution," *Ann. Math. Statist.* 27, 410–426.

Terry, M. E. (1952), "Some Rank Order Tests Which Are Most Powerful Against Specific Parametric Alternatives," *Ann. Math. Statist.* 23, 346–366.

Thomas, J. B. (1970), "Nonparametric Detection," *Proc. IEEE* 58(5), 623–631.

Tsao, C. K. (1957), "Approximations to the Power of Rank Tests," *Ann. Math. Statist.* 28, 159–172.

Walsh, J. E. (Vol. 1, 1962; Vol. 2, 1965; Vol. 3, 1968), *Handbook of Nonparametric Statistics*. Van Nostrand, Princeton, NJ.

Woinsky, M. N. and L. Kurz (1970), "Nonparametric Detection Using Dependent Samples," *IEEE Trans. Inf. Th.* IT-16, 355–358.

Wolff, S. S., J. L. Gastwirth, and H. Rubin (1967), "The Effects of Autoregressive Dependence on Nonparametric Tests," *IEEE Trans. Inf. Th.* IT-13, 311–313.

CHAPTER 8

SEQUENTIAL DETECTION

8.1 INTRODUCTION

In Chapters 4, 5, 6, and 7, we studied detection schemes under various degrees of a priori available information. All the schemes were formulated subject to the common assumption that the observation interval is fixed and not controlled by the analyst (or the system). In this chapter, we will eliminate this assumption. We will, instead, allow the analyst to decide *when* a set of observations is sufficiently reliable for decision making. In particular, the analyst will be allowed to observe data sequentially, to decide when data should stop being observed, and to determine which underlying hypothesis is then acting. The implication is that both a stopping rule and a decision rule are in the analyst's control, and the *sequential detection* model evolves, via a generalization of the Neyman-Pearson formalization. In particular, the following assets are then available to the analyst.

i. Two well known, or parametrically known, or nonparametrically described hypotheses H_0 and H_1 where H_1 is the emphatic hypothesis
ii. A realization $x = \{x^n; n \to \infty\}$ or $x(t); t \in (0, \infty)$ from the underlying hypothesis
iii. A power level β and a false alarm rate α where $0 < \alpha < 1$ and $0 < \beta \leq 1$.

Assuming, without losing generality, discrete-time hypotheses, the analyst then designs a stopping rule $\Delta_n(x)$ and a decision rule $\{\delta_n(x^n); n = 1, 2, \ldots\}$, where

$$\Delta_n(x) = \Pr\{\text{decide to stop just after the } n\text{th observed datum} | x \text{ observed}\} \quad (8.1.1)$$

$$\delta_n(x^n) = \Pr\{\text{hypothesis } H_1 \text{ decided} | \text{stopped just after the } n\text{th datum and } x^n \text{ observed}\} \qquad (8.1.2)$$

As we saw in Chapter 3, the stopping rule is such that $\sum_n \Delta_n(x) = 1$ for every observable sequence x. That is, for every observable x the collection of observed data stops with probability one. We note that, in contrast to the classical Neyman-Pearson formalization in Chapter 5, in addition to a false alarm rate α a power level β is also assumed given. The additional performance requirement represented by the power level β basically imposes a lower bound on the number n of observed data that are used for decision. We also note that in the model, the hypotheses may vary from well known to nonparametrically described. When the hypotheses are either well known or parametrically known, the *parametric sequential detection* schemes arise. The *nonparametric sequential detection* schemes evolve when the hypotheses are nonparametrically described. As we will show, the latter schemes are derived from a "robustification" of the parametric schemes via methods such as those in Chapter 6.

The sequential detection schemes are not generally the solutions of some well-defined optimization problem. Indeed, the formalization of an optimization problem, which involves both a stopping and a decision rule, is generally very complex, and the search for an "optimal" solution usually becomes intractable. Instead, the sequential detection procedures evolve as intuitively appealing modifications of the nonsequential Neyman-Pearson decision rules. The underlying objective is to minimize the expected number of data needed for reliable decision, where the decision reliability is determined by the given false alarm rate and power level. As we will see, some form of asymptotic optimality can generally be defined for the sequential procedures that so evolve.

In this chapter, we will consider only discrete-time stochastic processes. We will adopt the same notation as in Chapter 5 and will first present the parametric sequential detection procedures proposed and analyzed by Wald (1947). Then, we will discuss the "robustification" of Wald's procedures for nonparametrically described hypotheses. Finally, we will present some related sequential schemes for detecting a change in distribution. For some advanced results and proofs, we refer the reader to Wald (1947).

8.2 WALD'S PROCEDURES

To introduce the idea behind Wald's sequential detection schemes, we will refer to the Neyman-Pearson decision rules in Chapter 5. Let us initially consider noncomposite hypotheses H_1 and H_0 and let us then concentrate on the Neyman-Pearson decision rule in Theorem 5.2.1. Given a false alarm rate α and the size n of the observation vector, the Neyman-Pearson rule δ_n^* is then well-defined. For a fixed false alarm rate α, the rule δ_n^* induces a power $P_1(\delta_n^*)$ that increases monotonically with increasing sample size n—as can be easily seen, for example, from (5.2.20), (5.2.30), and (5.2.39). In other words, as the sample size n increases, the H_0 and H_1 hypotheses become more distinguishable; thus, as n increases, a smaller false alarm rate and a higher power can be attained. Furthermore, given n and the density functions $f_1(x^n)$ and $f_0(x^n)$ that correspond respectively to H_1 and H_0, the Neyman-Pearson rule tests the likelihood ratio $f_1(x^n)|f_0(x^n)$ against some threshold λ. If the likelihood ratio is higher than λ, H_1 is decided. Otherwise, H_0 is decided. Now let a false alarm rate α and a power level β be given, and let us demand that a detection scheme be devised such that its induced false alarm does not exceed α and its induced power is at least β. Let us also require that the minimum possible sample size n be used to satisfy the false alarm and power constraints. Then, due to the arguments we presented above, the following sequential scheme seems naturally applicable, to the latter requirements.

Given the density functions $f_1(x^n)$ and $f_0(x^n)$ $\forall$ $n \geq 1$ and given that those density functions reflect respectively the H_1 and H_0 hypotheses, select two thresholds λ_1 and λ_2 such that $\lambda_1 < \lambda_2$. Start observing data $x_1, x_2, \ldots$ sequentially. Given $x^n = \{x_1, x_2, \ldots, x_n; x \to \infty\}$, stop at the *first* n such that

$$\text{Either} \quad T(x^n) = \log \frac{f_1(x^n)}{f_0(x^n)} \leq \lambda_1 \qquad (8.2.1)$$

$$\text{Or} \quad T(x^n) = \log \frac{f_1(x^n)}{f_0(x^n)} \geq \lambda_2 \qquad (8.2.2)$$

If at the stopping instant n, (8.2.1) holds, decide in favor of the hypothesis H_0. If instead (8.2.2) holds, decide in favor of the hypothesis H_1.

The sequential detection scheme described above was proposed by Wald (1947); it clearly incorporates both a stopping and a decision rule. The Wald stopping rule is described by (8.2.3) and (8.2.4) below. The Wald decision rule is described by (8.2.5).

$$N_{\lambda_1,\lambda_2}(x) = \inf\{n: \text{either } T(x^n) \leq \lambda_1 \text{ or } T(x^n) \geq \lambda_2\} \qquad (8.2.3)$$

$$\Delta_n(x) = \begin{cases} 1 & \text{if } n = N_{\lambda_1,\lambda_2}(x) \\ 0 & \text{otherwise} \end{cases} \qquad (8.2.4)$$

$$\delta_{N_{\lambda_1,\lambda_2}(x)}(x) = \begin{cases} 1 & \text{if } T(x^{n^*}) \geq \lambda_2 \quad \text{where} \quad n^* = N_{\lambda_1,\lambda_2}(x) \\ 0 & \text{if } T(x^{n^*}) \leq \lambda_1 \quad \text{where} \quad n^* = N_{\lambda_1,\lambda_2}(x) \end{cases} \qquad (8.2.5)$$

The random variable $N_{\lambda_1,\lambda_2}(X)$ in (8.2.3) is called the *Wald stopping variable*. The thresholds λ_1 and λ_2 in the Wald test, computed via the false alarm and power constraints, represent two absorbing barriers. The test function $T(x^n)$ can be updated sequentially. Indeed, denoting $F(x_m|x_1^{m-1}) = \Pr\{X_m \leq x_m | x_1^{m-1}\}$; $m \geq 2$ and $f(x_m|x_1^{m-1})dx_m = dF(x_m|x_1^{m-1})$, we obtain

$$T(x^n) = T(x^{n-1}) + \log \frac{f_1(x_n|x_1^{n-1})}{f_0(x_n|x_1^{n-1})} \qquad (8.2.6)$$

8.2 Wald's Procedures

where,

$$x^{n^T} = [x_1, x_2, \ldots, x_n] \qquad f(x_1 | x_1^0) = f(x_1)$$

We note that the updating term, $\log[f_1(x_n|x_1^{n-1})/f_0(x_n|x_1^{n-1})]$, in (8.2.6) generally implies a memory size $n - 1$. If the density functions $f_1(\cdot)$ and $f_0(\cdot)$ are both induced by memoryless stochastic processes, then the updating term in (8.2.6) requires zero memory for every value n. We will now present a lemma, which relates the thresholds λ_1 and λ_2 in (8.2.1) and (8.2.2) to the given false alarm rate and power level.

LEMMA 8.2.1: Consider the Wald sequential test, and let the thresholds λ_1 and λ_2 in the test be respectively as in (8.2.1) and (8.2.2). Let a false alarm rate α and a power level β be given. Then λ_1 and λ_2 satisfy respectively

$$\lambda_1 \geq \log\left(\frac{1-\beta}{1-\alpha}\right) \tag{8.2.7}$$

$$\lambda_2 \leq \log\left(\frac{\beta}{\alpha}\right) \tag{8.2.8}$$

Without loss in false alarm and power performance, the conditions in (8.2.7) and (8.2.8) can be taken with equality.

Proof: Given α and β, we require that $\Pr\{\text{decide } H_1 | H_0\} \leq \alpha$ and $\Pr\{\text{decide } H_1 | H_1\} \geq \beta$. Thus, applying the theorem of total probability and using the Wald test in (8.2.1) and (8.2.2), we obtain, by defining $\mu_1 = \log^{-1} \lambda_1$ and $\mu_2 = \log^{-1} \lambda_2$,

$\Pr\{\text{decide } H_1 | H_0\}$

$$= \sum_{n=1}^{\infty} \Pr\{\text{stop at } n \text{ and decide } H_1 | H_0\}$$

$$= \sum_{n=1}^{\infty} \int f_0(x^n) \, dx^n$$

$$x^n \in \mathscr{A}^n = \{x^n: \mu_1 < f_1(x^k)/f_0(x^k) < \mu_2; \forall k < n, f_1(x^n)/f_0(x^n) \geq \mu_2\}$$

$$\leq \mu_2^{-1} \sum_{n=1}^{\infty} \int f_1(x^n) \, dx^n$$

$$x^n \in \mathscr{A}^n$$

$$= \mu_2^{-1}\{\Pr \text{ decide } H_1 | H_1\} \rightarrow \mu_2 \leq \frac{\Pr\{\text{decide } H_1 | H_1\}}{\Pr\{\text{decide } H_1 | H_0\}} \leq \frac{\beta}{\alpha} \rightarrow \mu_2$$

$$\leq \frac{\beta}{\alpha} \rightarrow \lambda_2 \leq \log\left(\frac{\beta}{\alpha}\right)$$

$\Pr\{\text{decide } H_0 | H_0\}$

$$= \sum_{n=1}^{\infty} \Pr\{\text{stop at } n \text{ and decide } H_0 | H_0\}$$

$$= \sum_{n=1}^{\infty} \int f_0(x^n) \, dx^n$$

$$x^n \in \mathscr{B}^n \cong \{x^n : \mu_1 < f_1(x^k)/f_0(x^k) < \mu_2; \, \forall \, k < n, \, f_1(x^n)/f_0(x^n) \leq \mu_1\}$$

$$\geq \mu_1^{-1} \sum_{n=1}^{\infty} \int f_1(x^n) \, dx^n$$

$$x^n \in \mathscr{B}^n$$

$$= \mu_1^{-1} \Pr\{\text{decide } H_0 | H_1\} \to \mu_1 \geq \frac{\Pr\{\text{decide } H_0 | H_1\}}{\Pr\{\text{decide } H_0 | H_0\}}$$

$$= \frac{1 - \Pr\{\text{decide } H_1 | H_1\}}{1 - \Pr\{\text{decide } H_1 | H_0\}} \geq \frac{1 - \beta}{1 - \alpha} \to \mu_1 \geq \frac{1 - \beta}{1 - \alpha} \to \lambda_1 \geq \log\left(\frac{1 - \beta}{1 - \alpha}\right) \quad \blacksquare$$

Due to the conditions in Lemma 8.2.1, the thresholds λ_1 and λ_2 in the Wald test are precisely determined, given α and β, and so then is the Wald test itself. If H_0 is instead composite, and if the conditions in Lemma 5.2.1 are satisfied, a sequential Wald test can be defined on the test function $h(x^n)$ in Lemma 5.2.1. Similarly, if the conditions in Lemma 5.3.1 are satisfied, a uniformly most powerful Wald test can be determined on the test function $h(x^n)$ therein. In both cases $h(x^n)$ substitutes for the test function $T(x^n)$ in (8.2.1) and (8.2.2). The thresholds λ_1 and λ_2 are then uniquely determined, via the given false alarm and power constraints α and β. In particular:

i. For the case in Lemma 5.2.1, the test function $h(x^n)$ is tested against

$$\lambda_1 = \inf_{\theta^m \in \mathscr{E}^m} \left(\frac{f[(1 - \beta)/(1 - \alpha)] - c(\theta^m)}{g(\theta^m)} \right)$$

$$\lambda_2 = \sup_{\theta^m \in \mathscr{E}^m} \left(\frac{f(\beta/\alpha) - c(\theta^m)}{g(\theta^m)} \right)$$

where the functions $f(\cdot)$, $g(\cdot)$, and $c(\cdot)$, and the space $\mathscr{E}^m$ are as in Lemma 5.2.1.

ii. For the case in Lemma 5.3.1, the test function $h(x^n)$ is tested against

$$\lambda_1 = \inf_{\theta_1^m \in \mathscr{E}_1^m, \, \theta_0^m \in \mathscr{E}_0^m} \left(\frac{f[(1 - \beta)/(1 - \alpha)] - c(\theta_1^m, \theta_0^m)}{g(\theta_1^m, \theta_0^m)} \right)$$

$$\lambda_2 = \sup_{\theta_1^m \in \mathscr{E}^m, \, \theta_0^m \in \mathscr{E}_0^m} \left(\frac{f(\beta/\alpha) - c(\theta_1^m, \theta_0^m)}{g(\theta_1^m, \theta_0^m)} \right)$$

where the spaces $\mathscr{E}_0^m$ and $\mathscr{E}_1^m$ and the functions $f(\cdot)$, $c(\cdot)$, and $g(\cdot)$ are as in Lemma 5.3.1.

8.2 Wald's Procedures

From this point on, we will consider only noncomposite hypotheses H_0 and H_1. We will also assume that the stochastic processes that generate H_0 and H_1 are both stationary. Given the corresponding density functions $f_1(x^n)$ and $f_0(x^n)$ for all n, we define

$$L_n = n^{-1} \log \frac{f_1(x^n)}{f_0(x^n)} \qquad I_{10} = \lim_{n \to \infty} L_n \qquad P_n(v) = P_{\mu_1}(L_n < v) \qquad (8.2.9)$$

We now state, without proof, a theorem proven in Wald (1947).

THEOREM 8.2.1: Let the quantities in (8.2.9) satisfy the following conditions: (1) I_{10} exists ($I_{10} < \infty$) and is equal to $E\{I_{10}|H_1\}$ almost everywhere. (2) For v in $(0, I_{10})$ we have

$$\lim_{n \to \infty} n P_n(v) = 0 \qquad \sum_n P_n(v) \to \infty \qquad (8.2.10)$$

Then, given a false alarm rate α and power level β as in Lemma 8.2.1, and denoting $N = N_{\lambda_1, \lambda_2}(X)$, where $N_{\lambda_1, \lambda_2}(X)$ is the Wald stopping variable in (8.2.3), we have

$$E\left\{\log \frac{f_1(X^N)}{f_0(X^N)} \bigg| H_1\right\} \geq (1-\beta) \log \frac{1-\beta}{\alpha} + \beta \log \frac{\beta}{1-\alpha} \geq (1-\beta)|\log \alpha| - \log 2$$

$$(8.2.11)$$

$$\lim_{\substack{\alpha \to 0 \\ \beta \to 1}} E\{N|H_1\} = I_{10}^{-1}|\log \lambda_2|$$

$$= \inf\{E\{N'|H_1\}, \text{ where } N' \text{ is some arbitrary stopping variable}\}$$

$$(8.2.12)$$

∎

Theorem 8.2.1 exhibits some very important properties of the Wald test. The conditions in (8.2.10) represent some ergodicity properties; they are satisfied, for example, when both the hypotheses are generated by memoryless processes, such that the Kullback-Leibler distance, $E \log\{f_1(X_1)/f_0(X_1)|H_1\} = \int du\, f_1(u) \log[f_1(u)/f_0(u)]$, is finite. Expression (8.2.12) represents, then, an asymptotic optimality property of the Wald test. It basically says that as the false alarm rate α decreases to values near zero and as the power level β increases to values near one, the Wald test attains the corresponding performance requirements with the minimum possible expected delay. That is, there exist then no other tests that can satisfy the levels α and β with smaller expected sample size. This asymptotic property, in conjunction with its sequential nature, are the important merits of the Wald test. We note that as $\alpha \to 0$ and $\beta \to 1$, λ_1 and λ_2 in the Wald test tend respectively to $-\infty$ and $+\infty$. The expected sample size $E\{N|H_1\}$ needed by the test when H_1 is active takes then asymptotic values. From (8.2.12), we notice that the asymptotic speed of the Wald test is controlled by the *information number*

$$I_{10} = \lim_{n \to \infty} E\left\{n^{-1} \log \frac{f_1(X^n)}{f_0(X^n)} \middle| H_1\right\}$$

As this number increases, the asymptotic speed of the Wald test increases as well (since $E\{N|H_1\}$ decreases). This is intuitively pleasing. It basically says that the further apart the two hypotheses are in Kullback-Leibler distance, the faster they can be distinguished from each other. We note that if the two hypotheses are generated by memoryless processes, then

$$I_{10} = E\left\{\log \frac{f_1(X_1)}{f_0(X_1)} \middle| H_1\right\} = \int du\, f_1(u) \log \frac{f_1(u)}{f_0(u)}$$

That is, the information number I_{10} is then the Kullback-Leibler distance between the first order density functions f_1 and f_0. We will conclude this section with two examples.

Example 8.2.1

Let the noncomposite hypothesis H_1 be described by a memoryless, unit variance, and mean θ Gaussian process, where $\theta > 0$. Let the noncomposite hypothesis H_0 be described by a memoryless, zero mean, unit variance Gaussian process. Then, given n, the test function $T(x^n)$ in the Wald test is

$$T(x^n) = \ln \frac{f_1(x^n)}{f_0(x^n)} = \sum_{i=1}^n \ln \frac{\phi(x_i - \theta)}{\phi(x_i)} = \sum_{i=1}^n \frac{\theta}{2}(2x_i - \theta) = \theta \sum_{i=1}^n x_i - \frac{\theta^2}{2}n \quad (8.2.13)$$

where ln denotes the natural logarithm and $\phi(x)$ is the zero mean, unit variance Gaussian density function at the point x.

In sequential terms, we have

$$T(x^n) = T(x^{n-1}) + \theta\left(x_n - \frac{\theta}{2}\right) \quad (8.2.14)$$

Thus, the updating step, at the collection of the nth datum, equals $\theta[x_n - (\theta/2)]$. The information number I_{10} is given here by

$$I_{10} = \int du\, f_1(u) \cdot \ln \frac{f_1(u)}{f_0(u)} = \int_{-\infty}^{\infty} \theta\left(u - \frac{\theta}{2}\right)\phi(u - \theta)\, du = \frac{\theta^2}{2} \quad (8.2.15)$$

Therefore, the asymptotic speed of the Wald test increases as the value of the parameter θ increases. Given a false alarm rate α and power level β, the thresholds in the Wald test are respectively selected as

$$\lambda_1 = \ln\left(\frac{1-\beta}{1-\alpha}\right) \qquad \lambda_2 = \ln\left(\frac{\beta}{\alpha}\right) \quad (8.2.16)$$

8.2 Wald's Procedures

The Wald test is then completely described by the stopping rule $N(x)$ and the decision rule $\delta_{N(x)}(x)$ below.

$$N(x) = \inf\left\{n: \begin{cases} \text{Either} & \theta \sum_{i=1}^{n} x_i - \frac{\theta^2}{2} n \leq \ln\left(\frac{1-\beta}{1-\alpha}\right) \\ \text{Or} & \theta \sum_{i=1}^{n} x_i - \frac{\theta^2}{2} n \geq \ln\left(\frac{\beta}{\alpha}\right) \end{cases}\right\}$$

$$\delta_{N(x)}(x) = \begin{cases} 1 & \text{If } \theta \sum_{i=1}^{N} x_i - \frac{\theta^2}{2} N \geq \ln\left(\frac{\beta}{\alpha}\right) \quad \text{where} \quad N = N(x) \\ 0 & \text{If } \theta \sum_{i=1}^{N} x_i - \frac{\theta^2}{2} N \leq \ln\left(\frac{1-\beta}{1-\alpha}\right) \quad \text{where} \quad N = N(x) \end{cases}$$

Alternatively, the Wald test is described as follows.

At time zero, set the test function T_0 at zero. Update the test function $T(x^n)$ sequentially, where

$$T(x^n) = T(x^{n-1}) + \theta\left(x_n - \frac{\theta}{2}\right)$$

Stop at the first n such that

$$\text{Either} \qquad T(x^n) \geq \ln\left(\frac{\beta}{\alpha}\right) \qquad (8.2.17)$$

$$\text{Or} \qquad T(x^n) \leq \ln\left(\frac{1-\beta}{1-\alpha}\right) \qquad (8.2.18)$$

If at the stopping time n, (8.2.17) occurs, decide that H_1 is active. If, instead, (8.2.18) occurs, decide that H_0 is active.

Example 8.2.2

Let both the noncomposite hypotheses H_0 and H_1 be described by Bernoulli processes that generate binary sequences with elements zero and one. Let the element one occur with probability q under H_1, and probability p under H_0. Then we obtain

$$\log \frac{f_1(x_i)}{f_0(x_i)} = \log \frac{q^{x_i}(1-q)^{1-x_i}}{p^{x_i}(1-p)^{1-x_i}} = x_i \log \frac{q}{p} + (1-x_i) \log \frac{1-q}{1-p}$$

$$= \log \frac{1-q}{1-p} + x_i \log \frac{q(1-p)}{p(1-q)} \qquad (8.2.19)$$

The Wald test function $T(x^n)$ is then tested against the thresholds $\lambda_1 = \log[(1-\beta)/(1-\alpha)]$ and $\lambda_2 = \log \beta/\alpha$ for given α and β, and it is described by

$$T(x^n) = \log\left(\frac{q(1-p)}{p(1-q)}\right) \sum_{i=1}^{n} x_i + n \log \frac{1-q}{1-p} \qquad (8.2.20)$$

$$T(x^n) = T(x^{n-1}) + \log \frac{1-q}{1-p} + x_n \log \frac{q(1-p)}{p(1-q)} \qquad (8.2.21)$$

Let us define,

$$\gamma(q, p) = \frac{\log \dfrac{1-q}{1-p}}{\log \dfrac{q(1-p)}{p(1-q)}} \qquad (8.2.22)$$

$$\lambda_1(q, p) = \frac{\log \dfrac{1-\beta}{1-\alpha}}{\log \dfrac{q(1-p)}{p(1-q)}} \qquad (8.2.23)$$

$$\lambda_2(q, p) = \frac{\log \dfrac{\beta}{\alpha}}{\log \dfrac{q(1-p)}{p(1-q)}} \qquad (8.2.24)$$

Then, the Wald test takes the following equivalent form.

Test the test function

$$T'(x^n) = \sum_{i=1}^{n} x_i + n\gamma(q, p)$$

$$T'(x^n) = T'(x^{n-1}) + [x_n + \gamma(q, p)]$$

against the thresholds, $\lambda_1(q, p)$ and $\lambda_2(q, p)$.

In particular, assuming $p \neq q$, $p \neq 0, 1$, and $q \neq 0, 1$, the Wald test is then described completely as follows:

i. *If $q > p$*

At time zero, set the test function T'_0 at zero. Update the test function sequentially by

$$T'(x^n) = T'(x^{n-1}) + [x_n + \gamma(q, p)]$$

Stop at the first n such that

Either $\quad T'(x^n) \geq \lambda_2(q, p)\quad$ and then decide $\quad H_1$

Or $\quad T'(x^n) \leq \lambda_1(q, p)\quad$ and then decide $\quad H_0$

ii. *If $q < p$*

At time zero, set the test function T'_0 at zero. Update the test function sequentially by

$$T'(x^n) = T'(x^{n-1}) + [x_n + \gamma(q, p)]$$

Stop at the first n such that

Either $\quad T'(x^n) \leq \lambda_2(q, p)\quad$ and then decide $\quad H_1$

Or $\quad T'(x^n) \geq \lambda_1(q, p)\quad$ and then decide $\quad H_0$

We observe that the quantity $\gamma(q, p)$ in (8.2.22) is negative, for any nonidentical p and q choices. It is also always absolutely less than the value one. Therefore, the updating step $x_n + \gamma(q, p)$ in the Wald test is positive and less than one if $x_n = 1$, it is negative and larger than -1 if $x_n = 0$. The information number I_{10} is given here by

$$I_{10} = P_1\{x = 1\} \log \frac{f_1(1)}{f_0(1)} + P_1\{x = 0\} \log \frac{f_1(0)}{f_0(0)}$$

$$= q\left[\log \frac{1-q}{1-p} + \log \frac{q(1-p)}{p(1-q)}\right] + (1-q) \log \frac{1-q}{1-p}$$

$$= q \log \frac{q}{p} + (1-q) \log \frac{1-q}{1-p} \qquad (8.2.25)$$

From (8.2.25) it can be easily seen that the information number I_{10} attains its minimum when the value of the ratio q/p is as close to the value one as possible, and its maximum when the value of the ratio q/p approaches either zero or infinity. In the latter case, the Wald test attains its maximum speed.

In addition to its asymptotic optimality, the Wald test has generally high speed characteristics, in the presence of relatively small sample sizes as well. The latter characteristics can be only studied numerically for specific choices of the H_1 and H_0. Some such studies can be found in Wald (1947).

8.3 ROBUST SEQUENTIAL TESTS

As in Chapter 6, let us consider the case where the hypotheses H_1 and H_0 are respectively described by two disjoint classes $\mathscr{F}_1$ and $\mathscr{F}_0$ of discrete-time and stationary stochastic processes. The question arising then is: Is there some sequential detection scheme that guarantees some minimum level of performance within the class $\mathscr{F}_1 \times \mathscr{F}_0$? In response to that question, the performance criterion must first be determined. This criterion must be related to some optimal characteristic of the parametric sequential schemes. But, as we saw in the previous section, the only known such characteristic is asymptotic. In particular, it is known that asymptotically the parametric Wald test requires the minimum possible expected sample size, and that this expected sample size is then inversely proportional to the corresponding Kullback-Leibler information number (8.2.12). We will thus select the Kullback-Leibler information number as the performance criterion for the definition of robust sequential detection procedures.

To avoid unnecessary complexity in our presentation, we will now consider classes $\mathscr{F}_1$ and $\mathscr{F}_0$ of memoryless, discrete-time, stationary processes. Let f_1 be some density function in $\mathscr{F}_1$, and let f_0 be some density function in $\mathscr{F}_0$. Consider then the Wald test at the density functions $f_1(x)$ and $f_0(x)$. Due to (8.2.12), the asymptotic speed of that test at the density function $f_1(x)$ is then determined by the information number

$$I(f_1, f_0) = \int dx\, f_1(x) \log \frac{f_1(x)}{f_0(x)} \tag{8.3.1}$$

That is, if the test function $T(x^n)$ of the Wald test was designed based on the assumption that the acting density functions are f_0 and f_1 and the true acting density function is indeed f_1 then the expected time for stopping is asymptotically ($\alpha \to 0$, $\beta \to 1$) proportional to $I^{-1}(f_1, f_0)$, where $I(f_1, f_0)$ is the information number in (8.3.1). If, however, the true acting density function is instead $f'_1 \neq f_1$ then the asymptotic expected time for stopping is instead proportional to $I^{-1}(f'_1, f_1, f_0)$ where

$$I(f'_1, f_1, f_0) = \int dx\, f'_1(x) \log \frac{f_1(x)}{f_0(x)} \tag{8.3.2}$$

$$I(f'_1, f_1, f_0) \leq I(f'_1, f_0) = \int dx\, f'_1(x) \log \frac{f'_1(x)}{f_0(x)} \tag{8.3.3}$$

The inequality in (8.3.3) is due to the concavity of the logarithm. In the quantity $I(f'_1, f_1, f_0)$ in (8.3.2), the density functions f_1 and f_0 are basically selected by the test designer, while the density function f'_1 is selected by Nature. In physical terms, the inequality in (8.3.3) states then that, if the designer knew Nature's selection f'_1, he should in his design select $f_1 = f'_1$ to ensure the fastest possible asymptotic stopping time in his sequential test. Since Nature's selection is unknown to him, the designer tries, instead, to optimize the case least favorable to him. That is, he designs his test at those density functions f_0^* and f_1^* such that

$$I(f_1^*, f_0^*) \leq I(f_1, f_0) \quad \forall\, f_1 \in \mathscr{F}_1 \quad \forall\, f_0 \in \mathscr{F}_0 \tag{8.3.4}$$

8.4 Tests for Detecting a Change in Distribution

If the pair (f_1^*, f_0^*) in (8.3.4) exists, it represents a least favorable pair of density functions in $\mathscr{F}_1 \times \mathscr{F}_0$, and minimizes the Kullback-Leibler information number in (8.3.1), in $\mathscr{F}_1 \times \mathscr{F}_0$. The *robust sequential test* is then completely described, for a given false alarm rate α and power level β, by

Given the classes $\mathscr{F}_1$ and $\mathscr{F}_0$ find the density functions f_1^* and f_0^* such that

$$I(f_1^*, f_0^*) = \int dx\, f_1^*(x) \log \frac{f_1^*(x)}{f_0^*(x)} = \inf_{f_1 \in \mathscr{F}_1,\, f_0 \in \mathscr{F}_0} I(f_1, f_0)$$

Start at time zero, by setting the test function T_0 at the value zero. Update the test function sequentially by

$$T(x^n) = T(x^{n-1}) + \log \frac{f_1^*(x_n)}{f_0^*(x_n)}$$

Stop at the first n such that

Either $\quad T(x^n) \geq \lambda_2 = \log \dfrac{\beta}{\alpha} \quad$ and decide then in favor of $\mathscr{F}_1$.

Or $\quad T(x^n) \leq \lambda_1 = \log \dfrac{1-\beta}{1-\alpha} \quad$ and decide then in favor of $\mathscr{F}_0$.

If $\mathscr{F}_0$ and $\mathscr{F}_1$ are Huber's classes in (6.3.31), the least favorable density functions f_1^* and f_0^* are given respectively by (6.3.33) and (6.3.32). If the nominal density functions f_0 and f_1 in the above classes are both Gaussian, and such that $f_0(x) = \phi(x)$ and $f_1(x) = \phi(x - \theta)$, where $\theta > 0$, and where $\phi(x)$ is the density function of the zero mean, unit variance, Gaussian random variable, then the updating step $\log[f_1^*(x_n)/f_0^*(x_n)]$ in the robust sequential test equals z_n, where z_n is the truncated form of the datum x_n as given by (6.3.41).

8.4 TESTS FOR DETECTING A CHANGE IN DISTRIBUTION

Let $f_0(x^n)$ and $f_1(x^n)$ denote the n-dimensional density functions of two well known, distinct, discrete-time, and mutually independent stochastic processes at the vector point x^n. Let it be known that the data sequence is initially generated by the density function f_0. Let it then be possible that, instead, at some point in time the density function f_1 may become active and remain so from that point on. Then, given an infinite data sequence $x = \{x_i;\, i \geq 1\}$ the possibilities are:

i. The f_0 to f_1 change never occurs. Thus, the total sequence x is generated by the density function f_0.

ii. The f_0 to f_1 change occurs before the sequence x starts being observed: Thus, the total sequence x is then generated by the density function f_1.
iii. The f_0 to f_1 change occurs just after the datum x_m; $m \geq 1$. Thus, the subsequence, x_1^m is then generated by the density function f_0 and the remaining sequence $x_{m+1}^\infty = \{x_k; k \geq m+1\}$, is then generated by the density function f_1.

Given the infinite sequence $x = \{x_i; i \geq 1\}$ and the density functions $f_0(x^n)$ and $f_1(x^n)$ $\forall$ $n \geq 1$ and $\forall$ x^n, the objective is to detect a possible f_0 to f_1 change as reliably and quickly as possible. Since the very occurrence of the change is uncertain, it is clear that only a sequential approach to the problem is meaningful. That is, to detect the possible occurrence of an f_0 to f_1 change, we must devise a sequential detection scheme. To gain insight, we first formalize a preliminary, nonsequential, optimal Bayesian detection scheme. In particular, we initially fix the size n of the observation sequence x_1^n and we test the occurrence of the change at all the possible points in the sequence x_1^n, assuming that those occurrences are equally probable. Then, the problem in Example 4.5.7 arises. From (4.5.71) we observe that, in this preliminary formalization, we conclude that the f_0 to f_1 change occurs just after the datum x_j; $0 \leq j \leq n$ (hypothesis H_j), iff

$$\sum_{i=j+1}^{n} \log \frac{f_1(x_i|x_{j+1}^{i-1})}{f_0(x_i|x_1^{i-1})} = \max_{0 \leq k \leq n} \left(\sum_{i=k+1}^{n} \log \frac{f_1(x_i|x_{k+1}^{i-1})}{f_0(x_i|x_1^{i-1})} \right) = T'(x^n) \quad (8.4.1)$$

where

$$f_0(x_1|x_1^0) = f_0(x_1)$$

$$f_1(x_{m+1}|x_{m+1}^m) = f_1(x_{m+1})$$

$$\sum_{i=n+1}^{n} \log \frac{f_1(x_i|x_{n+1}^{i-1})}{f_0(x_i|x_1^{i-1})} = 0$$

Studying the Bayesian rule in (8.4.1), we observe that its test function $T'(x^n)$ does not possess characteristics for convenient sequential transformation. We thus consider a variation of the test in (8.4.1), by replacing the conditional density function $f_1(x_i|x_{k+1}^{i-1})$ by $f_1(x_i|x_1^{i-1})$ $\forall$ i. Then the following variation of the test in (8.4.1) arises.

Given the data block x^n decide in favor of the hypothesis H_j, iff

$$\sum_{i=j+1}^{n} g_i(x^i) = \max_{0 \leq k \leq n} \left(\sum_{i=k+1}^{n} g_i(x^i) \right) \quad (8.4.2)$$

where

$$g_i(x^i) = \log \frac{f_1(x_i|x_1^{i-1})}{f_0(x_i|x_1^{i-1})} \quad (8.4.3)$$

An equivalent form of the test in (8.4.2) is given below.

8.4 Tests for Detecting a Change in Distribution

Given the data block x^n decide in favor of the hypothesis H_j, iff

$$\sum_{i=1}^{j} g_i(x^i) = \min_{0 \leq k \leq n} \left(\sum_{i=1}^{k} g_i(x^i) \right) \quad (8.4.4)$$

Let us now consider a sequential generalization of the test in (8.4.4). When data are observed sequentially, it seems natural to conclude that as the difference, $\sum_{i=1}^{n} g_i(x^i) - \min_{0 \leq k \leq n} (\sum_{i=1}^{k} g_i(x^i))$, increases, the odds that the f_0 to f_1 change has occurred increase as well. Thus, the following sequential test naturally evolves.

Select some positive threshold δ. Observe data sequentially, and decide that the f_0 to f_1 change has occurred at the first time n such that

$$T(x^n) = \sum_{i=1}^{n} g_i(x^i) - \min_{0 \leq k \leq n} \left(\sum_{i=1}^{k} g_i(x^i) \right) \geq \delta \quad (8.4.5)$$

It is easy to see that the sequential test in (8.4.5) can be expressed in the following equivalent form.

Select some positive threshold δ. Observe data sequentially, and decide that the f_0 to f_1 change has occurred at the first n such that $T(x^n) \geq \delta$, where

$$T(0) = 0$$
$$T(x^n) = \max\{0, T(x^{n-1}) + g_n(x^n); g: g_i(x^i) \text{ as in (8.4.3)}\} \quad (8.4.6)$$

Clearly, the test in (8.4.6) is sequential, with updating step $g_n(x^n)$. Also, it operates via the use of two thresholds, 0 and δ, where 0 represents a reflecting barrier and δ represents an absorbing barrier. When the two stochastic processes represented by the density functions f_0 and f_1 are not memoryless, then the memory x_1^{n-1} in the updating step $g_n(x^n)$ generally increases with increasing n. When the processes are instead memoryless, then the memory in the updating step is zero for all n. The test in (8.4.6) was originally proposed by Page (1954, 1955) for memoryless processes. Its asymptotic properties for such processes were studied by Lorden (1971). In its general form, the test in (8.4.6) was proposed by Bansal (1983) and Bansal et al (1986), who also studied its asymptotic properties for stationary and ergodic processes that also satisfy some general regularity conditions. The test generally consists of repeated Wald tests and is asymptotically optimal in the sense that it requires the minimum possible expected sample size for decision, subject to a false alarm constraint. We state this result in Theorem 8.4.1 below. The interested reader will find the proof of the theorem in Section 8.6.

Let us denote by $N_\delta(x)$ the stopping variable induced by the test in (8.4.6), and let $E_{f_0}\{N_\delta(X)\}$ and $E_{f_1}\{N_\delta(X)\}$ denote the expected value of the stopping variable when the density functions f_0 and f_1, respectively, are active. Then

$$N_\delta(x) = \inf\{n: T(x^n) \geq \delta\} \qquad (8.4.7)$$

THEOREM 8.4.1: Let the density functions f_0 and f_1 represent two ergodic and stationary stochastic processes satisfying some mixing conditions, and let (8.2.10) in Theorem 8.2.1 be satisfied. Consider the stopping variable $N_\delta(x)$ induced by the sequential test in (8.4.6). Then for $\delta \to \infty$ we have

$$E_{f_0}\{N_\delta(X)\} \geq \frac{\delta}{2} \qquad (8.4.8)$$

$$E_{f_1}\{N_\delta(X)\} \cong |\log \delta| E_{f_1}^{-1}\{I_{10}\} \qquad (8.4.9)$$

where I_{10} is the information number in (8.2.9).

If $N'(x)$ is the stopping variable of some arbitrary test, then, subject to $E_{f_0}\{N'(x)\} \geq \delta/2$, and for $\delta \to \infty$, we have $E_{f_1}\{N'(x)\} \geq |\log \delta| E_{f_1}^{-1}\{I_{10}\}$. ∎

Theorem 8.4.1 states that the test in (8.4.6) is optimal in an asymptotic sense. Specifically, when the threshold δ is asymptotically large, then the test requires the minimum possible expected time to cross that threshold, when the f_0 to f_1 change has occurred before the collection of data, subject to the constraint that when the f_0 to f_1 change never occurs then the expected time for threshold crossing is larger than $\delta/2$. Therefore, the test in (8.4.6) is asymptotically the fastest among all tests. In fact, due to the relationship between δ and $|\log \delta|$, we conclude that the speed of the test increases exponentially with δ. We note that Theorem 8.4.1 includes Lorden's (1971) result, for memoryless processes. Then the information number I_{10} equals $\int du f_1(u) \log[f_1(u)/f_0(u)]$.

Although the asymptotic optimality of the test in (8.4.6) is a significant virtue, its nonasymptotic properties are most important. Indeed, the very objective of the test is *fast* detection of change, which imposes the selection of relatively small δ, which selection is performed via the study of the false alarm and power characteristics induced by the test. This study varies with the nature of the processes. In the next section, we will fully analyze an example that finds applications in industrial quality control and in the dynamic monitoring of the performance of computer-communication networks. We point out that the algorithm in (8.4.6) has numerous applications, including automated image processing.

8.5 EXAMPLES

Example 8.5.1

Let us consider the two Bernoulli processes in Example 8.2.2. Let it be desirable to detect a possible change from the Bernoulli process whose parameter is p to

8.5 Examples

the Bernoulli process whose parameter is q. We then apply the sequential test in (8.4.6) while we assume that $q > p$. Due to the derivations in Example 8.2.2, we conclude that the test takes here the form

Select some positive threshold δ. Observe data sequentially and decide that the change has occurred at the first instant n such that $T(x^n) \geq \delta$, where

$$T(0) = 0 \qquad T(x^n) = \max(0, T(x^{n-1}) + z_n) \qquad (8.5.1)$$

$$z_n = x_n + \gamma(q, p) \qquad (8.5.2)$$

$$\gamma(q, p) = \frac{\log \dfrac{1-q}{1-p}}{\log \dfrac{q(1-p)}{p(1-q)}} \qquad (8.5.3)$$

where, due to the convexity of the logarithm, it is found that $p < -\gamma(q, p) < q$.

The asymptotic properties of the above algorithm have been discussed in Section 8.4. Here, we will study its nonasymptotic performance. In particular, given the threshold δ we will analyze the false alarm and the power characteristics induced by the algorithm. Let us define

$P_{r,\delta}(n)$ The probability that the threshold δ is first crossed at the time instant n, under the condition that the p to q change actually occurred just after the datum x_r.   (8.5.4)

The probabilities $\{P_{\infty,\delta}(n); n \geq 1\}$ represent then a false alarm set, while the probabilities $\{P_{0,\delta}(n); n \geq 1\}$ represent a power set. Indeed, the former probabilities show how probable is the crossing of the threshold at different time instants n when the p to q change never occurs. The latter probabilities correspond, instead, to threshold crossings at different time instants n when the "alarming," parameter-q Bernoulli process is active. We note that when the threshold δ is finite, it is eventually crossed with probability one, even when the p to q change never occurs. The false alarm curves $\{P_{\infty,\delta}(n); n \geq 1\}$ and the power curves $\{P_{0,\delta}(n); n \geq 1\}$ can be only computed numerically. For efficiency in such computations, we apply an approximation. In particular, we approximate the constant $\gamma(q, p)$ in (8.5.3) by the ratio of two integers. That is, we write

$$\gamma(q, p) \simeq -\frac{l}{s} \qquad (8.5.5)$$

where $l < s$ and l and s are both natural numbers.

Then we write the algorithm in (8.5.1) and (8.5.2) in the following equivalent form, where, without loss of generality, the threshold t is a natural number.

Select the threshold t. Decide the change has occurred at the first time n such that $T'(x^n) \geq t$, where

$$T'(0) = 0 \qquad T'(x^n) = \max(0, T'(x^{n-1}) + y_n) \qquad (8.5.6)$$

$$y_n = sx_n - l \qquad (8.5.7)$$

Clearly the updating step y_n in (8.5.7) is an integer, equal to $-l$ when $x_n = 0$, and equal to $s - l > 0$ when $x_n = 1$. Let us denote by $P_{r,t}(n)$ the probability in (8.5.4) as applied to the modified algorithm in (8.5.6) and (8.5.7). It is then possible to obtain recursive expressions for computing the probabilities $\{P_{r,t}(n); n \geq 1\}$ via a Markov chain formulation. The key element in the latter formulation is the probability $P_{r,t}(n, j)$ that $T'(x^k) < t \; \forall \; k < n$ and $T'(x^n) = j$, given that the p to q change occurs just after the datum, x_r. Indeed, via a method exhibited in Papantoni-Kazakos (1979a), we find:

i. If $t - 1 \geq s > l + 1$

$$
\begin{aligned}
P_{r,t}(n, 0) &= (1 - v) \sum_{i=0}^{l} P_{r,t}(n - 1, i) \\
P_{r,t}(n, j) &= (1 - v) P_{r,t}(n - 1, j + l) \qquad 1 \leq j < s - l \\
P_{r,t}(n, j) &= (1 - v) P_{r,t}(n - 1, j + l) + v P_{r,t}(n - 1, j - s + l) \\
&\qquad\qquad\qquad\qquad\qquad\qquad s - l \leq j \leq t - 1 - l \\
P_{r,t}(n, j) &= v P_{r,t}(n - 1, j - s + l) \qquad t - 1 - l < j \leq t - 1
\end{aligned}
\qquad (8.5.8)
$$

ii. If $t - 1 \geq s = l + 1$

$$
\begin{aligned}
P_{r,t}(n, 0) &= (1 - v) \sum_{i=0}^{l} P_{r,t}(n - 1, i) \\
P_{r,t}(n, j) &= (1 - v) P_{r,t}(n - 1, j + l) + v P_{r,t}(n - 1, j - 1) \\
&\qquad\qquad\qquad\qquad 1 \leq j \leq t - 1 - l \\
P_{r,t}(n, j) &= v P_{r,t}(n - 1, j - 1) \qquad t - 1 - l < j \leq t - 1
\end{aligned}
\qquad (8.5.9)
$$

where

$$v = \begin{cases} p & \text{if } n \leq r \\ q & \text{if } n > r \end{cases} \qquad (8.5.10)$$

$$P_{r,t}(n) = v \sum_{j=t-s+l}^{t-1} P_{r,t}(n - 1, j) \qquad t - 1 \geq s \geq l + 1 \qquad (8.5.11)$$

The recursive expressions in (8.5.8) and (8.5.9) allow for the efficient computation of the probability $P_{r,t}(n)$ in (8.5.11) for every r and every n.

8.5 Examples

The false alarm probabilities $\{P_{\infty,t}(n); n \geq 1\}$ and the power probabilities $\{P_{0,t}(n); n \geq 1\}$ evolve as a special case of the probabilities $\{P_{r,t}(n)\}$ and can also be computed recursively. Let us now define

$$\alpha(p, q, n, t) = \sum_{i=0}^{n} P_{\infty,t}(i) \qquad (8.5.12)$$

$$\beta(p, q, n, t) = \sum_{i=0}^{n} P_{0,t}(i) \qquad (8.5.13)$$

The quantities $\alpha(p, q, n, t)$ and $\beta(p, q, n, t)$ are the probabilities that the threshold t is crossed by the algorithm in (8.5.6) and (8.5.7) before or at the time instant n given that all the data are generated respectively by the p-parameter and q-parameter Bernoulli process. Thus, $\alpha(p, q, n, t)$ and $\beta(p, q, n, t)$ are respectively the false alarm curve and the power curve induced by the algorithm in (8.5.6) and (8.5.7). Those two curves represent the nonasymptotic performance of the algorithm. We computed those curves for $p = .01$ and $q = .05$ and for various choices of the threshold δ in the algorithm in (8.5.1) to (8.5.3). In our computations, we applied the rational approximations represented by (8.5.5) and the algorithm in (8.5.6) and (8.5.7), and we used the recursions in (8.5.8) and (8.5.9). Our results are shown in Figure 8.5.1, where P_0 stands for p and where P_1 stands for q. From this figure, the powerful nonasymptotic characteristics of the algorithm are evident. For relatively small sample size, it simultaneously attains both low false alarm rate and high power. The choice of the threshold δ will be determined from the specific false alarm and power requirements and will depend on where the emphasis (power versus false alarm) is

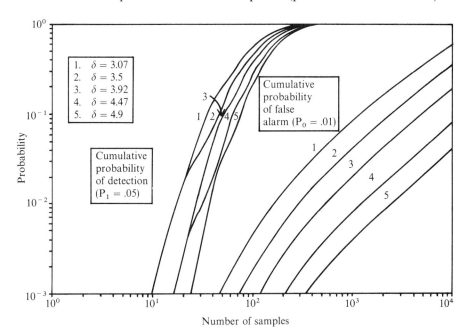

Figure 8.5.1

placed. When the probabilities p and q are further apart from each other than the pair, $p = .01$ and $q = .05$, then the false alarm and the power curves are drawn further apart than those in Figure 8.5.1. The performance characteristics of the algorithm are then further improved. When dependent Bernoulli trials are considered, instead, studies similar to those in this example, such as can be found in Bansal (1983), can be performed.

The algorithm in the present example can be applied to detect link failures in a computer-communication network. It attains this objective quickly and reliably, and can be generalized to perform its operations with minimal overhead [see Papantoni-Kazakos (1979b)].

Example 8.5.2

Let us consider two first order Markov Gaussian processes. Let the two processes be completely described by the following set of unconditional and conditional density functions, where $\theta > 0$, and $\mu: 0 < \mu < 1$.

Process #0

$$f_0(x_1) = \phi(x_1) = \frac{\exp\{-\frac{1}{2}x_1^2\}}{(2\pi)^{1/2}}$$

$$f_0(x_n|x_1^{n-1}) = f_0(x_n|x_{n-1}) = \frac{\exp\left\{-\frac{1}{2(1-\mu^2)}[x_n - \mu x_{n-1}]^2\right\}}{(2\pi)^{1/2}(1-\mu^2)^{1/2}} \quad (8.5.14)$$

Process #1

$$f_1(x_1) = \phi(x_1 - \theta) = \frac{\exp\left\{-\frac{(x_1 - \theta)^2}{2}\right\}}{(2\pi)^{1/2}} \quad (8.5.15)$$

$$f_1(x_n|x_1^{n-1}) = f_1(x_n|x_{n-1}) = \frac{\exp\left\{-\frac{1}{2(1-\mu^2)}[(x_n - \theta) - \mu(x_{n-1} - \theta)]^2\right\}}{(2\pi)^{1/2}(1-\mu^2)^{1/2}}$$

Let it be desirable to detect a possible change from Process #0 to Process #1. We will apply the sequential algorithm in Section 8.4, where here

$$\ln \frac{f_1(x_1)}{f_0(x_1)} = \theta x_1 - \frac{\theta^2}{2} = \frac{\theta}{1+\mu}\left[(1+\mu)\left(x_1 - \frac{\theta}{2}\right)\right] \quad (8.5.16)$$

$$n \geq 2 \quad \ln \frac{f_1(x_n|x_1^{n-1})}{f_0(x_n|x_1^{n-1})} = \frac{\theta}{1+\mu}\left[\left(x_n - \frac{\theta}{2}\right) - \mu\left(x_{n-1} - \frac{\theta}{2}\right)\right] \quad (8.5.17)$$

The conditions in Theorem 8.4.1 are all satisfied here, so the sequential test in (8.4.6) is asymptotically optimal, and its nonasymptotic form is

8.5 Examples

Select some positive threshold, δ. Observe data sequentially, and decide that the f_0 to f_1 change has occurred, at the first n, such that $T(x^n) \geq \delta$, where:

$$T(0) = 0 \qquad T(x^n) = \max(0, T(x^{n-1}) + g_n(x^n))$$

$$n \geq 2 \qquad g_n(x^n) = \left(x_n - \frac{\theta}{2}\right) - \mu\left(x_{n-1} - \frac{\theta}{2}\right) \qquad (8.5.18)$$

$$g_1(x_1) = (1 + \mu)\left(x_1 - \frac{\theta}{2}\right)$$

Let us now define the following probabilities:

$P_{\infty,\delta}(n, z)\,dz =$ Probability that $T(x^k) < \delta;\ \forall\ k \leq n - 1$, and
$n \geq 1$ $\qquad z < T(x^n) \leq z + dz$, conditioned on the
$\qquad\qquad$ event that the f_0 to f_1 change never occurs. $\qquad$ (8.5.19)

$P_{0,\delta}(n, z)\,dz:$ Probability that $T(x^k) < \delta;\ \forall\ k \leq n - 1$, and
$n \geq 1$ $\qquad z < T(x^n) \leq z + dz$, conditioned on the event
$\qquad\qquad$ that the f_0 to f_1 change occurred at time zero. $\qquad$ (8.5.20)

$P_{\infty,\delta}(n)$ $\qquad$ Probability that the threshold, δ, is first crossed
$n \geq 1$ $\qquad$ at time n, conditioned on the event that the
$\qquad\qquad$ f_0 to f_1 change never occurs. $\qquad\qquad\qquad\qquad$ (8.5.21)

$P_{0,\delta}(n)$ $\qquad$ Probability that the threshold, δ, is first crossed
$n \geq 1$ $\qquad$ at time n, conditioned on the event that the
$\qquad\qquad$ f_0 to f_1 change occurs at time zero. $\qquad\qquad\qquad$ (8.5.22)

The probabilities in (8.5.19) and (8.5.20) can be computed recursively, via the following expressions:

$$P_{\infty,\delta}(1, z) = \left[\Phi\left(\frac{\theta}{2}\right) - c\phi\left(\frac{\theta}{2}\right)\right]\delta(z) + c\phi\left(cz + \frac{\theta}{2}\right) \qquad z \geq 0 \quad c = \frac{1}{1+\mu}$$

(8.5.23)

$P_{\infty,\delta}(n, z) = \delta(z) \int_0^\delta dy\, P_{\infty,\delta}(n-1, y)$
$n \geq 2$

$$\times \left[\Phi\left(\frac{-y + \frac{1-\mu}{2}\theta}{(1-\mu^2)^{1/2}}\right) - \frac{\exp\left\{-\frac{\left[-y + \frac{(1-\mu)\theta}{2}\right]^2}{2(1-\mu^2)}\right\}}{(2\pi)^{1/2}(1-\mu^2)^{1/2}}\right]$$

$$+ \int_0^\delta dy\, P_{\infty,\delta}(n-1, y) \frac{\exp\left\{-\frac{\left[z - y + \frac{(1-\mu)\theta}{2}\right]^2}{2(1-\mu^2)}\right\}}{(2\pi)^{1/2}(1-\mu^2)^{1/2}} \qquad z \geq 0$$

(8.5.24)

$$P_{0,\delta}(1,z) = \left[\Phi\left(-\frac{\theta}{2}\right) - c\phi\left(\frac{\theta}{2}\right)\right]\delta(z) + c\phi\left(cz - \frac{\theta}{2}\right) \qquad z \geq 0 \quad c = \frac{1}{1+\mu}$$
(8.5.25)

$$\begin{aligned}P_{0,\delta}(n,z) = \delta(z)&\int_0^\delta dy\, P_{0,\delta}(n-1,y) \\ & n \geq 2\end{aligned}$$

$$\times \left[\Phi\left(\frac{-y - \frac{1-\mu}{2}\theta}{(1-\mu^2)^{1/2}}\right) - \frac{\exp\left\{-\frac{\left[y + \frac{1-\mu}{2}\theta\right]^2}{2(1-\mu^2)}\right\}}{(2\pi)^{1/2}(1-\mu^2)^{1/2}}\right]$$

$$+ \int_0^\delta dy\, P_{0,\delta}(n-1,y) \frac{\exp\left\{-\frac{\left[z - y - \frac{1-\mu}{2}\theta\right]^2}{2(1-\mu^2)}\right\}}{(2\pi)^{1/2}(1-\mu^2)^{1/2}} \qquad z \geq 0$$
(8.5.26)

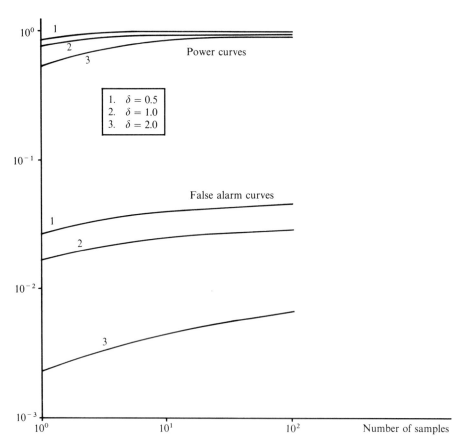

Figure 8.5.2 $\mu = 0.5$, $\theta = 3$.

8.5 Examples

where $\delta(x)$ denotes the delta function at point 0, and ϕ and Φ denote respectively the density function and the cumulative distribution of the unit variance, zero mean Gaussian random variable.

As in Example 8.5.1, the probabilities in (8.5.21) and (8.5.22) are given via the probabilities in (8.5.19) and (8.5.20) as

$$P_{\infty,\delta}(n) = \int_\delta^\infty P_{\infty,\delta}(n, z)\,dz \tag{8.5.27}$$

$$P_{0,\delta}(n) = \int_\delta^\infty P_{0,\delta}(n, z)\,dz \tag{8.5.28}$$

The false alarm curve $\alpha(\theta, \delta, n)$ and the power curve $\beta(\theta, \delta, n)$ are then provided by

$$\alpha(\theta, \delta, n) = \sum_{i=1}^n P_{\infty,\delta}(i) \qquad \beta(\theta, \delta, n) = \sum_{i=1}^n P_{0,\delta}(i) \tag{8.5.29}$$

We computed the false alarm and power curves for $\theta = 3$ and various values of the threshold δ. Our results are exhibited in Figure 8.5.2. We observe that proper choice of the threshold δ induces low false alarm rate and high power for small sample sizes n. As θ increases, the power of the algorithm is monotonically dramatized. Thus, in addition to being asymptotically ($\delta \to \infty$) optimal, the algorithm in (8.5.18) is also powerful for small sample sizes.

We point out that if the presence of outliers is suspected, the algorithm in (8.5.18) can be "robustified," by methods as in Chapter 6. The updating step $g_n(x^n)$ of the algorithm will then be truncated as

$$g_n(x^n) = \begin{cases} d & \left(x_n - \frac{\theta}{2}\right) - \mu\left(x_{n-1} - \frac{\theta}{2}\right) \geq d \\ \left(x_n - \frac{\theta}{2}\right) - \mu\left(x_{n-1} - \frac{\theta}{2}\right) & -d < \left(x_n - \frac{\theta}{2}\right) - \mu\left(x_{n-1} - \frac{\theta}{2}\right) < d \\ -d & \left(x_n - \frac{\theta}{2}\right) - \mu\left(x_{n-1} - \frac{\theta}{2}\right) \leq -d \end{cases}$$
$$n \geq 2 \tag{8.5.30}$$

$$g_1(x_1) = \begin{cases} d & (1+\mu)\left(x_1 - \frac{\theta}{2}\right) \geq d \\ (1+\mu)\left(x_1 - \frac{\theta}{2}\right) & -d < (1+\mu)\left(x_1 - \frac{\theta}{2}\right) < d \\ -d & (1+\mu)\left(x_1 - \frac{\theta}{2}\right) \leq -d \end{cases} \tag{8.5.31}$$

Given the frequency ε of the outliers, the constant d in (8.5.30) and (8.5.31) is found in a way parallel to those used in Chapter 6. The simultaneous selection of d and the threshold δ can be attained, however, via the study of the false alarm and power curves, computed for various choices of d and δ.

8.6 PROOF OF THEOREM 8.4.1

In this appendix, we present the proof of Theorem 8.4.1. We first consider a variation of the test in Theorem 8.4.1, and we prove its asymptotic optimality for processes that are more general than those considered in the theorem. Then we discuss when the two tests are asymptotically equivalent.

Consider the stopping variable $N'_\delta(w)$ given by

$$N'_\delta(w) = \inf\{n: T'(w_1^n) \geq \delta\} \qquad (8.6.1)$$

where

$$T'(w_1^n) = \max_{0 \leq k \leq n} \left(\sum_{i=k+1}^{n} \log \frac{f_1(w_i | w_{k+1}^{i-1})}{f_0(w_i | w_{k+1}^{i-1})} \right) \qquad (8.6.2)$$

We first express a theorem parallel to Theorem 2 in Lorden (1971).

THEOREM 8.6.1: Let f_0 and f_1 denote the density functions of two stationary and ergodic processes. Let $w = w_1, w_2, \ldots$ be an infinite data sequence, and let N be an extended stopping variable with respect to w, such that

$$P_{f_0}(N < \infty) \leq \alpha \qquad \alpha: 0 < \alpha < 1 \qquad (8.6.3)$$

For $k = 1, 2, \ldots$, let N_k denote the stopping variable obtained by applying N to $w_k, w_{k+1}, \ldots$, and define

$$N^* = \min\{N_k + k - 1 | k = 1, 2, \ldots,\} \qquad (8.6.4)$$

Then N^* is an extended stopping variable, and

$$E_{f_0}\{N^*\} \geq 2^{-1}(1 + \alpha^{-1}) \xrightarrow[\alpha \to 0]{} \frac{\alpha^{-1}}{2} \qquad (8.6.5)$$

$$E_{f_1}\{N^{*^+}\} = \bar{E}_{f_1}\{N^*\} \leq E_{f_1}\{N\} \qquad (8.6.6)$$

where $\bar{E}_{f_1}\{N^*\}$ is defined as in Lorden (1971).

Proof: The fact that N^* is an extended variable and (8.6.6) are proven exactly as in Theorem 2, Lorden (1971). Indeed, since the events $\{N_i \leq n - i + 1\}$; $1 \leq i \leq n$ are determined by the sequence w_1^n and the event $\{N^* \leq n\}$ is the union of the above, N^* is an extended stopping variable. Also, (8.6.6) holds since $E_{f_{1,m}}[(N^* - m + 1)^+ | w_1^{m-1}] \leq E_{f_{1,m}}[N_m | w_1^{m-1}] = E_{f_{1,m}} N_m = E_{f_1} N$. Here

8.6 Proof of Theorem 8.4.1

$f_{1,m}$ is the density, which is identical to f_0 on w_1^{m-1} and to f_1 on w_m^∞. The above inequality holds because f_0 and f_1 are assumed to be mutually independent—that is, for $n \geq m$, $f_{1,m}(x_1^n) = f_0(x_1^{m-1})f_1(x_m^n)$. Therefore, from the definition of $\bar{E}_{f_1}\{\cdot\}$ it follows that $\bar{E}_{f_1}\{N^*\} \leq E_{f_1}\{N\}$.

To prove (8.6.5), we proceed as follows. Consider a data sequence, $w = w_1, w_2, \ldots$. Let T be the shift transformation on R^∞ (one-sided infinite product of the real line), and let $\{\prod_j\}$ be the sequence of projections from R^∞ to R, where

$$\prod_j(Tw) = \prod_{j+1}(w)$$

Let $\sigma(w) = \{w, Tw, T^2w, \ldots\}$ denote the trajectory of w. Let $A = \{N < \infty\}$. Then $P_{\mu_0}(A) \leq \alpha$. Define the sequence $\{k_i(w)\}$ of reentries of w into A as follows.

$$k_1 = k_1(w) = \inf\{n: T^n w \in A; n = 1, 2, \ldots\}$$
$$k_i = k_i(w) = \inf\{n: T^{\sum_{1}^{i-1} k_j + n} w \in A; n = 1, 2, \ldots\} \quad i \geq 2 \qquad (8.6.7)$$

We then observe that

$$N(T^n w) = \infty \quad n \neq 0 \quad n \neq \sum_1^i k_j \quad \forall i \geq 1$$
$$N(w) = \infty \quad \text{if} \quad w \in A^c$$
$$N(w) < \infty \quad \text{if} \quad w \in A$$

Also, for $w \in A^c$, we have

$$N^*(T^i w) \geq k_1 - i + 1 \qquad 0 \leq i \leq k_1$$
$$N^*(T^{k_1 + i} w) \geq k_2 - i + 1 \qquad 1 \leq i \leq k_2$$
$$N^*(T^{\sum_1^i k_j + n} w) \geq k_{i+1} - n + 1 \qquad 1 \leq n \leq k_{i+1}$$

For $w \in A$, we have

$$N^*(w) \geq 1$$
$$N^*(T^i w) \geq k_1 - i + 1 \qquad 1 \leq i \leq k_1$$
$$N^*(T^{\sum_1^i k_j + n} w) \geq k_{i+1} - n + 1 \qquad 1 \leq n \leq k_{i+1}$$

Let us define

$$c_0 = 0 \qquad c_n = \sum_1^n k_j - 1 \qquad n \geq 1$$

Then from the above we conclude that

$$\sum_{i=c_{n-1}}^{c_n} N_{+1}^*(T^i w) \geq 1 + k_n + (k_n - 1) + \cdots + 2$$

$$= 2^{-1} k_n (k_n + 1) \qquad \forall w \quad \forall n \geq 1$$

$$\sum_{i=0}^{c_n} N^*(T^i w) \geq \sum_{j=1}^{n} 2^{-1} k_j (1 + k_j) \qquad \forall w \quad \forall n \geq 1$$

$$\rightarrow [1 + c_n]^{-1} \sum_{i=0}^{c_n} N^*(T^i w) \geq 2^{-1} \left(\sum_{j=1}^{n} k_j^2 \right) \left(\sum_{j=1}^{n} k_j \right)^{-1} + 2^{-1}$$

$$= 2^{-1} \left(n^{-1} \sum_{j=1}^{n} k_j^2 \right) \left(n^{-1} \sum_{j=1}^{n} k_j \right)^{-1} + 2^{-1} \qquad (8.6.8)$$

Applying in (8.6.8) the Jensen inequality, and due to the convexity of x^2, we obtain

$$[1 + c_n]^{-1} \sum_{i=0}^{c_n} N^*(T^i w) \geq 2^{-1} \left(n^{-1} \sum_{j=1}^{n} k_j \right)^2 \left(n^{-1} \sum_{j=1}^{n} k_j \right)^{-1} + 2^{-1}$$

$$= 2^{-1} n^{-1} \sum_{j=1}^{n} k_j + 2^{-1} \qquad (8.6.9)$$

But due to the ergodicity of the process represented by the density function f_0 we have

$$\lim n \left(\sum_{j=1}^{n} k_j \right)^{-1} = P_{f_0}(A) \leq \alpha$$

The left part of the inequality in (8.6.9) converges to $E_{f_0}\{N^*\}$ a.e. (f_0).
Thus, from (8.6.9) we obtain

$$E_{f_0}\{N^*\} \geq 2^{-1}(1 + \alpha^{-1}) \xrightarrow[\alpha \to 0]{} 2^{-1} \alpha^{-1}$$

We point out that for periodic ergodic processes, the bound $2^{-1}(1 + \alpha^{-1})$ is tight, and there exists a stopping variable N such that (8.6.5) is attained with equality. ∎

We point out here that if the extended stopping variable N is as in (8.6.1) and the threshold is α^{-1}, then (8.6.3) in the theorem is satisfied. Thus, Theorem 8.6.1 applies to the stopping variable $N'_\delta(x)$ in (8.6.1) if we put $\delta = \alpha^{-1}$. We thus have

$$E_{f_0}\{N'_\delta(x)\} \geq \frac{1}{2}(1 + \delta) \xrightarrow[\delta \to \infty]{} 2^{-1} \delta \qquad (8.6.10)$$

$$\bar{E}_{f_1}\{N'_\delta(x)\} \leq E_{f_1}\{N\} \qquad (8.6.11)$$

8.6 Proof of Theorem 8.4.1

for

$$N = \inf\left\{n: \sum_{i=0}^{n} g_i(x^i) \geq \delta\right\} \quad (8.6.12)$$

where $g_i(x^i)$ is as in (8.4.3).

We will now focus on showing that for $\delta \to \infty$ and under certain conditions, $\bar{E}_{f_1}\{N'_\delta(x)\}$ is the infimum among all the expected values $\bar{E}_{\mu_1}\{N^*\}$, where N^* is any extended stopping variable that satisfies the condition $E_{f_0}\{N^*\} \geq 2^{-1}\delta$. Thus, we will show that the stopping variable $N'_\delta(x)$ is then optimal in the above sense. Our approach will be as follows. We will first find an upper bound on $\bar{E}_{f_1}\{N'_\delta(x)\}$, for $\delta \to \infty$. Then, we will show that asymptotically ($\delta \to \infty$), this upper bound cannot be smaller than $\bar{E}_{f_1}\{N^*\}$; for any extended stopping variable satisfying $E_{f_0}\{N^*\} \geq \delta/2$, and under the appropriate assumptions. Let us consider the quantities in (8.2.9) and the conditions in (8.2.10). If the latter conditions are satisfied, we have from Berk (1973)

$$\lim_{\alpha \to 0} E_{f_1}\{N\} \cong \frac{|\log \alpha|}{E_{f_1}\{I_{10}\}}$$

or $\qquad (8.6.13)$

$$\lim_{\delta \to \infty} E_{f_1}\{N\} \cong \frac{|\log \delta|}{E_{f_1}\{I_{10}\}} \quad \text{for } \delta = \alpha^{-1}$$

where N is the extended stopping variable in (8.6.12).

We note that (8.6.13) is also satisfied if instead of (8.2.10), the martingale conditions of Chow and Robbins (1963), or the strong mixing conditions of Lai (1977) hold. Under conditions of Theorem 1 in Chow and Robbins (1963), the analysis remains unchanged. From (8.6.11) and (8.6.13) we thus conclude that, for $\delta \to \infty$ and if the conditions (8.2.10) hold, the expected value $\bar{E}_{f_1}\{N'_\delta(x)\}$ does not exceed $|\log \delta| E_{f_1}^{-1}\{I_{10}\}$. Therefore, $|\log \delta| E_{f_1}^{-1}\{I_{10}\}$ is then an upper bound on $\bar{E}_{f_1}\{N_\delta(x)\}$.

Let $\mathscr{C}_\delta^*$ be the class of all the extended stopping variables N^* satisfying $E_{f_0}\{N^*\} \geq \delta/2$. Let us define,

$$n^*(\delta) = \inf_{N_\delta^* \in \mathscr{C}_\delta^*} \bar{E}_{f_1}\{N^*\} \quad (8.6.14)$$

Our final objective will be to prove that for density functions f_0 and f_1 that satisfy (8.2.10) we have

$$\lim_{\delta \to \infty} n^*(\delta) \geq |\log \delta| E_{f_1}^{-1}\{I_{10}\} \quad (8.6.15)$$

We will now prove that for stationary processes that satisfy (8.2.10) the bound in (8.6.15) holds. We will do that via a theorem. In the proof of the theorem, Theorem 8.6.1 will be used. Note that if the limit I_{10} in (8.2.9) exists, and if the processes are also ergodic, then the limit I_{10} is the mismatch entropy of f_1 with respect to f_0.

THEOREM 8.6.2: Let $\mathscr{C}_\delta^*$ be the class of all extended stopping variables N^*, that satisfy the condition $E_{f_0}\{N^*\} \geq \delta/2$. Let $n^*(\delta)$ be as in (8.6.14), let (8.2.10) be satisfied, and let I_{10} be as in (8.2.9). Then

$$\lim_{\delta \to \infty} n^*(\delta) \cong \frac{\log \delta}{E_{f_1}\{I_{10}\}}$$

Proof: It suffices to show that for every ε in $(0, 1)$, there exists $c(\varepsilon) < \infty$ such that for any stopping variable N in $\mathscr{C}_\delta^*$ we have

$$E_{f_1}\{I_{10}\}\bar{E}_{f_1}\{N\} \geq (1 - \varepsilon) \log E_{f_0}\{N\} - c(\varepsilon) \qquad (8.6.16)$$

STEP 1: Given ε, let us define stopping variables T_i; $i \geq 0$ such that $T_0 \triangleq 0$, $T_i < T_{i+1} < \infty \; \forall \, i$, and T_{i+1}: the smallest n (or ∞ if such an n does not exist) such that $T_i < n$ and $f_1(w_{T_i+1}^n) \leq \varepsilon f_0(w_{T_i+1}^n)$.

From Wald's (1947) argument, we have $P_{f_1}(T_1 < \infty) = P_{f_1}(\text{decide } H_0) = \beta$. Then ε is the lowest threshold for deciding in favor of H_0, which in Wald's test is larger than $\beta(1 - \alpha)^{-1} > \beta$. Thus $P_{f_1}(T_1 < \infty) \leq \varepsilon$. Let us define

$$D_{rk} = \{w_1^k; T_{r-1} = k\} \qquad k < N \qquad (8.6.17)$$

Let us denote by P_{k+1} the probability induced by the measure f_1 and applied to data sequences $w_{k+1}, w_{k+2}, \ldots$. Then, provided that $P_1(D_{rk}) > 0$, we have, due to the arguments above, $P_{k+1}(T_r < \infty | D_{rk}) \leq \varepsilon$. On the set D_{rk} in (8.6.17), and for given N in $\mathscr{C}_\delta^*$, given T_r, we define a sequential test based on the sequence $w_{k+1}, w_{k+2}, \ldots$, as follows:

Stop at $\min(N, T_r)$

Decide $\begin{cases} P_{k+1} & \text{if } N \leq T_r \\ P_{\mu_0} & \text{if } N > T_r \end{cases}$

The number of observations taken for the above test is $\min(N, T_r) - k$, whose conditional expectation $P_{k+1}(\cdot | D_{rk})$ is at most $\bar{E}_{f_1}\{N\}$. To this point, our derivations are basically as in the proof of Theorem 3 in Lorden (1971), with the appropriate modifications.

STEP 2: Now, Wald's inequality (8.2.11) in Theorem 8.2.1 holds on D_{rk} with $\alpha = P_{f_0}(N \leq T_r | D_{rk})$ and $1 - \beta = P_{k+1}(N \leq T_r | D_{rk})$. So we have

$$E_{k+1}\left\{\log \frac{f_1(w_{k+1}^{N'})}{f_0(w_{k+1}^{N'})} \bigg| D_{rk}\right\} \geq P_{k+1}(N \leq T_r | D_{rk}) \log P_{f_0}(N \leq T_r | D_{rk}) - \log 2$$

$$\geq (1 - \varepsilon)|\log P_{f_0}(N \leq T_r | D_{rk})| - \log 2 \qquad (8.6.18)$$

where $N' = \min(N, T_r)$ and the last part in (8.6.18) is due to

$$P_{k+1}(N \leq T_r | D_{rk}) \geq P_{k+1}(T_r = \infty | D_{rk}) \geq 1 - \varepsilon$$

8.6 Proof of Theorem 8.4.1

Due to (8.2.10), given $\xi > 0$, $\exists N_0(\xi) < \infty$ such that

$$\left| E_{k+1}\left\{ (i-k)^{-1} \log \frac{f_1(w^i_{k+1})}{f_0(w^i_{k+1})} \right\} - E_{f_1}\{I_{10}\} \right| < \xi \qquad \forall\, i > k + N_0(\xi) \quad (8.6.19)$$

Due to (8.6.19), we now obtain

$$E_{k+1}\left\{ \log \frac{f_1(w^{N'}_{k+1})}{f_0(w^{N'}_{k+1})} \middle| D_{rk} \right\}$$

$$\leq \sum_{i=k+1}^{k+N_0(\xi)} P_{k+1}(N = i | D_{rk}) E_{k+1}\left\{ \log \frac{f_1(w^i_{k+1})}{f_0(w^i_{k+1})} \middle| N = i, D_{rk} \right\}$$

$$+ [\xi + E_{f_1}\{I_{10}\}] \operatorname*{sup\,ess\,sup}_{k} \sum_{i=k+N_0(\xi)+1}^{\infty} (i-k) P_{k+1}\{N = i | w^k_1\}$$

$$\leq \bar{E}_{f_1}\{N\}[\xi + E_{f_1}\{I_{10}\}] + \sum_{i=k+1}^{k+N_0(\xi)} P_{k+1}(N = i | D_{rk})$$

$$\times E_{k+1}\left\{ \log \frac{f_1(w^i_{k+1})}{f_0(w^i_{k+1})} \middle| N = i, D_{rk} \right\} \quad (8.6.20)$$

But due to (8.2.10) the expected values $E_{f_1}\{n^{-1}\log[f_1(w_1^n)/f_0(w_1^n)]\}$ are bounded for all n. If this bound is B, then $E_{f_1}\{\log[f_1(w_1^n)/f_0(w_1^n)]\} < nB < \infty$ $\forall\, n < \infty$. Therefore, $E_{f_1}\{\log[f_1(w^{k+n}_{k+1})/f_0(w^{k+n}_{k+1})] | w^k_1\}$ is bounded a.e. (f_1).

Due to the above arguments, and since $N_0(\xi)$ is finite, we conclude that there exists some finite constant $C(\xi)$ such that

$$\sum_{i=k+1}^{k+N_0(\xi)} P_{k+1}(N = i | D_{rk}) E_{k+1}\left\{ \log \frac{f_1(w^i_{k+1})}{f_0(w^i_{k+1})} \middle| N = i, D_{rk} \right\} < C(\xi) \quad (8.6.21)$$

for almost all w_1^k in measure f_1.

From (8.6.20) and (8.6.21) we obtain

$$E_{k+1}\left\{ \log \frac{f_1(w^{N'}_{k+1})}{f_0(w^{N'}_{k+1})} \middle| D_{rk} \right\} \leq \bar{E}_{f_1}\{N\}[\xi + E_{f_1}\{I_{10}\}] + C(\xi) \qquad \text{a.e. in } D_{rk}, \text{ in } f_1 \quad (8.6.22)$$

STEP 3: Let R be the smallest integer $r \geq 1$ such that $T_r > N$, where $\{T_i\}$ the sequence in step 1. If $P_{f_0}(R \geq r) > 0$, then $P_{f_0}(R < r+1 | R \geq r)$ is well defined and equals $P_{f_0}(N \leq T_r | T_{r-1} < N)$, which is an average over k of the probabilities $P_{f_0}(N \leq T_r | T_{r-1} = k < N)$ satisfying (8.6.18). Therefore, $P_{f_0}(R \geq r) > 0$ and the convexity of $-\log$ implies,

$$\sum_{k=0}^{\infty} P_{k+1}(D_{rk}) E_{k+1}\left\{ \log \frac{f_1(w^{N'}_{k+1})}{f_0(w^{N'}_{k+1})} \middle| D_{rk} \right\}$$

$$\geq (1-\varepsilon)|\log P_{f_0}(R < r+1 | R \geq r)| - \log 2 \quad (8.6.23)$$

If $P_{f_0}(R < r+1 | R \geq r) \geq Q$; $r \geq 1$ such that $P_{f_0}(R \geq r) > 0$, then $P_{f_0}(R \geq r+1) \leq (1-Q)^r$, hence $E_{f_0}\{R\} \leq Q^{-1}$. Thus, we obtain from (8.6.23)

$$\sum_{k=0}^{\infty} P_{k+1}(D_{rk}) E_{k+1} \left\{ \log \frac{f_1(w_{k+1}^{N'})}{f_0(w_{k+1}^{N'})} \bigg| D_{rk} \right\} \geq (1-\varepsilon) \log E_{f_0}\{R\} - \log 2 \quad (8.6.24)$$

We observe that due to (8.6.22) and (8.6.24), $E_{f_0}\{R\}$ is bounded—that is, $E_{f_0}\{R\} < \infty$. Also, from (8.6.22) and (8.6.24), we obtain

$$(1-\varepsilon) \log E_{f_0}\{R\} - \log 2 \leq \bar{E}_{f_1}\{N\}[\xi + E_{f_1}\{I_{10}\}] + C(\xi) \quad (8.6.25)$$

STEP 4: Given ε, let the stopping variable T_i be defined as in step 1. Given N, let R be defined as in step 3. Given the ergodic and stationary process with density function f_0, we then obtain the following expressions in a straightforward fashion, on $\{N = p\}$,

$E_{f_0}\{R | N = p\}$

$$= 1 + \sum_{l=1}^{p} \sum_{\substack{\sum_{j=1}^{l} k_j \leq p}} P_{f_0}\left(T_1 = k_1, T_2 = k_1 + k_2, \ldots, T_l = \sum_{j=1}^{l} k_j \bigg| N = p\right)$$

$$= 1 + \sum_{m=1}^{p} \sum_{l=1}^{m} \sum_{\substack{\sum_{j=1}^{l} k_j = m}} P_{f_0}\left(T_1 = k_1, T_2 = k_1 + k_2, \ldots, T_l = \sum_{j=1}^{l} k_j \bigg| N = p\right)$$

$$= 1 + \sum_{m=1}^{p} P_{f_0}(\exists\, k \leq m\colon T_k = m | N = p)$$

Let M_{m+1} denote the smallest n (or ∞ if such an n does not exist) such that $f_1(w_{m+1}^n) \leq \varepsilon f_0(w_{m+1}^n)$. Then

$E_{f_0}\{T_R | N = p\}$

$$= E_{f_0}\{T_1 | N = p\} + \sum_{l=1}^{p} \sum_{\substack{\sum_{j=1}^{l} k_j \leq p}} P_{f_0}\left(T_1 = k_1, T_2 = k_1 + k_2, \ldots, T_l\right.$$

$$= \sum_{j=1}^{l} k_j \bigg| N = p \bigg) \cdot E_{f_0}\left\{T_{l+1} - T_l \bigg| T_1 = k_1, \ldots, T_l = \sum_{j=1}^{l} k_j, N = p\right\}$$

$$= E_{f_0}\{T_1 | N = p\} + \sum_{m=1}^{p} \sum_{l=1}^{m} \sum_{\substack{\sum_{j=1}^{l} k_j = m}} P_{f_0}\left(T_1 = k_1, \ldots, T_l = \sum_{j=1}^{l} k_j \bigg| N = p\right)$$

$$\cdot E_{f_0}\left\{M_{m+1} \bigg| T_1 = k_1, \ldots, T_l = \sum_{j=1}^{l} k_j, N = p\right\}$$

$$= P_{f_0}\{T_1 | N = p\} + \sum_{m=1}^{p} P_{f_0}(\exists\, k \leq m\colon T_k = m | N = p)$$

$$\times E_{f_0}\{M_{m+1} | \exists\, k \leq m\colon T_k = m, N = p\} \quad (8.6.26)$$

8.6 Proof of Theorem 8.4.1

Define $A_m = (w: \exists\, k \leq m: T_k = m)$. Given m, and since the process with density function f_0 is stationary, we have

$$P_{f_0}(A_m)E_{f_0}\{M_{m+1}|A_m\} + P_{f_0}(A_m^c)E_{f_0}\{M_{m+1}|A_m^c\} = E_{f_0}\{T_1\} < \infty$$

Thus

$$\text{given } m: P_{f_0}(A_m) > 0 \quad \exists\, \beta(m) < \infty: E_{f_0}\{M_{m+1}|A_m\}$$
$$= E_{f_0}\{M_{m+1}|\exists\, k \leq m: T_k = m\} \leq \beta(m) \quad (8.6.27)$$

Let w denote an infinite data sequence generated by f_0. Let T be the shift transformation. Given w, let us define the sequence $\{k_i = k_i(w)\}$, as follows.

$$k_1 = \inf\{n + T_1(T^n w)\}$$
$$k_i = \inf\{n + T_1(T^{\sum_{j=1}^{i-1} k_j + n} w)\}$$

Then,

$$T_1(T^i w) \geq k_1 - i \quad 0 \leq i \leq k_1 - 1$$
$$T_1(T^{\sum_{j=1}^{i-1} k_j + l} w) \geq k_i - l \quad 0 \leq l \leq k_i - 1$$

$$\sum_{i=0}^{\sum_{j=1}^{n} k_j - 1} T_1(T^i w) \geq \sum_{j=2}^{n} \sum_{l=0}^{k_j - 1} (k_j - l) + \sum_{l=0}^{k_1 - 1} (k_1 - l)$$
$$= 2^{-1}\left(\sum_{j=1}^{n} k_j^2\right) + 2^{-1}\left(\sum_{j=1}^{n} k_j\right)$$

Thus, due to the ergodicity of the process with density function f_0, we obtain

$$\lim_{n \to \infty} \left(\sum_{j=1}^{n} k_j\right)^{-1} \sum_{i=0}^{\sum_{j=1}^{n} k_j - 1} T_1(T^i w)$$

$$= E_{f_0}\{T_1\} \geq \lim_{n \to \infty} 2^{-1}\left\{1 + \left(n^{-1}\sum_{j=1}^{n} k_j^2\right)\left(n^{-1}\sum_{j=1}^{n} k_j\right)^{-1}\right\}$$

$$\geq \text{(through Jensen's inequality)} \lim_{n \to \infty} 2^{-1}\left\{1 + \left(n^{-1}\sum_{j=1}^{n} k_j\right)^2 \left(n^{-1}\sum_{j=1}^{n} k_j\right)^{-1}\right\}$$

$$= 2^{-1}\left\{1 + \lim_{n \to \infty}\left(n^{-1}\sum_{j=1}^{n} k_j\right)\right\} \to \lim_{n \to \infty} n^{-1}\sum_{j=1}^{n} k_j \leq 2E_{f_0}\{T_1\} - 1 < 2E_{f_0}\{T_1\}$$

Consider $m \in \left[\sum_{j=1}^{i-1} k_j, \sum_{j=1}^{i} k_j - 1\right] \equiv B_i$. Then

$$T_1(T^{m+1}w) < \sum_{\sum_{j=1}^{i-1} k_j + 1 \leq l \leq \sum_{j=1}^{i} k_j} T_1(T^l w) \to \lim_{m \to \infty} E_{f_0}\{M_{m+1}|A_m\}$$

$$< \lim_{n \to \infty} n^{-1} \sum_{j=0}^{\sum_{j=1}^{n} k_j - 1} T_1(T^j w)$$

$$= \lim_{n \to \infty} \left(n^{-1} \sum_{j=1}^{n} k_j\right)\left(\sum_{j=1}^{n} k_j\right)^{-1} \sum_{j=0}^{\sum_{j=1}^{n} k_j - 1} T_1(T^j w) \leq 2E_{f_0}^2\{T_1\}$$

Thus, given $\zeta: 0 < \zeta$, $\exists N_0$:

$$E_{f_0}\{M_{m+1}|A_m\} < 2E_{f_0}^2\{T_1\} + \zeta = \beta E_{f_0}\{T_1\} \qquad \forall\, m > N_0 \quad (8.6.28)$$

From (8.6.26), (8.6.27), and (8.6.28), we thus obtain

$$E_{f_0}\{T_R|N=p\}$$

$$\leq E_{f_0}\{T_1|N=p\} + \max_{1 \leq m \leq p} \beta(m) \sum_{m=1}^{p} P_{f_0}(\exists\, k \leq m: T_k = m|N = p)$$

$$\leq E_{f_0}\{T_1|N=p\} + \max_{1 \leq m \leq p} \beta(m)\left[1 + \sum_{m=1}^{p} P_{f_0}(\exists\, k \leq m: T_k = m|N = p)\right]$$

$$\leq E_{f_0}\{T_1|N=p\} + \max_{m \leq N_0} \beta(m) E_{f_0}\{R|N=p\} \qquad p \leq N_0$$

$$E_{f_0}\{T_R|N=p\}$$

$$\leq E_{f_0}\{T_1|N=p\} + \max_{m \leq N_0} \beta(m) \sum_{m=1}^{N_0} P_{f_0}(\exists\, k \leq m: T_k = m|N = p)$$

$$+ \beta E_{f_0}\{T_1\} \sum_{m=N_0+1}^{} P_{f_0}(\exists\, k \leq m: T_k = m|N = p)$$

$$\leq E_{f_0}\{T_1|N=p\}$$

$$+ \max\left[\beta E_{f_0}\{T_1\}, \max_{m \leq N_0} \beta(m)\right]\left[1 + \sum_{m=1}^{p} P_{f_0}(\exists\, k \leq m: T_k = m|N = p)\right]$$

$$= E_{f_0}\{T_1|N=p\} + \max\left[\beta E_{f_0}\{T_1\}, \max_{m \leq N_0} \beta(m)\right] E_{f_0}\{R|N=p\} \qquad p > N_0$$

Thus, defining

$$\beta^* = E_{f_0}^{-1}\{T_1\} \max\left(E_{f_0}\{T_1\}, \max\left[\beta E_{f_0}\{T_1\}, \max_{m \leq N_0} \beta(m)\right]\right) \quad (8.6.29)$$

8.6 Proof of Theorem 8.4.1

we have

$$E_{f_0}\{T_R\} \leq \beta^* E_{f_0}\{T_1\} E_{f_0}\{R\} \qquad (8.6.30)$$

We note that if the process with density function f_0 is memoryless, then $E_{f_0}\{T_R\} = E_{f_0}\{T_1\} E_{f_0}\{R\}$.

From (8.6.30), and due to the definition of R in step 3, we obtain,

$$\log E_{f_0}\{N\} \leq \log E_{f_0}\{T_R\} \leq \log E_{f_0}\{R\} + \log E_{f_0}\{T_1\} - \log \beta^*$$

or $\qquad (8.6.31)$

$$\log E_{f_0}\{R\} \geq \log E_{f_0}\{N\} - \log E_{f_0}\{T_1\} + \log \beta^*$$

From (8.6.25) and (8.6.31), we now obtain directly

$$(1 - \varepsilon) \log E_{f_0}\{N\} - (1 - \varepsilon) \log E_{f_0}\{T_1\} - \log 2$$
$$\leq \bar{E}_{f_1}\{N\}[\xi + E_{f_1}\{I_{10}\}] + C(\xi) - (1 - \varepsilon) \log \beta^*$$
$$\underset{\xi \to 0}{\longrightarrow} \bar{E}_{f_1}\{N\} E_{f_1}\{I_{10}\} + C'(\xi) - (1 - \varepsilon) \log \beta^*$$

or

$$\bar{E}_{f_1}\{N\} E_{f_1}\{I_{10}\} \geq (1 - \varepsilon) \log E_{f_0}\{N\} - [(1 - \varepsilon) \log E_{f_0}\{T_1\}$$
$$+ \log 2 + C'(\xi) - (1 - \varepsilon) \log \beta^*] \qquad (8.6.32)$$

But $E_{f_0}\{T_1\}$ and β^* are finite, and they do not depend on N, but only on ε in step 1. Also $C'(\xi)$ is finite and independent of ε. We can thus write

$$c(\varepsilon) = (1 - \varepsilon) \log E_{f_0}\{T_1\} + \log 2 + C'(\xi) - (1 - \varepsilon) \log \beta^* < \infty$$

and we have proven inequality (8.6.16).

The proof of the theorem is now complete. ∎

Via the derivations in this section, we have basically proved the following theorem.

THEOREM 8.6.3: Let f_0 and f_1 be the density functions of two stationary, ergodic, and mutually independent stochastic processes with memory. Let (8.2.10) in Theorem 8.2.1 be satisfied. Consider the extended stopping variables N^* generated by testing an f_0 to f_1 change. Let $\mathscr{C}_\delta^*$ be the class of all such extended stopping variables N^* that also satisfy the condition $E_{f_0}\{N^*\} \geq \delta/2$. Let $n^*(\delta)$ be defined as

$$n^*(\delta) = \inf_{N^* \in \mathscr{C}_\delta^*} \bar{E}_{f_1}\{N^*\}$$

Let I_{10} be defined as

$$I_{10} = \lim_{n \to \infty} n^{-1} \log \frac{f_1(w_1^n)}{f_0(w_1^n)}$$

Then, due to (8.2.10), I_{10} exists a.e. (P_{f_1}) and is equal to $E_{f_1}\{I_{10}\}$. Futhermore,

$$\lim_{\delta \to \infty} n^*(\delta) \cong \frac{\log \delta}{E_{f_1}\{I_{10}\}}$$

and for the extended stopping variable $N'_\delta(w)$ in (8.6.1) and (8.6.2):

$$\lim_{\delta \to \infty} \bar{E}_{f_1}\{N'_\delta(w)\} \cong \frac{\log \delta}{E_{f_1}\{I_{10}\}}$$ ∎

What Theorem 8.6.3 basically says is that the sequential test described by the stopping variable $N'_\delta(w)$ in (8.6.1) and (8.6.2) is asymptotically ($\delta \to \infty$) optimal among all tests in class $\mathscr{C}^*_\delta$ in the theorem. The test then minimizes the expected time between the occurrence of a f_0 to f_1 change and its detection, under the constraint that if this change does not occur, then the expected time for a false alarm (exceeding the upper threshold δ) is no less than half the threshold value δ.

We note that conditions (8.2.10) hold for a large class of ergodic and stationary processes. As an example, let f_1 and f_0 be the density functions of two Gaussian processes with common spectral density and means respectively equal to θ and 0. Let R_n denote the n-dimensional covariance matrix induced by the common spectral density, and let R_n^{-1} be its inverse. Let $f^{-1}(\omega)$ denote the spectral density induced by R_n^{-1}, for $n \to \infty$. Then, if $f^{-1}(\omega)$ and $(2\pi)^{-1} \int_{-\pi}^{\pi} f^{-1}(\omega) \sin \omega [1 - \cos \omega]^{-1} d\omega$ exist, conditions (8.2.10) are satisfied.

Let us now consider the stopping variable $N_\delta(w)$ defined as

$$N_\delta(w) = \inf\{n: T(w_1^n) \geq \delta\} \qquad (8.6.33)$$

where

$$T(w_1^n) = \max_{0 \leq k \leq n} \left(\sum_{i=k+1}^n \log \frac{f_1(w_i/w_1^{i-1})}{f_0(w_i/w_1^{i-1})} \right)$$

The stopping variable in (8.6.33) is not as the extended stopping variables in Theorem 8.6.1. The two stopping variables are asymptotically ($\delta \to \infty$) equivalent, however, if the stochastic processes satisfy some mixing conditions. Such conditions may be the strong mixing conditions given by Lai (1976).

8.7 PROBLEMS

8.7.1 (a) Find the Wald sequential test, for

$$H_1: f_1(x) = \begin{cases} 1 & 0 < x < 1 \\ 0 & \text{otherwise} \end{cases} \qquad H_0: f_0(x) = \begin{cases} 2^{-1} & 0 < x < 2 \\ 0 & \text{otherwise} \end{cases}$$

Given α and β, find the two thresholds. Study the test's asymptotic behavior.

8.7 Problems

(b) Consider detection of change from f_0 to f_1. Express the sequential test, study its asymptotic performance, and study numerically the choice of thresholds based on false alarm and power curves behavior.

8.7.2 Repeat the studies in Problem 8.7.1, for

$$f_0(x) = \begin{cases} \frac{4}{5} & x = 0 \\ \frac{1}{5} & x = 1 \end{cases} \qquad f_1(x) = \begin{cases} \frac{2}{5} & x = 0 \\ \frac{2}{5} & x = 1 \\ \frac{1}{5} & x = 2 \end{cases}$$

8.7.3 Consider the case where a Poisson process of intensity λ_1 may shift to another Poisson process with intensity λ_2. Propose and study a sequential test for detecting this change. Study its asymptotic performance, and study the choice of thresholds based on small-sample behavior of the false alarm and power curves. Assume mutual independence between the two Poisson processes.

8.7.4 Consider two first order Markov chains, formed by a sequence of Bernoulli trials with dependence. Let one chain H_0 be characterized by the probabilities

$$P_0 = \Pr\{X_i = 1\} \qquad \lambda_0 = \Pr\{X_i = 1 | X_{i-1} = 1\}$$

Let the other chain H_1 be characterized by the probabilities,

$$P_1 = \Pr\{X_i = 1\} \qquad \lambda_1 = \Pr\{X_i = 1 | X_{i-1} = 1\}$$

Let H_0 and H_1 be mutually independent, and let an H_0 to H_1 shift be possible. Express the sequential test to detect this shift. Approximate the adaptation steps by rational numbers, and find recursive forms of the false alarm and power curves. Study the behavior of those curves numerically, for various choices of thresholds.

8.7.5 Consider Huber's robust test in Chapter 6, for Gaussian nominal processes, $H_0: \phi(x)$ vs. $H_1: \phi(x - \theta)$. Use a sequential form of this test, for detecting a possible change from class $\mathscr{F}_0$ to class $\mathscr{F}_1$. Study the asymptotic behavior of this test at $\phi(x)$ and $\phi(x - \theta)$, and compare it with the asymptotic behavior of the parametric test at $\phi(x)$ and $\phi(x - \theta)$. Express, study, and compare the false alarm and power curves of the above tests, for various ε_0 and ε_1 values. Perform the above comparisons, when the respective true densities are

$$f_1(x) = (1 - \varepsilon_1)\phi(x - \theta) + \varepsilon_1 \delta(x - y_1)$$
$$f_0(x) = (1 - \varepsilon_0)\phi(x) + \varepsilon_0 \delta(x - y_0)$$

where y_0 and y_1 are given. Compare, for various y_0 and y_1 choices.

REFERENCES

Bagshaw, M. and R. A. Johnson (1975), "The Effect of Serial Correlation in the Performance of CUSUM Tests II", *Technometrics* 17, No. 1, 73–80.

Balakrishnan, A. V. (1983), "Minimal-Time Detection Algorithm," An unpublished report, Optimization Software Inc., Los Angeles, CA.

Bansal, R. K. (1983), "An Algorithm for Detecting a Change in Stochastic Process," M. S. Thesis, University of Connecticut, EECS Dept.

Bansal, R. K. and P. Papantoni-Kazakos (1986) "An Algorithm for Detecting a Change in Stochastic Process," *IEEE Trans. Inform. Th.* IT-32, 227–235.

Berk, R. H. (1973), "Some Asymptotic Aspects of Sequential Analysis," *Ann. Statist.* 1, 1126–1136.

Billingsley, Patrick (1968), *Convergence of Probability Measures*, Wiley, New York.

Breiman, Leo (1968), *Probability*, Addison-Wesley, Reading, Mass.

Chow, Y. S. and H. Robbins (1963), "A Renewal Theorem for Random Variables which are Dependent or Non-Identically Distributed," *Ann. Math. Statist.* 34, 390–395.

Ibragimov, I. A. and Y. A. Rozanov (1978), *Gaussian Random Processes*, Springer-Verlag, New York.

Johnson, Richard A. and M. Bagshaw (1974), "The Effect of Serial Correlation on the Performance of CUSUM Tests," *Technometrics* 16, No. 1, 103–112.

Khan, Rasul A. (1978), "Wald's Approximations to the Average Run Length in CUSUM Procedures," *Journal of Statist. Planning and Inference* 2, 63–77.

——— (1979), "A Sequential Detection Procedure and the Related CUSUM Procedure," *Sankhya* 40, Series B. Pts. 3 and 4, pp. 146–162.

Kerr, T. H. (1983), "The Controversy over use of SPRT and GLR Techniques and other Losse-Ends in Failure Detections," Proc. of A.C.C. Conference, San Francisco, CA.

Kolmogorov, A. N. and Yu A. Rozanov (1960), "On Strong Mixing Conditions for Stationary Mixing Sequences," *Theory of Probability and Applications* V, No. 2, 204–208.

Lai, Tze Leung (1976), "On r-quick Convergence and a Conjecture of Strassen," *Ann. of Prob.* 4, No. 4, 612–617.

——— (1977), "Convergence Rates and r-quick Versions of the Strong Law for Stationary Mixing Sequences," *Ann. of Prob.* 5, No. 5, 693–706.

Loéve, M. (1963), *Probability Theory*, 3rd ed., Van Nostrand, Princeton, NJ.

References

Lorden, G. (1971), "Procedures for Reacting to a Change in Distribution," *Ann. Math. Statist.* 42, 1897–1908.

Page, E. S. (1954), "Continuous Inspection Schemes," *Biometrika* 41, 100–115.

——— (1955), "A Test for a Change in a Parameter Occurring at an Unknown Point," *Biometrika* 42, 523–527.

Papantoni-Kazakos, P. (1979a), "Algorithms for Monitoring Changes in Quality of Communication Links," *IEEE Trans. Comm.* COM-27, No. 4, 682–692.

——— (1979b), "The Potential of End-to-End Observations in Trouble Localization and Quality Control of Network Links," *IEEE Trans. Comm.* COM-27, No. 1, 16–30.

Shirayev, A. N. (1976), *Optimal Stopping Rules*, Springer-Verlag, New York.

Wald, A. (1947), *Sequential Analysis*, Wiley, New York.

Wijsman, R. A. (1971), "Exponentially Bounded Stopping Time of Sequential Probability Ratio Tests for Composite Hypotheses," *Ann. Math. Statist.* 42, 1859–1869.

Willsky, Alan S. (1976), "A Survey of Design Methods for Failure Detection in Dynamic Systems," *Automatica* 12, 601–611.

Willsky, Alan S. and M. L. Jones (1974), "A Generalized Likelihood Ratio Approach to the Detection and Estimation in Linear Systems Subject to Abrupt Changes," Proc. of 1974 IEEE Conf. on Decision and Control, Phoenix, Arizona.

Wong, S. P. (1968), "Asymptotically Optimum Properties of Certain Sequential Tests," *Ann. Math. Statist.* 39, 1244–1263.

Zacks, S. and Z. Kander (1966), "Test Procedures for Possible Changes in Parameters of Statistical Distribution Occurring at Unknown Time Points," *Ann. Math. Statist.* 37, 1196–1210.

CHAPTER 9

BAYESIAN AND MINIMAX PARAMETER ESTIMATION

9.1 INTRODUCTION

In Chapter 3, we outlined the various parameter estimation schemes as they evolve with the assets available to the analyst. In this chapter, we focus on those parameter estimation schemes whose formalization requires the maximum available set of assets; the *Bayesian* schemes. We also present and analyze the *minimax parameter estimation* schemes, where the latter evolve naturally from their Bayesian parallels. Both the Bayesian and the minimax parameter estimation schemes are presented as the optimal solution of appropriately formulated optimization problems.

Let us assume that there exists a parameter vector Θ^m of finite dimensionality m that takes values in some given m-dimensional space $\mathscr{E}^m$. Let each value θ^m in $\mathscr{E}^m$ determine a well known, discrete-time stochastic process. Let X^n denote an n-dimensional sequence of consecutive random variables generated by the above process, and let x^n denote some realization of the sequence X^n. When some realization x^n is observed, the crucial assumption in the formalization of all the parameter estimation schemes is that x^n has been generated by a single stochastic process. The latter assumption clearly implies that the value θ^m of the parameter vector Θ^m remains unchanged throughout the whole observation time period. Given the parameter space $\mathscr{E}^m$ and some realization x^n of sequences generated by the active stochastic process, an estimate $\hat{\theta}^m(x^n)$ of the true parameter value θ^m is sought. In the search for this estimate, the parameter space $\mathscr{E}^m$, the realization x^n, and the n-dimensional density function $f_{\theta^m}(x^n)$ induced by the process at the vector point x^n are all used. In the Bayesian and minimax formalizations, some given scalar penalty function $c[\theta^m, \hat{\theta}^m(x^n)]$ is also used, where

$$c[\theta^m, \hat{\theta}^m(x^n)] \geq 0 \qquad \forall \, \theta^m \in \mathscr{E}^m \quad \forall \, \hat{\theta}^m \quad \forall \, x^n \qquad (9.1.1)$$

9.2 The Bayesian Optimization Problem

In addition, when the Bayesian parameter estimation formalization is considered, then a density function $p(\theta^m)$; $\theta^m \in \mathscr{E}^m$ is also available. This density function represents the a priori probability with which some parameter value θ^m in $\mathscr{E}^m$ is selected, *before* the sequence realization x^n is observed. Via the full utilization of all the available assets, the objective function in both the minimax and the Bayesian formalizations is the expected penalty $c(\hat{\theta}^m, p)$ in (3.3.2), where Γ^n is the space where the realization x^n takes its value. In the Bayesian formalization, the estimate $\hat{\theta}_0^m$ that minimizes the expected penalty $c(\hat{\theta}^m, p)$ is sought. The minimax estimate is instead the result of a saddle-point game formalization, with payoff function $c(\hat{\theta}^m, p)$ and with variables the estimate vector function $\hat{\theta}^m$ and the a priori parameter density function p.

Given n, given the densities $f_{\theta^m}(x^n)$; $x^n \in \Gamma^n$, $\theta^m \in \mathscr{E}^m$ and $p(\theta^m)$; $\theta^m \in \mathscr{E}^m$, let us define the densities $f(x^n)$; $x^n \in \Gamma^n$ and $p(\theta^m | x^n)$ as

$$f(x^n) = \int_{\mathscr{E}^m} f_{\theta^m}(x^n) p(\theta^m) \, d\theta^m \qquad (9.1.2)$$

$$p(\theta^m | x^n) = \frac{f_{\theta^m}(x^n) p(\theta^m)}{f(x^n)} \qquad (9.1.3)$$

The expected penalty $c(\hat{\theta}^m, p)$ in (3.3.2) can be then expressed via (9.1.3) as

$$c(\hat{\theta}^m, p) = \int_{\Gamma^n} dx^n \, f(x^n) \int_{\mathscr{E}^m} c[\theta^m, \hat{\theta}^m(x^n)] p(\hat{\theta}^m | x^n) \, d\theta^m \qquad (9.1.4)$$

Given $\hat{\theta}^m$ and p, the integral

$$C_{\hat{\theta}^m, p}(x^n) = \int_{\mathscr{E}^m} c[\theta^m, \hat{\theta}^m(x^n)] p(\theta^m | x^n) \, d\theta^m \qquad (9.1.5)$$

in (9.1.4) is clearly strictly a function of the sequence realization x^n. Given x^n, the parameter estimate $\hat{\theta}^m(x^n)$ is an m-dimensional vector function of x^n. Minimization of $c(\hat{\theta}^m, p)$ in (9.1.4), with respect to $\hat{\theta}^m$ implies that for every value x^n in Γ^n, the m-dimensional vector $\hat{\theta}_0^m(x^n)$ in $\mathscr{E}^m$ that minimizes the nonnegative quantity $C_{\hat{\theta}^m, p}(x^n)$ in (9.1.5) should be selected (if this minimum exists and it can be attained). Then the optimal Bayesian parameter estimate $\hat{\theta}_0^m$ is equal to $\hat{\theta}_0^m(x^n)$ for each x^n in Γ^n.

9.2 THE BAYESIAN OPTIMIZATION PROBLEM

Due to the introductory discussion in Section 9.1, and directly from (9.1.2) and (9.1.5), the following optimization problem arises in the Bayesian parameter estimation formalization.

Given n, $\mathscr{E}^m$, p, and $c(\cdot, \cdot)$, as in (9.1.1), minimize

$$c(\hat{\theta}^m, p) = \int_{\Gamma^n} f(x^n) C_{\hat{\theta}^m, p}(x^n) \, dx^n \qquad (9.2.1)$$

subject to

$$\hat{\theta}^m(x^n) \in \mathscr{E}^m \qquad \forall \, x^n \in \Gamma^n$$

If it exists, a solution $\hat{\theta}_0^m(x^n)$; $x^n \in \Gamma^n$ of the optimization problem in (9.2.1) is an optimal Bayesian parameter estimate. We now express the general solution of the optimization problem in (9.2.1) in a theorem.

THEOREM 9.2.1: Consider the nonnegative function $C_{\hat{\theta}^m,p}(x^n)$; $x^n \in \Gamma^n$ in (9.1.5). Let the infimum

$$C_p(x^n) = \inf_{\hat{\theta}^m(x^n) \in \mathscr{E}^m} C_{\hat{\theta}^m,p}(x^n) \qquad (9.2.2)$$

exist for every x^n in Γ^n, and for a given x^n let this infimum be attained at $\hat{\theta}_0^m(x^n)$. Then $\hat{\theta}_0^m(x^n)$; $x^n \in \Gamma^n$, is a solution of the optimization problem in (9.2.1); thus, it is an optimal Bayesian parameter estimate. If $\hat{\theta}_0^m(x^n)$ is unique for every x^n in Γ^n, then the vector function $\hat{\theta}_0^m(x^n)$; $x^n \in \Gamma^n$ is the unique solution of the problem in (9.2.1); thus, this vector function is then the unique optimal Bayesian estimate.

Proof: Given some arbitrary vector function $\hat{\theta}^m(x^n)$; $x^n \in \Gamma^n$ such that $\hat{\theta}^m(x^n) \in \mathscr{E}^m$ $\forall \, x^n \in \Gamma^n$, we clearly obtain

$$c(\hat{\theta}^m, p) = \int_{\Gamma^n} dx^n \, f(x^n) C_{\hat{\theta}^m,p}(x^n) \geq \int_{\Gamma^n} dx^n \, f(x^n) C_p(x^n) \qquad (9.2.3)$$

assuming that the infimum $C_p(x^n)$ in (9.2.2) exists for each x^n in Γ^n. Given x^n in Γ^n, let the vector $\hat{\theta}_0^m(x^n)$ attain the infimum $C_p(x^n)$ in (9.2.2). Let $\hat{\theta}_0^m$ denote the function $\hat{\theta}_0^m(x^n)$; $x^n \in \Gamma^n$ generated by $\hat{\theta}_0^m(x^n)$, for x^n values varying in Γ^n. Due to (9.2.3) we then obtain

$$C(\hat{\theta}_0^m, p) \leq c(\hat{\theta}^m, p) \qquad \forall \, \hat{\theta}^m : \hat{\theta}^m(x^n) \in \mathscr{E}^m \quad \forall \, x^n \in \Gamma^n$$

The proof of the theorem is now complete, and the statement about uniquenes follows easily. ∎

From Theorem 9.2.1, we conclude that the search for an optimal Bayesian parameter estimate reduces to the search for the infimum $C_p(x^n)$ in (9.2.2) for every x^n in Γ^n. The existence of this infimum is clearly the initial important point, and it can be guaranteed if the penalty function $c[\hat{\theta}^m, \hat{\theta}^m(x^n)]$ and the density function $p(\theta^m|x^n)$ in (9.1.3) and (9.1.5), satisfy certain conditions. In Lemma 9.2.1 below we present some important such conditions, where we assume that θ^m and $\hat{\theta}^m(x^n)$ are column vectors, where * and T denote respectively conjugate and transpose, where Q is some $m \times m$ positive definite matrix, and where we denote

$$\|\theta^m - \hat{\theta}^m(x^n)\|_Q = [\theta^m - \hat{\theta}^m(x^n)]^T Q [\theta^m - \hat{\theta}^m(x^n)]^* \qquad (9.2.4)$$

$$E^m(x^n) = \int_{\mathscr{E}^m} \theta^m \, p(\theta^m|x^n) \, d\theta^m \qquad (9.2.5)$$

LEMMA 9.2.1: Let n, $\mathscr{E}^m$, p, and $c[\theta^m, \hat{\theta}^m(x^n)]$ be given. Let the space $\mathscr{E}^m$ be compact with respect to the metric in (9.2.4).

9.2 The Bayesian Optimization Problem

i. Let the penalty function, $c[\theta^m, \hat{\theta}^m(x^n)]$, be such that

> For every θ^m in $\mathscr{E}^m$, $c(\theta^m, \alpha^m)$ is strictly convex in $\alpha^m \in \mathscr{E}^m$ with respect to the metric in (9.2.4).

Then the infimum $C_p(x^n)$ in (9.2.2) exists and it is unique $\forall\ x^n \in \Gamma^n$.

ii. Let the density function $p(\theta^m|x^n)$ in (9.1.3) be symmetric around its expected value $E^m(x^n)$ in (9.2.5), and let $E^m = \mathbb{C}^m = \mathbb{C} \times \cdots \times \mathbb{C}$, where $\mathbb{C}$ signifies the complex plane. Let the penalty function $c[\theta^m, \hat{\theta}^m(x^n)]$, be such that

$$c[\theta^m, \hat{\theta}^m(x^n)] = c[\theta^m - \hat{\theta}^m(x^n)]$$

$$c(y^m) = c(-y^m) \qquad \forall\ y^m \in \mathbb{C}^m \tag{9.2.6}$$

$c(y^m)$ is strictly convex in $\mathbb{C}^m$, with respect to the metric in (9.2.4).

Then, given x^n in Γ^n, the infimum $C_p(x^n)$ in (9.2.2) exists, it is unique, and it is uniquely satisfied by the vector

$$\hat{\theta}_0^m(x^n) = E^m(x^n) \tag{9.2.7}$$

That is, the optimal Bayesian estimate $\hat{\theta}_0^m(x^n)$ is then the conditional expectation $E^m(x^n)$; $x^n \in \Gamma^n$ in (9.2.5).

iii. Let the penalty function $c[\theta^m, \hat{\theta}^m(x^n)]$ be such that

$$c[\theta^m, \hat{\theta}^m(x^n)] = 1 - \delta(\|\theta^m - \hat{\theta}^m(x^n)\|_Q) \tag{9.2.8}$$

for some positive definite $m \times m$ matrix Q, where $\delta(x)$ denotes the delta function at the point zero. Given x^n in Γ^n, let the density function $p(\theta^m|x^n)$ in (9.1.3) have a supremum

$$p(x^n) = \sup_{\theta^m \in \mathscr{E}^m} p(\theta^m|x^n) \tag{9.2.9}$$

attained at $\theta^m = \hat{\theta}_0^m(x^n)$. Then the infimum $C_p(x^n)$ in (9.2.2) exists and it equals $1 - p(x^n)$. If for every given $x^n \in \Gamma^n$ the parameter value $\hat{\theta}_0^m(x^n)$ that attains the supremum in (9.2.9) is unique, the optimal Bayesian estimate is also unique, and is equal to $\hat{\theta}_0^m(x^n)\ \forall\ x^n \in \Gamma^n$. This estimate is then called the *maximum a posteriori* estimate, since it maximizes the a posteriori density function $p(\theta^m|x^n)$ for every x^n in Γ^n.

Proof:

i. Let θ^m in $\mathscr{E}^m$ be given. Since $\mathscr{E}^m$ is compact, and $c(\theta^m, \alpha^m)$ is strictly convex in $\alpha^m \in \mathscr{E}^m$, both with respect to the metric in (9.2.4) the infimum $\inf_{\alpha^m \in \mathscr{E}^m} c(\theta^m, \alpha^m)$ exists, it is unique, and it is attained at some $\alpha^m(\theta^m)$ in $\mathscr{E}^m$.

Then, we clearly conclude that

$$C_p(x^n) = \int_{\mathscr{E}^m} c[\theta^m, \alpha^m(\theta^m)] p(\theta^m | x^n) \, d\theta^m$$

and that $C_p(x^n)$ is also unique $\forall\, x^n \in \Gamma^n$.

ii. We first note that the space $\mathbb{C}^m$ is compact with respect to the metric in (9.2.4). Given x^n in Γ^n let $\hat{\theta}_0^m(x^n)$ denote the conditional expected value $E^m(x^n)$ in (9.2.5). Let $\hat{\theta}^m(x^n)$ be some arbitrary m-dimensional vector parameter estimate, and let us then define the vector functions $f^m(x^n)$ and $z_{\theta^m}^m(x^n)$; $x^n \in \Gamma^n$, $\theta^m \in R^m$ as

$$\begin{aligned} f^m(x^n) &= \hat{\theta}^m(x^n) - \hat{\theta}_0^m(x^n) = \hat{\theta}^m(x^n) - E^m(x^n) \\ z_{\theta^m}^m(x^n) &= \theta^m - \hat{\theta}_0^m(x^n) = \theta^m - E^m(x^n) \end{aligned} \qquad (9.2.10)$$

Then, due to the conditions in (9.2.6) and the symmetry of $p(\theta^m | x^n)$ around $E^m(x^n)$, we can obtain

$$\begin{aligned} C_{\hat{\theta}^m, p}(x^n) &= \int_{\mathbb{C}^m} d\theta^m \, c[\theta^m, \hat{\theta}^m(x^n)] p(\theta^m | x^n) = \int_{\mathbb{C}^m} c[\theta^m - \hat{\theta}^m(x^n)] p(\theta^m | x^n) \, d\theta^m \\ &= \int_{\mathbb{C}^m} c[\theta^m - \hat{\theta}_0^m(x^n) - f^m(x^n)] p(\theta^m | x^n) \, d\theta^m \\ &= \int_{\mathbb{C}^m} c[z_{\theta^m}^m(x^n) - f^m(x^n)] p[z_{\theta^m}^m(x^n) + \hat{\theta}_0^m(x^n) | x^n] \, d\theta^m \qquad (9.2.11) \\ &= \int_{\mathbb{C}^m} c[-z_{\theta^m}^m(x^n) + f^m(x^n)] p[-z_{\theta^m}^m(x^n) + \hat{\theta}_0^m(x^n) | x^n] \, d\theta^m \\ &= \int_{\mathbb{C}^m} c[z_{\theta^m}^m(x^n) + f^m(x^n)] p[z_{\theta^m}^m(x^n) + \hat{\theta}_0^m(x^n) | x^n] \, d\theta^m \qquad (9.2.12) \end{aligned}$$

where the latter integral is derived via the change of variable $w_{\theta^m}^m(x^n) = -z_{\theta^m}^m(x^n)$.

From (9.2.11) and (9.2.12), we obtain

$$C_{\hat{\theta}^m, p}(x^n) = 2^{-1} \int_{\mathbb{C}^m} \{ c[z_{\theta^m}^m(x^n) - f^m(x^n)] + c[z_{\theta^m}^m(x^n) + f^m(x^n)] \}$$
$$\cdot p[z_{\theta^m}^m(x^n) + \hat{\theta}_0^m(x^n) | x^n] \, d\theta^m \qquad (9.2.13)$$

Due to the strict convexity of the function $c(y^m)$; $y^m \in \mathbb{C}^m$ we have for any x^n in Γ^n

$$\begin{aligned} 2^{-1} \{ c[z_{\theta^m}^m(x^n) - f^m(x^n)] &+ c[z_{\theta^m}^m(x^n) + f^m(x^n)] \} \\ &\geq c[z_{\theta^m}^m(x^n)] \qquad \forall\, z_{\theta^m}^m(x^n) \in \mathbb{C}^m \end{aligned} \qquad (9.2.14)$$

with equality in (9.2.14) if, and only if, $f^m(x^n) = 0$.

From (9.2.13), (9.2.14), and (9.2.10), we thus conclude

$$\begin{aligned} C_{\hat{\theta}^m, p}(x^n) &\geq \int_{\mathbb{C}^m} c[z_{\theta^m}^m(x^n)] p[z_{\theta^m}^m(x^n) + \hat{\theta}_0^m(x^n) | x^n] \, d\theta^m \\ &= \int_{\mathbb{C}^m} d\theta^m \, c[\theta^m - \hat{\theta}_0^m(x^n)] P(\theta^m | x^n) = C_{\hat{\theta}_0^m, p}(x^n) \qquad (9.2.15) \end{aligned}$$

9.2 The Bayesian Optimization Problem

with equality, if and only if, $\hat{\theta}^m(x^n) = \hat{\theta}_0^m(x^n) = E^m(x^n) \; \forall \; x^n \in \Gamma^n$. Thus $\hat{\theta}_0^m(x^n) = E^m(x^n) \; \forall \; x^n \in \Gamma^n$ is the unique Bayesian estimate.

iii. Let the penalty function $c[\theta^m, \hat{\theta}^m(x^n)]$ be as in (9.2.8). Then, from (9.1.5), we obtain by substitution

$$C_{\hat{\theta}^m, p}(x^n) = \int_{\mathscr{E}^m} d\theta^m \, p(\theta^m | x^n) - p[\hat{\theta}^m(x^n) | x^n] = 1 - p[\hat{\theta}^m(x^n) | x^n] \qquad (9.2.16)$$

Let the supremum in (9.2.9) exist, and let it be attained at $\theta^m = \hat{\theta}_0^m(x^n)$. Then, due to (9.2.16), we clearly obtain

$$\inf_{\hat{\theta}^m(x^n) \in \mathscr{E}^m} C_{\hat{\theta}^m, p}(x^n) = 1 - \sup_{\hat{\theta}^m(x^n) \in \mathscr{E}^m} p(\hat{\theta}^m(x^n) | x^n)$$

$$= 1 - p(x^n) = 1 - p(\hat{\theta}_0^m(x^n) | x^n) \qquad (9.2.17)$$

The statement about uniqueness follows trivially. ∎

As we will see further, case ii in the lemma is widely used, it describes an interesting class of penalty functions, and it specifies an interesting class of statistics. Case iii in the lemma defines the maximum a posteriori estimates as offsprings of certain Bayesian formalizations. At this point, we will relate, in the form of a corollary, some of the results in Lemma 9.2.1 with the sufficient statistics presented in Chapter 4.

COROLLARY 9.2.1: Given n, $\mathscr{E}^m$, p, and $\theta^m \in \mathscr{E}^m$, let the density function $f_{\theta^m}(x^n)$ satisfy the factorization criterion in (4.2.26). That is, let there exist a vector function $T^m(x^n)$ and scalar functions $h(x^n)$ and $g[\theta^m, T^m(x^n)]$ such that

$$f_{\theta^m}(x^n) = g[\theta^m, T^m(x^n)] h(x^n) \qquad \forall \; \theta^m \in \mathscr{E}^m \quad \forall \; x^n \in \Gamma^n \qquad (9.2.18)$$

Then the vector function $T^m(x^n)$ is the sufficient statistics at the data process. Let (9.2.18) hold. Then:

For every θ^m in $\mathscr{E}^m$, the density function $p(\theta^m | x^n)$ in (9.1.3) includes the sequence x^n strictly as a function of the sufficient statistics $T^m(x^n)$ and so does the expected value $E^m(x^n)$ in (9.2.5). Thus, if the conditions in either of cases ii and iii in Lemma 9.2.1 are satisfied, the resulting Bayesian estimates $\hat{\theta}_0^m(x^n)$; $x^n \in \Gamma^n$ include the data sequence x^n strictly as a function of the sufficient statistics $T^m(x^n)$.

Proof: Substituting (9.2.18) in (9.1.2) and (9.1.3), we trivially obtain

$$p(\theta^m | x^n) = \left[\int_{\mathscr{E}^m} d\theta^m \, g[\theta^m, T^m(x^n)] p(\theta^m) \right]^{-1} g[\theta^m, T^m(x^n)] p(\theta^m) \qquad (9.2.19)$$

which trivially includes x^n strictly as a function of $T^m(x^n)$. Everything else follows trivially. ∎

Let us now focus on part ii of Lemma 9.2.1. A special case of a penalty function $c[\theta^m, \hat{\theta}^m(x^n)]$ that clearly satisfies the conditions in (9.2.6) is the quadratic function $\|\theta^m - \hat{\theta}^m(x^n)\|_Q$ in (9.2.4), where Q is some $m \times m$ positive definite matrix. This case deserves special attention, since it has been widely used, it measures absolute deviation of the estimate $\hat{\theta}^m(x^n)$ from the true parameter value θ^m, and it even has its very own name. In addition, the optimal Bayesian estimate is then found explicitly, without any restrictions on the density function $p(\theta^m | x^n)$. We thus express the following lemma.

LEMMA 9.2.2: Let n, $\mathscr{E}^m = \mathbb{C}^m$, and p be given, and let $p(\theta^m | x^n)$; $\theta^m \in \mathbb{C}^m$; $x^n \in \Gamma^n$ be defined, as in (9.1.3). Let Q be some given $m \times m$ positive definite matrix, and let the penalty function $c[\theta^m, \hat{\theta}^m(x^n)]$ be defined as

$$c(\theta^m, \hat{\theta}^m(x^n)) = \|\theta^m - \hat{\theta}^m(x^n)\|_Q = [\theta^m - \hat{\theta}^m(x^n)]^T Q [\theta^m - \hat{\theta}^m(x^n)]^* \quad (9.2.20)$$

The optimal Bayesian estimate $\theta_0^m(x^n)$ is then called the *minimum mean-squared estimate*, it exists, it is unique, and it is equal to the expected value $E^m(x^n)$ in (9.2.5) for every x^n in Γ^n. This is true, without any imposed restrictions on the density function $p(\theta^m | x^n)$.

Proof: For the penalty function in (9.2.20), the quantity $C_{\hat{\theta}^m, p}(x^n)$ in (9.1.5) takes the form

$$C_{\hat{\theta}^m, p}(x^n) = \int_{\mathbb{C}^m} [\theta^m - \hat{\theta}^m(x^n)]^T Q [\theta^m - \hat{\theta}^m(x^n)]^* p(\theta^m | x^n) \, d\theta^m$$

The quadratic form in (9.2.20) is nonnegative and strictly convex with respect to $\hat{\theta}^m(x^n)$ in $\mathbb{C}^m$ and with respect to the metric in (9.2.4). Thus, due to part i in Lemma 9.2.1, the infimum $C_p(x^n) = \inf_{\hat{\theta}^m \in \mathbb{C}^m} C_{\hat{\theta}^m, p}(x^n)$ exists for every x^n in Γ^n and is unique. Furthermore, this infimum and the value $\hat{\theta}_0^m(x^n)$, where it is attained, can be found by setting the gradient $\nabla_{\hat{\theta}^m} C_{\hat{\theta}^m, p}(x^n)$ with respect to $\hat{\theta}^m(x^n)$ equal to zero. That is,

$$\hat{\theta}_0^m(x^n): \nabla_{\hat{\theta}^m} C_{\hat{\theta}^m, p}(x^n) = -2 \int_{\mathbb{C}^m} Q^* [\theta^m - \hat{\theta}_0^m(x^n)] p(\theta^m | x^n) \, d\theta^m = 0$$

which clearly gives

$$\hat{\theta}_0^m(x^n) = \int_{\mathbb{C}^m} d\theta^m \, \theta^m p(\theta^m | x^n) = E^m(x^n)$$

The solution is unique, due to the fact that the matrix Q is positive definite, and the proof of the lemma is now complete. ∎

From Lemma 9.2.2, we observe that different choices of positive definite matrices Q in the penalty function in (9.2.20) induce different weights on the components of the vector $\theta^m - \hat{\theta}^m(x^n)$. From the result in the lemma we con-

9.2 The Bayesian Optimization Problem

clude that the optimal Bayesian estimate is the expected value $E^m(x^n)$ independent of the particular selection of those weights. A special selection of the matrix Q corresponds to the $m \times m$ identity matrix. The latter weighs all the components of the vector $\theta^m - \hat{\theta}^m(x^n)$ equally. We will conclude this section, with some examples.

Example 9.2.1

Let us consider an observation time interval, that is represented by n discrete and scalar observation values. Given n, let s_i^n; $1 \leq i \leq m$, denote m distinct, real, deterministic, and known, n-dimensional column vectors, where m is known and finite. Let $\Theta^m = \{\Theta_i; 1 \leq i \leq m\}$ be a column, m-dimensional vector, whose components are scalar, real, unknown parameters. Let it be assumed that some value $\theta^m = \{\theta_i; 1 \leq i \leq m\}$, of the parameter vector Θ^m is a priori selected, and then, for given n, the signal $\sum_{i=1}^{m} \theta_i s_i^n$, is transmitted through an additive, noisy channel. Let N^n denote the n-dimensional random column vector induced by the channel, and let n^n denote some realization of N^n. Then the random, column, observation vector X^n and its realization x^n are given respectively by the following expressions, given that the value θ^m of the parameter vector Θ^m has been a priori selected.

$$X^n = \sum_{i=1}^{m} \theta_i s_i^n + N^n \qquad x^n = \sum_{i=1}^{m} \theta_i s_i^n + n^n \qquad (9.2.21)$$

Let us now define the $n \times m$ matrix S as

$$S = [s_1^n, s_2^n \cdots s_m^n] \qquad (9.2.22)$$

Then we obtain

$$\sum_{i=1}^{m} \theta_i s_i^n = S \cdot \theta^m \qquad \sum_{i=1}^{m} \Theta_i s_i^n = S \cdot \Theta^m \qquad (9.2.23)$$

where S is a deterministic and known matrix.

Due to (9.2.21) and (9.2.23) and for a specific value θ^m of the a priori selected parameter vector Θ^m we obtain,

$$X^n = S \cdot \theta^m + N^n \qquad (9.2.24)$$

For specific realizations of the random vectors in (9.2.24) we obtain

$$x^n = S \cdot \theta^m + n^n \qquad (9.2.25)$$

Let us now assume that the additive channel is zero mean Gaussian, and that the a priori value of the vector parameter Θ^m is also controlled by a zero mean, Gaussian distribution. In paraticular given n, let R_n be a given, known $n \times n$ autocovariance matrix, and let Λ_m be another known, $m \times m$ autocovariance matrix such that

N^n: Gaussian, zero mean vector, with autocovariance matrix R_n.

Θ^m: Gaussian, zero mean vector, with autocovariance matrix Λ_m.

Then, from the model in (9.2.23) and (9.2.25) and using the same notation previously used in this section, we clearly obtain

$$f_{\theta^m}(x^n) = (2\pi)^{-n/2}|R_n|^{-1/2} \exp\left\{-\frac{1}{2}[x^n - S\theta^m]^T R_n^{-1}[x^n - S\theta^m]\right\}$$

$$= (2\pi)^{-n/2}|R_n|^{-1/2} \exp\left\{-\frac{1}{2}x^{n^T} R_n^{-1} x^n\right\}$$

$$\cdot \exp\left\{\theta^{m^T} S^T R_n^{-1} x^n - \frac{1}{2}\theta^{m^T} S^T R_n^{-1} S\theta^m\right\} \quad (9.2.26)$$

$$p(\theta^m) = (2\pi)^{-m/2}|\Lambda_m|^{-1/2} \exp\left\{-\frac{1}{2}\theta^{m^T}\Lambda_m^{-1}\theta^m\right\} \quad (9.2.27)$$

where it is assumed that the selection of θ^m is independent from the channel statistics, and where $|\ |$ denotes determinant.

By direct substitution of the (9.2.26) and (9.2.27) in (9.1.2) and (9.1.3), we obtain

$$f(x^n) = \int_{R^m} d\theta^m\, f_{\theta^m}(x^n) p(\theta^m)$$

$$= (2\pi)^{-n/2}|R_n|^{-1/2} \exp\left\{-\frac{1}{2}x^{n^T} R_n^{-1} x^n\right\} \cdot \int_{R^m} (2\pi)^{-m/2}|\Lambda_m|^{-1/2}$$

$$\cdot \exp\left\{-\frac{1}{2}\theta^{m^T}[I_m + S^T R_n^{-1} S\Lambda_m]\Lambda_m^{-1}\theta^m + x^{n^T} R_n^{-1} S\theta^m\right\} d\theta^m \quad (9.2.28)$$

$$p(\theta^m|x^n) = \left[\int_{R^m} \exp\left\{-\frac{1}{2}\theta^{m^T}[I_m + S^T R_n^{-1} S\Lambda_m]\Lambda_m^{-1}\theta^m + x^{n^T} R_n^{-1} S\theta^m\right\} d\theta^m\right]^{-1}$$

$$\cdot \exp\left\{-\frac{1}{2}\theta^{m^T}[I_m + S^T R_n^{-1} S\Lambda_m]\Lambda_m^{-1}\theta^m + x^{n^T} R_n^{-1} S\theta^m\right\} \quad (9.2.29)$$

where I_m denotes the $m \times m$ identity matrix.

Let us now express

$$x^{n^T} R_n^{-1} S\theta^m = x^{n^T} R_n^{-1} S\Lambda_m[I_m + S^T R_n^{-1} S\Lambda_m]^{-1}[I_m + S^T R_n^{-1} S\Lambda_m]\Lambda_m^{-1}\theta^m \quad (9.2.30)$$

and let us denote

$$A^{m^T}(x^n) = x^{n^T} R_n^{-1} S\Lambda_m[I_m + S^T R_n^{-1} S\Lambda_m]^{-1}$$

$$M_m = \Lambda_m[I_m + S^T R_n^{-1} S\Lambda_m]^{-1} \quad (9.2.31)$$

9.2 The Bayesian Optimization Problem

where $A^m(x^n)$ is an m-dimensional column vector that is a function of the observation sequence x^n, and where M_m is an $m \times m$ constant nonsingular matrix. Via (9.2.31) and (9.2.30), and by substitution in (9.2.29), we then obtain

$$x^{nT} R_n^{-1} S \theta^m = A^{mT}(x^n) M_m^{-1} \theta^m \qquad (9.2.32)$$

$$p(\theta^m | x^n) = \left[\int_{R^m} \exp\left\{ -\frac{1}{2} [\theta^m - A^m(x^n)]^T M_m^{-1} [\theta^m - A^m(x^n)] \right\} d\theta^m \right]^{-1}$$

$$\cdot \exp\left\{ -\frac{1}{2} [\theta^m - A^m(x^n)]^T M_m^{-1} [\theta^m - A^m(x^n)] \right\}$$

$$= (2\pi)^{-m/2} |M_m|^{-1/2} \exp\left\{ -\frac{1}{2} [\theta^m - A^m(x^n)]^T M_m^{-1} [\theta^m - A^m(x^n)] \right\}$$

$$(9.2.33)$$

From (9.2.33) we conclude that, as expected, the conditional density function $p(\theta^m | x^n)$ is Gaussian. We also conclude that the expected value of this distribution and its autocovariance matrix are respectively equal to $A^m(x^n)$ and M_m in (9.2.31). From (9.2.26) and via the factorization criterion in Chapter 4, we conclude that the sufficient statistics at the Gaussian additive channel, in the present case, is the m-dimensional column vector $S^T R_n^{-1} x^n$. The vector $A^m(x^n)$ in (9.2.31) clearly includes the observation sequence x^n strictly as a function of the sufficient statistics $S^T R_n^{-1} x^n$; thus, the conditional distribution $p(\theta^m | x^n)$ includes x^n strictly in the same form as well. The expected value $E^m(x^n)$ in (9.2.5) is finally given here by

$$E^m(x^n) = A^m(x^n) = [I_m + S^T R_n^{-1} S \Lambda_m]^{-1} \Lambda_m S^T R_n^{-1} x^n = M_m^T S^T R_n^{-1} x^n$$

$$(9.2.34)$$

Since the density function $p(\theta^m | x^n)$ in (9.2.33) is symmetric around its mean $E^m(x^n)$ in (9.2.34), then for any penalty function $c[\theta^m, \hat{\theta}^m(x^n)]$ that satisfies conditions (9.2.6) in Lemma 9.2.1, the optimal Bayesian parameter estimate $\hat{\theta}_0^m(x^n)$ is the expected value $E^m(x^n)$. This is due to part ii in Lemma 9.2.1. This expected value is a linear transformation of the observation vector x^n. Therefore, the random vector $E^m(X^n)$ is Gaussian at every value θ^m of the parameter vector Θ^m. Given θ^m, the expected value $E_{\theta^m}^m = E\{E^m(X^n) | \theta^m\}$ and the covariance matrix C_m of the random vector $E^m(X^n)$ are respectively given by

$$E_{\theta^m}^m = M_m^T S^T R_n^{-1} S \theta^m = [I_m + S^T R_n^{-1} S \Lambda_m]^{-1} \Lambda_m S^T R_n^{-1} S \theta^m \qquad (9.2.35)$$

$$C_m = M_m^T S^T R_n^{-1} S M_m = [I_m + S^T R_n^{-1} S \Lambda_m]^{-1} \Lambda_m S^T R_n^{-1} S \Lambda_m [I_m + S^T R_n^{-1} S \Lambda_m]^{-1}$$

$$(9.2.36)$$

where, due to the symmetry of the matrices $S^T R_n^{-1} S$, R_n^{-1}, and Λ_m, we have

$$S^T R_n^{-1} S \Lambda_m = [S^T R_n^{-1} S \Lambda_m]^T = \Lambda_m^T S^T R_n^{-1} S = \Lambda_m S^T R_n^{-1} S \qquad (9.2.37)$$

Let us now assume that each of the diagonal components of the matrix $\Lambda_m S^T R_n^{-1} S$ is large compared to the value one. Then we can write

$$I_m + S^T R_n^{-1} S \Lambda_m \simeq S^T R_n^{-1} S \Lambda_m = \Lambda_m S^T R_n^{-1} S$$

and

$$[I_m + S^T R_n^{-1} S \Lambda_m]^{-1} \simeq [\Lambda_m S^T R_n^{-1} S]^{-1} = (S^T R_n^{-1} S)^{-1} \Lambda_m^{-1} \quad (9.2.38)$$

Substituting (9.2.38) in (9.2.35) and (9.2.36), we then obtain

$$E_{\hat{\theta}^m}^m \simeq \theta^m \qquad C_m \simeq (S^T R_n^{-1} S)^{-1} \quad (9.2.39)$$

We thus conclude that if the diagonal components of the matrix $\Lambda_m S^T R_n^{-1} S$ are all large as compared to the value one, then the conditional expectation $E^m(x^n)$ Bayesian estimate is unbiased, with variance $V_n(\hat{\theta}^m) = E\{\|E^m(X^n) - E_{\hat{\theta}^m}^m\|^2\} \simeq \operatorname{tr}(S^T R_n^{-1} S)^{-1}$, where tr denotes trace. Thus, if the diagonal components of the matrix $S^T R_n^{-1} S$ are also all large compared to the value one, then, the estimate $E^m(x^n)$ has also small variance, in addition to being unbiased.

We conclude this example, by observing that the maximum a posteriori estimate, in part iii, Lemma 9.2.1 is here the expected value $E^m(x^n)$ again. This is so, because the unique supremum of the density function $p(\theta^m | x^n)$ in (9.2.31) is attained at its expected value $A^m(x^n) = E^m(x^n)$.

Example 9.2.2

A special case of Example 9.2.1 deserves special attention. It corresponds to the estimation of a scalar real parameter, added to a sequence of independent and identically distributed random variables. This case has been widely studied in the literature, especially for Gaussian data and parameter statistics. Let s_I^n denote the column n-dimensional vector, whose components are all equal to one. Then the model in this example is given below, where N^n is a vector of independent, zero mean, and variance σ^2 Gaussian random variables, and Θ is a zero mean, variance v^2 Gaussian random variable that is independent of the noise vector N^n

$$X^n = \Theta s_I^n + N^n \quad (9.2.40)$$

The conditional density function $p(\theta | x^n)$ for the present model is clearly given via (9.2.33) in Example 9.2.1, if the following substitutions are made.

$$\begin{aligned} R_n &\to \sigma^2 I_n \\ \Lambda_m &\to v^2 \\ S &\to s_I^n \\ I_m &\to 1 \end{aligned} \quad (9.2.41)$$

9.2 The Bayesian Optimization Problem

We then trivially obtain,

$$p(\theta|x^n) = (2\pi)^{-1/2} \frac{(\sigma^2 + nv^2)^{1/2}}{v\sigma} \exp\left\{-\frac{1}{2}\frac{\sigma^2 + nv^2}{\sigma^2 v^2}\left[\theta - (\sigma^2 v^{-2} + n)^{-1}\sum_{i=1}^{n} x_i\right]^2\right\}$$
(9.2.42)

Both the optimal Bayesian and the maximum a posteriori estimates are now given by the following expression, where σ^2 and v^2 are both assumed finite.

$$\hat{\theta}_0(x^n) = \left(\frac{\sigma^2}{nv^2} + 1\right)^{-1} n^{-1} \sum_{i=1}^{n} x_i \simeq n^{-1} \sum_{i=1}^{n} x_i = m(x^n) \qquad (9.2.43)$$

$$n \gg \frac{\sigma^2}{v^2}$$

The estimate $\hat{\theta}_0(x^n)$ thus approaches the empirical mean $m(x^n) = n^{-1}\sum_{i=1}^{n} x_i$ as n increases. The empirical mean is trivially unbiased, since

$$E\{m(X^n)|\theta\} = n^{-1}\sum_{i=1}^{n} E\{X_i|\theta\} = n^{-1}(n\theta) = \theta \qquad (9.2.44)$$

The variance of the empirical mean, in the present model, is (due to the independence of the variables $\{X_i\}$) equal to $n^{-1}\sigma^2$. Thus, as n increases, the Bayesian and maximum a posteriori estimates become unbiased, and their variance around the true parameter value tends to zero. Those nice qualities hold strictly under the Gaussian assumption. When deviations from the Gaussian model occur, the performance of the empirical mean deteriorates drastically. We will discuss this point further, in Chapter 10.

Example 9.2.3

Let $X^n = \{X_i; 1 \leq i \leq n\}$ be a sequence of independent, and uniformly distributed in $[0, \theta]$ random variables, where θ is some positive parameter. Let the parameter θ be a priori selected uniformly in an interval $[a, b]$, where $a > 0$ and $b > a$. That is,

$$f_\theta(x^n) = \theta^{-n} U\left(\theta - \max_i x_i\right) U\left(\min_i x_i\right) \qquad (9.2.42)$$

$$p(\theta) = [b-a]^{-1} \qquad a \leq \theta \leq b \qquad (9.2.46)$$

where $\max_i x_i$ is the sufficient statistics and

$$U(x) = \begin{cases} 1 & x \geq 0 \\ 0 & x < 0 \end{cases}$$

From (9.2.45) and (9.2.46) we obtain

$$f(x^n) = \int_a^b d\theta\, f_\theta(x^n) p(\theta)$$

$$= [b-a]^{-1} U\left(\min_i x_i\right) \left[\int_{\max(a,\max_i x_i)}^b \theta^{-n}\, d\theta\right] U\left[b - \max\left(a, \max_i x_i\right)\right]$$

$$= \frac{U\left(\min_i x_i\right)}{(b-a)(n-1)} U\left[b - \max\left(a, \max_i x_i\right)\right]$$

$$\cdot \left\{\max^{-(n-1)}\left(a, \max_i x_i\right) - b^{-(n-1)}\right\} \tag{9.2.47}$$

$$p(\theta|x^n) = \frac{f_\theta(x^n) p(\theta)}{f(x^n)}$$

$$= (n-1) \frac{\theta^{-n} U\left(\theta - \max_i x_i\right)}{U\left[b - \max\left(a, \max_i x_i\right)\right]\left\{\max^{-(n-1)}\left(a, \max_i x_i\right) - b^{-(n-1)}\right\}} \tag{9.2.48}$$

$$E(x^n) = \int_a^b d\theta\, \theta p(\theta|x^n)$$

$$= \frac{n-1}{U\left[b - \max\left(a, \max_i x_i\right)\right]\left\{\max^{-(n-1)}\left(a, \max_i x_i\right) - b^{-(n-1)}\right\}}$$

$$\cdot U\left[b - \max\left(a, \max_i x_i\right)\right] \int_{\max(a,\max_i x_i)}^b \theta^{-(n-1)}\, d\theta$$

$$= \frac{n-1}{n-2} \frac{\max^{-(n-2)}\left(a, \max_i x_i\right) - b^{-(n-2)}}{\max^{-(n-1)}\left(a, \max_i x_i\right) - b^{-(n-1)}} \tag{9.2.49}$$

The density function $p(\theta|x^n)$ in (9.2.48) has a unique supremum, attained at

$$\theta = \max\left(a, \max_i x_i\right)$$

Thus,

$$\hat{\theta}_0(x^n) = \max\left(a, \max_i x_i\right)$$

9.2 The Bayesian Optimization Problem

is the maximum a posteriori estimate here. If the penalty function $c[\theta, \hat{\theta}(x^n)] = [\theta - \hat{\theta}(x^n)]^2$ is given, the optimal Bayesian estimate is, instead, the expected value $E(x^n)$ in (9.2.49) (via Lemma 9.2.2). For asymptotically large sample sizes n the expected value $E(x^n)$ converges to

$$\max\left(a, \max_i x_i\right)$$

Thus, asymptotically $(n \to \infty)$, the mean-squared optimal Bayesian estimate and the maximum a posteriori estimate become identical. Let us now consider the expected value of the estimate,

$$\hat{\theta}_0(x^n) = \max\left(a, \max_i x_i\right)$$

conditioned on some value θ in $[a, b]$. We first derive some useful probabilities.

$$\Pr\left(\max_i X_i \leq y;\ 1 \leq i \leq n \big| \theta\right) = \Pr(X_i \leq y;\ \forall i;\ 1 \leq i \leq n | \theta)$$

$$= \prod_{i=1}^{n} \Pr(X_i \leq y | \theta)$$

$$= \begin{cases} 0 & y \leq 0 \\ y^n \theta^{-n} & 0 \leq y \leq \theta \\ 1 & y \geq \theta \end{cases} \quad (9.2.50)$$

$$\frac{\partial}{\partial y} \Pr\left(\max_i X_i \leq y;\ 1 \leq i \leq n \big| \theta\right) = n y^{n-1} \theta^{-n} U(\theta - y) U(y) \quad (9.2.51)$$

Due to (9.2.50) and (9.2.51), we now obtain the expected value of the estimate

$$\hat{\theta}_0(x^n) = \max\left(a, \max_i x_i\right)$$

conditioned on some value θ as

$$E\left\{\max\left(a, \max_i X_i\right) \big| \theta\right\} = a \Pr\left(\max_i X_i \leq a;\ 1 \leq i \leq n | \theta\right)$$

$$+ \int_{u \geq a} u\, d\Pr\left(\max_i X_i \leq u;\ 1 \leq i \leq n | \theta\right)$$

$$= \frac{a^{n+1}}{\theta^n} + \int_a^\theta n\theta^{-n} u^n\, du$$

$$= \frac{a^{n+1}}{\theta^n} + \frac{n}{n+1}\theta - \frac{n}{n+1}\frac{a^{n+1}}{\theta^n}$$

$$= a\left(\frac{a}{\theta}\right)^n (n+1)^{-1} + \frac{n}{n+1}\theta \quad (9.2.52)$$

From (9.2.52) we easily conclude that as n increases, the expected value

$$E\left\{\max\left(a, \max_i X_i\right)\middle|\theta\right\}$$

converges to the value θ. Thus, asymptotically ($n \to \infty$) the maximum a posteriori estimate becomes unbiased and so then does the optimal mean-squared Bayesian estimate since the two estimates become asymptotically indentical. We similarly obtain,

$$E\left\{\max^2\left(a, \max_i X_i\right)\middle|\theta\right\} = a^2 \Pr\left(\max_i X_i \leq a; 1 \leq i \leq n\middle|\theta\right)$$

$$+ \int_{u \geq a} u^2 d \Pr\left(\max_i X_i \leq u; 1 \leq i \leq n\middle|\theta\right)$$

$$= \frac{a^{n+2}}{\theta^n} + \int_a^\theta n\theta^{-n} u^{n+1} \, du$$

$$= 2a^2 \left(\frac{a}{\theta}\right)^n (n+2)^{-1} + \frac{n}{n+2}\theta^2 \qquad (9.2.53)$$

$$\text{Var}\left\{\max\left(a, \max_i X_i\right)\middle|\theta\right\} = E\left\{\max^2\left(a, \max_i X_i\right)\middle|\theta\right\}$$

$$- E^2\left\{\max\left(a, \max_i X_i\right)\middle|\theta\right\}$$

$$= 2a^2 \left(\frac{a}{\theta}\right)^n (n+2)^{-1} + \frac{n}{n+2}\theta^2$$

$$= \left[a\left(\frac{a}{\theta}\right)^n (n+1)^{-1} + \frac{n}{n+1}\theta\right]^2$$

$$= \frac{n\theta^2}{(n+2)(n+1)^2} - \frac{a^2}{(n+1)^2}\left(\frac{a}{\theta}\right)^{2n}$$

$$- \frac{2na\theta}{(n+1)^2}\left(\frac{a}{\theta}\right)^n \left[1 - \frac{(n+1)^2}{n(n+2)}\frac{a}{\theta}\right] \xrightarrow[n \to \infty]{} 0$$

$$(9.2.54)$$

Thus, asymptotically ($n \to \infty$), the variance of the maximum a posteriori and optimal mean-squared Bayesian estimates, converges to zero.

9.3 THE LINEAR MEAN-SQUARED SCHEME

As expressed by Lemma 9.2.2, when the penalty function is the quadratic form in (9.2.20), then, without any restrictions on the distribution $p(\theta^m|x^n)$, the optimal Bayesian estimate is the expected value $E^m(x^n)$ in (9.2.5). When the par-

9.3 The Linear Mean-Squared Scheme

ameter vector Θ^m and the observation random vector X^n are jointly Gaussian, the expected value $E^m(x^n)$ is linear, as a function of the vector realization x^n. This is a familiar property of the Gaussian statistics that can be observed in Examples 9.2.1 and 9.2.2. When the joint statistics of the random vectors Θ^m and X^n is other than Gaussian, however, the minimum mean-squared Bayesian estimate $E^m(x^n)$ is, in general, a nonanalytic function of the data vector x^n. Thus, while $E^n(x^n)$ is well known in principle, its evaluation at each given vector point x^n is, in general, practically cumbersome and computationally inefficient, especially for large sample sizes n. To resolve this problem, a compromise is reached via the introduction of a suboptimal Bayesian scheme for the quadratic penalty function in (9.2.20). In particular, only the class of linear parameter estimates is considered, and a member of this class is sought that minimizes the expected quadratic penalty. The latter estimate is called the linear mean-squared estimate; it is introduced formally in Definition 9.3.1.

DEFINITION 9.3.1: Let n, $\mathscr{E}^m = \mathbb{C}^m$, and $p(\theta^m)$; $\theta^m \in \mathbb{C}^m$ be given. Let Q be some given $m \times m$ positive definite matrix, and let the penalty function $c[\theta^m, \hat{\theta}^m(x^n)] = [\theta^m - \hat{\theta}^m(x^n)]^T Q [\theta^m - \hat{\theta}^m(x^n)]^*$ be adopted, where $x^n \in R^n$. Consider the class $\mathscr{A}_{m,n}$ of all the $m \times n$ constant matrices A whose elements take values in the complex plane $\mathbb{C}$ and the class $\mathscr{C}_m$ of all the m-dimensional, constant column vectors C whose components take values in $\mathbb{C}$. Let A^0 and C^0 be such that, for $C^0 \in \mathscr{C}_m$, $A^0 \in \mathscr{A}_{m,n}$,

$$E\{[\Theta^m - A^0 X^n - C^0]^T Q [\Theta^m - A^0 X^n - C^0]^*\}$$
$$= \int_{\mathbb{C}^m} d\theta^m\, p(\theta^m) \int_{R^n} dx^n\, f_{\theta^m}(x^n) [\theta^m - A^0 x^n - C^0]^T Q [\theta^m - A^0 x^n - C^0]^*$$
$$= \inf_{A \in \mathscr{A}_{m,n},\, C \in \mathscr{C}_m} E\{[\Theta^m - A X^n - C]^T Q [\Theta^m - A X^n - C]^*\} \quad (9.3.1)$$

Then, the linear, with respect to x^n, vector $A^0 x^n + C^0$; $x^n \in R^n$ is called the *optimal linear mean-squared estimate*, of the true vector parameter θ^m. ∎

If $m = 1$, Θ is scalar. Then, the matrix A^0 and the vector C^0 in Definition 9.3.1, reduce, respectively, to an n-dimensional row vector and to a scalar. For each given sequence x^n the optimal linear mean-squared estimate is then a scalar. When the parameter vector Θ^m and the random observation vector X^n are jointly Gaussian, the optimal linear mean-squared estimate and the optimal mean-squared Bayesian estimate coincide. We now express a lemma that provides the matrix A^0 and the vector C^0 for the optimal linear minimum mean-squared estimate.

LEMMA 9.3.1: The matrix A^0 and the vector C^0 in Definition 9.3.1 are found from the solution of the following linear system.

$$A^0 E\{X^n\} + C^0 = E\{\Theta^m\} \quad (9.3.2)$$

$$E\{[\Theta^m - A^0 X^n - C^0]^* X^{n^T}\} = 0 \quad (9.3.3)$$

The equation in (9.3.3) is called the *orthogonality principle*. The solution of the above linear system provides the unique optimal linear mean-squared estimate.

Proof: Since the space $\mathscr{A}_{m,n} \times \mathscr{C}_m$ is compact with respect to the quadratic metric, and since the quadratic penalty function is strictly convex, in the same space, with respect to both C and A the infimum in (9.3.1) is unique, and it is attained in $A_{m,n} \times \mathscr{C}_m$. The values C^0 and A^0 that attain this infimum are then found, by setting the corresponding gradients of the expected value $E\{[\theta^m - AX^n - C]^T Q[\theta^m - AX^n - C]\}$ equal to zero. Setting the gradient with respect to the vector C equal to zero, results in (9.3.2), via familiar results in linear algebra [Bellman (1970)]. Similarly, setting the matrix gradient with respect to the matrix A equal to zero results in (9.3.3). ∎

Now let M_p^m and $M_{f,p}^n$ denote respectively the expected values of the parameter vector Θ^m and the observation vector X^n as they are induced by the density functions f and p. Let $R_{m,p}$ and $R_{n,f,p}$ denote respectively the autocovariance matrices of the vectors Θ^m and X^n as induced by f and p. Let $R_{m,n,f,p}$ denote the cross covariance matrix of the vectors Θ^m and X^n. That is,

$$M_p^m = \int_{\mathscr{C}^m} \theta^m p(\theta^m) \, d\theta^m$$

$$M_{f,p}^n = \int_{\mathscr{C}^m} d\theta^m \, p(\theta^m) \int_{\mathscr{C}^n} x^n f_{\theta^m}(x^n) \, dx^n$$

$$R_{m,p} = \int_{\mathscr{C}^m} [\theta^m - M_p^m][\theta^m - M_p^m]^{*T} p(\theta^m) \, d\theta^m \qquad (9.3.4)$$

$$R_{n,f,p^m} = \int_{\mathscr{C}^m} d\theta^m \, p(\theta^m) \int_{\mathscr{C}^n} [x^n - M_{f,p}^n][x^n - M_{f,p}^n]^{*T} f_{\theta^m}(x^n) \, dx^n$$

$$R_{m,n,f,p} = \int_{\mathscr{C}^m} d\theta^m \, p(\theta^m) \int_{\mathscr{C}^n} [\theta^m - M_p^m][x^n - M_{f,p}^n]^{*T} f_{\theta^m}(x^n) \, dx^n$$

The solution of the linear system in (9.3.2) and (9.3.3) is then easily found to be given by the following expressions, where (-1) denotes inverse, and where the autocovariance matrix $R_{n,f,p}$ is nonsingular.

$$A^0 = R_{m,n,f,p} R_{n,f,p}^{-1} \qquad (9.3.5)$$

$$C^0 = M_p^m - R_{m,n,f,p} R_{n,f,p}^{-1} M_{f,p}^n \qquad (9.3.6)$$

Due to (9.3.5) and (9.3.6), we conclude easily that the expected value of the optimal linear mean-squared estimate $A^0 X^n + C^0$ equals the expected value M_p^m of the parameter vector Θ^m. The autocovariance matrix of the estimate $A^0 X^n + C^0$ is easily found by $A^0 R_{n,f,p} A^{0*T} = R_{m,n,f,p} \cdot [R_{n,f,p}^{-1}]^{*T} R_{m,n,f,p}^{*T}$. We note that if the vector parameter Θ^m and the observation vector X^n are statistically independent, then the distribution $f_{\theta^m}(x^n)$ is independent of θ^m; thus, the cross covariance matrix $R_{m,n,f,p}$ in (9.3.4) is then equal to the zero matrix. Then the matrix A^0 in (9.3.5) is also the zero matrix, and the vector, C^0 in (9.3.6), equals the expected value M_p^m of the vector parameter Θ^m. Therefore, if the

9.3 The Linear Mean-Squared Scheme

parameter θ^m and the vector X^n are mutually independent, the optimal linear mean-squared parameter estimate is independent of the observation vector x^n and equals the mean vector M_p^m. Indeed, the observation vector x^n provides then no information about the true value θ^m of the vector parameter, and the best parameter guess is then represented by the expected value M_p^m. We will complete this section with two examples.

Example 9.3.1

Let us consider the case, where the vector parameter Θ^m and the observation sequence X^n are jointly Gaussian. Then, the optimal mean-squared Bayesian estimate is the expected value $E^m(x^n)$ in (9.2.5), and it is known in advance to be linear as a function of the vector value x^n. Therefore, $E^m(x^n)$ is then identical to the optimal linear mean-squared estimate, and it is convenient that $E^m(x^n)$ be then found directly from (9.3.5) and (9.3.6). To quantify the above statements, let us revisit Example 9.2.1. Clearly, the parameter Θ^m and the observation vector X^n are jointly Gaussian there. So, the optimal mean-squared Bayesian estimate $\hat{\theta}_0^m(x^n)$ is then given by

$$\hat{\theta}_0^m(x^n) = A^0 x^n + C^0 \tag{9.3.7}$$

where,

$$A^0 = R_{m,n,f,p} \cdot R_{n,f,p}^{-1}$$
$$C^0 = M_p^m - R_{m,n,f,p} \cdot R_{n,f,p}^{-1} \cdot M_{f,p}^n \tag{9.3.8}$$
$$M_p^m = 0$$

$$M_{f,p}^n = E\left\{\sum_{i=1}^m \Theta_i s_i^n + N^n\right\} = \sum_{i=1}^m E\{\Theta_i\} s_i^n + E\{N^n\} = 0 \tag{9.3.9}$$

$$R_{n,f,p} = E\{X^n X^{n*T}\} = E\{[S\Theta^m + N^n][S\Theta^m + N^n]^{*T}\}$$
$$= SE\{\Theta^m \Theta^{m*T}\}S^{*T} + E\{N^n N^{n*T}\}$$
$$= S\Lambda_m S^{*T} + R_n = S\Lambda_m S^T + R_n \tag{9.3.10}$$

$$R_{m,n,f,p} = E\{\Theta^m X^{n*T}\} = E\{\Theta^m [S\Theta^m + N^n]^{*T}\}$$
$$= E\{\Theta^m \Theta^{m*T}\}S^{*T} = \Lambda_m S^T \tag{9.3.11}$$

Substituting (9.3.9), (9.3.10), and (9.3.11) in (9.3.8) and (9.3.7), we obtain

$$C^0 = 0$$
$$A^0 = \Lambda_m S^T [S\Lambda_m S^T + R_n]^{-1} \tag{9.3.12}$$
$$= [I_m + S^T R_n^{-1} S\Lambda_m]^{-1} \Lambda_m S^T R_n^{-1}$$

and by substitution of (9.3.12) in (9.3.7), we obtain, as in (9.2.32),

$$\hat{\theta}_0^m(x^n) = \Lambda_m S^T [S\Lambda_m S^T + R_n]^{-1} x^n = [I_m + S^T R_n^{-1} S\Lambda_m]^{-1} \Lambda_m S^T R_n^{-1} x^n \tag{9.3.13}$$

Let us revisit Example 9.2.2, and find the optimal mean-squared Bayesian estimate, following the same method as in Example 9.3.1. Then we only need to substitute the values in (9.2.41) into (9.3.13) above. The result of this substitution is trivially the expression in (9.2.43).

Example 9.3.2

Let us consider Example 9.2.3, and let us then search for the optimal linear mean-squared estimate. Equivalently, we will compute the quantities in (9.3.4), as they evolve in the present example. We first compute the expected values

$$E\{\Theta\} = \frac{1}{b-a}\int_a^b \theta\,d\theta = \frac{b+a}{2} \quad (9.3.14)$$

$$E\{X_i|\theta\} = \frac{1}{\theta}\int_0^\theta u\,du = \frac{\theta}{2} \quad (9.3.15)$$

$$E\{X_i\} = E\left\{\frac{\Theta}{2}\right\} = \frac{b+a}{4} \quad (9.3.16)$$

$$E\left\{\left[X_i - \frac{b+a}{4}\right]\left[X_j - \frac{b+a}{4}\right]\right\} = 0 \quad i \neq j \quad (9.3.17)$$

$$E\{X_i^2|\theta\} = \frac{1}{\theta}\int_0^\theta u^2\,du = \frac{\theta^2}{3}$$

$$E\{X_i^2\} = E\left\{\frac{\Theta^2}{3}\right\} = \frac{1}{3(b-a)}\int_a^b \theta^2\,d\theta = \frac{b^2+ab+a^2}{9}$$

$$E\left\{\left[X_i - \frac{b+a}{4}\right]^2\right\} = \frac{b^2+ab+a^2}{9} - \frac{(b+a)^2}{16} = \frac{7b^2+7a^2-2ab}{144} \quad (9.3.18)$$

$$E\{\Theta X_i\} = E\left\{\frac{\Theta^2}{2}\right\} = \frac{b^2+ab+a^2}{6}$$

$$E\left\{\left[\Theta - \frac{b+a}{2}\right]\left[X_i - \frac{b+a}{4}\right]\right\} = \frac{b^2+ab+a^2}{6} - \frac{(b+a)^2}{8} = \frac{(b-a)^2}{24} \quad (9.3.19)$$

Let us denote by I_n the n-dimensional identity matrix, and let s_l^n be the n-dimensional column vector whose components are all equal to one. For notational simplicity, let us change the notation in (9.3.4) to

$$M_p^m \to m$$

$$M_{f,p}^n \to M^n$$

$$R_{n,f,p} \to R_n$$

$$R_{m,n,f,p} \to R_{m,n}$$

9.3 The Linear Mean-Squared Scheme

From the above notation, and from (9.3.14) through (9.3.19), we then find

$$m = \frac{b+a}{2}$$

$$M^n = \frac{b+a}{4} s_I^n$$

$$R_n = \frac{7b^2 + 7a^2 - 2ab}{144} I_n \qquad (9.3.20)$$

$$R_{m,n} = \frac{(b-a)^2}{24} s_I^{n^T}$$

Substituting the expressions in (9.3.20), in (9.3.5) and (9.3.6), we obtain,

$$A^0 = \frac{6(b-a)^2}{7b^2 + 7a^2 - 2ab} s_I^{n^T}$$

$$C^0 = \frac{b+a}{2} - \frac{3}{2}\frac{(b+a)(b-a)^2}{7b^2 + 7a^2 - 2ab} n \qquad (9.3.21)$$

$$A^0 x^n + C^0 = \frac{6(b-a)^2}{7b^2 + 7a^2 - 2ab} \sum_{i=1}^{n} x_i + \frac{b+a}{2} - \frac{3}{2}\frac{(b+a)(b-a)^2}{7b^2 + 7a^2 - 2ab} n$$

$$= n\frac{6(b-a)^2}{7b^2 + 7a^2 - 2ab}\left[n^{-1}\sum_{i=1}^{n} x_i - \frac{b+a}{4}\right] + \frac{b+a}{2} \qquad (9.3.22)$$

From (9.3.22), we observe that the linear mean-squared parameter estimate includes the data sequence strictly in the form of the empirical mean, $n^{-1}\sum_{i=1}^{n} x_i$. Strictly from (9.3.15), and due to (9.3.22), we obtain

$$E\{A^0 X^n + C^0 | \theta\} = n\frac{3(b-a)^2}{7b^2 + 7a^2 - 2ab}\left[\theta - \frac{b+a}{2}\right] + \frac{b+a}{2} \qquad (9.3.23)$$

From (9.3.23), we observe that in contrast to the optimal mean-squared Bayesian estimate studied in Example 9.2.3, the bias of the linear mean-squared estimate increases monotonically with the sample size n unless $\theta = 2^{-1}(b+a)$. In the latter case the bias equals $|\theta - (b+a)/2|$ for every value n. This poor performance of the linear mean-squared estimate, in terms of bias, is due to the fact that its deviation from the optimal mean-squared Bayesian estimate increases monotonically with the sample size n. Indeed, the Bayesian estimate converges asymptotically ($n \to \infty$) to the statistics,

$$\lim_{n\to\infty} \max(a, \max_i x_i) \to b$$

while the linear mean-squared estimate converges then to the value $2^{-1}(b+a)$, since due to ergodicity $n^{-1}\sum_{i=1}^{n} x_i$ converges then to the expected value in

(9.3.16), and $\max_i x_i$ converges to the value b. The asymptotic deviation of the two estimates increases with increasing difference $b - a$.

As is clear from the results and the discussion in Example 9.3.2, the linear mean-squared estimate, although computationally efficient, may deviate substantially from the optimal mean-squared Bayesian estimate. In general, the further from Gaussian the statistics are, the larger this deviation is.

9.4 UNBIASED MEAN-SQUARED ESTIMATES

As it should be evident from the examples in Sections 9.2 and 9.3, both the optimal mean-squared Bayesian, and the optimal linear mean-squared estimates are not necessarily unbiased. On the other hand, the design of unbiased parameter estimates is frequently desirable. Then, if Θ^m represents a vector parameter that takes values in some space, $\mathscr{E}^m$, and if, $\hat{\theta}^m(x^n)$; $x^n \in \Gamma^n$, is some estimate of Θ^m it is required that the following condition be satisfied.

$$E_{\theta^m}\{\hat{\theta}^m(X^n)\} = \int_{\Gamma^n} \hat{\theta}^m(x^n) f_{\theta^m}(x^n) \, dx^n = \theta^m \qquad \forall \, \theta^m \in \mathscr{E}^m \qquad (9.4.1)$$

The condition in (9.4.1) represents a constraint. Thus, the general Bayesian optimization problem in (9.2.1) now takes the modified form

Given n, $\mathscr{E}^m$, p, and $c(\cdot, \cdot)$ as in (9.1.1), minimize

$$c(\theta^m, p) = \int_{\Gamma^n} f(x^n) C_{\hat{\theta}_m, p}(x^n) \, dx^n$$

Subject to
$$\hat{\theta}^m(x^n) \in \mathscr{E}^m \qquad \forall \, x^n \in \Gamma^n \qquad (9.4.2)$$
$$E_{\theta^m}\{\hat{\theta}^m(X^n)\} = \int_{\Gamma^n} \hat{\theta}^m(x^n) f_{\theta^m}(x^n) \, dx^n = \theta^m \qquad \forall \, \theta^m \in \mathscr{E}^m$$

The optimization problem in (9.4.2) may not have a solution. If it does, its solution $\hat{\theta}^m_*(x^n)$; $x^n \in \Gamma^n$ is called the *unbiased optimal Bayesian estimate*. To avoid complicated derivations, we will limit ourselves to quadratic penalty functions, and we will thus search for unbiased optimal mean-squared estimates. That is, given some $m \times m$ positive definite matrix Q we will adopt the penalty function

$$c[\theta^m, \hat{\theta}^m(x^n)] = [\theta^m - \hat{\theta}^m(x^n)]^T Q [\theta^m - \hat{\theta}^m(x^n)]^* \qquad (9.4.3)$$

We now express a theorem, where $\mathbb{C}$ and $\mathbb{C}^k$ signify respectively the complex plane and k multiples of this plane.

THEOREM 9.4.1: Let n, $\mathscr{E}^m = \mathbb{C}^m$, $\Gamma^n = \mathbb{C}^n$, and $c[(\theta^m, \hat{\theta}^m(x^n)]$, as in (9.4.3), be given. Then, if the optimization problem in (9.4.2) has a solution $\hat{\theta}_*(x^n)$; $x^n \in \mathbb{C}^n$ this

9.4 Unbiased Mean-Squared Estimates

solution is given by

$$\hat{\theta}_*(x^n) = E^m(x^n) + g^m(x^n) \qquad x^n \in \mathbb{C}^n \qquad (9.4.4)$$

where

$$E^m(x^n) = \int_{\mathbb{C}^m} \theta^m p(\theta^m|x^n) \, d\theta^m \qquad (9.4.5)$$

$$g^m(x^n): \int_{\mathbb{C}^n} g^m(x^n) f_{\theta^m}(x^n) \, dx^n = \theta^m - \int_{\mathbb{C}^n} E^m(x^n) f_{\theta^m}(x^n) \, dx^n \qquad \forall \, \theta^m \in \mathbb{C}^m \qquad (9.4.6)$$

Proof: Due to the strict convexity of the penalty function in (9.4.3) with respect to the same metric as in Lemma 9.2.2, and using differential calculus, we conclude that, given x^n, we are searching for a vector $\hat{\theta}_*^m(x^n)$ such that

$$\int_{\mathbb{C}^m} d\theta^m p(\theta^m|x^n)\{2Q^*[\theta^m - \hat{\theta}_*^m(x^n)] + Q^* f^m(\theta^m)\} = 0 \qquad (9.4.7)$$

where $f^m(\theta^m)$ is some vector function of the parameter θ^m that will satisfy the constraint in (9.4.1). Directly from (9.4.7), and due to the fact that the matrix, Q is positive definite, we obtain,

$$\hat{\theta}_*^m(x^n) = \int_{\mathbb{C}^m} d\theta^m \, \theta^m p(\theta^m|x^n) + \int_{\mathbb{C}^m} 2^{-1} f^m(\theta^m) p(\theta^m|x^n) \, d\theta^m$$

$$= E^m(x^n) + \int_{\mathbb{C}^m} 2^{-1} f^m(\theta^m) p(\theta^m|x^n) \, d\theta^m \qquad (9.4.8)$$

Defining,

$$g^m(x^n) = \int_{\mathbb{C}^m} 2^{-1} f^m(\theta^m) p(\theta^m|x^n) \, d\theta^m \qquad (9.4.9)$$

we obtain from (9.4.8), by substitution,

$$\hat{\theta}_*(x^n) = E^m(x^n) + g^m(x^n) \qquad (9.4.10)$$

Finally, imposing the constraint in (9.4.1) on (9.4.10), we obtain the condition in (9.4.6). ∎

From (9.4.4), (9.4.5), and (9.4.9), we observe that if the factorization criterion in Corollary 9.2.1 holds, then the observation sequence x^n appears in both the vectors $E^m(x^n)$ and $g^m(x^n)$ strictly as a function of the sufficient statistic $T^m(x^n)$. Thus, the same is then true, for the unbiased Bayesian mean-squared estimate $\hat{\theta}_*(x^n)$ in (9.4.4). In general, from Theorem 9.4.1 we observe that, if it exists, the unbiased optimal Bayesian mean-squared estimate is formed by adding a term $g^m(x^n)$ to the optimal Bayesian mean-squared estimate $E^m(x^n)$, where the additive term $g^m(x^n)$ is such that the constraint in (9.4.1) is satisfied. Due to this observation, and since the linear mean-squared estimates comprise a subclass of all the mean-squared estimates, we can express directly from Theorem 9.4.1 the following lemma.

LEMMA 9.4.1: Given n, $\mathscr{E}^m$, Γ^n, and $c[\theta^m, \hat{\theta}^m(x^n)]$ as in Theorem 9.4.1, and the classes $\mathscr{A}_{m,n}$ and $\mathscr{C}_m$ in Definition 9.3.1, then, if an optimal unbiased linear mean-squared estimate $\hat{\theta}^m_{*L}(x^n)$; $x^n \in \mathbb{C}^n$ exists, it is given by

$$\hat{\theta}^m_{*L}(x^n) = [A^0 x^n + C^0] + [Ax^n + C] \qquad x^n \in \mathbb{C}^n \qquad (9.4.11)$$

where A^0 and C^0 are exactly as in Lemma 9.3.1, and where, $A \in \mathscr{A}_{m,n}$, $C \in \mathscr{C}_m$, and for,

$$E^n\{X^n | \theta^m\} = \int_{\mathbb{C}^n} x^n f_{\theta^m}(x^n)\, dx^n \qquad (9.4.12)$$

we also have

$$AE^n\{X^n | \theta^m\} + C = \theta^m - A^0 E^n\{X^n | \theta^m\} - C^0 \qquad (9.4.13)$$

∎

We will conclude this section with an example.

Example 9.4.1

Let us consider again Example 9.2.1. As we discussed in Section 9.3, the optimal Bayesian mean-squared estimate and the optimal linear mean-squared estimate are then identical. Furthermore, they are both given by the vector function $\hat{\theta}^m_0(x^n)$; $x^n \in R^n$ where from (9.2.34) we have

$$\hat{\theta}^m_0(x^n) = E^m(x^n) = [I_m + S^T R_n^{-1} S \Lambda_m]^{-1} \Lambda_m S^T R_n^{-1} x^n = D_{m,n} x^n \qquad (9.4.14)$$

and where, due to (9.2.35), we have,

$$E^m_{\theta^m} = E\{\hat{\theta}^m_0(x^n) | \theta^m\} = [I_m + S^T R_n^{-1} S \Lambda_m]^{-1} \Lambda_m S^T R_n^{-1} S \theta^m \qquad (9.4.15)$$

The matrix $D_{m,n} S = [I_m + S^T R_n^{-1} S \Lambda_m]^{-1} \Lambda_m S^T R_n^{-1} S$ in (9.4.15) is not, in general, equal to the identity matrix I_m; thus, the estimate in (9.4.14) is not, in general, unbiased. Due to Lemma 9.4.1, searching for an unbiased linear estimate is equivalent to searching for some $m \times n$, real matrix $A_{m,n}$ such that

$$[D_{m,n} + A_{m,n}] S = I_m$$

Another equivalent form of the problem is the following.
Find an $m \times n$ real matrix $B_{m,n}$ such that

$$B_{m,n} S = I_m \qquad (9.4.16)$$

For dimensionality n larger than m, the equation in (9.4.16) defines an overdetermined linear system, with unknowns the components of the matrix $B_{m,n}$. Thus, the system in (9.4.16) has then several solutions. Each such solution $B^*_{m,n}$ provides an unbiased linear mean-squared estimate $B^*_{m,n} x^n$; $x^n \in R^n$. Let us now

9.4 Unbiased Mean-Squared Estimates

define the $m \times n$ real matrix $B^0_{m,n}$ by

$$B^0_{m,n} = (S^T R_n^{-1} S)^{-1} S^T R_n^{-1} \tag{9.4.17}$$

Clearly, the matrix $B^0_{m,n}$, in (9.4.17) satisfies the condition in (9.4.16). That is,

$$B^0_{m,n} S = I_m \tag{9.4.18}$$

Also, every other matrix $B_{m,n}$ that satisfies (9.4.16) can be written as

$$B_{m,n} = B^0_{m,n} + D_{m,n} \tag{9.4.19}$$

where $D_{m,n}$ is some $m \times n$ real matrix such that

$$D_{m,n} S = 0 \tag{9.4.20}$$

Let us now consider the expected quadratic error, $e(B_{m,n}) = E\{[\Theta^m - B_{m,n} X^n]^T Q [\Theta^m - B_{m,n} X^n]\}$, induced by the linear estimate $B_{m,n} x^n$, where $B_{m,n}$ is as in (9.4.19). Then, if tr denotes trace, and using linear algebra results, we obtain

$$e(B_{m,n}) = e(B^0_{m,n}) - \operatorname{tr} Q D_{m,n} E\{X^n [\Theta^m - B^0_{m,n} X^n]^T\}$$
$$- \operatorname{tr} Q E\{[\Theta^m - B^0_{m,n} X^n] X^{nT}\} D^T_{m,n} + E\{(D_{m,n} X^n)^T Q (D_{m,n} X^n)\} \tag{9.4.21}$$

Due to the model $X^n = S\Theta^m + N^n$, the independence of Θ^m from N^n, and condition (9.4.18), we find

$$E\{X^n X^{nT}\} = S\Lambda_m S^T + R_n$$
$$E\{X^n [\Theta^m - B^0_{m,n} X^n]^T\} = -R_n B^{0T}_{m,n} \tag{9.4.22}$$
$$E\{[\Theta^m - B^0_{m,n} X^n] X^{nT}\} = -B^0_{m,n} R_n$$

Substituting (9.4.22) in (9.4.21), we find

$$e(B_{m,n}) = e(B^0_{m,n}) + \operatorname{tr} Q D_{m,n} R_n B^{0T}_{m,n} + \operatorname{tr} Q B^0_{m,n} R_n D^T_{m,n} + E\{(D_{m,n} X^n)^T Q (D_{m,n} X^n)\} \tag{9.4.23}$$

Due to the form of the matrix, $B^0_{m,n}$ in (9.4.17), due to (9.4.20), and by substitution, we also find

$$Q D_{m,n} R_n B^{0T}_{m,n} = Q D_{m,n} R_n R_n^{-1} S (S^T R_n^{-1} S)^{-1} = Q (D_{m,n} S)(S^T R_n^{-1} S)^{-1} = 0$$
$$Q B^0_{m,n} R_n D^T_{m,n} = Q (S^T R_n^{-1} S)^{-1} S^T R_n^{-1} R_n D^T_{m,n} = Q (S^T R_n^{-1} S)^{-1} (D_{m,n} S)^T = 0$$

Thus, (9.4.23) can be finally written as

$$e(B_{m,n}) = e(B^0_{m,n}) + E\{(D_{m,n} X^n)^T Q (D_{m,n} X^n)\} \tag{9.4.24}$$

Due to the fact that the matrix Q is positive definite, the quadratic errors $e(B_{m,n})$ and $e(B_{m,n}^0)$ in (9.4.24) can be equal to each other only if $D_{m,n} = 0$. Therefore, the error $e(B_{m,n})$ attains its minimum at $B_{m,n} = B_{m,n}^0$, where the matrix $B_{m,n}^0$ is given in (9.4.17). As a conclusion, the optimal unbiased linear mean-squared estimate $\hat{\theta}_{*L}^m(x^n)$ is here unique and given by

$$\hat{\theta}_{*L}^m(x^n) = (S^T R_n^{-1} S)^{-1} S^T R_n^{-1} x^n \qquad x^n \in R^n \qquad (9.4.25)$$

The optimal unbiased linear mean-squared estimate in (9.4.25) is identical to the optimal mean-squared Bayesian estimate if each of the diagonal components of the matrix, $\Lambda_m S^T R_n^{-1} S$ is large compared to one—that is, if $I_m + S^T R_n^{-1} S \Lambda_m \simeq S^T R_n^{-1} S \Lambda_m$. This was found in Example 9.2.1. Then, the optimal unbiased linear estimate also attains the minimum possible expected quadratic error.

9.5 THE MINIMAX OPTIMIZATION PROBLEM

In the Bayesian parameter estimation formalizations, the assets available to the analyst are the space $\mathscr{E}^m$ where the vector parameter Θ^m takes its values; for every value θ^m in $\mathscr{E}^m$, the stochastic process that generates the observation sequences, some observation sequence x^n; a penalty function, $c[\theta^m, \hat{\theta}^m(x^n)]$; and the a priori distribution $p(\theta^m)$; $\theta^m \in \mathscr{E}^m$ of the vector parameter θ^m. The objective function is then the expected penalty $c(\hat{\theta}^m, p)$ in (9.1.4), and the estimate vector function $\hat{\theta}_0^m(x^n)$; $x^n \in \Gamma^n$ that minimizes the expected penalty is sought. Let us now assume, that the a priori distribution $p(\theta^m)$; $\theta^m \in \mathscr{E}^m$ is unknown, while it is still assumed that the total observation sequence x^n, is generated by a single stochastic process—that is, a single parameter value θ^m in $\mathscr{E}^m$ is acting. Then the minimax parameter estimation formalization arises. As with the minimax detection in Chapter 6, the minimax parameter estimation is formulated as a saddle-point game, played between the analyst and Nature. Nature selects the a priori distribution $p(\theta^m)$; $\theta^m \in \mathscr{E}^m$ while the analyst independently selects the estimate $\hat{\theta}^m(x^n)$; $x^n \in \Gamma^n$. The payoff function is the expected penalty $c(\hat{\theta}^m, p)$. A pair $(\hat{\theta}_*^m, p_*)$ is then sought such that

$$\forall\; p(\theta^m) \quad \theta^m \in \mathscr{E}^m \qquad c(\hat{\theta}_*^m, p) \leq c(\hat{\theta}_*^m, p_*) \leq c(\hat{\theta}^m, p_*) \qquad \forall\; \hat{\theta}^m(x^n) \in \mathscr{E}^m \;\; \forall\; x^n \in \Gamma^n$$
(9.5.1)

If a pair $(\hat{\theta}_*^m, p_*^m)$ that satisfies the conditions in (9.5.1) exists, it is a saddle-point solution of the game. The distribution $p_*(\theta^m)$; $\theta^m \in \mathscr{E}^m$ is then called *least favorable*, and the vector function $\hat{\theta}_*^m(x^n)$; $x^n \in \Gamma^n$ is then the *minimax estimate*. We observe that the part

$$c(\hat{\theta}_*^m, p_*) \leq c(\hat{\theta}^m, p_*) \qquad \forall\; \hat{\theta}^m(x^n) \in \mathscr{E}^m \;\; \forall\; x^n \in \Gamma^n$$

of the double inequality in (9.5.1) represents a Bayesian formalization at the least favorable distribution $p_*(\theta^m)$; $\theta^m \in \mathscr{E}^m$. That is, if the saddle-point $(\hat{\theta}_*^m, p_*)$ exists, the minimax estimate $\hat{\theta}_*^m(x^n)$; $x^n \in \Gamma^n$ is then the optimal Bayesian estimate,

9.5 The Minimax Optimization Problem

when the a priori distribution of the vector parameter Θ^m is $p_*(\theta^m)$; $\theta^m \in \mathscr{E}^m$. Furthermore, the existence of the pair $(\hat{\theta}_*^m, p_*)$ implies the existence of the optimal Bayesian estimate at the distribution p_*. In addition, from the theory of saddle-point games, if a saddle point $(\hat{\theta}_*^m, p_*)$ exists, then

$$c(\hat{\theta}_*^m, p_*) = \inf_{\hat{\theta}^m} \sup_p c(\hat{\theta}^m, p) = \sup_p \inf_{\hat{\theta}^m} c(\hat{\theta}^m, p) \qquad (9.5.2)$$

As in Chapter 6, the search for a saddle-point pair $(\hat{\theta}_*^m, p_*)$ is tractable if the distribution $p(\theta^m)$; $\theta^m \in \mathscr{E}^m$ is first fixed, and the Bayesian estimate $\hat{\theta}_p^m$ at this distribution is then found. If this Bayesian problem has a solution, for each selected p this solution induces a minimum expected penalty $c(\hat{\theta}_p^m, p) = \inf_{\hat{\theta}^m} c(\hat{\theta}^m, p)$ at p. The final step is the search for the distribution p_* that realizes the supremum $\sup_p c(\hat{\theta}_p^m, p)$. If such a p_* exists, it is the least favorable distribution. The estimate $\hat{\theta}_{p_*}^m$ is then the minimax estimate. We will now present a useful definition:

DEFINITION 9.5.1: Given Γ^n, $\mathscr{E}^m$, and $c[\theta^m, \hat{\theta}^m(x^n)]$, the estimate $\hat{\theta}^m(x^n)$; $x^n \in \Gamma^n$ is called *admissible* if it is the optimal Bayesian estimate at some a priori distribution $p(\theta^m)$; $\theta^m \in \mathscr{E}^m$. ∎

From Definition 9.5.1 and the earlier discussion, we conclude that the search for a minimax estimate can be limited within the class of admissible estimates. This conclusion is parallel to a similar conclusion in Chapter 6, where minimax detection is studied. Let $c(\theta^m, \hat{\theta}^m)$ be the conditional expected penalty in (3.3.1). We will now express a lemma which basically describes a special case, for the existence of a minimax estimate.

LEMMA 9.5.1: Given, Γ^n, $\mathscr{E}^m$, and $c[\theta^m, \hat{\theta}^m(x^n)]$, let $\hat{\theta}_0^m(x^n)$; $x^n \in \Gamma^n$ be some admissible estimate such that the conditional expected penalty $c(\theta^m, \hat{\theta}_0^m)$ is a constant with respect to the parameter value θ^m in $\mathscr{E}^m$. Then, the estimate $\hat{\theta}_0^m(x^n)$; $x^n \in \Gamma^n$ is also a minimax estimate.

Proof: Since the conditional expected penalty $c(\theta^m, \hat{\theta}_0^m)$ is a constant with respect to the parameter value θ^m, given any a priori distribution $p(\theta^m)$; $\theta^m \in \mathscr{E}^m$ of the vector parameter θ^m, we obtain

$$c(\hat{\theta}_0^m, p) = \int_{\mathscr{E}^m} c(\theta^m, \hat{\theta}_0^m) p(\theta^m) \, d\theta^m = c(\theta^m, \hat{\theta}_0^m) = c(\hat{\theta}_0^m) \qquad (9.5.3)$$

Let $p_0(\theta^m)$; $\theta^m \in \mathscr{E}^m$ denote the a priori parameter distribution at which the estimate $\hat{\theta}_0^m(x^n)$; $x^n \in \Gamma^n$ is optimal Bayesian. Then, in conjunction with (9.5.3), we obtain

$$c(\hat{\theta}_0^m, p) = c(\hat{\theta}_0^m, p_0) \qquad \forall \ p(\theta^m) \quad \theta^m \in \mathscr{E}^m \qquad (9.5.4)$$

Thus, the left part of the double inequality in (9.5.1) is satisfied with equality by the pair $(\hat{\theta}_0^m, p_0)$. Since the estimate $\hat{\theta}_0^m(x^n)$; $x^n \in \Gamma^n$ is the optimal Bayesian estimate at the distribution $p_0(\theta^m)$; $\theta^m \in \mathscr{E}^m$, we also have

$$c(\hat{\theta}_0^m, p_0) \leq c(\hat{\theta}^m, p_0) \qquad \forall \ \hat{\theta}^m(x^n) \in \mathscr{E}^m \quad \forall \ x^n \in \Gamma^n \qquad (9.5.5)$$

Thus, the right part of the double inequality in (9.5.1) is satisfied by the pair $(\hat{\theta}_0^m, p_0)$. As a conclusion, the pair $(\hat{\theta}_0^m, p_0)$ satisfies the saddle-point game in (9.5.1); thus $\hat{\theta}_0^m(x^n)$; $x^n \in \Gamma^n$ is a minimax estimate, not necessarily unique.

We will complete this section with two examples.

Example 9.5.1

Let us consider the problem in Example 9.2.1. Let us assume that an a priori distribution $p(\theta^m)$; $\theta^m \in R^m$ of the vector parameters θ^m is not available. Let us then search for a minimax estimate. In Example 9.2.1 we found that if the a priori distribution $p(\theta^m)$; $\theta^m \in \mathscr{E}^m$ is zero mean Gaussian with autocovariance matrix Λ_m, then the optimal Bayesian mean-squared estimate is given by (9.2.34). Let now the matrix $I_m + S^T R_n^{-1} S \Lambda_m$ in (9.2.34) be such that

$$I_m + S^T R_n^{-1} S \Lambda_m \simeq \Lambda_m S^T R_n^{-1} S$$

Then, as we found in Example 9.2.1, the optimal Bayesian estimate $\hat{\theta}_0^m(x^n)$; $x^n \in R^n$ at the Gaussian distribution (as above), takes the form

$$\hat{\theta}_0^m(x^n) = (S^T R_n^{-1} S)^{-1} S^T R_n^{-1} x^n = B_{m,n} x^n \qquad x^n \in R^n \qquad (9.5.6)$$

where the estimate in (9.5.6) is then clearly admissible, and where

$$B_{m,n} = (S^T R_n^{-1} S)^{-1} S^T R_n^{-1} \qquad (9.5.7)$$

$$B_{m,n} S = (S^T R_n^{-1} S)^{-1} S^T R_n^{-1} S = I_m \qquad (9.5.8)$$

The penalty function in our example is given by the quadratic expression

$$c[\theta^m, \hat{\theta}^m(x^n)] = [\theta^m - \hat{\theta}^m(x^n)]^T Q [\theta^m - \hat{\theta}^m(x^n)] \qquad (9.5.9)$$

The conditional expected penalty $c(\theta^m, \hat{\theta}_0^m)$ for $\hat{\theta}_0^m$ as in (9.5.6), and for the penalty function as in (9.5.9) is then computed as follows, where results from linear algebra and (9.5.7) and (9.5.8) are used, and where tr signifies trace.

$$\begin{aligned}
c(\theta^m, \hat{\theta}_0^m) &= E\{[\theta^m - B_{m,n} X^n]^T Q [\theta^m - B_{m,n} X^n] | \theta^m\} \\
&= E\{[\theta^m - B_{m,n} S \theta^m - B_{m,n} N^n]^T Q [\theta^m - B_{m,n} S \theta^m - B_{m,n} N^n] | \theta^m\} \\
&= E\{[B_{m,n} N^n]^T Q [B_{m,n} N^n]\} = \operatorname{tr} E\{Q [B_{m,n} N^n][B_{m,n} N^n]^T\} \\
&= \operatorname{tr} Q B_{m,n} E\{N^n N^{nT}\} B_{m,n}^T = \operatorname{tr} Q B_{m,n} R_n B_{m,n}^T \\
&= \operatorname{tr} Q (S^T R_n^{-1} S)^{-1} S^T R_n^{-1} R_n R_n^{-1} S (S^T R_n^{-1} S)^{-1} = \operatorname{tr} Q (S^T R_n^{-1} S)^{-1}
\end{aligned}$$
$$(9.5.10)$$

From (9.5.10), we observe that the conditional expected penalty $c(\theta^m, \hat{\theta}_0^m)$ is a constant with respect to θ^m in R^m. Applying Lemma 9.5.1, we conclude then that the parameter estimate $\hat{\theta}_0^m(x^n)$; $x^n \in R^n$ in (9.5.6) is a minimax estimate for the present model.

Example 9.5.2

Let us revisit Example 9.2.2. Let us assume that the variance σ^2 is finite and that no a priori distribution $p(\theta)$; $\theta \in R$ of the scalar parameter Θ is available. We then search for a minimax estimate. In example 9.2.2, we found that if the distribution $p(\theta)$; $\theta \in R$ is zero mean, variance v^2 Gaussian, then the optimal Bayesian mean-squared estimate is given by (9.2.43). Let us now denote by $p_0(\theta)$; $\theta \in R$ the zero mean Gaussian distribution whose variance equals σ^2 and let us select a large enough dimensionality n of the observation sequence so that

$$1 + n^{-1} \simeq 1 \qquad (9.5.11)$$

Then, as concluded by (9.2.43), the optimal Bayesian mean-squared estimate $\hat{\theta}_0(x^n)$; $x^n \in R^n$ at the a priori parameter distribution $p_0(\theta)$; $\theta \in R$ is given by

$$\hat{\theta}_0(x^n) = n^{-1} \sum_{i=1}^{n} x_i \qquad x^n \in R^n \qquad (9.5.12)$$

Given some parameter value θ in R, the conditional expected mean-squared penalty induced by the estimate in (9.5.12) is computed as

$$E\{[\theta - \hat{\theta}_0(X^n)]^2 | \theta\} = E\left\{\left[\theta - n^{-1} \sum_{i=1}^{n} X_i\right]^2 \Big| \theta\right\} = n^{-1}\sigma^2 \qquad (9.5.13)$$

since, given the value θ, the expected value of the random variable $n^{-1}\sum_{i=1}^{n} X_i$ equals θ and its variance equals $n^{-1}\sigma^2$.

From (9.5.13), we conclude that the conditional expected mean-squared penalty induced by the estimate in (9.5.12) is a constant with respect to the parameter value θ in R. In addition, for n such that (9.5.11) is satisfied, the estimate in (9.5.12) is also admissible, since it is the optimal Bayesian mean-squared estimate at the Gaussian parameter distribution $p_0(\theta)$; $\theta \in R$. Applying Lemma 9.5.1, we then conclude that for $1 + n^{-1} \simeq 1$, the estimate in (9.5.12) is also minimax.

9.6 LOWER BOUNDS IN PARAMETER ESTIMATION

As we established earlier, given the vector parameter Θ^m, that takes its values θ^m in the product space $\mathbb{C}^m$, where $\mathbb{C}$ signifies the complex plane, given the density function $f_{\theta^m}(x^n)$ $\forall$ n, and given the data realization $x^n \in \mathbb{C}^n$, the parameter estimate $\hat{\theta}^m(x^n)$ estimates the acting vector parameter value $\theta^m \in \mathbb{C}^m$ where it is assumed that this value remains unchanged throughout the observation sequence x^n. It is thus desirable that the estimate $\hat{\theta}^m(x^n)$ be as "close" to the true vector parameter value θ^m as possible. Given the acting vector parameter value θ^m, this closeness is represented by the difference $d_{\theta m}^m(X^n) = \hat{\theta}^m(X^n) - \theta^m$, where the stochastic observation vector X^n is generated by the density function $f_{\theta m}(x^n)$. The vector $d_{\theta m}^m(X^n)$ is then clearly stochastic, and for given estimate $\hat{\theta}^m(x^n)$ its statistics is controlled by $f_{\theta m}(x^n)$. The first order statistics of $d_{\theta m}^m(X^n)$ is then given by the expected value $E_{f_{\theta m}}\{d_{\theta m}^m(X^n)\} = E_{f_{\theta m}}\{\hat{\theta}^m(X^n)\} - \theta^m$, and represents the bias of the estimate $\hat{\theta}^m(x^n)$; $x^n \in \mathbb{C}^n$. Even order statistics of the vector $d_{\theta m}^m(X^n)$ is then provided by the expected value of powers of some quadratic expression on the stochastic matrix $d_{\theta m}^{m*}(X^n)d_{\theta m}^{mT}(X^n) = [\hat{\theta}^m(X^n) - \theta^m]^*[\hat{\theta}^m(X^n) - \theta^m]^T$. In particular, the latter statistics is, in general, given by the expected value $E_{f_{\theta m}}\{A^{*T}M^*(\theta^m)[\hat{\theta}^m(X^n) - \theta^m]^*[\hat{\theta}^m(X^n) - \theta^m]^T M^T(\theta^m)A\}^p$ where p is some natural number, A is some m-dimensional complex and constant column vector, and $M(\theta^m)$ is some $m \times m$ complex matrix whose components are functions of the acting vector parameter value θ^m. The dependence of the matrix $M(\theta^m)$ on the vector parameter value θ^m allows for flexible weighing of the difference $\hat{\theta}^m(X^n) - \theta^m = d_{\theta m}^m(X^n)$. Indeed, the even order statistics of this difference can be evaluated differently, for vector parameter values θ^m of high versus low importance.

In this section, we are concerned with the derivation of lower bounds for the even order statistics of the stochastic difference $d_{\theta m}^m(X^n) = \hat{\theta}^m(X^n) - \theta^m$. In particular, we will derive such lower bounds for the expected value

$$E_{f_{\theta m}}\{A^{*T}M^*(\theta^m)[\hat{\theta}^m(X^n) - \theta^m]^*[\hat{\theta}^m(X^n) - \theta^m]^T M^T(\theta^m)A\}^p$$
$$= E_{f_{\theta m}}\{\|[\hat{\theta}^m(X^n) - \theta^m]^T M^T(\theta^m)A\|^{2p}\} \quad (9.6.1)$$

where the matrix $M(\theta^m)$, the vector A, and the natural number p are arbitrary.

In (9.6.1) we have denoted $b*b = \|b\|^2$, where b is any complex scalar. Given θ^m, let $f_{\theta m}(x^n)$ denote the n-dimensional density function induced by the data generating process at the vector point x^n when θ^m is acting. Let $x^n \in \mathbb{C}^n$, and let $\hat{\theta}^m(x^n)$; $x^n \in \mathbb{C}^n$ be some vector parameter estimate of the acting vector parameter value θ^m. Let A be some arbitrary constant, m-dimensional column vector, and let $M(\theta^m)$ be some $m \times m$ matrix that is a function of the acting vector parameter value θ^m. We will then develop lower bounds on the expected value in (9.6.1) for an arbitrary natural number p. We will denote by ∇ the m-dimensional column gradient with respect to θ^m. That is, if $f(\theta^m)$ is some scalar function of θ^m, then

$$\nabla^T f(\theta^m) = \left[\frac{\partial}{\partial \theta_1} f(\theta^m), \ldots, \frac{\partial}{\partial \theta_m} f(\theta^m)\right]$$

where T denotes transpose and $\theta^{mT} = [\theta_1, \ldots, \theta_m]$.

We will now express a theorem that provides the major result of this section.

THEOREM 9.6.1: Given n, let the density function of the observation vector X^n at the vector point x^n in $\mathbb{C}^n$, be $f_{\theta m}(x^n)$. Let $\hat{\theta}^m(x^n)$; $x^n \in \mathbb{C}^n$ be some estimate of the acting vector parameter value θ^m. Let the following two conditions be satisfied.

9.6 Lower Bounds in Parameter Estimation

$$\nabla \int_{\mathbb{C}^n} dx^n \, f_{\theta^m}(x^n) = \int_{\mathbb{C}^n} dx^n \, \nabla f_{\theta^m}^n(x^n) \tag{9.6.2}$$

$$\nabla E_{f_{\theta^m}}^T \{\hat{\theta}^m(X^n)\} = \nabla \int_{\mathbb{C}^n} \hat{\theta}^{mT}(x^n) f_{\theta^m}(x^n) \, dx^n = \int_{\mathbb{C}^n} dx^n \, \nabla f_{\theta^m}(x^n) \hat{\theta}^{mT}(x^n) \tag{9.6.3}$$

Then, for an arbitrary vector A and matrix $M(\theta^m)$, as in the beginning of this section, and for an arbitrary natural number p, the following inequality holds, where ln denotes natural logarithm.

$$\frac{E_{f_{\theta^m}}\{\|A^T M(\theta^m)[\hat{\theta}^m(X^n) - \theta^m]\|^{2p}\}}{\|A^T M(\theta^m)[\nabla E_{f_{\theta^m}}^T \{\hat{\theta}^m(X^n)\}]^T A\|^{2p}}$$

$$\geq \left\{ \int_{\mathbb{C}^n} \|A^T \nabla \ln[f_{\theta^m}(x^n)]\|^{2p/(2p-1)} f_{\theta^m}(x^n) \, dx^n \right\}^{1-2p} \tag{9.6.4}$$

The above inequality becomes an equality if and only if the scalar $A^T M(\theta^m)[\hat{\theta}^m(x^n) - \theta^m] A^T \nabla \ln[f_{\theta^m}(x^n)]$ is real and positive, $\forall \, x^n \in \mathbb{C}^n$ and $\forall \, \theta^m \in \mathbb{C}^m$, and there exists some real, positive, scalar function $c(\theta^m)$; $\theta^m \in \mathbb{C}^m$ such that

$$\|A^T \nabla \ln(f_{\theta^m}(x^n))\| = c(\theta^m) \|A^T M(\theta^m)[\hat{\theta}^m(x^n) - \theta^m]\|^{2p-1} \qquad \forall \, x^n \in \mathbb{C}^n \tag{9.6.5}$$

Proof: Since $\int_{\mathbb{C}^n} dx^n \, f_{\theta^m}(x^n) = 1 \, \forall \, \theta^m \in \mathbb{C}^m$, then $\nabla \int_{\mathbb{C}^n} dx^n \, f_{\theta^m}(x^n) = 0 \, \forall \, \theta^m \in \mathbb{C}^m$. Thus, if (9.6.2) is satisfied, we obtain

$$\int_{\mathbb{C}^n} dx^n \, \nabla f_{\theta^m}(x^n) = 0 \qquad \forall \, \theta^m \in \mathbb{C}^m \tag{9.6.6}$$

From (9.6.3) we obtain

$$[\nabla E_{f_{\theta^m}}^T \{\hat{\theta}^m(X^n)\}]^T = \int_{\mathbb{C}^n} \hat{\theta}^m(x^n) \nabla^T f_{\theta^m}(x^n) \, dx^n$$

$$A^T M(\theta^m)[\nabla E^T f_{\theta^m}\{\hat{\theta}^m(X^n)\}]^T A = \int_{\mathbb{C}^n} A^T M(\theta^m) \hat{\theta}^m(x^n) \nabla^T f_{\theta^m}(x^m) \, A \, dx^n$$

$$= \int_{\mathbb{C}^n} A^T M(\theta^m) \hat{\theta}^m(x^n) A^T \nabla f_{\theta^m}(x^n) \, dx^n \tag{9.6.7}$$

Due to (9.6.6), we can write (9.6.7) as

$$A^T M(\theta^m)[\nabla E_{f_{\theta^m}}^T \{\hat{\theta}^m(X^n)\}]^T A$$

$$= \int_{\mathbb{C}^n} dx^n \, A^T M(\theta^m) \hat{\theta}^m(x^n) A^T \nabla f_{\theta^m}(x^n) - A^T M(\theta^m) \theta^m A^T \int_{\mathbb{C}^n} dx^n \, \nabla f_{\theta^m}(x^n)$$

$$= \int_{\mathbb{C}^n} A^T M(\theta^m)[\hat{\theta}^m(x^n) - \theta^m] A^T \nabla f_{\theta^m}(x^n) \, dx^n \tag{9.6.8}$$

We can now express trivially, the equality

$$\nabla f_{\theta^m}(x^n) = f_{\theta^m}(x^n) \nabla \ln[f_{\theta^m}(x^n)] = [f_{\theta^m}(x^n)]^{1/2p} [f_{\theta^m}(x^n)]^{(2p-1)/2p} \nabla \ln[f_{\theta^m}(x^n)] \tag{9.6.9}$$

Substituting the equality in (9.6.9) to (9.6.8), we obtain

$$\|A^T M(\theta^m)[\nabla E^T_{f_{\theta^m}}\{\hat{\theta}^m(X^n)\}]^T A\| = \left\| \int_{\mathbb{C}^n} dx^n \{A^T M(\theta^m)[\hat{\theta}^m(x^n) - \theta^m][f_{\theta^m}(x^n)]^{1/2p}\} \right.$$
$$\left. \cdot \{A^T \nabla \ln[f_{\theta^m}(x^n)][f_{\theta^m}(x^n)]^{(2p-1)/2p}\} \right\|$$
$$\leq \int_{\mathbb{C}^n} \{\|A^T M(\theta^m)[\hat{\theta}^m(x^n) - \theta^m]\|[f_{\theta^m}(x^n)]^{1/2p}\}$$
$$\cdot \|A^T \nabla \ln[f_{\theta^m}(x^n)]\|[f_{\theta^m}(x^n)]^{(2p-1)/2p} dx^n \quad (9.6.10)$$

Applying now the Hölder inequality on the right hand side expression of the inequality in (9.6.10), we obtain

$$\|A^T M(\theta^m)[\nabla E^T_{f_{\theta^m}}\{\hat{\theta}^m(X^n)\}]^T A\|$$
$$\leq \left\{ \int_{\mathbb{C}^n} \|A^T M(\theta^m)[\hat{\theta}^m(x^n) - \theta^m]\|^{2p} f_{\theta^m}(x^n) dx^n \right\}^{1/2p}$$
$$\cdot \left\{ \int_{\mathbb{C}^n} \|A^T \nabla \ln[f_{\theta^m}(x^n)]\|^{2p/(2p-1)} f_{\theta^m}(x^n) dx^n \right\}^{(2p-1)/2p}$$
$$= E^{1/2p}_{f_{\theta^m}}\{\|A^T M(\theta^m)[\hat{\theta}^m(X^n) - \theta^m]\|^{2p}\}$$
$$\cdot \left\{ \int_{\mathbb{C}^n} \|A^T \nabla \ln(f_{\theta^m}(x^n)\|^{2p/(2p-1)} f_{\theta^m}(x^n) dx^n \right\}^{(2p-1)/2p} \quad (9.6.11)$$

The inequality in (9.6.11) becomes an equality, iff,

i. The scalar $A^T M(\theta^m)[\hat{\theta}^m(x^n) - \theta^m] A^T \nabla \ln(f_{\theta^m}(x^n))$ is real and positive $\forall\, x^n \in \mathbb{C}^n \,\forall\, \theta^m \in \mathbb{C}^m$
ii. $c(\theta^m)\|A^T M(\hat{\theta}^m)[\hat{\theta}^m(x^n) - \theta^m]\|^{2p-1} = \|A^T \nabla \ln(f_{\theta^m}(x^n))\| \,\forall\, x^n \in \mathbb{C}^n, \,\forall\, \theta^m \in \mathbb{C}^m$

where $c(\theta^m)$ is some positive real scalar function of the vector parameter θ^m
From (9.6.11) we finally obtain

$$\frac{E_{f_{\theta^m}}\{\|A^T M(\theta^m)[\hat{\theta}^m(X^n) - \theta^m]\|^{2p}\}}{\|A^T M(\theta^m)[\nabla E^T_{\mu_{\theta^m}}\{\hat{\theta}^m(X^n)\}]^T A\|^{2p}}$$
$$\geq \left\{ \int_{\mathbb{C}^n} \|A^T \nabla \ln[f_{\theta^m}(x^n)]\|^{2p/(2p-1)} f_{\theta^m}(x^n) dx^n \right\}^{1-2p} \quad \blacksquare$$

We note that if both the vector parameter Θ^m and the data sequence X^n represent real stochastic processes, then the conditions in Theorem 9.6.1, for satisfying (9.6.4) with equality, reduce to

$$A^T \nabla \ln[f_{\theta^m}(x^n)] = c(\theta^m)\{A^T M(\theta^m)[\hat{\theta}^m(x^n) - \theta^m]\}^{2p-1} \quad \forall\, x^n \;\forall\, \theta^m \quad (9.6.12)$$

for some real and positive scalar function $c(\theta^m)$.

9.6 Lower Bounds in Parameter Estimation

The condition in (9.6.12) specifies a density function $f_{\theta^m}(x^n)$ that has exponential form. Furthermore, the estimate $\hat{\theta}^m(x^n)$ is then the sufficient statistics that correspond to the density function $f_{\theta^m}(x^n)$. In general, given n, some arbitrary density function $f_{\theta^m}(x^n)$, and p, any estimate $\hat{\theta}^m(x^n)$ satisfies the inequality in (9.6.4). If there exists some estimate $\hat{\theta}_0^m(x^n)$ that satisfies (9.6.4) with equality, this estimate is called, p-efficient at the density function $f_{\theta^m}(x^n)$. A p-efficient estimate $\hat{\theta}_0^m(x^n)$ exists only for stochastic processes that induce density functions $f_{\theta^m}(x^n)$ of exponential form, as in (9.6.12). This estimate induces then the minimum value of the ratio $E_{f_{\theta^m}}\{\|A^T M(\theta^m)[\hat{\theta}^m(X^n) - \theta^m]\|^{2p}\}/\{A^T M(\theta^m) \cdot [\nabla E_{f_{\theta^m}}^T\{\hat{\theta}^m(X^n)\}]^T A\}^{2p}$ among all possible estimates $\hat{\theta}^m(x^n)$. The denominator of the latter ratio represents a measure on the bias induced by the estimate $\hat{\theta}^m(x^n)$. Thus, the ratio reflects $2p$-order deviation of the estimate $\hat{\theta}^m(x^n)$ from the acting vector parameter value θ^m normalized by some bias measure. To see this point clearly, let us consider a real and scalar parameter Θ and a real data process, where R represents the real line. Then, for any estimate $\hat{\theta}(x^n)$ of the parameter value θ we have directly from Theorem 9.6.1,

$$\frac{E_{f_\theta}\{[\hat{\theta}(X^n) - \theta]^{2p}\}}{\left[\frac{\partial}{\partial \theta} E_{f_\theta}\{\hat{\theta}(X^n)\}\right]^{2p}} \geq \left\{\int_{R^n} \left[\frac{\partial}{\partial \theta} \ln f_\theta(x^n)\right]^{2p/(2p-1)} f_\theta(x^n)\, dx^n\right\}^{1-2p} \quad (9.6.13)$$

satisfied with equality if and only if there exists a positive scalar function $c(\theta)$ such that

$$\frac{\partial}{\partial \theta} \ln f_\theta(x^n) = c(\theta)[\hat{\theta}(x^n) - \theta]^{2p-1} \quad \forall\, x^n \in R^n \quad \theta \in R \quad (9.6.14)$$

If the estimate $\hat{\theta}(x^n)$; $x^n \in R^n$ in (9.6.13) is unbiased, then $E_{f_\theta}\{\hat{\theta}(X^n)\} = \theta$ and $(\partial/\partial\theta)E_{f_\theta}\{\hat{\theta}(X^n)\} = 1$. In this case, the expected $2p$-order deviation $E_{f_\theta}\{[\hat{\theta}(X^n) - \theta]^{2p}\}$ of the estimate $\hat{\theta}(x^n)$; $x^n \in R^n$ from the acting parameter θ is normalized by the constant one. If $\hat{\theta}(x^n)$; $x^n \in R^n$ is not unbiased, then the scalar $(\partial/\partial\theta)E_{f_\theta}\{\hat{\theta}(X^n)\}$ is a function of θ and so is then the scalar that normalizes the expected value $E_{f_\theta}\{[\hat{\theta}(X^n) - \theta]^{2p}\}$. From the condition in (9.6.14) we conclude that exponential density functions of the form $f_\theta(x^n) = a \exp\{-b[g(x^n) - \theta]^{2p}\}$; $x^n \in R^n$, where a and b are positive and real scalars and $g(x^n)$ is a real scalar function of the vector x^n, induce p-efficient estimates. In particular, the p-efficient estimate $\hat{\theta}(x^n)$; $x^n \in R^n$ for the density function expressed above is such that $\hat{\theta}(x^n) = g(x^n)\ \forall\, x^n \in R^n$.

In Theorem 9.6.1, we developed lower bounds for $2p$-order moments of scalar linear transformations of the vector difference $\hat{\theta}^m(x^n) - \theta^m$ where θ^m is the acting vector parameter value and $\hat{\theta}^m(x^n)$ is its estimate. At this point, we concentrate on the special case where $p = 1$. We express the general form of this case in a theorem that is a special case of Theorem 9.6.1.

THEOREM 9.6.2: Consider vector parameter values θ^m, data density functions $f_{\theta^m}(x^n)$, vector estimates $\hat{\theta}^m(x^n)$; $x^n \in \mathcal{C}^n$, a vector A, and a matrix $M(\theta^m)$ as in

Theorem 9.6.1. Let the conditions in (9.6.2) and (9.6.3) be satisfied, and let us denote

$$R_{f_{\theta^m},\hat{\theta}^m,n} = E_{f_{\theta^m}}\{[\hat{\theta}^m(X^n) - \theta^m]* [\hat{\theta}^m(X^n) - \theta^m]^T\}$$

Then

$$\frac{A^T M(\theta^m) R_{f_{\theta^m},\hat{\theta}^m,n} M^{*T}(\theta^m) A^*}{\|A^T M(\theta^m)[\nabla E^T_{f_{\theta^m}}\{\hat{\theta}^m(X^n)\}]^T A\|^2} \geq \left\{\int_{\mathbb{C}^n} \|A^T \nabla \ln(f_{\theta^m}(x^n))\|^2 f_{\theta^m}(x^n) dx^n\right\}^{-1}$$

(9.6.15)

with equality if and only if there exists some positive scalar function $c(\theta^m)$ such that

$$A^T \nabla \ln(f_{\theta^m}(x^n)) = c(\theta^m) A^{*T} M^*(\theta^m) [\hat{\theta}^m(x^n) - \theta^m]^* \quad \forall\, x^n \in \mathbb{C}^n \quad (9.6.16)$$

The bound $\{\int_{\mathbb{C}^n} \|A^T \nabla \ln[f_{\theta^m}(x^n)]\|^2 f_{\theta^m}(x^n) dx^n\}^{-1}$ in (9.6.15) is called the Rao-Cramér lower bound, and estimates that satisfy (9.6.15) with equality are called efficient. ∎

The Rao-Cramér bound in Theorem 9.6.2 took its name from the two statisticians who first developed it. The condition in (9.6.16) for satisfying the bound in (9.6.15) with equality is trivially derived from the parallel conditions in Theorem 9.6.1 when $p = 1$. We observe that the inequality in (9.6.15) represents a lower bound on the quadratic expression, $A^T M(\theta^m) R_{f_{\theta^m},\hat{\theta}^m,n} M^{*T}(\theta^m) A^*$. The latter involves the autocorrelation matrix, $R_{f_{\theta^m},\hat{\theta}^m,n}$ in (9.6.14), whose trace is the expected norm $E_{f_{\theta^m}}\{\|\hat{\theta}^m(X^n) - \theta^m\|^2\}$.

Let the parameter θ be real and scalar. Then, the Rao-Cramér bound in Theorem 9.6.2 takes the following form for real data sequences x^n.

$$\frac{E_\theta\{[\hat{\theta}(X^n) - \theta]^2\}}{\left[\frac{\partial}{\partial \theta} E_{f_\theta}\{\hat{\theta}(X^n)\}\right]^2} \geq \left\{\int_{R^n} \left[\frac{\partial}{\partial \theta} \ln[f_\theta(x^n)]\right]^2 f_\theta(x^n) dx^n\right\}^{-1} \quad (9.6.17)$$

This is satisfied with equality if and only if there exists some positive scalar function $c(\theta)$ such that

$$\frac{\partial}{\partial \theta} \ln[f_\theta(x^n)] = c(\theta)[\hat{\theta}(x^n) - \theta] \quad \forall\, x^n \in R^n \quad (9.6.18)$$

So, if there exists some scalar function $\hat{\theta}(x^n)$; $x^n \in R^n$ such that the density function $f_\theta(x^n)$ is as in (9.6.18), then $\hat{\theta}(x^n)$ is also an efficient estimate of the acting parameter value θ. Furthermore, from (9.6.18) we then obtain

$$E_{f_\theta}\left\{\frac{\partial}{\partial \theta} \ln[f_\theta(X^n)]\right\} = c(\theta)[E_{f_\theta}\{\hat{\theta}(X^n)\} - \theta]$$

9.6 Lower Bounds in Parameter Estimation

or

$$\int_{R^n} \frac{\partial}{\partial \theta} \ln[f_\theta(x^n)] f_\theta(x^n)\, dx^n = \int_{R^n} \frac{\partial}{\partial \theta} f_\theta(x^n)\, dx^n$$
$$= c(\theta)[E_{f_\theta}\{\hat{\theta}(X^n)\} - \theta] \qquad (9.6.19)$$

But, due to (9.6.2) we have

$$\int_{R^n} \frac{\partial}{\partial \theta} f_\theta(x^n)\, dx^n = \frac{\partial}{\partial \theta} \int_{R^n} f_\theta(x^n)\, dx^n = \frac{\partial}{\partial \theta}(1) = 0 \qquad (9.6.20)$$

and from (9.6.19) and (9.6.20) we thus obtain

$$E_{f_\theta}\{\hat{\theta}(X^n)\} = \theta$$

Therefore, the efficient estimate $\hat{\theta}(x^n)$; $x^n \in R^n$ in (9.6.18) is also unbiased. That is, if the density function $f_\theta(x^n)$ is as in (9.6.18), then the scalar $\hat{\theta}(x^n)$ is an unbiased estimate of the acting parameter value θ and also attains the minimum possible variance among all the unbiased estimates of θ at $f_\theta(x^n)$. The density function $f_\theta(x^n)$ that satisfies (9.6.18) is exponential with respect to θ, and $\hat{\theta}(x^n)$ is a sufficient statistic at $f_\theta(x^n)$. Indeed, integrating (9.6.18), with respect to θ, we obtain

$$\ln[f_\theta(x^n)] - \ln[f_0(x^n)] = \int_0^\theta c(u)[\hat{\theta}(x^n) - u]\, du$$
$$= \hat{\theta}(x^n) \int_0^\theta c(u)\, du - \int_0^\theta u c(u)\, du$$

and

$$f_\theta(x^n) = f_0(x^n) \exp\left\{\hat{\theta}(x^n) \int_0^\theta c(u)\, du - \int_0^\theta u c(u)\, du\right\} \qquad (9.6.21)$$

From (9.6.21), we observe that the factorization criterion in Chapter 4 is satisfied; thus, $\hat{\theta}(x^n)$ is a sufficient statistic at $f_\theta(x^n)$. Also, the parameter θ appears only in the exponent in (9.6.21); thus, the density $f_\theta(x^n)$ is exponential with respect to θ.

We will complete this section with an example.

Example 9.6.1

Let θ^m be some m-dimensional vector parameter value in R^m and let the density function $f_{\theta^m}(x^n)$ be Gaussian. That is, there exists an $n \times m$ real matrix S_{nm} such that,

$$f_{\theta^m}(x^n) = (2\pi)^{-n/2} |R_n|^{-1/2} \exp\left\{-\frac{1}{2}[x^n - S_{nm}\theta^m]^T R_n^{-1}[x^n - S_{nm}\theta^m]\right\} \qquad (9.6.22)$$

where R_n is the autocovariance matrix of the Gaussian distribution. From (9.6.22) we obtain

$$\nabla \ln[f_{\theta^m}(x^n)] = -\frac{1}{2}\nabla\{[x^n - S_{nm}\theta^m]^T R_n^{-1}[x^n - S_{nm}\theta^m]\}$$

$$= -\frac{1}{2}\nabla\{x^{n^T}R_n^{-1}x^n - 2x^{n^T}R_n^{-1}S_{nm}\theta^m + \theta^{m^T}S_{nm}^T R_n^{-1}S_{nm}\theta^m\}$$

$$= S_{nm}^T R_n^{-1} x^n - S_{nm}^T R_n^{-1} S_{nm}\theta^m$$

$$= S_{nm}^T R_n^{-1} S_{nm}[(S_{nm}^T R_n^{-1} S_{nm})^{-1} S_{nm}^T R_n^{-1} x^n - \theta^m] \quad (9.6.23)$$

Let us now define

$$\hat{\theta}^m(x^n) = (S_{nm}^T R_n^{-1} S_{nm})^{-1} S_{nm}^T R_n^{-1} x^n \quad (9.6.24)$$

$$M(\theta^m) = S_{nm}^T R_n^{-1} S_{nm} \quad (9.6.25)$$

Then we obtain

$$\nabla E_{f_{\theta^m}}^T\{\hat{\theta}^m(X^n)\} = \nabla \theta^{m^T} S_{nm}^T R_n^{-1} S_{nm}(S_{nm}^T R_n^{-1} S_{nm})^{-1} = \nabla \theta^{m^T} = I_m \quad (9.6.26)$$

where I_m denotes the $m \times m$ identity matrix.

For the estimate in (9.6.24) and the matrix in (9.6.25) and for any real vector A, the condition (9.6.16), theorem 9.6.2, is clearly satisfied for $c(\theta^m) = 1$. Thus, the estimate in (9.6.24) is then efficient at the Gaussian distribution in (9.6.22). Due to (9.6.26), the estimate is also unbiased. Also, for the estimate in (9.6.24), we have

$$R_{f_{\theta^m},\hat{\theta}^m,n} = E_{f_{\theta^m}}\{[\hat{\theta}(X^n) - \theta^m][\hat{\theta}(X^n) - \theta^m]^T\}$$

$$= (S_{nm}^T R_n^{-1} S_{nm})^{-1} S_{nm}^T R_n^{-1}$$

$$\times E_{f_{\theta^m}}\{[X^n - S_{nm}\theta^m][X^n - S_{nm}\theta^m]^T\} R_n^{-1} S_{nm}(S_{nm}^T R_n^{-1} S_{nm})^{-1}$$

$$= (S_{nm}^T R_n^{-1} S_{nm})^{-1} S_{nm}^T R_n^{-1} R_n R_n^{-1} S_{nm}(S_{nm}^T R_n^{-1} S_{nm})^{-1}$$

$$= (S_{nm}^T R_n^{-1} S_{nm})^{-1} \quad (9.6.27)$$

Now let A be some arbitrary, constant, real, m-dimensional column vector. Then, for the matrix $M(\theta^m)$ in (9.6.25), and due to (9.6.26), and (9.6.27), we have

$$r = \frac{A^T M(\theta^m) R_{f_{\theta^m},\hat{\theta}^m,n} M^{*T}(\theta^m) A^*}{\|A^T M(\theta^m)[\nabla E_{f_{\theta^m}}^T\{\hat{\theta}^m(X^n)\}]^T A\|^2} = \frac{A^T S_{nm}^T R_n^{-1} S_{nm} A}{[A^T S_{nm}^T R_n^{-1} S_{nm} A]^2}$$

$$= [A^T S_{nm}^T R_n^{-1} S_{nm} A]^{-1} \quad (9.6.28)$$

Given the vector A and the matrix $M(\theta^m)$ in (9.6.25), the minimum value of the ratio r is $[A^T S_{nm}^T R_n^{-1} S_{nm} A]^{-1}$ and is attained by the efficient estimate in (9.6.24). If the dimensionality m of the vector parameter is one, then the efficient estimate in (9.6.24) takes the following form, where S_n is now a real, constant, n-dimensional column vector.

$$\hat{\theta}(x^n) = \frac{S_n^T R_n^{-1} x^n}{S_n^T R_n^{-1} S_n} \qquad (9.6.29)$$

If the data are independent, with variance one, and θ is their common mean, then R_n is the identity matrix and S_n is the vector whose components are all equal to one. The efficient estimate in (9.6.29) becomes then equal to the data average $n^{-1} \sum_{i=1}^{n} x_i$. This latter estimate is unbiased, and its conditional variance, $\sigma_\theta^2 = E\{[n^{-1} \sum_{i=1}^{n} X_i - \theta]^2\}$ equals n^{-1} for every value θ. Asymptotically ($n \to \infty$), the variance σ_θ^2 approaches the value zero. Therefore, asymptotically, the estimate $n^{-1} \sum_{i=1}^{n} x_i$ equals the acting parameter value θ almost everywhere.

9.7 PROBLEMS

9.7.1 Consider the Bayesian estimate $\hat{\theta}_0(x^n)$ in (9.2.43), Example 9.2.2. Given n and parameter value θ, find its mean and variance. With what rate does the estimate converge to an unbiased estimate as n increases?

9.7.2 For the problem in Example 9.2.2, what is the optimal unbiased mean-squared estimate, and how does it compare to that in Problem 9.7.1?

9.7.3 Let S be an n-vector with mean zero and covariance matrix R_s. Let $V = HS + N$ be a p observation vector where H is known and N is independent noise with mean zero and covariance R_N. Find the estimate of S that minimizes the quadratic form

$$I = S^T R_s^{-1} S + (V - HS)^T R_N^{-1} (V - HS)$$

What is $E(I)$ for this estimate?

9.7.4 Let $\hat{\theta}(x)$ be an estimate of θ, and X a random variable governed by the density function $f(x, \theta)$. If $\hat{\theta}(x)$ is unbiased, the variance satisfies the Rao-Cramér bound

$$E\{(\hat{\theta}(x) - \theta)^2\} \geq \frac{1}{E\{[(\partial/\partial\theta) \ln f(x, \theta)]^2\}}$$

Show that this bound is also given by

$$E\{(\hat{\theta}(x) - \theta)^2\} \geq \frac{-1}{E\{[(\partial^2/\partial\theta^2) \ln f(x, \theta)]\}}$$

9.7.5 Suppose the observation x is exponentially distributed with location parameter μ.

$$p_\mu(x) = e^{-|x-\mu|} \qquad x \geq \mu$$

and μ is uniformly distributed on $[0, \theta]$. Find the minimum variance estimate of μ. Is this estimate biased?

9.7.6 Let $V = \theta S + N$ be observed where $s_i = 1$, $i = 1, \ldots, n$ and N is a vector of Gaussian random variables with mean zero and covariance matrix $R = \{\sigma^{-2}\rho^{|i-j|}; i, j = 1, \ldots, n\}$, $0 < \rho < 1$. Find the minimum variance unbiased estimate of θ explicitly. Evaluate the variance. Find the efficiency of the sample mean. What happens as $n \to \infty$?

9.7.7 Let $V = \{v_i; i = 1, \ldots, n\}$ be observed with density

$$f(v|\theta) = \begin{cases} \dfrac{1}{\theta} & 0 \le v \le \theta \\ 0 & v > \theta \text{ or } v < 0 \end{cases}$$

where θ has uniform a priori density as well on $[0, \theta_0]$.

$$p(\theta) = \begin{cases} \dfrac{1}{\theta_0} & 0 \le \theta \le \theta_0 \\ 0 & \text{otherwise} \end{cases}$$

Find the minimum variance estimate of θ. Evaluate the variance that results.

9.7.8 Let $A_1, A_2, \ldots$ be possible outcomes of an experiment with unknown a priori probabilities $p_1, p_2, \ldots, \sum p_i = 1$. In m independent trials, suppose A_i occurs X_i times, $\sum X_i = m$, $i = 1, 2, \ldots$. Define

$$\phi_i = \begin{cases} 1 & X_i = 0 \\ 0 & X_i \ne 0 \end{cases}$$

Then the random variable, $u = \sum_i p_i \phi_i$ represents the sum of probabilities of unobserved outcomes—that is, u is unobservable. Suppose one more independent trial is made. Now A_i occurs Y_i times; $\sum Y_i = m + 1$. Define

$$\psi_i = \begin{cases} 1 & Y_i = 1 \\ 0 & Y_i \ne 1 \end{cases}$$

The random variable $(m + 1)v = \sum_i \psi_i$ is the number of single outcomes in $(m + 1)$ trials. (Note that v can be observed.) Show that v is an unbiased estimate of u.

9.7.9 Let X have a binomial distribution with n trials and probability θ; $0 < \theta < 1$ of success. Using the penalty function

$$C(\theta, a) = (\theta - a)^2 / [\theta(1 - \theta)]$$

show that $d(X) = X/n$ is a minimax estimate of θ with constant risk $1/n$.

9.7.10 Let θ be the half-open interval $[0, 1)$ (that is, $0 \leq \theta < 1$) let α be the closed interval $[0, 1]$, and let $C(\theta, \alpha) = (\theta - \alpha)^2/(1 - \theta)$. Let the observable X have the geometric distribution

$$f(x|\theta) = (1 - \theta)\theta^x \qquad x = 0, 1, 2, \ldots$$

Find the minimax rule.

REFERENCES

Blackwell, D. (1947), "Conditional Expectation and Unbiased Sequential Estimation," *Ann. Math. Statist.* 18, 105–110.

——— (1951), "On the Translation Parameter Problem for Discrete Variables," *Ann. Math. Statist.* 22, 393–399.

De Groot, M. H. (1959), "Unbiased Sequential Estimation for Binomial Populations," *Ann. Math. Statist.* 30, 80–101.

Ferguson, T. S., (1967), *Mathematical Statistics: A Decision Theoretical Approach*, Academic Press, New York.

Girshick, M. A., F. Mosteller, and L. J. Savage (1946), "Unbiased Estimates for Certain Binomial Sampling Problems with Applications," *Ann. Math. Statist.* 17, 13–23.

James, W., and C. Stein (1961), "Estimation with Quadratic Loss," *Proc. 4th Berkeley Symp. on Math. Statist. and Prob.* 1, Univ. of Calif. Press, Berkeley, 361–380.

McKinsey, J. C. C. (1952), *Introduction to the Theory of Games*. The Rand Corp., McGraw-Hill, New York.

Parzen, E. (1960), *Modern Probability Theory and its Applications*. Wiley, New York.

Pitman, E. J. G. (1939), "The Estimation of Location and Scale Parameters of a Continuous Population of Any Given Form," *Biometrika* 30, 391–421.

Rao, C. R. (1945), "Information and Accuracy Attainable in Estimation of Statistical Parameters," *Bull. Calcutta Math. Soc.* 37, 81–91.

Ray, S. N. (1965), "Bounds on the Maximum Sample Size of a Bayes Sequential Procedure," *Ann. Math. Statist.* 36, 859–878.

Savage, L. J. (1947), "A Uniqueness Theorem for Unbiased Sequential Estimation," *Ann. Math. Statist.* 18, 295–297.

Scheffé, H. (1959), *The Analysis of Variance*, Wiley, New York.

Stein, C. (1956), "Inadmissibility of the Usual Estimator for the Mean of a Multivariate Normal Distribution," *Proc. 3rd Berkeley Symp. on Math. Statist. and Prob.* 1, the University of California Press, Berkeley, 197–206.

Steinhaus, H. (1957), "The Problem of Estimation," *Ann. Math. Statist.* 28, 633–648.

Tucker, H. G. (1962), *An Introduction to Probability and Mathematical Statistics.* Academic Press, New York.

Van Trees, H. L., (1968), *Detection, Estimation, and Modulation Theory*, Wiley, New York.

Von Neumann, J., and Morgenstern, O., (1944), *Theory of Games and Economic Behavior*, 3d. ed. Princeton Univ. Press, Princeton, NJ.

Wolfowitz, J., (1946), "On Sequence Binomial Estimation," *Ann. Math. Statist.* 17, 489–493.

Wolfowitz, J., (1950), "Minimax Estimates of the Mean of a Normal Distribution with Known Variance," *Ann. Math. Statist.* 21, 218–230.

CHAPTER 10

MAXIMUM LIKELIHOOD PARAMETER ESTIMATION

10.1 INTRODUCTION

As we saw in Chapter 9, the Bayesian parameter estimation schemes evolve when an a priori parameter distribution is available. Such a distribution is frequently unknown, however, and the a priori choice of a reasonable penalty function may likewise be impossible. Then, the *maximum likelihood* parameter estimates arise. In particular, if Θ^m is an m-dimensional parameter vector taking values in $\mathscr{E}^m$, it is uniformly assumed that a single parameter value θ^m is acting throughout the observation sequence x^n. If $f_{\theta^m}(x^n)$ denotes the density function induced by the data generating process at the vector point x^n in Γ^n, when the value θ^m is acting, and assuming that the density $f_{\theta^m}(x^n)$ is well known for every given θ^m in $\mathscr{E}^m$, then given x^n, the maximum likelihood parameter estimate $\hat{\theta}^m(x^n)$, is defined as

$$\hat{\theta}^m(x^n): f_{\hat{\theta}^m(x^n)}(x^n) = \sup_{\theta^m \in \mathscr{E}^m} f_{\theta^m}(x^n) \qquad (10.1.1)$$

That is, given x^n, the maximum likelihood estimate is the parameter value that attains the supremum of the conditional density function $f_{\theta^m}(x^n)$ in $\mathscr{E}^m$. We note that if sufficient statistics $T^m(x^n)$ exists such that the factorization in (9.2.18) holds, then the maximum likelihood estimate clearly satisfies the supremum $\sup_{\theta^m \in \mathscr{E}^m} g[\theta^m, T^m(x^n)]$; thus it is then strictly a function of the sufficient statistics, $T^m(x^n)$. If the function $g(\cdot, \cdot)$ in (9.2.18) is in addition such that $\sup_{y^m} g(y^m, w^m) = g(w^m, w^m) \ \forall \ w^m$, then the supremum $\sup_{\theta^m \in \mathscr{E}^m} g[\theta^m, T^m(x^n)]$ is attained at $\theta^m = T^m(x^n)$; therefore, the maximum likelihood estimate $\hat{\theta}^m(x^n)$ equals then the sufficient statistics $T^m(x^n)$ for every x^n in Γ^n.

Example 10.1.1

Let the density function $f_{\theta^m}(x^n)$ correspond to a memoryless, real, unit variance and stationary Gaussian process with unknown mean θ in R, where R denotes the real line. Then

$$f_{\theta^m}(x^n) = f_\theta(x^n) = (2\pi)^{-n/2} \exp\left\{-\frac{1}{2}\sum_{i=1}^{n}(x_i - \theta)^2\right\}$$

$$= (2\pi)^{-n/2} \exp\left\{-n\left[\frac{\theta^2}{2} - \theta n^{-1}\sum_{i=1}^{n} x_i\right]\right\} \exp\left\{-\frac{1}{2}\sum_{i=1}^{n} x_i^2\right\} \quad (10.1.2)$$

Due to (10.1.2), we conclude that the sufficient statistics here is $n^{-1}\sum_{i=1}^{n} x_i$. Also, given $x^n = [x_1, \ldots, x_n]^T$, the exponential $\exp\{-n[(\theta^2/2) - \theta n^{-1}\sum_{i=1}^{n} x_i]\}$ attains its unique maximum at $\hat{\theta}(x^n) = n^{-1}\sum_{i=1}^{n} x_i$. Thus, the maximum likelihood estimate $\hat{\theta}(x^n)$ of the mean θ equals here the sufficient statistics $n^{-1}\sum_{i=1}^{n} x_i$.

Example 10.1.1 brings up two important special cases in maximum likelihood parameter estimation—density functions generated by stationary and memoryless processes, and parameters that represent the mean values of such processes. The latter parameters are called *location parameters* in the statistical literature, since they represent shifts in density functions. The majority of the statistical results on maximum likelihood estimates concentrate around parameters in memoryless and stationary processes. We will devote Section 10.2 to parametrically known stationary and memoryless processes.

Given a parametrically known stochastic process, given its density function, $f_{\theta^m}(x^n) \,\forall\, \theta^m \in \mathscr{E}^m, \forall n, \forall x^n \in \Gamma^n$, the maximum likelihood estimate $\hat{\theta}^m(x^n)$ in (10.1.1) is unique if the supremum $\sup_{\theta^m \in \mathscr{E}^m} f_{\theta^m}(x^n)$ exists and is unique, for every x^n in Γ^n, and for every n. The maximum likelihood estimate $\hat{\theta}^m(x^n)$ is called *consistent* if it converges asymptotically to the true parameter value θ^m—that is, if $\lim_{n\to\infty} \hat{\theta}^m(x^n) = \theta^m$ almost everywhere in f_{θ^m}. Various consistent estimates can be evaluated and compared, based on their asymptotic variance around the true parameter value θ^m. We note that consistency is a stronger property than the concept of unbias in Chapter 9. That is, a consistent parameter estimate is also asymptotically unbiased, but the inverse is generally untrue.

10.2 PARAMETRICALLY KNOWN STATIONARY AND MEMORYLESS PROCESSES

In this section, we focus on the special case where the data process is parametrically known, memoryless, and stationary, and the components of the unknown parameter vector θ^m are all included in the first order density function of the process. Then, given n and observation vector $x^n = [x_1, \ldots, x_n]^T$, we have

$$f_{\theta^m}(x^n) = \prod_{i=1}^{n} f_{\theta^m}(x_i) \quad (10.2.1)$$

10.2 Parametrically Known Stationary and Memoryless Processes

where T denotes transpose and $\prod$ denotes product, where $f_{\theta^m}(x^n)$ is the n-dimensional density function of the data process conditioned on the parameter value θ^m, and where $f_{\theta^m}(x^n)$ is well known for every given value θ^m.

To avoid complications in our presentation, we will assume that the data process and the vector parameter are both real. Therefore, $x^n \in R^n$ and $\theta^m \in \mathscr{E}^m$, where $\mathscr{E}^m \subset R^m$, R denotes the real line, and where R^k denotes k multiples of R. Given x^n, let us now search for the maximum likelihood parameter estimate $\hat{\theta}^m(x^n)$ that maximizes the density function $f_{\theta^m}(x^n)$ in (10.2.1). Due to the strict monotonicity of the logarithmic function, we may equivalently search for the vector parameter value that maximizes the function $\log f_{\theta^m}(x^n)$. That is

$$\hat{\theta}^m(x^n): \log f_{\hat{\theta}^m(x^n)}(x^n) = \sum_{i=1}^n \log f_{\hat{\theta}^m(x^n)}(x_i) = \sup_{\theta^m \in \mathscr{E}^m} \sum_{i=1}^n \log f_{\theta^m}(x_i) \quad (10.2.2)$$

Due to (10.2.2), it is clear that the properties of $\sum_{i=1}^n \log f_{\theta^m}(x_i)$ must be studied before the maximum likelihood estimate is sought. If, for example, the latter function is strictly concave with respect to θ^m for every n and every x^n in R^n, then a unique maximum likelihood estimate $\hat{\theta}^m(x^n)$ exists. If, as in Chapter 9, ∇ denotes the gradient with respect to the vector parameter θ^m, then $\hat{\theta}^m(x^n)$ is found by setting $\nabla \sum_{i=1}^n \log f_{\theta^m}(x_i) = 0$, where

$$\nabla \sum_{i=1}^n \log f_{\hat{\theta}^m}(x_i) = \sum_{i=1}^n \nabla \log f_{\theta^m}(x_i) = \sum_{i=1}^n \frac{\nabla f_{\theta^m}(x_i)}{f_{\theta^m}(x_i)} \quad (10.2.3)$$

It turns out that asymptotically ($n \to \infty$) the maximum likelihood estimate can be found by setting the gradient in (10.2.3) equal to zero, under conditions more general than strict concavity. Before we discuss such conditions, however, we will present an example.

Example 10.2.1

Let us assume that the data generating process is stationary, memoryless, and Gaussian, with unknown mean θ and unknown variance σ^2. The unknown parameter vector has thus dimensionality two, and its components are contained in the first order density function of the process. Let us denote,

$$f_{\theta^2}(x) = f_{\theta,\sigma}(x) = (2\pi)^{-1/2}\sigma^{-1}\exp\left\{-\frac{1}{2\sigma^2}(x-\theta)^2\right\} \quad (10.2.4)$$

Then, given x^n, we have

$$\log f_{\theta^2}(x^n) = \sum_{i=1}^n \log f_{\theta,\sigma}(x_i) \quad (10.2.5)$$

The expression in (10.2.5) is strictly concave with respect to both θ and σ, for every given x^n. Thus, due to (10.2.3), the maximum likelihood estimate

$\hat{\theta}^2(x^n) = [\hat{\theta}(x^n), \hat{\sigma}(x^n)]^T$ is found by setting

$$\sum_{i=1}^{n} \frac{\nabla f_{\theta,\sigma}(x_i)}{f_{\theta,\sigma}(x_i)} = 0 \tag{10.2.6}$$

where

$$\nabla f_{\theta,\sigma}(x_i) = \left[\frac{\partial}{\partial \theta} f_{\theta,\sigma}(x_i), \frac{\partial}{\partial \sigma} f_{\theta,\sigma}(x_i)\right]^T \tag{10.2.7}$$

$$\frac{\partial}{\partial \theta} f_{\theta,\sigma}(x_i) = \sigma^{-2}(x_i - \theta) f_{\theta,\sigma}(x_i) \tag{10.2.8}$$

$$\frac{\partial}{\partial \sigma} f_{\theta,\sigma}(x_i) = \sigma^{-3}[(x_i - \theta)^2 - \sigma^2] f_{\theta,\sigma}(x_i) \tag{10.2.9}$$

$$\frac{\nabla f_{\theta,\sigma}(x_i)}{f_{\theta,\sigma}(x_i)} = \sigma^{-2}[x_i - \theta, \sigma^{-1}\{(x_i - \theta)^2 - \sigma^2\}]^T \tag{10.2.10}$$

Substituting (10.2.10) in (10.2.6), we obtain

$$\left[\sum_{i=1}^{n} x_i - n\theta, \sigma^{-1}\left\{\sum_{i=1}^{n}(x_i - \theta)^2 - n\sigma^2\right\}\right]^T = 0 \tag{10.2.11}$$

or

$$\hat{\theta}(x^n), \hat{\sigma}(x^n): \begin{cases} \sum_{i=1}^{n} x_i - n\hat{\theta}(x^n) = 0 & (10.2.12) \\ \sum_{i=1}^{n} [x_i - \hat{\theta}(x^n)]^2 - n\hat{\sigma}^2(x^n) = 0 & (10.2.13) \end{cases}$$

Expression (10.2.12) gives

$$\hat{\theta}(x^n) = n^{-1} \sum_{i=1}^{n} x_i \tag{10.2.14}$$

Expression (10.2.13) then gives

$$\hat{\sigma}^2(x^n) = n^{-1} \sum_{i=1}^{n} \left[x_i - n^{-1} \sum_{j=1}^{n} x_j\right]^2 \tag{10.2.15}$$

The maximum likelihood estimates in (10.2.14) and (10.2.15) are unique for all x^n in R^n. In addition (due to the strong law of large numbers) they both converge asymptotically ($n \to \infty$) to the true acting parameters [$\hat{\theta}(x^n)$ to the true mean θ, and $\hat{\sigma}^2(x^n)$ to the true variance σ^2] almost everywhere in $f_{\theta,\sigma}$. They are thus consistent. If X^n denotes the n-dimensional random vector generated by the memoryless Gaussian process whose first order density is $f_{\theta,\sigma}(x)$ in (10.2.4), then $\hat{\theta}(X^n)$ is a Gaussian random variable for every n. The random variable $\hat{\sigma}^2(X^n)$ is only asymptotically ($n \to \infty$) Gaussian instead. The *rate of convergence* of

10.2 Parametrically Known Stationary and Memoryless Processes

the two estimates to the corresponding true parameter values is determined by the asymptotic variances of the random variables $\hat{\theta}(X^n)$ and $\hat{\sigma}^2(X^n)$, which both converge to zero inversely proportionally to n. The rate of convergence in both cases is thus n^{-1}.

Let us now consider a memoryless and stationary data generating process whose first order density function $f_{\theta^m}(x); x \in R$ exists for every $\theta^m \in \mathscr{E}^m \subset R^m$ and satisfies the conditions

$$0 = \nabla \int_R f_{\theta^m}(u)\, du = \int_R \nabla f_{\theta^m}(u)\, du \qquad \theta^m \in \mathscr{E}^m \qquad (10.2.16)$$

$$\int_R f_{\theta^m}(u) \left[\frac{\nabla f_{\theta^m}(u)}{f_{\theta^m}(u)}\right]^2 du < \infty \qquad \forall\, \theta^m \in \mathscr{E}^m \qquad (10.2.17)$$

$$H(f) = H(f_{\theta^m}) = -\int_R f_{\theta^m}(u) \log f_{\theta^m}(u)\, du < \infty \qquad \forall\, \theta^m \in \mathscr{E}^m \qquad (10.2.18)$$

$$\int_R f_{\theta^m}(u)[\log f_{\theta^m}(u) + H(f)]^2\, du < \infty \qquad \forall\, \theta^m \in \mathscr{E}^m \qquad (10.2.19)$$

where given θ^m, the quantity $H(f_{\theta^m})$ in (10.2.18) is the entropy of the data generating process, where it is assumed that the gradient $\nabla f_{\theta^m}(u)$ exists everywhere in $\mathscr{E}^m$ for every u in R, and where $H(f_{\theta^m}) = H(f)$ means that the entropy is not a function of the parameter value θ^m.

We can now express the following theorem.

THEOREM 10.2.1: If conditions (10.2.16)–(10.2.19) are satisfied, then given $x^n = [x_1, \ldots, x_n]^T$, the maximum likelihood parameter estimate $\hat{\theta}^m(x^n)$ is asymptotically ($n \to \infty$) found as a solution (with respect to θ^m) of the following equation.

$$T_f(\hat{\theta}^m, x^n) = n^{-1} \sum_{i=1}^{n} \frac{\nabla f_{\hat{\theta}^m}(x_i)}{f_{\hat{\theta}^m}(x_i)} = 0 \qquad (10.2.20)$$

If the solution of the equation in (10.2.20) is unique, for every given x^n, then the maximum likelihood estimate $\hat{\theta}^m(x^n)$ is asymptotically unique. ∎

Proof: As we have already established, given x^n, the maximum likelihood estimate is the parameter value that attains the supremum of $\sum_{i=1}^{n} \log f_{\theta^m}(x_i)$ or $n^{-1} \sum_{i=1}^{n} \log f_{\theta^m}(x_i)$ in $\mathscr{E}^m$. Now assume that the true vector parameter value is θ^m and consider the random variable

$$Y = n^{-1} \sum_{i=1}^{n} \log f_{\theta^m}(X_i) \qquad (10.2.21)$$

The random variables $W_i = \log f_{\theta^m}(X_i);\ 1 \le i \le n$, are then independent and identically distributed, with distribution determined by the density function $f_{\theta^m}(x)$. For $n \to \infty$, the strong law of large numbers then applies on the random variable Y in (10.2.21). That is, for $n \to \infty$, the random variable Y is Gaussian, with mean m and variance σ_n^2, given as

$$m = n^{-1} \sum_{i=1}^{n} E_{f_{\theta^m}}\{\log f_{\theta^m}(X_i)\} = \int_R f_{\theta^m}(u) \log f_{\theta^m}(u) \, du = -H(f) \quad (10.2.22)$$

$$\sigma_n^2 = n^{-2} \sum_{i=1}^{n} E_{f_{\theta^m}}\{\log f_{\theta^m}(X_i) + H(f)\}^2 = n^{-1} \int_R f_{\theta^m}(u) [\log f_{\theta^m}(u) + H(f)]^2 \, du$$
$$(10.2.23)$$

If conditions (10.2.18) and (10.2.19) are satisfied, then the entropy $H(f)$ is finite and the value of the random variable Y converges asymptotically ($n \to \infty$) to $-H(f)$, with rate inversely proportional to n, as determined by the variance in (10.2.23). Then, the gradient ∇Y exists asymptotically ($n \to \infty$) and it converges to $-\nabla H(f) = 0$. But from (10.2.21) we obtain

$$Z = \nabla Y = n^{-1} \sum_{i=1}^{n} \nabla \log f_{\theta^m}(X_i) = n^{-1} \sum_{i=1}^{n} \frac{\nabla f_{\theta^m}(X_i)}{f_{\theta^m}(X_i)} \quad (10.2.24)$$

Following the same arguments as with the random variable Y we conclude that the random variable Z in (10.2.24) is asymptotically ($n \to \infty$) Gaussian, with mean μ and variance ρ_n^2 given as

$$\mu = n^{-1} \sum_{i=1}^{n} E_{f_{\theta^m}}\left\{\frac{\nabla f_{\theta^m}(X_i)}{f_{\theta^m}(X_i)}\right\} = \int_R f_{\theta^m}(u) \frac{\nabla f_{\theta^m}(u)}{f_{\theta^m}(u)} \, du = \int_R \nabla f_{\theta^m}(u) \, du \quad (10.2.25)$$

$$\rho_n^2 = n^{-3} \sum_{i=1}^{n} E_{f_{\theta^m}}\left\{\frac{\nabla f_{\theta^m}(X_i)}{f_{\theta^m}(X_i)} - \mu\right\}^2 = n^{-1} \int_R f_{\theta^m}(u) \left[\frac{\nabla f_{\theta^m}(u)}{f_{\theta^m}(u)} - \mu\right]^2 du$$
$$(10.2.26)$$

But due to (10.2.16), (10.2.25) gives $\mu = 0$ and (10.2.26) gives

$$\rho_n^2 = n^{-1} \int_R f_{\theta^m}(u) \left[\frac{\nabla f_{\theta^m}(u)}{f_{\theta^m}(u)}\right]^2 du \quad (10.2.27)$$

If (10.2.17) is satisfied, then the variance in (10.2.27) converges (for $n \to \infty$) to zero with rate inversely proportional to n. Therefore, the quantity $n^{-1} \sum_{i=1}^{n} [\nabla f_{\theta^m}(x_i)/f_{\theta^m}(x_i)]$ converges asymptotically ($n \to \infty$) to $\mu = 0$ with rate n^{-1}, which are exactly the convergence characteristics of $\nabla n^{-1} \sum_{i=1}^{n} \log f_{\theta^m}(x_i)$ as established above. Thus, the solution $\hat{\theta}^m(x^n)$ of the equation in (10.2.20) is then such that

$$\lim_{n \to \infty} n^{-1} \sum_{i=1}^{n} \log f_{\hat{\theta}^m(x^n)}(x_i) = -H(f)$$

and $\hat{\theta}^m(x^n)$ then qualifies asymptotically as a maximum likelihood estimate. ∎

10.2 Parametrically Known Stationary and Memoryless Processes

An interesting application of Theorem 10.2.1 evolves when the unknown parameter is a scalar *location parameter*. Then, $f_\theta(x) = f_0(x - \theta) = f(x - \theta)$ where $f(x)$ denotes the first order density function of the data generating process at the point x, when the value of the location parameter equals zero. The gradient ∇ is here represented by the partial derivative, $\partial/\partial\theta$ and we obtain

$$\frac{\partial}{\partial\theta} \int_R f_\theta(u)\,du = \frac{\partial}{\partial\theta}(1) = 0 \tag{10.2.28}$$

$$\int_R \frac{\partial}{\partial\theta} f_\theta(u)\,du = \int_R \frac{\partial}{\partial\theta} f(u - \theta)\,du = -\int_R \frac{\partial}{\partial u} f(u)\,du = f(-\infty) - f(\infty) = 0 \tag{10.2.29}$$

$$\int_R f_\theta(u) \left[\frac{\frac{\partial}{\partial\theta} f_\theta(u)}{f_\theta(u)}\right]^2 du = \int_R f(u - \theta) \left[\frac{\frac{\partial}{\partial\theta} f(u - \theta)}{f(u - \theta)}\right]^2 du$$

$$= \int_R f(u) \left[\frac{-\frac{\partial}{\partial u} f(u)}{f(u)}\right]^2 du$$

$$= \int_R f(u) \left[\frac{\dot{f}(u)}{f(u)}\right]^2 du \tag{10.2.30}$$

$$-\int_R f_\theta(u) \log f_\theta(u)\,du = -\int_R f(u - \theta) \log f(u - \theta)\,du$$

$$= -\int_R f(u) \log f(u)\,du = H(f) \tag{10.2.31}$$

$$\int_R f_\theta(u)[\log f_\theta(u) + H(f)]^2\,du = \int_R f(u)[\log f(u) + H(f)]^2\,du \tag{10.2.32}$$

where $\dot{f}(u) = (\partial/\partial u)f(u)$.

Due to (10.2.28) and (10.2.29), (10.2.16) is always satisfied for the location parameter case. The entropy in (10.2.31) and the expression in (10.2.30) are also automatically not functions of the unknown parameter. The conditions in Theorem 10.2.1 are thus reduced, in the present case, to

$$I(f) = \int_R f(u) \left[\frac{\dot{f}(u)}{f(u)}\right]^2 du < \infty$$

$$H(f) = -\int_R f(u) \log f(u)\,du < \infty \tag{10.2.33}$$

$$\int_R f(u)[\log f(u) + H(f)]^2\,du < \infty$$

At the same time $T_f(\theta^m, x^n)$ in (10.2.20) takes here the form

$$T_f(\theta, x^n) = n^{-1} \sum_{i=1}^n \frac{\frac{\partial}{\partial\theta} f_\theta(x_i)}{f_\theta(x_i)} = n^{-1} \sum_{i=1}^n \frac{\frac{\partial}{\partial\theta} f(x_i - \theta)}{f(x_i - \theta)} \tag{10.2.34}$$

The quantity $I(f)$ in (10.2.33) is the Fisher information measure of the density f that we first saw in Chapter 7. From (10.2.27) in the proof of Theorem 10.2.1, we conclude that the variance of the random variable $T_f(\theta, X^n)$ in (10.2.34) when x^n is replaced by the random vector X^n equals $n^{-1}I(f)$. Thus, the Fisher information measure determines the rate with which the variable $T_f(\theta, X^n)$ converges (for $n \to \infty$) to its asymptotic value zero. From the above derivations, we can express the following corollary of Theorem 10.2.1.

COROLLARY 10.2.1: Let us consider the asymptotic maximum likelihood estimate of a scalar location parameter in the presence of data sequences x^n generated by a memoryless and stationary process whose first order density function is denoted f. Let the process satisfy the conditions in (10.2.33). That is, let the entropy $H(f)$ of the process and the Fisher information measure $I(f)$ exist ($< \infty$), and let $\int_R f(u)[\log f(u) + H(f)]^2 \, du < \infty$. Then, given x^n, the maximum likelihood estimate $\hat{\theta}(x^n)$ of the location parameter θ is asymptotically ($n \to \infty$) found from the solution of the following equation if, in addition, $(\partial/\partial x)f(x)$ exists everywhere.

$$T_f(\hat{\theta}, x^n) = n^{-1} \sum_{i=1}^{n} \frac{\left.\frac{\partial}{\partial \theta} f(x_i - \theta)\right|_{\theta=\hat{\theta}}}{f(x_i - \hat{\theta})} = 0 \qquad (10.2.35)$$

If θ is the true value of the location parameter, the mean of the random variable $T_f(\theta, X^n)$ equals zero and its variance equals $n^{-1}I(f)$. In addition, $T_f(\theta, X^n)$ is asymptotically ($n \to \infty$) Gaussian. ∎

It is now time to examine the properties of the maximum likelihood estimate $\hat{\theta}(x^n)$, which evolves as a solution of the equation in (10.2.35). Considering location parameter θ, we denote by $f(x)$; $x \in R$ the first order density function of the stationary and memoryless data generating process when $\theta = 0$, and we denote $\dot{f}(x) = (\partial/\partial x)f(x)$ and $\ddot{f}(x) = (\partial^2/\partial x^2)f(x)$, assuming that the latter derivatives exist. We subsequently express the following theorem.

THEOREM 10.2.2: Let us consider the maximum likelihood estimate of a location parameter in the presence of data sequences x^n generated by a stationary and memoryless real process, whose first order density is $f(u)$; $u \in R$. Let $f(u)$ be such that the derivatives $\dot{f}(u)$ and $\ddot{f}(u)$ exist for every u in R. Let in addition, $\dot{f}(\infty) = \dot{f}(-\infty) = 0$, $\int_R du\, \dot{f}(u)(\partial/\partial u)(\dot{f}(u)/f(u)) < \infty$, let the conditions in (10.2.33) be satisfied, and let the function $\psi(u) = -\dot{f}(u)/f(u)$ be strictly monotonically increasing in R and such that $\psi(-\infty) < 0$ and $\psi(\infty) > 0$. Then, the maximum likelihood estimate $\hat{\theta}(x^n)$ of the location parameter θ is asymptotically the solution of (10.2.35), and it is then unique. In addition, the random variable $n^{1/2}\hat{\theta}(X^n)$ is then asymptotically ($n \to \infty$) Gaussian with variance $I^{-1}(f)$, where $I(f)$ is the Fisher information measure in (10.2.33). If the true value of the parameter is θ, the asymptotic mean of the random variable $\hat{\theta}(X^n)$ is θ. The estimate $\hat{\theta}(x^n)$ is thus consistent. ∎

10.2 Parametrically Known Stationary and Memoryless Processes

Proof: That the maximum likelihood estimate is asymptotically a solution of the equation in (10.2.35), has been established in Corollary 10.2.1. Since $f(u)$ is doubly differentiable, $\psi(u)$ is continuous and differentiable everywhere in R. Since

$$\frac{\frac{\partial}{\partial \theta} f(x_i - \theta)}{f(x_i - \theta)} = -\frac{\frac{\partial}{\partial x_i} f(x_i - \theta)}{f(x_i - \theta)} = -\frac{\dot{f}(x_i - \theta)}{f(x_i - \theta)} = \psi(x_i - \theta)$$

and due to the continuity and the assumed strict monotonicity of $\psi(u)$, we conclude that given x^n, the solution $\hat{\theta}(x^n)$ of (10.2.35) is unique. In addition, due to the continuity and strict monotonicity of $\psi(\cdot)$, we also then conclude

$$\hat{\theta}(x^n) < \theta + n^{-1/2} y \Rightarrow n^{-1} \sum_{i=1}^{n} \psi(x_i - \theta - n^{-1/2} y) < 0$$

$$\Rightarrow \hat{\theta}(x^n) \leq \theta + n^{-1/2} y \quad (10.2.36)$$

where θ is the true value of the location parameter, and where y is some arbitrary constant in R. Let P_θ denote probability conditioned on the event that the true value of the parameter is θ. From (10.2.36) we then obtain,

$$P_\theta\{\hat{\theta}(X^n) \leq \theta + n^{-1/2} y\} \geq P_\theta\left\{n^{-1} \sum_{i=1}^{n} \psi(X_i - \theta - n^{-1/2} y) < 0\right\}$$

$$\geq P_\theta\{\hat{\theta}(X^n) < \theta + n^{-1/2} y\} \quad (10.2.37)$$

Given θ, the random variables $\psi(X_i - \theta - n^{-1/2} y); 1 \leq i \leq n$ in (10.2.37) are independent and identically distributed. Due to the strong law of large numbers, the random variable $n^{-1} \sum_{i=1}^{n} \psi(X_i - \theta - n^{-1/2} y)$, is then asymptotically ($n \to \infty$) Gaussian, with mean m and variance σ_n^2 given by

$$m = E_{f_\theta}\{\psi(X - \theta - n^{-1/2} y)\} = \int_R \psi(u - \theta - n^{-1/2} y) f_\theta(u) \, du$$

$$= -\int_R \frac{\dot{f}(u - \theta - n^{-1/2} y)}{f(u - \theta - n^{-1/2} y)} f(u - \theta) \, du = -\int_R \frac{\dot{f}(u - n^{-1/2} y)}{f(u - n^{-1/2} y)} f(u) \, du$$

$$= \int_R \psi(u - n^{-1/2} y) f(u) \, du \quad (10.2.38)$$

$$\sigma_n^2 = n^{-1} E_{f_\theta}\{\psi(X - \theta - n^{-1/2} y) - m\}^2$$

$$= n^{-1}\left\{\int_R \psi^2(u - \theta - n^{-1/2} y) f(u - \theta) \, du - m^2\right\}$$

$$= n^{-1}\left\{\int_R \psi^2(u - n^{-1/2} y) f(u) \, du - m^2\right\} \quad (10.2.39)$$

For fixed y, for asymptotically large values of n, and due to the differentiability of $\psi(\cdot)$, we use the following first order approximations.

$$\psi(u - n^{-1/2}y) \underset{n \to \infty}{\simeq} \psi(u) - n^{-1/2}y\dot{\psi}(u)$$
$$\psi^2(u - n^{-1/2}y) \underset{n \to \infty}{\simeq} \psi^2(u) - n^{-1/2}y[2\psi(u)\dot{\psi}(u)] \qquad (10.2.40)$$

Then,

$$\int_R \psi(u - n^{-1/2}y)f(u)\,du \underset{n \to \infty}{\simeq} \int_R \psi(u)f(u)\,du - n^{-1/2}y\int_R \dot{\psi}(u)f(u)\,du$$

$$= -\int_R \frac{\dot{f}(u)}{f(u)} f(u)\,du + n^{-1/2}y$$

$$\times \int_R \left[\frac{\ddot{f}(u)}{f(u)} - \left(\frac{\dot{f}(u)}{f(u)}\right)^2\right] f(u)\,du$$

$$= -\int_R \dot{f}(u)\,du + n^{-1/2}y \int_R \ddot{f}(u)\,du$$

$$- n^{-1/2}y \int_R \left[\frac{\dot{f}(u)}{f(u)}\right]^2 f(u)\,du$$

$$= -n^{-1/2}yI(f) \qquad (10.2.41)$$

$$\int_R \psi^2(u - n^{-1/2}y)f(u)\,du \underset{n \to \infty}{\simeq} \int_R \psi^2(u)f(u)\,du - 2n^{-1/2}y \int_R \psi(u)\dot{\psi}(u)f(u)\,du$$

$$= \int_R \left[\frac{\dot{f}(u)}{f(u)}\right]^2 f(u)\,du - 2n^{-1/2}y \int_R du\, \dot{f}(u)\frac{\partial}{\partial u}\left(\frac{\dot{f}(u)}{f(u)}\right)$$

$$= I(f) - 2n^{-1/2}y \int_R du\, \dot{f}(u) \frac{\partial}{\partial u}\left(\frac{\dot{f}(u)}{f(u)}\right) \qquad (10.2.42)$$

From (10.2.38) and (10.2.41) we obtain

$$m \underset{n \to \infty}{\simeq} -n^{-1/2}yI(f) \qquad (10.2.43)$$

From (10.2.39) and (10.2.42) we obtain

$$\sigma_n^2 \underset{n \to \infty}{\simeq} n^{-1}\left\{I(f) - 2n^{-1/2}y \int_R du\, \dot{f}(u)\frac{\partial}{\partial u}\left(\frac{\dot{f}(u)}{f(u)}\right) - m^2\right\} \qquad (10.2.44)$$

Due to (10.2.43) and the assumed boundedness of $\int_R du\, \dot{f}(u)(\partial/\partial u)[\dot{f}(u)/f(u)]$, we further simplify (10.2.44) to obtain

$$\sigma_n^2 \underset{n \to \infty}{\simeq} n^{-1}I(f) \qquad (10.2.45)$$

Denoting by $\Phi(x)$ the cumulative distribution of the Gaussian, zero mean and unit variance random variable at the point x, and due to (10.2.43) and

10.2 Parametrically Known Stationary and Memoryless Processes

(10.2.45), we obtain, for $n \to \infty$,

$$P_\theta\left\{n^{-1}\sum_{i=1}^n \psi(X_i - \theta - n^{-1/2}y) < 0\right\} = \Phi\left(-\frac{m}{\sigma_n}\right)$$

$$= \Phi[yI^{1/2}(f)] \quad (10.2.46)$$

Directly from (10.2.37) and (10.2.46), we then conclude, for $n \to \infty$,

$$P_\theta\{\hat{\theta}(X^n) - \theta < n^{-1/2}y\} \leq \Phi[yI^{1/2}(f)] \leq P_\theta\{\hat{\theta}(X^n) - \theta \leq n^{-1/2}y\}$$

which implies that

$$P_\theta\{\hat{\theta}(X^n) - \theta \leq n^{-1/2}y\} \underset{n \to \infty}{\simeq} \Phi\left(\frac{y}{I^{-1/2}(f)}\right) \quad (10.2.47)$$

Expression (10.2.47) basically says that the random variable $n^{1/2}\hat{\theta}(X^n)$ is asymptotically ($n \to \infty$) Gaussian, with mean $n^{1/2}\theta$ (if the true parameter value is θ) and with variance $I^{-1}(f)$. Since $I^{-1}(f)$ is bounded, the estimate $\hat{\theta}(x^n)$ converges then to the true parameter value θ with rate n^{-1}. The estimate is thus consistent. ∎

We will complete this section with an example.

Example 10.2.2

Let us revisit the problem in Example 10.2.1, now assuming that $\sigma = 1$ and known, and where the location parameter θ must be only estimated. Here, the first order density function f is the zero mean, unit variance Gaussian density function, $\phi(u)$; $u \in R$, which satisfies all the conditions in Theorem 10.2.2. Indeed, the derivatives $\dot{\phi}(u)$ and $\ddot{\phi}(u)$ exist for all u in R, $\dot{\phi}(-\infty) = \dot{\phi}(\infty) = 0$, and

$$\psi(u) = -\frac{\dot{\phi}(u)}{\phi(u)} = u \quad \text{(strictly monotonic, } \psi(-\infty) < 0, \psi(\infty) > 0,\text{ continuous and differentiable)}$$

$$\int_R du\, \phi(u) \frac{\partial}{\partial u}\left(\frac{\dot{\phi}(u)}{\phi(u)}\right) = \int_R \dot{\phi}(u)\, du = 0$$

$$I(\phi) = \int_R \left[\frac{\dot{\phi}(u)}{\phi(u)}\right]^2 \phi(u)\, du = \int_R u^2 \phi(u)\, du = 1 < \infty$$

$$H(f) = -\int_R \phi(u) \ln \phi(u)\, du = +\frac{1}{2}\ln 2\pi + \frac{1}{2}\int_R u^2 \phi(u)\, du$$

$$= \frac{1}{2}[1 + \ln 2\pi] < \infty$$

$$\int_R \phi(u)[\ln \phi(u) + H(f)]^2\, du = \int_R \phi(u)\left[-\frac{u^2}{2} + \frac{1}{2}\right]^2 du < \infty$$

where ln denotes natural logarithm.

Thus, the maximum likelihood estimate is here found asymptotically by (10.2.35), it is unique, and it is asymptotically Gaussian with variance $n^{-1}I(\phi) = n^{-1}$. It is also consistent.

Since we have

$$\frac{\frac{\partial}{\partial \theta} \phi(x_i - \theta)}{\phi(x_i - \theta)} = x_i - \theta$$

the equation in (10.2.35) takes here the form

$$\sum_{i=1}^{n} (x_i - \hat{\theta}) = 0 \rightarrow \hat{\theta}(x^n) = n^{-1} \sum_{i=1}^{n} x_i \qquad (10.2.48)$$

The estimate in (10.2.48) is the unique maximum likelihood estimate for every dimensionality n, as established in Example 10.2.1.

10.3 CONSISTENT LOCATION PARAMETER ESTIMATES

In this section we will present and analyze some estimates of location parameters that do not necessarily evolve from maximum likelihood formalizations but are consistent. In particular, assuming real, stationary and ergodic data generating processes, and given a data vector x^n, we first consider the following two estimates for location parameters, where f denotes the first order density function of the data process when the location parameter is zero, and where

$$F(x) = \int_{-\infty}^{x} f(u)\,du \quad \text{and} \quad F^{-1}(y) = x \rightarrow y = F(x).$$

$$\hat{\theta}_A(x^n) = n^{-1} \sum_{i=1}^{n} x_i \qquad (10.3.1)$$

$$\hat{\theta}_S(x^n) = -F^{-1}\left(n^{-1} \sum_{i=1}^{n} \operatorname{sgn} x_i\right) \quad \operatorname{sgn} x = \begin{cases} 0 & x \geq 0 \\ 1 & x < 0 \end{cases}$$

For real stationary and memoryless data generating processes, we also consider the following class of location parameter estimates.

$$\text{Given} \quad x^n, \hat{\theta}_M(x^n): n^{-1} \sum_{i=1}^{n} \psi[x_i - \hat{\theta}_M(x^n)] = 0 \qquad (10.3.2)$$

where $\psi(u)$ is a continuous, scalar, real, and monotonically increasing function such that $\psi(-\infty) < 0$ and $\psi(\infty) > 0$.

The estimate $\hat{\theta}_A(x^n)$ in (10.3.1) is the empirical mean estimate. The estimate $\hat{\theta}_S(x^n)$ in (10.3.1) is called the median estimate. The class of estimates in (10.3.2) generated by various choices of the function $\psi(\cdot)$ has been called M-estimates

10.3 Consistent Location Parameter Estimates

by Huber (1963). We now express in a lemma, the properties of the estimates $\hat{\theta}_A(x^n)$ and $\hat{\theta}_S(x^n)$ in (10.3.1).

LEMMA 10.3.1: Let the data vector x^n be generated by a real, stationary, and ergodic process. Then $\hat{\theta}_A(x^n)$ is a consistent estimate of the mean of the process. If the first order cumulative distribution function F of the data process with mean zero is also continuous and strictly monotone, then $\hat{\theta}_S(x^n)$ is also a consistent estimate of the mean. If the data generating process is stationary and memoryless, then the random variable $\hat{\theta}_A(X^n)$ is also asymptotically Gaussian. The same is then true for the random variable $F[-\hat{\theta}_S(X^n)]$, if in addition F is continuous and strictly monotone. ∎

Proof:

i. Let the data generating process be stationary and ergodic (not necessarily memoryless). Then, due to the ergodic theorem, $n^{-1}\sum_{i=1}^{n} x_i$ converges to the mean of the process almost everywhere and in probability. Thus, $\hat{\theta}_A(x^n)$ is a consistent estimate of the mean of the process (or location parameter). Due to the ergodic theorem, we also conclude that the statistic $n^{-1}\sum_{i=1}^{n} \text{sgn } x_i$ is a consistent estimate of the probability that a single datum from the process attains negative values. Given mean value θ, the latter probability equals $F(-\theta)$. If F is also continuous and strictly monotone, then clearly $F^{-1}(n^{-1}\sum_{i=1}^{n} \text{sgn } x_i)$ is a consistent estimate of $-\theta$, and $\hat{\theta}_S(x^n)$ is a consistent estimate of θ.

ii. Let the data generating process be stationary and memoryless. Since ergodicity then holds, the conclusions in part i are still true. In addition, if the process has mean θ and variance σ^2, the strong law of large numbers applies to both the random variables, $n^{-1}\sum_{i=1}^{n} X_i = \hat{\theta}_A(X^n)$ and $n^{-1}\sum_{i=1}^{n} \text{sgn } X_i$. Thus, they are both asymptotically Gaussian. The mean of the variable $\hat{\theta}_A(X^n)$ is θ and its variance is $n^{-1}\sigma^2$. The mean of the variable $n^{-1}\sum_{i=1}^{n} \text{sgn } X_i$ is then $F(-\theta)$ and its variance is $n^{-1}[F(-\theta) - F^2(-\theta)]$. If F is continuous and strictly monotone, then $F[-\hat{\theta}_S(X^n)] = n^{-1}\sum_{i=1}^{n} \text{sgn } X_i$, which completes the proof of the lemma. We note that $F(-\theta) - F^2(-\theta) \leq 4^{-1} \; \forall \; \theta \in R$. Thus, if $\sigma^2 > 4^{-1}$ the estimate $\hat{\theta}_S(x^n)$ converges faster than the estimate $\hat{\theta}_A(x^n)$, assuming that F is continuous and strictly monotone. In the latter case, the median estimate is asymptotically better than the empirical mean estimate, for every true parameter value θ. ∎

As we pointed out in the proof of Lemma 10.3.1, the median estimate $\hat{\theta}_S(x^n)$ may be asymptotically better than the empirical mean estimate $\hat{\theta}_A(x^n)$. The above property does not generally extend to include small sample sizes n, however. This is due to the discontinuities of the small-sample distribution of the variable $\hat{\theta}_S(X^n)$ and its relatively heavy tails. The effects of the above discontinuities will be further discussed in Chapter 12. We now turn to the M-estimates. Let $f(u); u \in R$ be the first order density function of the process that generates the data sequences x^n, and let this process be stationary and memoryless. Let $\psi(u)$; $u \in R$ be a real scalar function, and let us then define

$$\left.\begin{array}{l} m(\lambda) = \int_R \psi(u - \lambda) f(u)\, du \\ \sigma^2(\lambda) = \int_R \psi^2(u - \lambda) f(u)\, du \end{array}\right\} \quad \lambda \in R \quad (10.3.3)$$

We can then express a theorem, regarding the asymptotic properties of the M-estimates.

THEOREM 10.3.1: Let us consider a real, memoryless, and stationary data generating process, whose first order density function is $f(u)$; $u \in R$. Let $\psi(u)$; $u \in R$ be a scalar real function that is monotonically increasing and is such that $\psi(-\infty) < 0$ and $\psi(\infty) > 0$. In addition, let some λ_0 exist such that $m(\lambda_0) = 0$, where the function $m(\lambda)$ is given in (10.3.3). Let $m(\lambda)$ and $\sigma^2(\lambda)$ [also in (10.3.3)] both be differentiable at $\lambda = \lambda_0$, let $-\infty < \dot{m}(\lambda_0) < 0$, and let $|(\partial/\partial\lambda)\sigma^2(\lambda)|_{\lambda=\lambda_0}| < \infty$, $\sigma^2(\lambda_0) < \infty$. Then, the M-estimate $\hat{\theta}_M(x^n)$ in (10.3.2) is such that $n^{1/2}\hat{\theta}_M(X^n)$ is asymptotically Gaussian with mean $n^{1/2}\lambda_0$ and variance $\sigma^2(\lambda_0)/[\dot{m}(\lambda_0)]^2$. If λ_0 is the mean value of the data generating process, then $\hat{\theta}_M(x^n)$ is a consistent location parameter estimate.

Proof: Given a data sequence x^n, the M-estimate $\hat{\theta}_M(x^n)$ satisfies

$$n^{-1} \sum_{i=1}^{n} \psi[x_i - \hat{\theta}_M(x^n)] = 0$$

Let y be some arbitrary real scalar. Then, due to the monotonicity of the function $\psi(u)$ and the assumption that $\psi(-\infty) < 0$ and $\psi(\infty) > 0$, and exactly as in (10.2.36) and (10.2.37) in the proof of Theorem 10.2.2, we obtain

$$P\{\hat{\theta}_M(X^n) \leq \lambda_0 + n^{-1/2}y\} \geq P\left\{n^{-1}\sum_{i=1}^{n}\psi(X_i - \lambda_0 - n^{-1/2}y) < 0\right\}$$

$$\geq P\{\hat{\theta}(X^n) < \lambda_0 + n^{-1/2}y\} \quad (10.3.4)$$

Since the variables $\psi(X_i - \lambda_0 - n^{-1/2}y)$; $1 \leq i \leq n$ are independent and identically distributed, the strong law of large numbers gives

$$\lim_{n \to \infty} P\left\{n^{-1}\sum_{i=1}^{n}\psi(X_i - \lambda_0 - n^{-1/2}y) < 0\right\}$$

$$= \Phi\left(-n^{-1/2} \frac{m(\lambda_0 + n^{-1/2}y)}{[\sigma^2(\lambda_0 + n^{-1/2}y) - m^2(\lambda_0 + n^{-1/2}y)]^{1/2}}\right) \quad (10.3.5)$$

where $m(\lambda)$ and $\sigma^2(\lambda)$ are given in (10.3.3), and where $\Phi(\cdot)$ is the cumulative distribution of the zero mean, unit variance Gaussian random variable.

10.3 Consistent Location Parameter Estimates

Due to the differentiability of the functions $m(\lambda)$ and $\sigma^2(\lambda)$ at $\lambda = \lambda_0$, due to the boundness of $|\dot{m}(\lambda_0)|$ and $|(\partial/\partial\lambda)\sigma^2(\lambda)|_{\lambda=\lambda_0}|$, and using first order approximations, we obtain, for $n \to \infty$,

$$m(\lambda_0 + n^{-1/2}y) \simeq m(\lambda_0) + n^{-1/2}y\dot{m}(\lambda_0) = n^{-1/2}y\dot{m}(\lambda_0)$$

$$\sigma^2(\lambda_0 + n^{-1/2}y) \simeq \sigma^2(\lambda_0) + n^{-1/2}y\frac{\partial}{\partial\lambda}\sigma^2(\lambda)\Big|_{\lambda=\lambda_0} \simeq \sigma^2(\lambda_0)$$

$$\sigma^2(\lambda_0 + n^{-1/2}y) - m^2(\lambda_0 + n^{-1/2}y) \simeq \sigma^2(\lambda_0) + n^{-1/2}y\frac{\partial}{\partial\lambda}\sigma^2(\lambda)\Big|_{\lambda=\lambda_0}$$

$$- n^{-1}y^2[\dot{m}(\lambda_0)]^2 \simeq \sigma^2(\lambda_0)$$

(10.3.6)

Due to the expressions in (10.3.6), (10.3.5) gives by substitution

$$\lim_{n\to\infty} P\left\{n^{-1}\sum_{i=1}^{n}\psi(X_i - \lambda_0 - n^{-1/2}y) < 0\right\} = \Phi\left(y\frac{|\dot{m}(\lambda_0)|}{\sigma(\lambda_0)}\right) \quad (10.3.7)$$

From (10.3.4) and (10.3.7), we finally obtain

$$\lim_{n\to\infty} P\{\hat{\theta}_M(X^n) \leq \lambda_0 + n^{-1/2}y\} = \lim_{n\to\infty} P\{(\hat{\theta}_M(X^n) - \lambda_0)n^{1/2} \leq y\}$$

$$= \Phi\left(y\frac{|\dot{m}(\lambda_0)|}{\sigma(\lambda_0)}\right) \quad (10.3.8)$$

From (10.3.8) we conclude that the random variable $n^{1/2}(\hat{\theta}_M(X^n) - \lambda_0)$ is asymptotically Gaussian, with zero mean and variance $\sigma^2(\lambda_0)/[\dot{m}(\lambda_0)]^2$, which proves the theorem. ∎

If the density function $f(u)$; $u \in R$, in Theorem 10.3.1 has finite variance σ^2 then the function $\psi(u) = u$; $u \in R$ satisfies all the conditions in the theorem and provides a consistent M-estimate of the mean value of $f(u)$. In fact, the latter estimate is then identical to the empirical mean estimate in (10.3.1). We note that every continuous function $\psi(\cdot)$ provides a continuous M-estimate, with respect to x^n as is clear from (10.3.2). A special case of an M-estimate for a location parameter is provided by Theorem 10.2.2, where $\psi(u) = -\dot{f}(u)/f(u)$. If the density function $f(u)$ in Theorem 10.3.1 is symmetric around its mean, then any $\psi(u)$ function that is odd, differentiable at zero, and bounded in R, provides a consistent M-estimate for the mean of f, as is clearly concluded from Theorem 10.3.1.

Subject to the conditions in Lemma 10.3.1 and Theorem 10.3.1, the empirical mean, median, and M-estimates are all consistent, converging to the true value of the location parameter with rate n^{-1} times a constant. The asymptotic ($n \to \infty$) comparison of those estimates is thus based on the values of the respective constants. In that spirit, we conclude this section with two examples.

Example 10.3.1

Let us consider the case where the mean of a stationary and memoryless Gaussian process must be estimated. Let the variance of the process be σ^2 and let the data sequence x^n be given. Let us consider the empirical mean and the median estimates, and the M-estimate with

$$\psi(u) = \begin{cases} c & u \geq c \\ u & -c < u < c \\ -c & u \leq -c \end{cases} \qquad c > 0 \qquad (10.3.9)$$

All three estimates are consistent. As we have seen, the empirical mean estimate converges to the mean with rate $n^{-1}\sigma^2$. If the true value of the mean is θ, the median estimate converges to θ with rate $n^{-1}[\Phi(-\theta/\sigma) - \Phi^2(-\theta/\sigma)] = n^{-1}\Phi(-\theta/\sigma)\Phi(\theta/\sigma)$, where $\Phi(u)$ is the cumulative distribution of the zero mean, unit variance Gaussian random variable at the point u. The M-estimate converges to θ with rate $n^{-1}\sigma^2(\theta)/[\dot{m}(\theta)]^2$, where

$$m(\lambda) = \int_R \psi(u - \lambda)f(u)\,du = \int_R \psi(x)f(x + \lambda)\,dx$$

$$m(\theta) = 0$$

$$\dot{m}(\lambda) = \int_R \psi(x)\frac{\partial}{\partial \lambda}f(x + \lambda)\,dx = \int_R \psi(x)\frac{\partial}{\partial x}f(x + \lambda)\,dx$$

$$= \int_R \psi(x)\frac{\partial}{\partial x}\frac{\exp\{-(x + \lambda - \theta)^2/2\sigma^2\}}{(2\pi)^{1/2}\sigma}\,dx$$

$$= -\int_R \sigma^{-2}(x + \lambda - \theta)\psi(x)\frac{\exp\{-(x + \lambda - \theta)^2/2\sigma^2\}}{(2\pi)^{1/2}\sigma}\,dx$$

$$\dot{m}(\theta) = -\int_R \sigma^{-2}x\psi(x)\frac{\exp\{-x^2/2\sigma^2\}}{(2\pi)^{1/2}\sigma}\,dx = -\sigma^{-1}\int_{-\infty}^{\infty} u\psi(\sigma u)\phi(u)\,du$$

$$= -\int_{-c/\sigma}^{c/\sigma} u^2\phi(u)\,du - \frac{c}{\sigma}\int_{c/\sigma}^{\infty} u\phi(u)\,du + \frac{c}{\sigma}\int_{-\infty}^{-c/\sigma} u\phi(u)\,du$$

$$= 1 - 2\Phi\left(\frac{c}{\sigma}\right) \qquad (10.3.10)$$

$$\sigma^2(\theta) = \int_R \psi^2(u - \theta)f(u)\,du = \int_R \psi^2(x)f(x + \theta)\,dx = \int_R \psi^2(x)\frac{\exp\{-x^2/2\sigma^2\}}{(2\pi)^{1/2}\sigma}\,dx$$

$$= \int_R \psi^2(\sigma u)\phi(u)\,du = 2\int_{-\infty}^{0} \psi^2(\sigma u)\phi(u)\,du$$

$$= 2c^2 \int_{-\infty}^{-c/\sigma} \phi(u)\,du + 2\sigma^2 \int_{-c/\sigma}^{0} u^2\phi(u)\,du$$

$$= \sigma^2 + 2(c^2 - \sigma^2)\Phi\left(-\frac{c}{\sigma}\right) - 2\sigma c\phi\left(\frac{c}{\sigma}\right) \qquad (10.3.11)$$

10.3 Consistent Location Parameter Estimates

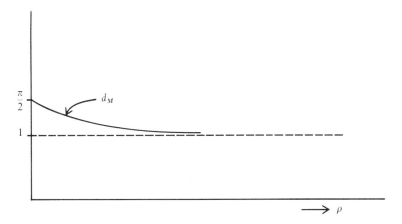

Figure 10.3.1

where $\phi(u)$ is the density of the zero mean unit variance Gaussian random variable at the point u.

Therefore, denoting the constants that control the convergence rate of the empirical mean, median, and M-estimates, respectively by d_A, d_S, and d_M, we have

$$d_A = \sigma^2 \qquad d_S = \Phi\left(-\frac{\theta}{\sigma}\right)\Phi\left(\frac{\theta}{\sigma}\right) \qquad (10.3.12)$$

$$d_M = \left[\sigma^2 + 2(c^2 - \sigma^2)\Phi\left(-\frac{c}{\sigma}\right) - 2\sigma c\phi\left(\frac{c}{\sigma}\right)\right]\left[1 - 2\Phi\left(\frac{c}{\sigma}\right)\right]^{-2}$$

$$= \sigma^2[1 + 2(\rho^2 - 1)\Phi(-\rho) - 2\rho\phi(\rho)][1 - 2\Phi(\rho)]^{-2}$$

where

$$\rho = c/\sigma$$

We observe that the constants d_A and d_M are not functions of the true mean value θ while the constant d_S is. In Figure 10.3.1, we plot the constant d_M as a function of ρ for $\sigma = 1$. We observe that the constant d_M attains its minimum, one, for $c \to \infty$. Then the M-estimate becomes identical to the empirical mean estimate, and we conclude that for Gaussian data, the empirical mean location parameter estimate is generally superior to the M-estimate, where in the latter the function in (10.3.9) is considered. From (10.3.12), we observe that $d_S \leq 4^{-1}$ for every θ value. Considering $\sigma = 1$, we then conclude that the Median estimate is superior to the empirical mean estimate, for every true value θ.

Example 10.3.2

Let us consider the case where the density function $f(u)$; $u \in R$ of the memoryless and stationary process that generates the data sequence is given by the following mixture, where $0 < \varepsilon < 1$, $m > 0$, and $\phi(u)$ denotes the density function of the zero mean unit variance Gaussian random variable at the point u.

$$f(u) = (1 - \varepsilon)\phi(u) + \frac{\varepsilon}{2}[\phi(u - m) + \phi(u + m)] \tag{10.3.13}$$

The mean of the density function in (10.3.13) is clearly zero, while its variance σ^2 is given by

$$\sigma^2 = (1 - \varepsilon)\int_R u^2 \phi(u)\, du + \frac{\varepsilon}{2}\int_R u^2 \phi(u - m)\, du + \frac{\varepsilon}{2}\int_R u^2 \phi(u + m)\, du$$

$$= (1 - \varepsilon) + \frac{\varepsilon}{2}\int_R (x + m)^2 \phi(x)\, dx + \frac{\varepsilon}{2}\int_R (x - m)^2 \phi(x)\, dx$$

$$= (1 - \varepsilon) + \frac{\varepsilon}{2}[2 + 2m^2] = 1 + \varepsilon m^2 \tag{10.3.14}$$

The empirical mean estimate $\hat{\theta}_A(x^n) = n^{-1}\sum_{i=1}^n x_i$ is thus consistent with rate of convergence

$$r_A = n^{-1}[1 + \varepsilon m^2] \tag{10.3.15}$$

Let us now consider the M-estimate with $\psi(u)$; $u \in R$ as in (10.3.9), as a candidate for the estimation of the mean zero of the data process. From (10.3.3), (10.3.9), and (10.3.13), we obtain

$$m(\lambda) = \int_R \psi(u - \lambda) f(u)\, du$$

$$= (1 - \varepsilon)\int_R \psi(u - \lambda)\phi(u)\, du + \frac{\varepsilon}{2}\int_R \psi(u - \lambda)\phi(u - m)\, du$$

$$+ \frac{\varepsilon}{2}\int_R \psi(u - \lambda)\phi(u + m)\, du$$

$$m(0) = \frac{\varepsilon}{2}\int_R \psi(u)\phi(u - m)\, du + \frac{\varepsilon}{2}\int_R \psi(u)\phi(u + m)\, du = 0 \tag{10.3.16}$$

$$\sigma^2(0) = \int_R \psi^2(u) f(u)\, du$$

$$= (1 - \varepsilon)\int_R \psi^2(u)\phi(u)\, du + \frac{\varepsilon}{2}\int_R \psi^2(u)\phi(u - m)\, du + \frac{\varepsilon}{2}\int_R \psi^2(u)\phi(u + m)\, du$$

$$= (1 - \varepsilon)[2c^2 - 1 - 2(c^2 - 1)\Phi(c) - 2c\phi(c)]$$
$$+ \varepsilon\{2c^2 - m^2 - 1 + (m^2 - c^2 + 1)[\Phi(c + m) + \Phi(c - m)]$$
$$- (c + m)\phi(c - m) + (m - c)\phi(c + m)\} \tag{10.3.17}$$

10.3 Consistent Location Parameter Estimates

where $\Phi(u)$ denotes the cumulative distribution function of the zero mean unit variance Gaussian random variable at the point u. As in (10.3.10), we also find

$$\dot{m}(0) = -(1-\varepsilon)\int_R x\psi(x)\phi(x)\,dx$$

$$-\frac{\varepsilon}{2}\left\{\int_R (x-m)\psi(x)\phi(x-m)\,dx + \int_R (x+m)\psi(x)\phi(x+m)\,dx\right\}$$

$$= (1-\varepsilon)[1 - 2\Phi(c)] + \varepsilon[\Phi(c+m) + \Phi(c-m) - 1] \qquad (10.3.18)$$

The M-estimate defined via the function $\psi(u)$ in (10.3.9) satisfies the conditions in Theorem 10.3.1. Due to (10.3.16), this estimate is consistent, regarding the mean zero of the data generating process. In addition, the estimate converges with rate $r_M = n^{-1}\sigma^2(0)/[\dot{m}(0)]^2$, where $\dot{m}(0)$ and $\sigma^2(0)$ are given respectively by (10.3.18) and (10.3.17).

Let us now consider the case where the parameter m in (10.3.13) is asymptotically large. Then the density in (10.3.13) represents the occurrence (with probability ε) of extreme, symmetric around zero outliers, when the nominal data generating process is represented by the zero mean Gaussian density $\phi(u)$. For such extreme m values, we will use the following approximations (for fixed c).

$$\Phi(c+m) \underset{m\gg 1}{\simeq} \Phi(m) \qquad \Phi(c-m) \underset{m\gg 1}{\simeq} \Phi(-m)$$

$$c+m \underset{m\gg 1}{\simeq} m \qquad m-c \underset{m\gg 1}{\simeq} m$$

Applying the above approximations in (10.3.18) and (10.3.17), we easily obtain

$$d_M(c) = \sigma^2(0)/[\dot{m}(0)]^2 \underset{m\gg 1}{\simeq} (1-\varepsilon)^{-1}[1-2\Phi(c)]^{-2}$$

$$\times \left[\frac{2-\varepsilon}{1-\varepsilon}c^2 - 1 - 2(c^2-1)\Phi(c) - 2c\,\phi(c)\right] \qquad (10.3.19)$$

It can be found that $d_M(0) = \pi/2(1-\varepsilon)$ and $\lim_{c\to\infty} d_M(c) = \infty$. For $\varepsilon < 0.5$, we plot the function $d_m(c)$ in Figure 10.3.2. We observe that the function is then convex, attaining minimum value $d_{\min}$ that is strictly less than $\pi/2(1-\varepsilon)$. Thus, if the constant c in (10.3.9) is selected equal to $c_{\min}$ (Fig. 10.3.2), then the M-estimate will converge to the value zero with rate

$$r_M = n^{-1}d_{\min} < n^{-1}\frac{\pi}{2(1-\varepsilon)} \qquad (10.3.20)$$

Let us now compare the rates in (10.3.15) and (10.3.20). Clearly, when m is asymptotically large, the rate in (10.3.15) becomes asymptotically slow, while the rate in (10.3.20) remains below the value $n^{-1}[\pi/2(1-\varepsilon)]$. Thus, in the case where extreme outliers may occur with nonzero probability, the empirical mean estimate tends to nonconvergence while the M-estimate can be designed to provide high-quality performance (consistency with a relatively fast rate of convergence).

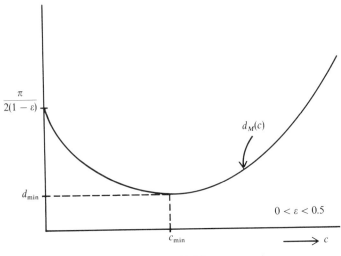

Figure 10.3.2

The median estimate $\hat{\theta}_S(x^n) = -\Phi^{-1}(n^{-1}\sum_{i=1}^{n}\operatorname{sgn} x_i)$ is also a fast converging location parameter estimate for the present problem. Its rate of convergence never exceeds $n^{-1}4^{-1}$, but it is generally nonconsistent.

10.4 ROBUST LOCATION PARAMETER ESTIMATION—MEMORYLESS PROCESSES

In Section 10.2, we studied maximum likelihood location parameter estimates when the data generating process is well known (at the zero value location parameter). When the process is a member of a whole class of processes, the robust parameter estimation formalizations arise. In loose terms, a location parameter estimate $\hat{\theta}(x^n)$ is *robust* at a class $\mathscr{C}$ of data generating processes if its performance remains satisfactory at each member in $\mathscr{C}$. For example, the median estimate $-\Phi^{-1}(n^{-1}\sum_{i=1}^{n}\operatorname{sgn} x_i)$ in Example 10.3.2 converges with rate not exceeding $n^{-1}4^{-1}$ at every memoryless and stationary data generating process, (at the expense in general of consistency). Thus, its asymptotic convergence is satisfactory, subject to a consistency tradeoff. On the other hand, the M-estimate with function $\psi(u)$ as in (10.3.9) is a consistent and fast converging location parameter estimate for all the density functions that are symmetric around their mean and also satisfy some additional general conditions. To define robustness in precise terms, we may adopt a saddle-point game formalization, as we did in Chapter 6. Such a formalization implies the utilization of an appropriate payoff function and the precise knowledge of a class $\mathscr{C}$ of data generating processes. If such a class is unknown, local performance stability around some given process may be sought instead, and then the qualitative robustness in Chapter 12 evolves. In this section, we will consider a saddle-point game approach for location parameter estimation.

10.4 Robust Location Parameter Estimation—Memoryless Processes

Let $\mathscr{F}$ be a given class of first order, zero mean density functions, generated by real, scalar, memoryless, and stationary processes. Let $f(u);\ u \in R$ denote some density function in $\mathscr{F}$ and let each f in $\mathscr{F}$ satisfy the conditions in Corollary 10.2.1. That is, let the entropy $H(f)$ and the Fisher information measure $I(f)$ exist, and let $|\dot{f}(x)| < \infty\ \forall\ x \in R$, $\int_R f(u)[\log f(u) + H(f)]^2\, du < \infty$. Then, if the density f were given, the maximum likelihood location parameter estimate would be given as a solution of (10.4.1), it would be consistent, and would converge with rate $n^{-1}I(f)$.

$$T_f(\theta, x^n) = n^{-1}\sum_{i=1}^{n}\psi_f(x_i - \theta) = 0 \qquad (10.4.1)$$

where

$$\psi_f(x - \theta) = \frac{\dfrac{\partial}{\partial \theta}f(x - \theta)}{f(x - \theta)} = -\frac{\dfrac{\partial}{\partial x}f(x - \theta)}{f(x - \theta)}$$

We observe that given f in $\mathscr{F}$, the maximum likelihood location parameter estimate is an M-estimate. We also observe that if $f(x) = f(-x)\ \forall\ x \in R$, then $\psi_f(x) = -\psi_f(-x)\ \forall\ x \in R$. Let us now assume that the class $\mathscr{F}$ contains densities that are symmetric around their mean zero. Then, considering the whole class, and due to the above observations, we are led to the initial game formalization:

Knowing the class $\mathscr{F}$ described above, the analyst (or system designer) considers an M-location parameter estimate, utilizing a function $\psi(u);\ u \in R$, such that $\psi(u) = -\psi(-u)$. Independently, Nature selects some density f in $\mathscr{F}$.

Let $\mathscr{D}$ denote the class of real, scalar, bounded, monotonically increasing, continuous, functions $\psi(\cdot)$, differentiable almost everywhere, that are also such that $\psi(u) = -\psi(-u)\ \forall\ u \in R$, $\psi(\infty) > 0$. Given some density f in the above class $\mathscr{F}$, let f_θ denote the same density when its mean is instead θ. We then obtain $\forall\ \psi \in \mathscr{D}$

$$m_{f_\theta}(\lambda) = \int_R \psi(u - \lambda)f_\theta(u)\, du \qquad m_f(0) = \int_R \psi(u)f(u)\, du = 0$$

$$\dot{m}_{f_\theta}(\lambda) = \frac{\partial}{\partial \lambda}\int_R \psi(u - \lambda)f_\theta(u)\, du = \frac{\partial}{\partial \lambda}\int_R \psi(u - \lambda)f(u - \theta)\, du$$

$$= \frac{\partial}{\partial \lambda}\int_R \psi(u + \theta - \lambda)f(u)\, du = -\int_R \frac{\partial}{\partial u}\psi(u - \theta - \lambda)f(u)\, du \qquad (10.4.2)$$

$$\dot{m}_{f_\theta}(\theta) = -\int_R \dot{\psi}(u)f(u)\, du < 0 \quad \text{and bounded}$$

$$\sigma_f^2(\theta) = \int_R \psi^2(u - \theta)f_\theta(u)\, du = \int_R \psi^2(u - \theta)f(u - \theta)\, du = \int_R \psi^2(u)f(u)\, du < \infty$$

where $f(x)$ does not need to be differentiable.

From the expressions in (10.4.2), we observe that the conditions in Theorem 10.3.1 are all satisfied for every f in $\mathscr{F}$. Due to that, and directly from Theorem 10.3.1, we can now express the following lemma.

LEMMA 10.4.1: Consider a class $\mathscr{F}$ of density functions whose entropies and Fisher information measures exist. In addition, for every f in $\mathscr{F}$ let

$$f(x) = f(-x) \qquad \dot{f}(x) < \infty \qquad \forall\, x \in R \quad \forall\, x \in R$$

$$\int_R f(u)[\log f(u) + H(f)]^2\, du < \infty$$

Let $\psi(u);\ u \in R$ be some function in the class $\mathscr{D}$, and consider the M-estimate defined via ψ. That is,

$$\hat{\theta}_M(x^n): n^{-1} \sum_{i=1}^{n} \psi[x_i - \hat{\theta}_M(x^n)] = 0$$

Then, the estimate $\hat{\theta}_M(x^n)$ is consistent at every f in $\mathscr{F}$. Given f in $\mathscr{F}$, the convergence rate of the estimate is $n^{-1}R^{-1}(\psi, f)$, where

$$R(\psi, f) = \left[\int_R \dot{\psi}(u) f(u)\, du \right]^2 \Big/ \int_R \psi^2(u) f(u)\, du \qquad (10.4.2)$$

∎

Let us now consider a game whose payoff function is the quantity $R(\psi, f)$ in (10.4.3). That is, the system designer selects the function ψ in $\mathscr{D}$, Nature selects independently some f in $\mathscr{F}$, and the reward paid to the system designer is then $R(\psi, f)$. This reward is asymptotic ($n \to \infty$), and a saddle-point solution (ψ^*, f^*) of the game exists if

$$\forall\, \psi \in \mathscr{D} \qquad R(\psi, f^*) \le R(\psi^*, f^*) \le R(\psi^*, f) \qquad \forall\, f \in \mathscr{F} \qquad (10.4.3)$$

where the classes $\mathscr{D}$ and $\mathscr{F}$ are as in Lemma 10.4.1.

In the search for a saddle-point solution (ψ^*, f^*) in (10.4.3), we will use some useful relationships. First, since every ψ in $\mathscr{D}$ is bounded, and since $f(\infty) = f(-\infty) = 0\ \forall\, f \in \mathscr{F}$, we have

$$\int_R \dot{\psi}(u) f(u)\, du = \psi(u) f(u)\Big|_{-\infty}^{\infty} - \int_R \psi(u) \dot{f}(u)\, du$$

$$= -\int_R \psi(u) \dot{f}(u)\, du \qquad \forall\, \psi \in \mathscr{D}\ \ \forall\, f \in \mathscr{F} \qquad (10.4.4)$$

Due to (10.4.4), and using the Schwartz inequality, we obtain

$$\left[\int_R \dot{\psi}(u) f(u)\, du \right]^2 = \left[\int_R \psi(u) \dot{f}(u)\, du \right]^2 = \left(\int_R [\psi(u) f^{1/2}(u)] \left[\frac{\dot{f}(u)}{f(u)} f^{1/2}(u) \right] du \right)^2$$

$$\le \left\{ \int_R \psi^2(u) f(u)\, du \right\} \left\{ \int_R \left[\frac{\dot{f}(u)}{f(u)} \right]^2 f(u)\, du \right\}$$

10.4 Robust Location Parameter Estimation—Memoryless Processes

Thus, $\forall \psi \in \mathcal{D}$ and $\forall f \in \mathcal{F}$

$$R(\psi, f) = \left[\int_R \dot{\psi}(u) f(u)\, du\right]^2 \Big/ \int_R \psi^2(u) f(u)\, du \leq \int_R \left[\frac{\dot{f}(u)}{f(u)}\right]^2 f(u)\, du = I(f) \tag{10.4.5}$$

where $I(f)$ is the Fisher information measure of the density f.

We now observe that $R(\psi, f)$ and $I(f)$ are both strictly convex with respect to f. Thus, to show that there exist f_ψ and f^* in $\mathcal{F}$ such that

$$R(\psi, f_\psi) \leq R(\psi, f) \qquad \forall f \in \mathcal{F} \tag{10.4.6}$$

$$I(f^*) \leq I(f) \qquad \forall f \in \mathcal{F} \tag{10.4.7}$$

it suffices to show that

$$J_{f_\psi f}(0) = \frac{\partial}{\partial \varepsilon} R(\psi, (1-\varepsilon) f_\psi + \varepsilon f)\Big|_{\varepsilon=0} \geq 0 \qquad \forall f \in \mathcal{F} \tag{10.4.8}$$

$$I_{f^* f}(0) = \frac{\partial}{\partial \varepsilon} I[(1-\varepsilon) f^* + \varepsilon f]\Big|_{\varepsilon=0} \geq 0 \qquad \forall f \in \mathcal{F} \tag{10.4.9}$$

Performing the operations in (10.4.8) and (10.4.9) for $\dot{\psi}(u) = -\dot{f}^*(u)/f^*(u)$ we obtain, from (10.4.8), and for $\psi^*(u) = -\dot{f}^*(u)/f^*(u)$,

$$J_{f^* f}(0) = \int_R \left\{ 2 \frac{\dot{f}^*(u)}{f^*(u)} [\dot{f}(u) - \dot{f}^*(u)] - \left(\frac{\dot{f}^*(u)}{f^*(u)}\right)^2 [f(u) - f^*(u)] \right\} du$$

and from (10.4.9),

$$I_{f^* f}(0) = J_{f^* f}(0) \qquad \forall f \in \mathcal{F} \tag{10.4.10}$$

Due to (10.4.10) we conclude that if there exists some f^* in $\mathcal{F}$ such that $I_{f^* f}(0) \geq 0 \;\forall f \in \mathcal{F}$, then $J_{f^* f}(0) \geq 0 \;\forall f \in \mathcal{F}$. The density f^* is then such that $I(f^*) \leq I(f) \;\forall f \in \mathcal{F}$, and $R(\psi^*, f^*) \leq R(\psi^*, f) \;\forall f \in \mathcal{F}$, where $\psi^*(u) = -\dot{f}^*(u)/f^*(u)$. In addition, due to (10.4.5), we also have $R(\psi, f^*) \leq I(f^*) \;\forall \psi \in \mathcal{D}$. We now summarize the above conclusions in a lemma.

LEMMA 10.4.2: Consider the classes $\mathcal{F}$ and $\mathcal{D}$ in Lemma 10.4.1. Let f^* be some density in $\mathcal{F}$ such that

$$I(f^*) \leq I(f) \qquad \forall f \in \mathcal{F}$$

$$-\dot{f}^*/f^* \in \mathcal{D}$$

Then the pair (ψ^*, f^*), where $\psi^*(u) = -\dot{f}^*(u)/f^*(u)$, is a saddle-point solution of the game in (10.4.3). ∎

Lemma 10.4.2 says that if there exists some density function f^* in $\mathscr{F}$ whose Fisher information measure is the minimum in $\mathscr{F}$, and if f^* is then such that the function $-\dot{f}^*(u)/f^*(u) = \psi^*(u)$ is bounded, monotonically increasing, differentiable almost everywhere, and such that $\psi^*(u) = -\psi^*(-u) \; \forall \; u \in R$, then the pair (ψ^*, f^*) satisfies the game in (10.4.3). The M-estimate $\hat{\theta}^*_M(x^n)$ determined via the function $\psi^*(u)$ is then called *robust in* $\mathscr{F}$, and it is such that

$$\hat{\theta}^*_M(x^n): n^{-1} \sum_{i=1}^{n} \psi^*[x_i - \hat{\theta}^*_M(x^n)] = 0 \qquad (10.4.11)$$

The estimate in (10.4.11) is consistent at every f in $\mathscr{F}$. At f (the sequence x^n generated by f), its rate of convergence is $n^{-1}R^{-1}(\psi^*, f)$. At f^*, its rate of convergence is the minimum in $\mathscr{F}$ and equals $n^{-1}I^{-1}(f^*)$. Since the function $\psi^*(u)$ is continuous everywhere in R, the estimate $\hat{\theta}^*(x^n)$ is a continuous function of x^n for every dimensionality n. We will now proceed with an example, first studied by Huber (1963), which corresponds to an outliers model (as in Chapter 6).

Example 10.4.1

Let us consider the possible occurrence of extreme independent outliers on independent, unit variance, and identically distributed Gaussian data. Let the outliers appear with frequency ε and let their distribution be represented by some unknown density function g whose entropy and Fisher information measure exist, and which is such that $g(x) = g(-x) \; \forall \; x \in R$; $\int_R g(u)[\log g(u) + H(g)]^2 \, du < \infty$; $\dot{g}(x) < \infty \; \forall \; x \in R$. Under zero mean, the data sequences are thus generated by a density function f as below, where $\phi(u)$ denotes the density function of the unit variance, zero mean Gaussian random variable at the point u, and where $\mathscr{G}$ denotes the class of densities g as above.

$$f(u) = (1 - \varepsilon)\phi(u) + \varepsilon g(u) \qquad u \in R, \quad g \in \mathscr{G} \qquad (10.4.12)$$

If $\mathscr{F}$ denotes the class of densities f as in (10.4.12), when the densities g vary in $\mathscr{G}$, then $\mathscr{F}$ satisfies the conditions in Lemmas 10.4.1 and 10.4.2. Given f in $\mathscr{F}$, let $I(f)$ denote its Fisher information measure. Via calculus of variation, we find f^* in $\mathscr{F}$ such that

$$I(f^*) = \inf_{f \in \mathscr{F}} I(f) \qquad (10.4.13)$$

In particular, for $0 < \varepsilon < 0.03$, f^* is as follows.

$$f^*(x) = \begin{cases} \phi(a)[\cos 2^{-1}ca]^{-2}[\cos 2^{-1}cx]^2 & -a \leq x \leq a \\ \phi(x) & \begin{cases} a < x < b \\ -b < x < -a \end{cases} \\ \phi(b)e^{-b|x-b|} & |x| \geq b \end{cases} \qquad (10.4.14)$$

10.4 Robust Location Parameter Estimation—Memoryless Processes

where the constants a, b, c are such that

$$c \tan(2^{-1}ca) = a$$

$$\phi(b) = b[\Phi(-b) + \varepsilon] \tag{10.4.15}$$

$$\phi(a)[\cos 2^{-1}ca]^{-2} = 2c[ca + \sin ca]^{-1}[\Phi(a) - 2^{-1} - \varepsilon]$$

where $\Phi(\cdot)$ is the cumulative distribution function of the zero mean, unit variance Gaussian random variable.

Directly from (10.4.14), we then conclude, for $x \geq 0$,

$$\psi^*(x) = \begin{cases} c \tan(2^{-1}cx) & 0 \leq x \leq a \\ x & a < x < b \\ b & x \geq b \end{cases} \tag{10.4.16}$$

The function $\psi^*(x)$ in (10.4.16) is clearly monotonically increasing, bounded, and differentiable almost everywhere, and $\psi^*(x) = -\psi^*(-x)$. It thus satisfies the conditions in Lemma 10.4.2 and defines a robust in $\mathscr{F}$ location parameter estimate for $0 < \varepsilon < 0.03$. The bound, 0.03, of ε is called the *breakdown point* of the robust estimate $\hat{\theta}^*(x^n)$, which, for $0 < \varepsilon < 0.03$, is given by (10.4.11) with ψ^* as in (10.4.16). For ε values above 0.03, neither consistency nor good convergence rate can be guaranteed by the estimate $\hat{\theta}^*(x^n)$. For $0 < \varepsilon < 0.03$, the estimate is consistent at every f in $\mathscr{F}$, with convergence rate not exceeding $n^{-1}I^{-1}(f^*)$, where f^* is given by (10.4.14) and

$$I(f^*) = \int_R \left[\frac{\dot{f}^*(x)}{f^*(x)}\right]^2 f^*(x)\,dx = 2\int_0^\infty [\psi^*(x)]^2 f^*(x)\,dx$$

$$= 2c^2\phi(a)[\cos 2^{-1}ca]^{-2} \int_0^a [\tan 2^{-1}cx]^2 [\cos 2^{-1}cx]^2\,dx$$

$$+ 2\int_a^b x^2\phi(x)\,dx + 2b^2\phi(b)\int_b^\infty e^{-b(x-b)}\,dx$$

$$= 2\Phi(b) - 2\Phi(a) + ac^2\phi(a)[\cos 2^{-1}ca]^2 \tag{10.4.17}$$

where the constants a, b, c are as in (10.4.15).

Table 10.4.1

ε	a	b	c
0.001	0.65	2.44	1.37
0.002	0.75	2.23	1.35
0.005	0.91	1.95	1.32
0.01	1.06	1.72	1.29
0.02	1.24	1.49	1.26
0.03	1.34	1.36	1.23

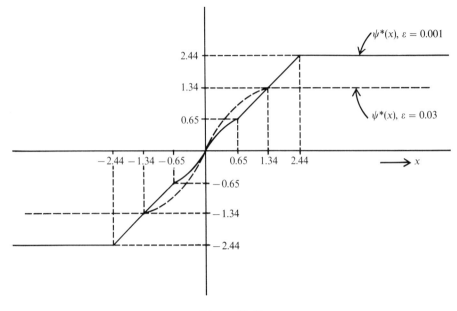

Figure 10.4.1

In Table 10.4.1 below, we list the values of the constants a, b, and c, in (10.4.15), for various values of ε. This table is from Huber (1963). In Figure 10.4.1, we plot the function $\psi^*(u)$ in (10.4.16) for $\varepsilon = 0.001$ and $\varepsilon = 0.03$.

When data sequences $x^n = [x_1, \ldots, x_n]$ are observed, an arbitrary parameter θ is first subtracted from each datum. Then, each difference $x_i - \theta$; $i = 1, \ldots, n$ is truncated as determined by the function $\psi^*(x)$ in (10.4.16). The truncated data are summed and the parameter θ is adjusted to make the truncated sum equal to zero. The parameter value that attains the zero value of the truncated sum is the location parameter estimate. The above data truncation is actually the operation that reduces the effect of extreme outliers; its quantitative effects are parallel to those discussed in Example 10.3.2. From Figure 10.4.1 we observe that as the frequency ε of the outliers increases, the upper bound of the data truncation decreases, to provide better protection against wrong or slowly converging estimates.

The outliers model in Example 10.4.1 is very important in many applications. It represents lack of confidence regarding histograms, which do not measure low-probability distribution tails. If the histogram indicates a central distribution mass that looks Gaussian, and if the frequency of outliers can be guessed and is less than 0.03, then the robust M-estimate in Example 10.4.1 may be adopted. If the frequency of the outliers is completely unknown, however, but protection against their possible occurrence is still needed, then ad hoc but intuitively pleasing estimates may be adopted. In the latter case, some performance criteria are needed for the comparison between such ad hoc schemes. Such performance criteria are the *breakdown point* and the *influence function*—

10.4 Robust Location Parameter Estimation—Memoryless Processes

both defined for the outliers model. We provide the definition of those two performance criteria below.

DEFINITION 10.4.1: Given a parameter estimate $\hat{\theta}(x^n)$, its breakdown point ε^* is the highest frequency of extreme amplitude ($\to \pm \infty$) outliers that the estimate can tolerate, while still converging to a function of the true parameter value. Considering outliers of amplitude x that occur with frequency ε, the influence function $IF(x)$ of the estimate at the point x is defined as follows:

$$IF(x) = \frac{\partial}{\partial \varepsilon} \left[\lim_{n \to \infty} \hat{\theta}(x^n) \right]_{\varepsilon = 0}$$ ■

From Definition 10.4.1, we conclude that the influence function represents the gradient, at frequency zero, of outliers with varying amplitudes. The breakdown point and influence function criteria will perhaps be better understood if their computation for some location parameter estimates is undertaken. We will compute the above criteria for the empirical mean, the median, and the M estimates, where in the latter we will adopt the function $\psi(u)$ in (10.3.9), Example 10.3.1.

The Empirical Mean Estimate

Let us consider the estimate $\hat{\theta}_A(x^n) = n^{-1} \sum_{i=1}^{n} x_i$ in (10.3.1). Let us assume that with probability $(1 - \varepsilon)$ the data sequence x^n is generated by a stationary and ergodic process whose mean value is θ. Let independent outliers of constant amplitude x occur with probability ε. The estimate $\hat{\theta}_A(x^n)$ then converges asymptotically ($n \to \infty$) to the value $F_A(x) = (1 - \varepsilon)\theta + \varepsilon x$. When the outlier amplitude x tends to $\pm \infty$, then $F_A(x) \to \varepsilon x$, which is not a function of θ, for every nonzero value of ε. The breakdown point ε_A^*, of the empirical mean estimate is thus zero. Since

$$\lim_{n \to \infty} \hat{\theta}_A(x^n) = F_A(x) = (1 - \varepsilon)\theta + \varepsilon x,$$

we have

$$\frac{\partial}{\partial \varepsilon} \left[\lim_{n \to \infty} \hat{\theta}_A(x^n) \right]_{\varepsilon = 0} = \frac{\partial}{\partial \varepsilon} F_A(x) \bigg|_{\varepsilon = 0} = x - \theta.$$

Thus, the influence function $IF_A(x)$ of the empirical mean estimate is linear in x. In summary,

$$\varepsilon_A^* = 0 \qquad IF_A(x) = x - \theta \tag{10.4.18}$$

The Median Estimate

Let us consider the estimate $\hat{\theta}_S(x^n) = -\Phi^{-1}(n^{-1} \sum_{i=1}^{n} \operatorname{sgn} x_i)$ in (10.3.1), where instead of the cumulative distribution F we now have Φ, the cumulative distribution of the zero mean, unit variance Gaussian random variable. Let each element in the data sequence x^n be generated by a unit variance, mean θ, Gaussian and memoryless process with probability $1 - \varepsilon$. Let each datum be instead an independent, amplitude x outlier with probability ε. Then,

$$\lim_{n \to \infty} \hat{\theta}_S(x^n) = -\Phi^{-1}[(1 - \varepsilon)\Phi(-\theta) + \varepsilon \text{ sgn } x]$$

and

$$\frac{\partial}{\partial \varepsilon} \lim_{n \to \infty} \hat{\theta}_S(x^n)\bigg|_{\varepsilon=0} = -\frac{\partial}{\partial \varepsilon}\Phi^{-1}[(1 - \varepsilon)\Phi(-\theta) + \varepsilon \text{ sgn } x]\bigg|_{\varepsilon=0}$$

$$= [\Phi(-\theta) - \text{sgn } x]/\phi(\theta) = IF_S(x)$$

where ϕ is the density function of the zero mean, unit variance Gaussian random variable, and where $IF_S(x)$ is the influence function of the Median estimate. Let us now consider $x \to -\infty$. Then, $\lim_{n \to \infty} \hat{\theta}_S(x^n) = -\Phi^{-1}[(1 - \varepsilon)\Phi(-\theta) + \varepsilon]$, since sgn $x = \{0; x > 0 \text{ and } 1; x \leq 0\}$. We observe that if $\varepsilon \geq 0.5$, then $-\Phi^{-1}[(1 - \varepsilon)\Phi(-\theta) + \varepsilon] < 0; \forall \theta$. That is, if $\varepsilon \geq 0.5$, then even a large positive θ value will be estimated as a negative value. On the other hand, if $\varepsilon < 0.5$, then the sign of θ will be preserved, if $|\theta|$ is large enough. The breakdown point ε_S^* of the Median estimate is thus 0.5. In summary,

$$\varepsilon_S^* = 0.5 \qquad IF(x) = [\Phi(-\theta) - \text{sgn } x]/\phi(\theta) \qquad (10.4.19)$$

where sgn $x = \{0; x > 0 \text{ and } 1; x \leq 0\}$.

The M-Estimate

Let us consider the M-estimate $\hat{\theta}_M(x^n)$ in (10.3.2), with function $\psi(u); u \in R$, as in (10.3.9), Example 10.3.1. We consider this estimate, rather than the one in Example 10.4.1, for simplicity. Now let each datum in x^n be generated by a stationary, memoryless, unit variance, mean θ Gaussian process with probability $1 - \varepsilon$. With probability ε, let instead each datum be an independent, amplitude x outlier. Then, the estimate $\hat{\theta}_M(x^n)$ converges asymptotically to some value λ_0 such that $m_x(\lambda_0) = 0$, where

$$m_x(\lambda) = (1 - \varepsilon) \int_R \psi(u - \lambda)\phi(u - \theta)\,du + \varepsilon \int_R \psi(u - \lambda)\delta(x - u)\,du$$

$$= (1 - \varepsilon) \int_R \psi(u)\phi(u + \lambda - \theta)\,du + \varepsilon\psi(x - \lambda)$$

$$= \varepsilon\psi(x - \lambda) + (1 - \varepsilon)\left\{-c\int_{-\infty}^{-c} \phi(u + \lambda - \theta)\,du + c\int_c^{\infty} \phi(u + \lambda - \theta)\,du \right.$$

$$\left. + \int_{-c}^{c} u\phi(u + \lambda - \theta)\,du\right\}$$

$$= \varepsilon\psi(x - \lambda) + (1 - \varepsilon)G(\lambda) \qquad (10.4.20)$$

10.4 Robust Location Parameter Estimation—Memoryless Processes

where $\delta(u)$ is the delta function at the point u and

$$G(\lambda) = c + \phi(c - \lambda + \theta) - \phi(c + \lambda - \theta) - (c + \lambda - \theta)\Phi(c + \lambda - \theta)$$
$$+ (\lambda - c - \theta)\Phi(-c + \lambda - \theta) \quad (10.4.21)$$

It can easily be found that the function $G(\lambda)$ in (10.4.21) is monotone, and such that $G(-\infty) = c$ and $G(\infty) = -c$. Given x, the function $\psi(x - \lambda)$ is also monotone in λ and linear, with $\psi[x - (-\infty)] = c$, and $\psi(x - \infty) = -c$. We thus conclude that, given absolutely bounded x, the function $m_x(\lambda)$ in (10.4.20) is monotone with $m_x(-\infty) = c$ and $m_x(\infty) = -c$. A value λ_0 such that $m_x(\lambda_0) = 0$, then exists such that

$$G(\lambda_0) = -\frac{\varepsilon}{1-\varepsilon}\psi(x - \lambda_0) \qquad |x| < \infty \quad (10.4.22)$$

where

$$\psi(x - \lambda) = \begin{cases} c & \lambda \leq x - c \\ x - \lambda & x - c < \lambda < x + c \\ -c & \lambda \geq x + c \end{cases} \quad (10.4.23)$$

Let us denote by λ^* the unique value such that $G(\lambda^*) = 0$, where $G(\lambda)$ is given in (10.4.21), and let $\dot{G}(\lambda^*) = (\partial/\partial\lambda)G(\lambda)|_{\lambda=\lambda^*} = -\Phi(c + \lambda^* - \theta) + \Phi(-c + \lambda^* - \theta)$. Then $\lambda^* = \theta$, $G(\theta) = 0$, $\dot{G}(\theta) = 1 - 2\Phi(c)$, and from (10.4.22) we find

$$\frac{\partial \lambda_0}{\partial \varepsilon}\bigg|_{\varepsilon=0} = -\frac{\psi(x - \lambda^*)}{\dot{G}(\lambda^*)} - \frac{G(\lambda^*)}{[\dot{G}(\lambda^*)]^2}$$
$$= -\psi(x - \theta)[1 - 2\Phi(c)]^{-1} = IF_S(x) \quad (10.4.24)$$

The expression in (10.4.24) gives the influence function of the M-estimate considered here. Now let $x \to \infty$. From (10.4.20), we then obtain $m_\infty(\lambda) = c\varepsilon + (1 - \varepsilon)G(\lambda)$, where $m_\infty(\lambda)$ is strictly monotone, with $m_\infty(\infty) = c\varepsilon + (1 - \varepsilon)G(\infty) = c\varepsilon - c(1 - \varepsilon) = -c[1 - 2\varepsilon]$ and $m_\infty(-\infty) = c\varepsilon + (1 - \varepsilon)G(-\infty) = c\varepsilon + c(1 - \varepsilon) = c$. We observe that for $\varepsilon > 0.5$, there is no value λ_0 such that $m_\infty(\lambda_0) = 0$, since then $m_\infty(\lambda) > 0 \; \forall \; \lambda$. If, on the other hand, $\varepsilon < 0.5$, then there exists a λ_0 such that $m_\infty(\lambda_0) = 0$, and it is a function of θ. Therefore, the breakdown point ε_S^* of the median estimate is 0.5. So

$$\varepsilon_S^* = 0.5 \quad (10.4.25)$$

Comparing now the empirical mean, median, and M-estimates with ψ as in (10.3.9), we first note that the empirical mean estimate can tolerate no extreme outliers (breakdown point zero). In fact, this is the reason why its variance in Example 10.3.2 was found to be unbounded. The influence function of the empirical mean estimate is linear, exhibiting again the nonresistance of the estimate to even infrequently occurring outliers. Since outliers occur in most real applications, we can safely characterize the empirical mean estimate as useless.

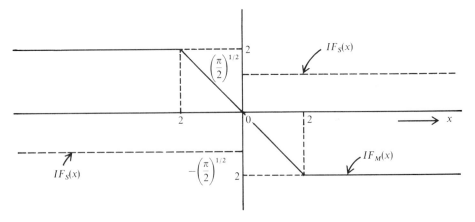

Figure 10.4.2

Turning now to the median and M-estimates, we observe that they both have breakdown point 0.5. That is, they can both tolerate up to 50% extreme outliers while still converging to some function of the true parameter value. Of course, consistency is lost then, but fast convergence to some function of the true parameter is maintained. Assuming $\theta = 0$ in both (10.4.19) and (10.4.24), the influence functions $IF_S(x)$ and $IF_M(x)$ represent the gradients of the corresponding estimates, at the zero frequency of outliers occurrence, and they are both bounded. In Figure 10.4.2, we plot the influence functions $IF_S(x)$ and $IF_M(x)$ for $\theta = 0$, and $c = 2$ (in (10.3.9)). We observe that for x values in $[-\sqrt{(\pi/2)} + 1, \sqrt{\pi/2}]$, the values of the function $IF_M(x)$ are absolutely smaller than those of the function $IF_S(x)$. Thus, the M-estimate is better in $[-\sqrt{(\pi/2)} + 1, \sqrt{\pi/2}]$, than the Median-estimate is. As the constant c in (10.3.9) decreases, the region where $IF_M(x)$ is absolutely smaller than $IF_S(x)$ increases, at the expense of reduced convergence rate of the M-estimate in the absence of outliers (see Example 10.3.1, Section 10.3).

10.5 DISCUSSION AND EXAMPLES

To this point, most of our studies in this chapter have concentrated around the estimation of a location parameter from data generated by stationary and memoryless processes. We have mentioned, however, that the empirical mean and the median estimates are consistent estimates of a location parameter, when the data are generally generated by some stationary and ergodic process. If this latter process is also Markovian, then the random variables $n^{-1} \sum_{i=1}^{n} X_i$ and $n^{-1} \sum_{i=1}^{n} \operatorname{sgn} X_i$ are in addition asymptotically Gaussian [see Doob (1967)]. As we saw in Section 10.4, the empirical mean estimate is extremely sensitive to extreme outliers; it is thus totally nonrobust; and it should generally be excluded from consideration. In the case of data generated by some stationary process with memory, variations of the M-estimates in Section 10.3 qualify as location parameter estimates. Indeed, if $f(x^n)$ denotes the density function of

10.5 Discussion and Examples

the data sequence at the point x^n when the location parameter is zero, and if $f_\theta(x^n)$ is the same density when the location parameter equals θ, then

$$f_\theta(x^n) = \prod_{i=1}^{n} f_\theta(x_i|x_1^{i-1}) = \prod_{i=1}^{n} f(x_i - \theta|x_1 - \theta, \ldots, x_{i-1} - \theta) \quad (10.5.1)$$

where, $f_\theta(x_1|x_0) = f_\theta(x_1) = f(x_1 - \theta)$, where $\prod$ denotes product, and where $f(x_i - \theta|x_1 - \theta, \ldots, x_{i-1} - \theta)$ denotes conditional density.

Therefore,

$$\log f_\theta(x^n) = \sum_{i=1}^{n} \log f(x_i - \theta|x_1 - \theta, \ldots, x_{i-1} - \theta) \quad (10.5.2)$$

$$\frac{\partial}{\partial \theta} \log f_\theta(x^n) = \sum_{i=1}^{n} \psi_i(x_1 - \theta, \ldots, x_i - \theta)$$

where

$$\psi_i(x_1 - \theta, \ldots, x_i - \theta) = \frac{\frac{\partial}{\partial \theta} f(x_i - \theta|x_1 - \theta, \ldots, x_{i-1} - \theta)}{f(x_i - \theta|x_1 - \theta, \ldots, x_{i-1} - \theta)} \quad (10.5.3)$$

Given $f(x^n) \; \forall \; x^n \in R^n$, the maximum likelihood estimate of the location parameter θ is found by setting the expression in (10.5.2) equal to zero. This induces a generalized M-estimate, with generally n functions, ψ_i; $1 \leq i \leq n$. If the density function $f(x^n) \; x^n \in R^n$ is Gaussian, the function in (10.5.3) takes the form

$$\psi_i(x_1 - \theta, \ldots, x_i - \theta) = \psi\left[x_i - \theta - \sum_{j=1}^{i-1} a_{ij}(x_j - \theta)\right]$$

$$= -\sigma^{-2}\left[x_i - \theta - \sum_{j=1}^{i-1} a_{ij}(x_j - \theta)\right]$$

The latter expression leads to the consideration of the following location parameter M-estimates for processes with memory that are "close" to Gaussian.

$$\hat{\theta}_M(x^n): n^{-1} \sum_{i=1}^{n} \psi(x_i - \hat{\theta}_M(x^n) - \sum_{j=1}^{i-1} a_{ij}[x_j - \hat{\theta}_M(x^n)]) = 0 \quad (10.5.4)$$

where $\psi(u)$; $u \in R$ is a monotone and continuous function such that $\psi(-\infty) < 0$ and $\psi(\infty) > 0$. For protection against occasional extreme outliers, and using the philosophy in Section 10.4, the function $\psi(\cdot)$ should also be bounded and differentiable almost everywhere on R. Such a function, for example, is the one in (10.3.9), Section 10.3. For K-order Markov processes, the M-estimate in (10.5.4) takes the asymptotic form

$$\hat{\theta}_M(x^n): T[x^n, \hat{\theta}_M(x^n)] = n^{-1} \sum_{i=K+1}^{K+n} \psi\left(x_i - \hat{\theta}_M(x^n) - \sum_{j=i-K}^{i-1} a_j[x_j - \hat{\theta}_M(x^n)]\right) = 0$$

$$(10.5.5)$$

From the Markovian theorem [see Doob (1967)] we then conclude that the random variable $T(X^n, \theta)$ is asymptotically Gaussian for every value θ, and the convergence properties of the estimate in (10.5.5) can be then found via methods parallel to those in Sections 10.3 and 10.4. If the function $\psi(\cdot)$ in (10.5.5) is also monotone, bounded, continuous, and differentiable almost everywhere in R, then, via methods as in the proof of Theorem 10.3.1, we conclude that the estimate $\hat{\theta}_M(x^n)$ in (10.5.5) is such that $n^{1/2}\hat{\theta}_M(X^n)$ is asymptotically Gaussian. If $f(x^n)$; $x^n \in R^n$ is the density function that generates the data sequence, then the mean of the random variable $n^{1/2}\hat{\theta}_M(X^n)$ is λ_0, and its variance is $\sigma_f^2(\lambda_0)|[\dot{m}_f(\lambda_0)]^2$, where

$$m_f(\lambda) = \int_{R^{K+1}} \psi(x_{K+1} - \lambda - \sum_{j=1}^{K} a_j[x_j - \lambda]) f(x_1^{K+1}) dx_1^{K+1}$$

$$m_f(\lambda_0) = 0 \qquad \dot{m}_f(\lambda_0) = \frac{\partial}{\partial \lambda} m_f(\lambda)\bigg|_{\lambda = \lambda_0}$$

$$\sigma_f^2(\lambda) = n^{-1} \sum_{i=K+1}^{K+n} \sum_{l=K+1}^{K+n} E_f\left\{\psi\left(x_i - \lambda - \sum_{j=i-K}^{i-1} a_j[x_j - \lambda]\right) \times \psi\left(x_l - \lambda - \sum_{j=l-K}^{l-1} a_j[x_j - \lambda]\right)\right\}$$

(10.5.6)

where $E_f\{\cdot\}$ denotes expectation with respect to the density function $f(\cdot)$.

Getting away from location parameter estimates, in the next two examples we will consider the estimation of autoregressive parameters in autoregressive models of known order. We note that if the latter order is known, then the estimation of the autoregressive parameters is the best method for spectral estimation as well.

Example 10.5.1

Let us consider a data generating process represented by the m-order autoregressive model

$$X_i = \sum_{k=1}^{m} \alpha_k X_{i-k} + W_i \qquad (10.5.7)$$

We assume that the random variables $\{W_i\}$ are known to be independent, identically distributed, and zero mean, unit variance Gaussian. Let the autoregressive parameters $\{\alpha_k\}$ be real and unknown. Then, given a realization x^n from the autoregressive process, we wish to estimate the parameters α_k; $1 \le k \le m$. We will adopt the maximum likelihood criterion. Specifically, given x^n, we will derive the estimate $\{\hat{\alpha}_k(x^n); 1 \le k \le m\}$ of the autoregressive parameters, by finding the supremum with respect to $\{\alpha_k\}$ of the density function

$$f_{\{\alpha_k\}}(x^n) = \prod_i \phi\left(x_i - \sum_{k=1}^{m} \alpha_k x_{i-k}\right) \qquad (10.5.8)$$

10.5 Discussion and Examples

where Π denotes product, and where $\phi(u)$ is the density function of the zero mean unit variance Gaussian random variable, at the point u.

Equivalently, the supremum of $\ln f_{\{\alpha_k\}}(x^n)$ in (10.5.8) can be sought, where $\ln$ denotes natural logarithm. That is, the supremum with respect to $\{\alpha_k\}$ of the following function is now sought.

$$T(x^n, \{\alpha_k\}) = \sum_i \ln \phi\left(x_i - \sum_{k=1}^m \alpha_k x_{i-k}\right)$$

$$= -\frac{n}{2}\ln 2\pi - 2^{-1}\sum_i\left[x_i - \sum_{k=1}^m \alpha_k x_{i-k}\right]^2$$

The function $T(x^n, \{\alpha_k\})$ is strictly concave and differentiable with respect to $\{\alpha_k\}$. Its supremum is thus found by taking the gradient ∇ with respect to $\{\alpha_k\}$ of $T(x^n, \{\alpha_k\})$, and setting it equal to zero. Thus, the maximum likelihood estimate $\{\hat{\alpha}_k = \hat{\alpha}_k(x^n); 1 \le k \le m\}$ is such that

$$\sum_i x_{i-l}\left[x_i - \sum_{k=1}^m \hat{\alpha}_k x_{i-k}\right] = 0 \qquad 1 \le l \le m \qquad (10.5.9)$$

where (10.5.9) determines a system of m equations.

Let $M(x^n)$ denote the $m \times m$ square matrix, whose k-row and l-column component q_{kl} equals $n^{-1}\sum_i x_{i-l}x_{i-k}$. Then, if, given x^n, the matrix $M(x^n)$ is invertible, (10.5.9) gives

$$\begin{bmatrix}\hat{\alpha}_1(x^n) \\ \vdots \\ \hat{\alpha}_m(x^n)\end{bmatrix} = M^{-1}(x^n)\begin{bmatrix}n^{-1}\sum_i x_i x_{i-1} \\ \vdots \\ n^{-1}\sum_i x_i x_{i-m}\end{bmatrix} \qquad (10.5.10)$$

To analyze the maximum likelihood estimate in (10.5.10), and to avoid unnecessary complications, let us assume that the order of the autoregressive model in (10.5.7) is one. We then denote by α the autoregressive parameter α_1, and we obtain from (10.5.10) the following maximum likelihood estimate $\hat{\alpha}(x^n)$.

$$\hat{\alpha}(x^n) = \left[n^{-1}\sum_i x_i^2\right]^{-1} n^{-1}\sum_i x_{i+1} x_i \qquad (10.5.11)$$

Let us denote by $F_{\alpha,n}(y)$ the cumulative distribution of the estimate in (10.5.11) at the point y where α is the value of the autoregressive parameter that generates the data sequence x^n. Then we easily obtain

$$F_{\alpha,n}(y) = \Pr\left\{n^{-1}\sum_{i=l+1}^{l+n} X_i(X_{i+1} - yX_i) \le 0 \Big| \alpha\right\} \qquad (10.5.12)$$

The autoregressive process is first order Markov. Applying the Markovian theorem [Doob (1967)], we conclude that the random variable $n^{-1}\sum_{i=l+1}^{l+n}$

$X_i(X_{i+1} - yX_i)$ is asymptotically Gaussian. For $l \to \infty$, the mean m and the variance σ_n^2 of the latter variable can be easily computed, and they are given below. We assume that $\alpha: 0 < \alpha < 1$.

$$m = \lim_{i \to \infty} E\{X_i(X_{i+1} - yX_i)|\alpha\} = \frac{\alpha - y}{1 - \alpha^2} \tag{10.5.13}$$

$$\sigma_n^2 = \lim_{l \to \infty} n^{-2} \sum_{i=l+1}^{l+n} \sum_{j=l+1}^{l+n} E\{X_i X_j(X_{i+1} - yX_i)(X_{j+1} - yX_j)|\alpha\} - m^2$$
$$= n^{-1}(1-\alpha^2)^{-1}\{1 + (y-\alpha)^2(3+\alpha^2)(1-\alpha^4)^{-1}[4 - (1-\alpha^2)^2]\} \tag{10.5.14}$$

Straightforward application of (10.5.13) and (10.5.14) in (10.5.12) provides the following expression for the asymptotic $(n \to \infty)$ distribution $F_{\alpha,n}(y)$, where $\Phi(\cdot)$ denotes the cumulative distribution of the zero mean, unit variance Gaussian random variable.

$$\lim_{n \to \infty} F_{\alpha,n}(y) = \Phi[n^{1/2}\{(3+\alpha^2)(1-\alpha^4)^{-1}[4-(1-\alpha^2)^2] + (y-\alpha)^{-2}\}^{-1/2} \operatorname{sgn}(\alpha - y)] \tag{10.5.15}$$

where

$$\operatorname{sgn} x = \begin{cases} 1 & x > 0 \\ 0 & x = 0 \\ -1 & x < 0 \end{cases}$$

From (10.5.15), we easily conclude that the asymptotic $(n \to \infty)$ density function $f_{\alpha,n}(y)$ of the estimate in (10.5.11) is symmetric around the value α and is not Gaussian. In fact, from (10.5.15) we obtain

$$\lim_{n \to \infty} f_{\alpha,n}(y) = n^{1/2}|y - \alpha|^{-3}[A(\alpha) + (y-\alpha)^{-2}]^{-3/2} \phi(n^{1/2}[A(\alpha) + (y-\alpha)^{-2}]^{-1/2}) \tag{10.5.16}$$

where $\phi(\cdot)$ is the density function of the zero mean, unit variance Gaussian random variable, and where

$$A(\alpha) = (3+\alpha^2)(1-\alpha^4)^{-1}[4 - (1-\alpha^2)^2] \tag{10.5.17}$$

From (10.5.16), it can easily be found that the variance of the estimate in (10.5.11) is asymptotically n^{-1}. Therefore, if the variables $\{W_i\}$ in (10.5.7) are indeed i.i.d. and Gaussian, then the maximum likelihood estimate in (10.5.11) is consistent, and its convergence rate is inversely proportional to n.

Example 10.5.2

Let us consider the autoregressive model in (10.5.7), where now the random variables $\{W_k\}$ are generally non-Gaussian, but still independent, identically dis-

10.5 Discussion and Examples

tributed, and zero mean. If the density function of each random variable W_k is $f(u)$; $u \in R$, then the expression in (10.5.8) takes the form

$$f_{\{\alpha_k\}}(x^n) = \prod_i f\left(x_i - \sum_{k=1}^m \alpha_k x_{i-k}\right)$$

and

$$T_f(x^n, \{\alpha_k\}) = \log f_{\{\alpha_k\}}(x^n) = \sum_i \log f\left(x_i - \sum_{k=1}^m \alpha_k x_{i-k}\right) \quad (10.5.18)$$

If the density function is known, and such that the function $T_f(x^n, \{\alpha_k\})$ in (10.5.18) is strictly concave with respect to $\{\alpha_k\}$, then the maximum likelihood estimate $\{\hat{\alpha}_k\}$ of the autoregressive parameters is found by setting the gradient (w.r.t. $\{\alpha_k\}$) of $T_f(x^n, \{\alpha_k\})$ equal to zero. That is,

$$\{\hat{\alpha}_k\}: \nabla T_f(x^n, \{\hat{\alpha}_k\}) = \sum_i \frac{\nabla f\left(x_i - \sum_{k=1}^m \hat{\alpha}_k x_{i-k}\right)}{f\left(x_i - \sum_{k=1}^m \hat{\alpha}_k x_{i-k}\right)} = 0 \quad (10.5.19)$$

where

$$\frac{\left[\nabla f\left(x_i - \sum_{k=1}^m \hat{\alpha}_k x_{i-k}\right)\right]^T}{f\left(x_i - \sum_{k=1}^m \hat{\alpha}_k x_{i-k}\right)} = -\frac{\dot{f}\left(x_i - \sum_{k=1}^m \hat{\alpha}_k x_{i-k}\right)}{f\left(x_i - \sum_{k=1}^m \hat{\alpha}_k x_{i-k}\right)} [x_{i-1}, \ldots, x_{i-m}]^T \quad (10.5.20)$$

$$\dot{f}(u) = \frac{\partial}{\partial x} f(x) \bigg|_{x=u}$$

Due to the above, and denoting

$$\psi\left(x_i - \sum_{k=1}^m \hat{\alpha}_k x_{i-k}\right) = \frac{\dot{f}\left(x_i - \sum_{k=1}^m \hat{\alpha}_k x_{i-k}\right)}{f\left(x_i - \sum_{k=1}^m \hat{\alpha}_k x_{i-k}\right)}$$

we observe that the vector function in (10.5.19), which determines the maximum likelihood estimate, can be written in the form

$$\{\hat{\alpha}_k\}: \sum_i x_{i-l} \psi\left(x_i - \sum_{k=1}^m \hat{\alpha}_k x_{i-k}\right) = 0 \quad 1 \leq l \leq m \quad (10.5.21)$$

The form in (10.5.21) is an M-estimate-type, for autoregressive parameters. To analyze it, avoiding unnecessary complications, let us consider the special case of an order one autoregressive model, and let us then denote the single autoregressive parameter α. From (10.5.21), we then conclude

$$\text{For} \quad T_0(x^n, \alpha) = n^{-1} \sum_i x_{i-1} \psi(x_i - \alpha x_{i-1}) \tag{10.5.22}$$

$$\hat{\alpha}: T_0(x^n, \hat{\alpha}) = 0$$

Let the autoregressive process that generates the data x^n have autoregressive parameter β, where $0 < \beta < 1$, and let the density of the variables $\{W_k\}$ be $f(u)$; $u \in R$. Then, due to the Markovian theorem, the random variable $T_0(X^n, \alpha)$ in (10.5.22) is asymptotically Gaussian. Its mean and variance depend on the choice of the function $\psi(\cdot)$ that controls the convergence properties of the variable $T_0(X^n, \alpha)$. Let us consider a whole class $\mathscr{F}$ of density functions $f(u); u \in R$ whose variance is one, and which are symmetric around zero. Let us then select the function $\psi(\cdot)$ as follows, where c is some positive constant.

$$\psi(u) = \begin{cases} c & u \geq c \\ u & -c < u < c \\ -c & u \leq -c \end{cases} \tag{10.5.23}$$

Let f be some density function in class $\mathscr{F}$, and let us then denote by $m_{f,\beta}$ and $\sigma^2_{n,f,\beta}$ the mean and the variance of the random variable $T_0(X^n, \alpha)$, as induced by f and the autoregressive parameter β. Then

$$m_{f,\beta} = E\{X_{l-1}\psi(X_l - \alpha X_{l-1})|_{f,\beta}\} \tag{10.5.24}$$

$$\sigma^2_{n,f,\beta} = n^{-2} \sum_{i=l+1}^{l+n} \sum_{j=l+1}^{l+n} E\{[X_{i-1}\psi(X_i - \alpha X_{i-1}) - m_{f,\beta}]$$
$$\cdot [X_{j-1}\psi(X_j - \alpha X_{j-1}) - m_{f,\beta}]|_{f,\beta}\} \tag{10.5.25}$$

For $l \to \infty$, the variance in (10.5.25) is bounded from above by $n^{-1}c^2(1-\beta^2)^{-1}$. So, if $c < \infty$, then the random variable $T_0(X^n, \alpha)$ converges asymptotically ($n \to \infty$) to its mean $m_{f,\beta}$ with rate inversely proportional to n. To study the mean $m_{f,\beta}$, we first observe that given f in the class $\mathscr{F}$, and for $l \to \infty$, the random variable X_{l-1} becomes zero mean Gaussian, with variance $(1-\beta^2)^{-1}$. Also, $X_l = \beta X_{l-1} + W_l$. Therefore, $X_l - \alpha X_{l-1} = (\beta - \alpha)X_{l-1} + W_l$, where X_{l-1} and W_l are mutually independent and the density function of W_l is f. From the above observations, we conclude that

$$m_{f,\beta} = E\{X_{l-1}\psi([\beta - \alpha]X_{l-1} + W_l)|_{f,\beta}\}$$

$$= \int_R du\, u \frac{\exp\{-u^2(1-\beta^2)/2\}}{(2\pi)^{1/2}} \sqrt{1-\beta^2} \int_R \psi([\beta - \alpha]u + w)f(w)\, dw$$

$$= \frac{1}{\sqrt{1-\beta^2}} \int_R du\, u\phi(u) \int_R \psi\left(\frac{\beta - \alpha}{(1-\beta^2)^{1/2}}u + w\right) f(w)\, dw \tag{10.5.26}$$

10.5 Discussion and Examples

where $\phi(u)$ is the density function of the zero mean, unit variance Gaussian random variable at the point u.

For the function $\psi(\cdot)$ in (10.5.23), let us now define

$$m_f(\lambda) = \int_R \psi(x - \lambda) f(x)\, dx \qquad (10.5.27)$$

Then, defining $F(x) = \int_{-\infty}^x f(u)\, du$ and $\dot{m}_f(\lambda) = (\partial/\partial y) m_f(y)|_{y=\lambda}$ and by substitution, we obtain

$$\dot{m}_f(\lambda) = F(-c + \lambda) - F(c + \lambda) \qquad (10.5.28)$$

In addition, (10.5.26) gives

$$m_{f,\beta} = \frac{1}{(1-\beta^2)^{1/2}} \int_R du\, u\phi(u) m_f\!\left(-\frac{\beta-\alpha}{(1-\beta^2)^{1/2}} u\right)$$

$$= \frac{-1}{(1-\beta^2)^{1/2}} \int_R du\, \dot{\phi}(u) m_f\!\left(-\frac{\beta-\alpha}{(1-\beta^2)^{1/2}} u\right)$$

$$= -\frac{\beta-\alpha}{1-\beta^2} \int_R du\, \phi(u) \dot{m}_f\!\left(-\frac{\beta-\alpha}{(1-\beta^2)^{1/2}} u\right)$$

Substituting (10.5.28) in the last expression, we obtain,

$$m_{f,\beta} = \frac{\beta-\alpha}{1-\beta^2} \int_R du\, \phi(u) \left[F\!\left(c - \frac{\beta-\alpha}{(1-\beta^2)^{1/2}} u\right) - F\!\left(-c - \frac{\beta-\alpha}{(1-\beta^2)^{1/2}} u\right) \right]$$
$$(10.5.29)$$

It can be easily verified that for symmetric density functions f the mean in (10.5.29) attains a unique zero at $\beta = \alpha$. Therefore, the estimate in (10.5.22) is consistent for every symmetric density function of the sequence $\{W_l\}$ if the function $\psi(\cdot)$ is as in (10.5.23).

Let us now consider independent, symmetric, and Gaussian outliers, occurring with probability ε, that are superimposed on zero mean, unit variance, Gaussian variables $\{W_k\}$. Then,

$$f(u) = (1-\varepsilon)\phi(u) + \frac{\varepsilon}{2}[\phi(u+m) + \phi(u-m)] \qquad u \in R \qquad (10.5.30)$$

where m is an asymptotically large positive value.

Carrying out computations parallel to the ones performed before, we conclude that the M-estimate in (10.5.22), with function $\psi(\cdot)$ as in (10.5.23), is again consistent at f in (10.5.30). The variance in (10.5.25) is then analogous to $n^{-1} c^2 m^2 (1-\beta^2)^{-1}$, however. Therefore, for large values of m, the convergence of the estimate is very slow. The convergence rate of the estimate in (10.5.11), Example 10.5.1, on the other hand, is $n^{-1} m^4 (1-\beta^2)^{-1}$. We thus conclude that in the presence of outliers as in (10.5.30), the M-estimate is superior to the estimate in (10.5.11) by a magnitude of m^2. Superior protection against extreme outliers will be provided by a variation of the M-estimate in (10.5.22), as below.

$$T_*(x^n, \alpha) = n^{-1} \sum_i \psi_1(x_{i-1})\psi_2(x_i - \alpha x_{i-1})$$

$$\hat{\alpha}: T_*(x^n, \hat{\alpha}) = 0$$

(10.5.31)

where $\psi_1(\cdot)$ and $\psi_2(\cdot)$ are both continuous, monotone, and bounded functions.

If both functions $\psi_1(\cdot)$ and $\psi_2(\cdot)$ are bounded absolutely by a constant c, then the convergence rate of the corresponding estimate does not exceed $n^{-1}c^4(1-\beta^2)^{-1}$, even in the presence of extreme outliers, where β is the true value of the autoregressive parameter. In addition, if the functions $\psi_1(\cdot)$ and $\psi_2(\cdot)$ are as in (10.5.23), then the estimate in (10.5.31) is also consistent for every symmetric around zero density function of the sequence $\{W_k\}$.

10.6 PROBLEMS

10.6.1 Assume that only events A or B can occur and that the probability of event A is p.
  (a) What is the probability of event A occurring n times in N independent trials?
  (b) Suppose p is unknown and that the event A is observed to occur y times in N independent trials. Find the maximum likelihood estimate of p.
  (c) Compare the maximum likelihood estimate to the empirical frequency estimate of p.

10.6.2 Let X be a binomial random variable with n trials and probability of success θ. Find the maximum likelihood estimate of θ and show that this is Bayes with respect to the loss function $L(\theta, \hat{\theta}) = (\theta - \hat{\theta})^2/\theta(1-\theta)$ when the a priori distribution on $(0, 1)$ is uniform.
  (Hint: $\int_0^1 y^k(1-y)^i \, dy = k!i!/(k+i+1)!$)

10.6.3 The a posteriori maximum likelihood estimate of a parameter θ, based on the observation y, is given by the solution

$$\hat{\theta}(y) = \max_\theta p(\theta|y)$$

where $p(\theta|y)$ is the conditional probability density of θ given the observation y. Show that if $p(\theta|y)$ is unimodal and symmetric for every y, the above estimate is equivalent to the minimum mean-square estimate.

10.6.4 Let $\{X_i, i = 1, \ldots, N\}$ be exponentially distributed.

$$p(x) = \alpha e^{-\alpha x} \qquad x \geq 0 \quad \alpha > 0$$

The mean of X is then $\mu = 1/\alpha$. It is proposed to estimate α by one over the sample mean.

$$\hat{\alpha} = 1 \bigg/ \frac{1}{N}\sum_{i=1}^N X_i$$

Is this convergent in (a) mean square? (b) probability?

10.6.5 Consider the hypothesis testing problem

$$H_0: v = n \qquad \pi_0 = \tfrac{1}{2}$$
$$H_1: v = \theta s + n \qquad \pi_1 = \tfrac{1}{2}$$

Where s is a known signal vector and n is a zero mean Gaussian random vector with covariance matrix

$$E\{nn^T\} = R.$$

Obtain the test that minimizes the probability of error and calculate its probability as a function of θ.

Suppose θ above is an unknown random variable with mean $\bar{\theta}$ and second moment $\bar{\theta}^2$, and that the vector v is observed.

(a) Calculate the linear, minimum variance estimate of θ. Show that this Bayes estimate is biased, with the bias $\to 0$ as $\bar{\theta}^2 \to \infty$.

(b) Use the Lagrange multiplier technique to obtain the linear, unbiased, minimum variance estimate of θ.

(c) Let $p(v|\theta)$ be the conditional p.d.f. of the observation v given θ. The maximum likelihood estimate $\hat{\theta}^*$ is the estimate that maximizes the p.d.f..

$$p(v|\hat{\theta}^*) > p(v|\hat{\theta})$$

Calculate $\hat{\theta}^*$ and compare it (in terms of mean-square error) to the estimates obtained in parts (a) and (b).

10.6.6 Suppose a sequence of independent random variables $\{n_i\}$ is observed with Poisson density

$$\Pr[n_i = k] = \frac{e^{-\lambda}\lambda^k}{k!} \qquad k = 0, 1, \ldots.$$

The parameter λ is a priori exponentially distributed.

$$f(\lambda) = e^{-\alpha\lambda}$$

(a) Find a sufficient statistic for λ.
(b) Find the maximum likelihood estimate of λ.
(c) Find the minimum variance estimate of λ.
(d) Propose a robust estimate of λ. Is it consistent? What is its breakdown point? Study its influence function.

10.6.7 Let $\{v_i = \theta + n_i; i = 1, \ldots, N\}$ where $\{n_i\}$ is an independent Cauchy sequence

$$p(n_i) = 1/\pi(1 + n_i^2)$$

and θ is uniformly distributed on $[0, a]$. Give the equation satisfied by the maximum likelihood estimate of θ. Evaluate as $a \to 0$.

Can you propose a robust estimate of θ that is also consistent? Find its breakdown point, and study its influence function.

10.6.8 Consider the location parameter estimate

$$\hat{\theta}_M(x^n): n^{-1} \sum_{i=2}^{n+1} \psi(x_i - \hat{\theta}_M(x^n) - a[x_{i-1} - \hat{\theta}_M(x^n)]) = 0$$

$$\psi(u) = \begin{cases} c & u \geq c \\ u & -c < u < c \\ -c & u \leq -c \end{cases}$$

If X^n is Gaussian such that $E\{X_i X_j\} = e^{-\beta|i-j|}$ under zero mean, find that the estimate is consistent, and find its asymptotic variance. Also find its breakdown point and its influence function at the above Gaussian distribution. Discuss the selection of the constants a and c.

10.6.9 In Example 10.5.2, find the breakdown point and the influence function of the estimate in (10.5.31), with $\psi_1 = \psi_2 = \psi$ as in (10.5.23).

REFERENCES

Andrews, D. F. et al (1972), *Robust Estimates of Location: Survey and Advances*, Princeton Univ. Press, Princeton, NJ.

Beran, R. (1974), "Asymptotically Efficient Adaptive Rank Estimates in Location Models," *Ann. Statist.* 2, 63–74.

―――― (1977), "Robust Location Estimates," *Ann. Statist.* 5, 431–444.

―――― (1977), "Minimum Hellinger Distance Estimates for Parametric Models," *Ann. Statist.* 5, 445–463.

―――― (1978), "An Efficient and Robust Adaptive Estimator of Location," *Ann. Statist.* 6, 292–313.

Berkson, J. (1955), "Maximum Likelihood and Minimum χ^2 Estimates of the Logistic Function," *J. Amer. Statist. Assoc.* 50, 130–162.

Bickel, P. J. and J. L. Hodges (1967), "The Asymptotic Theory of Galton's Test and a Related Simple Estimate of Location," *Ann. Math. Statist.* 4, pp. 68–85.

Chernoff, H., J. L. Gastwirth, and M. V. Johns (1967), "Asymptotic Distribution of Linear Combinations of Functions of Order Statistics with Applications to Estimation," *Ann. Math. Statist.* 38, 52–72.

Collins, J. R. (1976), "Robust Estimation of a Location Parameter in the Presence of Asymmetry," *Ann. Statist.* 4, 68–85.

Denby, L. and C. L. Mallows (1977), "Two Diagnostic Displays for Robust Regression Analysis," *Technometrics* 19, 1–13.

References

Devlin, S. J., R. Gnanadesikan, and J. R. Kettenring (1975), "Robust Estimation and Outlier Detection with Correlation Coefficients," *Biometrika* 62, 531–545.

Doob, J. L. (1953), *Stochastic Processes*, Wiley, New York.

Dudley, R. M. (1969), "The Speed of Mean Glivenko-Cantelli Convergence," *Ann. Math. Statist.* 40, 40–50.

Dudley, R. M. (1969), "The Speed of Mean Glivenko-Cantelli Convergence," *Ann. Math. Statist.* 40, 40–50.

Dutter, R. (1975), *Robust Regression: Different Approaches to Numerical Solutions and Algorithms*, Res. Rep. no. 6, Fachgruppe fur Statistik, Eidgen. Technische Hochschule, Zurich.

Ferguson, T. S. (1967), *Mathematical Statistics. A Decision Theoretic Approach*, Academic Press, New York.

Fisher, R. A. (1920), "A Mathematical Examination of the Methods of Determining the Accuracy of an Observation by the Mean Error and the Mean Square Error," *Monthly Not. Roy. Astron. Soc.* 80, 758–770.

Gnanadesikan, R. and Kettenring, J. R. (1972), "Robust Estimates, Residuals and Outlier Detection with Multiresponse Data," *Biometrics* 28, 81–124.

Hampel, F. R. (1968), "Contributions to the theory of robust estimation," Ph.D. Thesis, Univ. of Calif., Berkeley.

────── (1974), "The Influence Curve and Its Role in Robust Estimation," *J. Amer. Statist. Ass.* 62, 1179–1186.

────── (1976), "On the Breakdown Point of Some Rejection Rules with Mean," *Res. Rep. No. 11, Fachgruppe für Statistik, Eidgen.* Technische Hochschule, Zurich.

Hill, R. W. (1977), "Robust Regression When There Are Outliers in the Carriers," Ph.D. Thesis, Harvard Univ., Cambridge, MA.

Hogg, R. V. (1967), "Some Observations on Robust Estimation," *J. Amer. Statist. Ass.* 62, 1179–1186.

Holland, P. W. and R. E. Welsch (1977), "Robust Regression Using Iteratively Reweighted Least Squares," *Comm. Statist.* A6, 813–827.

Huber, P. J. (1964), "Robust Estimation of a Location Parameter," *Ann. Math. Statist.* 35, 73–101.

────── (1973), "Robust Regression: Asymptotics, Conjectures and Monte Carlo," *Ann. Statist.* 1, 799–821.

────── (1981), *Robust Statistics*, Wiley, New York.

Jaeckel, L. A. (1971), "Robust Estimates of Location: Symmetry and Asymmetric Contamination," *Ann. Math. Statist.* 42, 1020–1034.

Jureckova, J. (1971), "Nonparametric Estimates of Regression Coefficients," *Ann. Math. Statist.* 42, 1328–1338.

Kleiner, B. R. D. Martin, and D. J. Thompson, (1979), "Robust Estimation of Power Spectra," *J. Roy. Statist. Soc.* B41, No. 3, 313–351.

LeCam, L. (1953), "On Some Asymptotic Properties of Maximum Likelihood Estimates and Related Bayes' Estimates," *Univ. Calif. Publ. Statist.* 1, 277–330.

Maronna, R. A. (1976), "Robust M-estimators of Multivariate Location and Scatter," *Ann. Statist.* 4, 51–67.

Martin, R. D. (1980), "Robust Estimation of Autoregressive Models," in D. R. Brillinger and G. C. Tias, eds., *Direction in Time* series, Institute of Mathematical Statistics Publication, Haywood CA.

CHAPTER 11

STOCHASTIC DISTANCE MEASURES

11.1 INTRODUCTION

To this point, we have mainly concentrated on the presentation and evaluation of statistical schemes for hypothesis testing and parameter estimation when the data generating processes are given. A question naturally arising at this point is: How do the various statistical schemes perform as functions of the data generating processes involved? The answer is provided by the stochastic distance measures, which measure how close two stochastic processes are in some specific sense, and which generally relate this closeness to the performance of statistical procedures. We will discuss the latter relationships after we present various stochastic distances and discuss their properties. We point out that the term distance is here used loosely, since the triangular and symmetry properties may not be always satisfied.

Let us consider two real stochastic processes whose density functions $f_1(x^n)$ and $f_2(x^n)$ exist for every dimensionality n where, given n, x^n takes values on R^n. Given n, the variational distance $V_n(f_1, f_2)$, the Chernoff distance $C_{n,t}(f_1, f_2)$, the Kullback-Leibler or I-divergence distance $I_n(f_1, f_2)$, and the J-divergence distance $J_n(f_1, f_2)$ are then respectively defined as

$$V_n(f_1, f_2) = \int_{R^n} |f_1(x^n) - f_2(x^n)| \, dx^n \tag{11.1.1}$$

$$C_{n,t}(f_1, f_2) = -n^{-1} \ln\left(\int_{R^n} [f_1(x^n)]^t [f_2(x^n)]^{1-t} \, dx^n\right) \quad 0 \le t \le 1 \tag{11.1.2}$$

$$I_n(f_1, f_2) = n^{-1} \int_{R^n} dx^n \, f_1(x^n) \log \frac{f_1(x^n)}{f_2(x^n)} \tag{11.1.3}$$

$$J_n(f_1, f_2) = I_n(f_1, f_2) + I_n(f_2, f_1)$$

$$= n^{-1} \int_{R^n} dx^n [f_1(x^n) - f_2(x^n)] \log \frac{f_1(x^n)}{f_2(x^n)} \tag{11.1.4}$$

where ln denotes natural logarithm, R is the real line, and R^n denotes n multiples of R.

The Chernoff distance in (11.1.2) is defined for every constant t in $[0, 1]$. Given t, the quantity $[\int_{R^n}[f_1(x^n)]^t[f_2(x^n)]^{1-t}dx^n]^{1/n}$ in the logarithm of (11.1.2) is called the *Chernoff coefficient*. If $t = 0.5$, then the distance in (11.1.2) is instead called the *Bhattacharyya distance* and it is denoted $B_n(f_1, f_2)$, while the quantity in the logarithm is then called the *Bhattacharyya coefficient*. That is,

$$B_n(f_1, f_2) = -\ln c_n(f_1, f_2) \tag{11.1.5}$$

where

$$c_n(f_1, f_2) = \left[\int_{R^n}[f_1(x^n)f_2(x^n)]^{1/2} dx^n\right]^{1/n} \tag{11.1.6}$$

From the definition of the distances in (11.1.1) to (11.1.6), it is easily concluded that when the two processes considered are both memoryless and stationary, then the Chernoff, Bhattacharyya, Kullback-Leibler, and J-divergence distances are independent of the dimensionality n, and they can be then uniquely defined via the corresponding first order density functions. When the two processes are instead stationary with memory, then varying dimensionalities of n induce varying numbers $C_{n,t}(f_1, f_2)$, $B_n(f_1, f_2)$, $I_n(f_1, f_2)$ and $J_n(f_1, f_2)$, with generally no specific trends or monotonicities as functions of n. The above distances can be then defined as limits. That is, the Chernoff, Bhattacharyya, Kullback-Leibler, and J-divergence distances between two stationary processes with memory are respectively defined as follows.

$$C_t(f_1, f_2) = \lim_{n \to \infty} C_{n,t}(f_1, f_2) \qquad B(f_1, f_2) = \lim_{n \to \infty} B_n(f_1, f_2)$$
$$I(f_1, f_2) = \lim_{n \to \infty} I_n(f_1, f_2) \qquad J(f_1, f_2) = \lim_{n \to \infty} J_n(f_1, f_2) \tag{11.1.7}$$

The variational distance in (11.1.1) is used for fixed n, and it is a quite strong distance, as we will see later. We note that the Chernoff (for $t \neq 0.5$) and the Kullback-Leibler distances are not symmetric with respect to the involved densities, while the variational, J-divergence, and Bhattacharyya distances are. Also, all the above distances, with the exception of the variational distance, do not generally satisfy the triangular property. They all satisfy the identity property, however. Indeed, directly from (11.1.1), applying the Hölder inequality in (11.1.2) and (11.1.5), and using the inequality $\log x \geq 1 - x^{-1}$ in (11.1.3) and (11.1.4), we conclude that all the above distances equal zero if and only if the two densities are identical almost everywhere in R^n.

The variational, Chernoff, Bhattacharyya, Kullback-Leibler, and J-divergence distances involve only the density functions of the corresponding stochastic processes, and they can be considered as comprising a class. Another class consists of stochastic distances that involve distortion measures on data sequences and joint probability measures whose marginals are the densities of the considered stochastic processes. This latter class includes the Prohorov,

11.1 Introduction

Lèvy, Vasershtein, and Rho-Bar stochastic distances, whose definition will be given in the sequel. Given two real data sequences x^n and y^n, a distortion measure $\rho_n(x^n, y^n)$ is a real scalar function that represents the distance between the two sequences. A single-letter distortion measure is such that

$$\rho_n(x^n, y^n) = n^{-1} \sum_{i=1}^{n} \rho(x_i, y_i) \qquad (11.1.8)$$

where $\rho(\cdot, \cdot)$ is a given scalar real function. Candidates for the latter function may be $\rho(x_i, y_i) = |x_i - y_i|$, $\rho(x_i, y_i) = (x_i - y_i)^2$, or $\rho(x_i, y_i) = \max(x_i, y_i)$. Let us now consider two stochastic processes whose density functions of every dimensionality exist. Let us then denote the arbitrary dimensionality densities respectively f_1 and f_2, and let $f_1(x^n)$ and $f_2(x^n)$ denote respectively the n-dimensional density functions at the vector point x^n. Let $\mathscr{P}^n$ denote the class of all the $2n$-dimensional density functions whose marginal densities are f_1 and f_2. That is if g is some density function in $\mathscr{P}^n$ then

$$f_1(x^n) = \int_{R^n} g(x^n, y^n)\,dy^n \qquad \forall\, x^n \in R^n$$
$$f_2(y^n) = \int_{R^n} g(x^n, y^n)\,dx^n \qquad \forall\, y^n \in R^n \qquad (11.1.9)$$

Given f_1, f_2, and n, given some distortion measure $\rho_n(x^n, y^n)$ and the class $\mathscr{P}^n$ above, the Prohorov distance $\Pi_{n,\rho_n}(f_1, f_2)$, and the Vasershtein distance $V_{n,\rho_n}(f_1, f_2)$ are defined, due to Strassen (1965) and Dudley (1968), as

$$\Pi_{n,\rho_n}(f_1, f_2) = \inf_{g \in \mathscr{P}^n} \inf\{\gamma : \Pr(x^n, y^n : \rho_n(x^n, y^n) > \gamma \mid g) \leq \gamma\} \qquad (11.1.10)$$

$$V_{n,\rho_n}(f_1, f_2) = \inf_{g \in \mathscr{P}^n} E\{\rho_n(X^n, Y^n) \mid g\} \qquad (11.1.11)$$

When $\rho_n(x^n, y^n)$ is a single-letter distortion measure as in (11.1.8), then the Vasershtein distance in (11.1.11) is denoted instead $\bar{\rho}_n(f_1, f_2)$, and it represents the smallest possible expected single-letter distortion between random vectors X^n and Y^n that are respectively generated by f_1 and f_2. The Rho-Bar distance between the two stochastic processes whose arbitrary dimensionality densities are f_1 and f_2 is then defined as

$$\bar{\rho}(f_1, f_2) = \sup_n \bar{\rho}_n(f_1, f_2) \qquad (11.1.12)$$

The Rho-Bar distance represents the smallest possible expected single-letter distortion between random vectors generated by f_1 and f_2 that is "worst" in terms of dimensionality. Furthermore, it can be shown [see Gray et al (1975)], that the supremum in (11.1.12) exists and equals the limit $\lim_{n \to \infty} \bar{\rho}_n(f_1, f_2)$. Thus, the Rho-Bar distance represents asymptotic ($n \to \infty$), smallest possible expected single-letter distortion, and it bounds from above the smallest possible expected such distortions that correspond to data vectors of arbitrary dimensionality.

Now let the two processes whose generalized densities are f_1 and f_2 be stationary, and let $\mathscr{P}_s^n$ denote the class of all $2n$-dimensional stationary density functions whose marginal densities are f_1 and f_2. Then $\mathscr{P}_s^n \subset \mathscr{P}^n$, where $\mathscr{P}^n$ is as in (11.1.11), and

$$\bar{\rho}_n(f_1, f_2) = \inf_{g \in \mathscr{P}^n} E\left\{n^{-1} \sum_{i=1}^n \rho(X_i, Y_i) \Big| g\right\} \leq \inf_{g \in \mathscr{P}_s^n} E\left\{n^{-1} \sum_{i=1}^n \rho(X_i, Y_i) \Big| g\right\}$$

$$= n^{-1} \sum_{i=1}^n \inf_{g \in \mathscr{P}^1} E\{\rho(X_i, Y_i) | g\} = \inf_{g \in \mathscr{P}^1} E\{\rho(X, Y) | g\}$$

$$= V_{1,\rho}(f_1, f_2) \tag{11.1.13}$$

$$\bar{\rho}_n(f_1, f_2) = \inf_{g \in \mathscr{P}^n} E\left\{n^{-1} \sum_{i=1}^n \rho(X_i, Y_i) \Big| g\right\} \geq n^{-1} \sum_{i=1}^n \inf_{g \in \mathscr{P}^n} E\{\rho(X_i, Y_i) | g\}$$

$$= n^{-1} \sum_{i=1}^n \inf_{g \in \mathscr{P}^1} E\{\rho(X_i, Y_i) | g\} = \inf_{g \in \mathscr{P}^1} E\{\rho(X, Y) | g\}$$

$$= V_{1,\rho}(f_1, f_2) \tag{11.1.14}$$

From (11.1.13) and (11.1.14), we conclude that if the two processes are stationary, then $\bar{\rho}_n(f_1, f_2) = V_{1,\rho}(f_1, f_2) \,\forall\, n$. Therefore, $\bar{\rho}_n(f_1, f_2) = V_{1,\rho}(f_1, f_2)$ then, where $V_{1,\rho}(f_1, f_2)$ is the one-dimensional Vasershtein distance in (11.1.11), which represents the smallest possible expected ρ distortion between single data from the two processes.

Let us now focus on the Prohorov distance in (11.1.10). Given n and the distortion measure $\rho_n(x^n, y^n)$, the Prohorov distance γ_n^* between f_1 and f_2 is the smallest possible probability with which the ρ_n distortion between sequences from the two processes is higher than γ_n^*. The number γ_n^* is clearly between zero and one, and the Prohorov distance represents closeness of processes in a weak sense, since it is in probability. Given two processes, the Prohorov distances of various dimensionalities do not generally converge in any sense; that is, the $\sup_n \Pi_{n,\rho_n}(f_1, f_2)$ and the $\lim_{n \to \infty} \Pi_{n,\rho_n}(f_1, f_2)$ do not generally exist. This is a disadvantage of the Prohorov distance that makes it generally inappropriate as a stochastic distance between processes with memory. The weak closeness represented by this distance is an advantage, however, for some problems, as we will see later. Given n and the density functions $f_1(x^n)$ and $f_2(x^n)$; $x^n \in R^n$, let $F_1(x^n)$ and $F_2(x^n)$ be the corresponding cumulative distributions at the vector point x^n. Given a distortion measure $\rho_n(x^n, y^n)$, x^n and some real scalar positive number δ, let $s(x^n, \delta)$ denote all the sequences y^n whose ρ_n distortion from x^n is less than or equal to δ; that is, $s(x^n, \delta) = \{y^n: \rho_n(x^n, y^n) \leq \delta\}$. Let then x_δ^n denote the sequence in $s(x^n, \delta)$ that attains the maximum value of the cumulative distribution $F_2(\cdot)$; that is, $F_2(x_\delta^n) = \sup_{y^n \in s(x^n, \delta)} F_2(y^n)$. Then the Lèvy distance $L_{n,\rho_n}(f_1, f_2)$ is defined as

$$L_{n,\rho_n}(f_1, f_2) = \inf\{\delta: F_1(x^n) \leq F_2(x_\delta^n) + \delta \,\forall\, x^n \in R^n\} \tag{11.1.15}$$

It can be shown that the Lèvy distance is symmetric; that is, $L_{n,\rho_n}(f_1, f_2) = L_{n,\rho_n}(f_2, f_1)$. It is also weaker than the Prohorov distance, representing closeness of cumulative distributions. In fact, it arises as a special case of the Prohorov distance, and thus has similar properties with the latter. For $n = 1$ and $\rho(x, y) = |x - y|$, the Lèvy distance takes the form $L_{1,\rho}(f_1, f_2) = \inf\{\delta : F_1(x) \leq F_2(x + \delta) + \delta \, \forall \, x \in R\}$.

The properties of the Prohorov, Vasershtein, and Rho-Bar distances are controlled by the distortion measure on data sequences included in their definition. If this distortion measure is a metric (satisfying the triangular property), so are the distances [see Dobrushin (1970) and Gray et al (1975)]. On the other hand, the Prohorov distance takes values strictly in [0, 1] for every distortion measure, while the Vasershtein and the Rho-Bar distances may be instead unbounded. Those distances are sometimes equivalent, as we will discuss in the next section, and the choice of the appropriate distortion measure in their definition depends on the particular application.

11.2 PROPERTIES AND RELATIONSHIPS

In the previous section, we presented some important stochastic distance measures and discussed some of their fundamental properties. Here we will present some relationships between various distances that are important in selecting the appropriate distance for different applications. We will begin by developing relationships between the variational, Bhattacharyya, Kullback-Leibler, and J-divergence distances. First, we define a generalized version of the variational distance that includes a priori probabilities. Consider densities f_1 and f_2, occurring respectively with a priori probabilities p_1 and p_2, where $p_1 + p_2 = 1$. The n-dimensional variational distance $V_n(f_1, f_2, p_1, p_2)$ is then defined as

$$V_n(f_1, f_2, p_1, p_2) = 2^{-1} \int_{R^n} |p_1 f_1(x^n) - p_2 f_2(x^n)| dx^n \qquad (11.2.1)$$

We can now express the following lemma.

LEMMA 11.2.1: Given two density functions f_1 and f_2, given n, the variational distance in (11.2.1), the Bhattacharyya coefficient in (11.1.6), the Bhattacharyya distance in (11.1.5), the Kullback-Leibler distance in (11.1.3), and the J-divergence distance in (11.1.4) are related as follows, where in the definition of the last two distances, the natural logarithm is used.

$$[1 - 4p_1 p_2 C_n^{2n}(f_1, f_2)]^{1/2} \geq 2V_n(f_1, f_2, p_1, p_2) \geq 1 - 2[p_1 p_2]^{1/2} c_n^n(f_1, f_2)$$
$$(11.2.2)$$

$$2B_n(f_1, f_2) = -2 \ln c_n(f_1, f_2) \leq I_n(f_1, f_2) \qquad (11.2.3)$$

$$2B_n(f_1, f_2) \leq I_n(f_2, f_1) \qquad (11.2.4)$$

$$4B_n(f_1, f_2) \leq J_n(f_1, f_2) \text{ or } c_n(f_1, f_2) \geq \exp\{-J_n(f_1, f_2)/4\} \qquad (11.2.5)$$

Proof: To prove the right part of (11.2.2), we use

$$2V_n(f_1, f_2, p_1, p_2) = \int_{R^n} |p_1 f_1(x^n) - p_2 f_2(x^n)| \, dx^n$$

$$\geq \int_{R^n} |[p_1 f_1(x^n)]^{1/2} - [p_2 f_2(x^n)]^{1/2}|^2 \, dx^n$$

$$= 1 - 2[p_1 p_2]^{1/2} \int_{R^n} [f_1(x^n) f_2(x^n)]^{1/2} \, dx^n$$

$$= 1 - 2[p_1 p_2]^{1/2} c_n^n(f_1, f_2)$$

To prove the left part of (11.2.2), we use the Schwartz inequality.

$$2V_n(f_1, f_2, p_1, p_2) = \int_{R^n} |p_1 f_1(x^n) - p_2 f_2(x^n)| \, dx^n$$

$$= \int_{R^n} |[p_1 f_1(x^n)]^{1/2} - [p_2 f_2(x^n)]^{1/2}|$$
$$\times |[p_1 f_1(x^n)]^{1/2} + [p_2 f_2(x^n)]^{1/2}| \, dx^n$$

$$\leq \left\{ \int_{R^n} |[p_1 f_1(x^n)]^{1/2} - [p_2 f_2(x^n)]^{1/2}|^2 \, dx^n \right\}^{1/2}$$

$$\times \left\{ \int_{R^n} |[p_1 f_1(x^n)]^{1/2} + [p_2 f_2(x^n)]^{1/2}|^2 \, dx^n \right\}^{1/2}$$

$$= \left\{ 1 - 2[p_1 p_2]^{1/2} \int_{R^n} [f_1(x^n) f_2(x^n)]^{1/2} \, dx^n \right\}^{1/2}$$

$$\times \left\{ 1 + 2[p_1 p_2]^{1/2} \int_{R^n} [f_1(x^n) f_2(x^n)]^{1/2} \, dx^n \right\}^{1/2}$$

$$= \{1 - 2[p_1 p_2]^{1/2} c_n^n(f_1, f_2)\}^{1/2} \{1 + 2[p_1 p_2]^{1/2} c_n^n(f_1, f_2)\}^{1/2}$$

$$= [1 - 4 p_1 p_2 c_n^{2n}(f_1, f_2)]^{1/2}$$

To prove (11.2.3), and identically (11.2.4), we use Jensen's inequality.

$$-\frac{n}{2} I_n(f_1, f_2) = \int_{R^n} dx^n \, f_1(x^n) \ln\left[\frac{f_2(x^n)}{f_1(x^n)}\right]^{1/2} = E\left\{\ln\left[\frac{f_2(x^n)}{f_1(x^n)}\right]^{1/2} \bigg/ f_1\right\}$$

$$\leq \ln E\left\{\left[\frac{f_2(x^n)}{f_1(x^n)}\right]^{1/2} \bigg/ f_1\right\}$$

$$= \ln\left[\int_{R^n} dx^n \, f_1(x^n) [f_1(x^n)]^{-1/2} [f_2(x^n)]^{1/2}\right]$$

$$= \ln\left[\int_{R^n} [f_1(x^n) f_2(x^n)]^{1/2} \, dx^n\right] = \ln c_n^n(f_1, f_2) = n \ln c_n(f_1, f_2)$$

$$= -n B_n(f_1, f_2) \quad \text{or} \quad 2 B_n(f_1, f_2) \leq I_n(f_1, f_2)$$

and equivalently

$$c_n(f_1, f_2) \geq \exp\{-J_n(f_1, f_2)/4\}$$

11.2 Properties and Relationships

The inequality $4B_n(f_1, f_2) \leq J_n(f_1, f_2)$ follows directly from (11.2.3) and (11.2.4). ∎

Setting $p_1 = p_2 = 0.5$ in (11.2.2), we obtain, in a straightforward fashion the following relationship between the variational distance in (11.1.1) and the Bhattacharyya coefficient in (11.1.6).

$$2[1 - c_n^{2n}(f_1, f_2)]^{1/2} \geq V_n(f_1, f_2) \geq 2[1 - c_n^n(f_1, f_2)] \tag{11.2.6}$$

Due to inequalities (11.2.3), (11.2.4), and (11.2.5), we conclude that the Kullback-Leibler and the J-divergence distances are both stronger than the Bhattacharyya distance. That is, when the former are bounded, so is the latter. We note that the Kullback-Leibler and the J-divergence distances are related to the information measure of Shannon. At this point, it is perhaps appropriate to exhibit a special relationship between the Bhattacharyya coefficient, and the Fisher information measure encountered in parameter estimation. Without lack in generality, let us consider some one-dimensional density function $f(x); x \in R$, such that $\dot{f}(\infty) = \dot{f}(-\infty) = 0$ and $f(\infty) = f(-\infty) = 0$, where $\dot{f}$ denotes first order derivative. Let $\ddot{f}$ denote second order derivative, let θ be some location parameter value, and let us consider the Bhattacharyya coefficient $c(f)$ between the densities $f(x - \theta)$ and $f(x - \theta - \varepsilon)$, for $\varepsilon \ll 1$. Applying Taylor expansion, and partial differential assuming that all the derivatives of f exist, we obtain

$$\frac{\partial}{\partial \varepsilon} [f(x - \theta) f(x - \theta - \varepsilon)]^{1/2} \big|_{\varepsilon = 0}$$

$$= [-2^{-1} f^{1/2}(x - \theta) f^{-1/2}(x - \theta - \varepsilon) \dot{f}(x - \theta - \varepsilon)] \big|_{\varepsilon = 0} = -2^{-1} \dot{f}(x - \theta)$$

$$\frac{\partial^2}{\partial \varepsilon^2} [f(x - \theta) f(x - \theta - \varepsilon)]^{1/2} \big|_{\varepsilon = 0}$$

$$= [-4^{-1} f^{1/2}(x - \theta) f^{-3/2}(x - \theta - \varepsilon) [\dot{f}(x - \theta - \varepsilon)]^2$$
$$+ 2^{-1} f^{1/2}(x - \theta) f^{-1/2}(x - \theta - \varepsilon) \ddot{f}(x - \theta - \varepsilon)] \big|_{\varepsilon = 0}$$

$$= -4^{-1} \frac{[\dot{f}(x - \theta)]^2}{f(x - \theta)} + 2^{-1} \ddot{f}(x - \theta)$$

$$[f(x - \theta) f(x - \theta - \varepsilon)]^{1/2} \underset{\varepsilon \ll 1}{\simeq} f(x - \theta) - \frac{\varepsilon}{2} \dot{f}(x - \theta)$$

$$+ \frac{\varepsilon^2}{2} \left\{ -4^{-1} \frac{[\dot{f}(x - \theta)]^2}{f(x - \theta)} + 2^{-1} \ddot{f}(x - \theta) \right\}$$

$$c(f) = \int_R [f(x - \theta) f(x - \theta - \varepsilon)]^{1/2} dx$$

$$\underset{\varepsilon \ll 1}{\simeq} \int_R f(x - \theta) dx - \frac{\varepsilon^2}{8} \int_R \left[\frac{\dot{f}(x - \theta)}{f(x - \theta)} \right]^2 f(x - \theta) dx$$

$$= 1 - \frac{\varepsilon^2}{8} \int_R \left[\frac{\dot{f}(x)}{f(x)} \right]^2 f(x) dx \tag{11.2.7}$$

Expression (11.2.7) relates the Bhattacharyya coefficient to the Fisher information measure, for density functions that are location parameter close. To this point, we presented relationships within the class of distances that do not involve distortion measures on data sequences. Making the transition to the latter class smooth, we will first relate the variational distance in (11.1.1), with the Vasershtein distance in (11.1.11), and the Lèvy distance in (11.1.15). We thus present the following lemma.

LEMMA 11.2.2: Given two densities f_1 and f_2, whose cumulative distributions are respectively denoted F_1 and F_2, then

i. Given n, and any distortion measure $\rho_n(x^n, y^n)$, we have

$$L_{n,\rho_n}(f_1, f_2) \leq V_n(f_1, f_2) \qquad (11.2.8)$$

ii. Given n and $\rho_n(x^n, y^n) = \{1; x^n \neq y^n \text{ and } 0; x^n = y^n\}$, we have

$$V_{n,\rho_n}(f_1, f_2) = V_n(f_1, f_2) \qquad (11.2.9)$$

Proof:

i. Parthasarathy (1968) has shown that

$$V_n(f_1, f_2) \geq \sup_{x^n \in R^n} |F_1(x^n) - F_2(x^n)|$$

Then, if $V_n(f_1, f_2) = \delta$, we have

$$F_1(x^n) - F_2(x^n) \leq |F_1(x^n) - F_2(x^n)| \leq \delta \leq F_2(x^n_\delta) - F_2(x^n) + \delta \qquad \forall\, x^n \in R^n$$

and thus

$$F_1(x^n) \leq F_2(x^n_\delta) + \delta \qquad \forall\, x^n \in R^n \qquad (11.2.10)$$

From (11.2.10), we conclude that the infimum in (11.1.15) is generally less than the δ in (11.2.10). Therefore, $L_{n,\rho_n}(f_1, f_2) \leq V_n(f_1, f_2)$.

ii. Dobrushin (1970) showed that the variational distance in (11.1.1) equals the infimum, among all joint densities with marginals f_1 and f_2, probability that the sequences x^n and y^n in (11.1.1) are unequal. But if

$$\rho_n(x^n, y^n) = \begin{cases} 1 & x^n \neq y^n \\ 0 & x^n = y^n \end{cases}$$

then the Vasershtein distance equals the above probability. Thus, (11.2.9) follows. ∎

11.2 Properties and Relationship

We will now focus on the Prohorov and Vasershtein distances, where the former is given by (11.1.10); where given n the $\bar{\rho}_n(f_1, f_2)$ distance is the Vasershtein distance with the single-letter distortion measure in (11.1.8); and where $\bar{\rho}(f_1, f_2) = \sup \{\bar{\rho}_n(f_1, f_2): n\}$. We express two important relationships between the Prohorov and the Rho-Bar distances in Lemma 11.2.3 below.

LEMMA 11.2.3: Consider two density functions f_1 and f_2 and a single-letter distortion measure $\rho_n(x^n, y^n) = n^{-1} \sum_{i=1}^{n} \rho(x_i, y_i)$, where $\rho(x, y) \geq 0 \; \forall \; x, y \in R$. Then,

i. Given n,
$$[\Pi_{n,\rho_n}(f_1, f_2)]^2 \leq \bar{\rho}_n(f_1, f_2) \leq \bar{\rho}(f_1, f_2) \qquad (11.2.11)$$

ii. If $\rho(x, y)$ is bounded by some positive finite constant c for every x and y in R, and given n,
$$\bar{\rho}_n(f_1, f_2) \leq [1 + c]\Pi_{n,\rho_n}(f_1, f_2) \qquad (11.2.12)$$

Proof:

i. Let $\bar{\rho}_n(f_1, f_2) = \varepsilon$, and let p be the joint density function that attains $\bar{\rho}_n(f_1, f_2)$. Then, since $\rho_n(x^n, y^n)$ is nonnegative, we can apply Lemma 4.6.1 to obtain

$$\Pr\{x^n, y^n: \rho_n(x^n, y^n) > \varepsilon^{1/2} | p\} \leq \varepsilon^{-1/2} E\{\rho_n(X^n, Y^n) | p\}$$
$$= \varepsilon^{-1/2} \bar{\rho}_n(f_1, f_2) = \varepsilon^{-1/2} \varepsilon = \varepsilon^{1/2} \qquad (11.2.13)$$

From (11.2.13), and in conjunction with the definition of the Prohorov distance in (11.1.10), we conclude

$$\Pi_{n,\rho_n}(f_1, f_2) \leq \varepsilon^{1/2}$$

or

$$[\Pi_{n,\rho_n}(f_1, f_2)]^2 \leq \varepsilon = \bar{\rho}_n(f_1, f_2) \leq \sup_n \bar{\rho}_n(f_1, f_2) = \bar{\rho}(f_1, f_2)$$

ii. Let $\rho(x, y) \leq c < \infty \; \forall \; x$ and y, let $\Pi_{n,\rho_n}(f_1, f_2) = \varepsilon$, and let p be the joint density function that attains the above Prohorov distance. Then

$$\bar{\rho}_n(f_1, f_2) \leq E\{\rho_n(X^n, Y^n) | p\}$$
$$\leq \varepsilon \Pr\{x^n, y^n: \rho_n(x^n, y^n) \leq \varepsilon | p\} + c \Pr\{x^n, y^n: \rho_n(x^n, y^n) > \varepsilon | p\}$$
$$\leq \varepsilon + c \Pr\{x^n, y^n: \rho_n(x^n, y^n) > \varepsilon | p\} \leq \varepsilon + c\varepsilon$$
$$= (1 + c)\varepsilon = (1 + c)\Pi_{n,\rho_n}(f_1, f_2) \qquad \blacksquare$$

The results in Lemma 11.2.3 are significant. Condition (11.2.11) says that the Rho-Bar distance is a strong distance that bounds from above the Prohorov distance of any dimensionality. Conditions (11.2.11) and (11.2.12) say that if the distortion measure $\rho(\cdot,\cdot)$ is bounded, then the n-dimensional Rho-Bar and Prohorov distances are equivalent, in the sense that they generate the same topology. In the latter case, the supremum $\sup_n \Pi_{n,\rho_n}(f_1, f_2)$ exists, it equals the limit $\lim_{n\to\infty} \Pi_{n,\rho_n}(f_1, f_2)$, and it is equivalent to $\bar{\rho}(f_1, f_2)$.

11.3 RELATION TO STATISTICAL PROCEDURES

In this section, we will present relationships between various stochastic distances and some performance criteria in statistical procedures. We will start by relating the variational distance in (11.2.1), the Bhattacharyya coefficient in (11.1.6), and the J-divergence distance in (11.1.4), with the probability of error in single hypothesis testing. Consider two hypotheses H_1 and H_2, represented respectively by the densities f_1 and f_2. Let p_1 and p_2 be the a priori probabilities of H_1 versus H_2. Given an observation vector x^n and via the Bayesian procedures in Chapter 4, H_1 is decided if $p_1 f_1(x^n) \geq p_2 f_2(x^n)$. Otherwise, H_2 is decided. Given n, let us define $A_1 = \{x^n : p_1 f_1(x^n) \geq p_2 f_2(x^n)\}$ and $A_2 = \{x^n : p_1 f_1(x^n) < p_2 f_2(x^n)\}$, where $A_1 \cup A_2 = R^n$. The probability of error $P_e(n)$ induced by the test is then given by

$$P_e(n) = p_1 \int_{A_2} f_1(x^n) dx^n + p_2 \int_{A_1} f_2(x^n) dx^n$$
$$= \int_{R_n} \min[p_1 f_1(x^n), p_2 f_2(x^n)] dx^n \qquad (11.3.1)$$

The variational distance in (11.2.1), the Bhattacharyya coefficient in (11.1.6), and the J-divergence distance in (11.1.4) all provide bounds for the probability of error in (11.3.1). Those bounds are included in the statement of Lemma 11.3.1 below.

LEMMA 11.3.1: Given densities f_1 and f_2, given the a priori probabilities p_1 and p_2, given n, the following expressions hold.

$$P_e(n) = 2^{-1} - V_n(f_1, f_2, p_1, p_2) \qquad (11.3.2)$$
$$2^{-1} - 2^{-1}[1 - 4p_1 p_2 c_n^{2n}(f_1, f_2)]^{1/2} \leq P_e(n) \leq [p_1 p_2]^{1/2} c_n^n(f_1, f_2) \qquad (11.3.3)$$
$$P_e(n) \geq p_1 p_2 \exp\{-n J_n(f_1, f_2)/2\} \qquad (11.3.4)$$

where the logarithm in the definition of the J-divergence distance is the natural logarithm.

Proof: From the definition of the variational distance in (11.2.1), the probability of error in (11.3.1), and A_1 and A_2 as in (11.3.1), we prove (11.3.2) as follows.

11.3 Relation to Statistical Procedures

$$2V_n(f_1, f_2, p_1, p_2) = \int_{A_1} [p_1 f_1(x^n) - p_2 f_2(x^n)] dx^n + \int_{A_2} [p_2 f_2(x^n) - p_1 f_1(x^n)] dx^n$$

$$= \int_{A_1} p_1 f_1(x^n) dx^n + \int_{A_2} p_2 f_2(x^n) dx^n$$

$$\quad - \left\{ p_2 \int_{A_1} f_2(x^n) dx^n + p_1 \int_{A_2} f_1(x^n) dx^n \right\}$$

$$= p_1 - p_1 \int_{A_2} f_1(x^n) dx^n + p_2 - p_2 \int_{A_1} f_2(x^n) dx^n$$

$$\quad - \left\{ p_2 \int_{A_1} f_2(x^n) dx^n + p_1 \int_{A_2} f_1(x^n) dx^n \right\}$$

$$= 1 - 2 \left\{ p_2 \int_{A_1} f_2(x^n) dx^n + p_1 \int_{A_2} f_1(x^n) dx^n \right\} = 1 - 2P_e(n)$$

We obtain (11.3.3) by straightforward substitution of (11.3.2) in (11.2.2), Lemma 11.2.1. To prove (11.3.4), we use inequality (11.2.5) in Lemma 11.2.1, in conjunction with the following easily proven inequality.

$$2^{-1} - 2^{-1}[1 - 4p_1 p_2 c_n^{2n}(f_1, f_2)]^{1/2} \geq p_1 p_2 c_n^{2n}(f_1, f_2) \qquad \blacksquare$$

Among the expressions in Lemma 11.3.1, (11.3.3) is perhaps the most significant, since it expresses both upper and lower bounds on the probability of error as functions of the Bhattacharyya coefficient. Applying the Schwartz inequality, we find

$$c_n(f_1, f_2) = \left[\int_{R^n} [f_1(x^n) f_2(x^n)]^{1/2} dx^n \right]^{1/n}$$

$$\leq \left\{ \left[\int_{R^n} f_1(x^n) dx^n \right]^{1/2} \left[\int_{R^n} f_2(x^n) dx^n \right]^{1/2} \right\}^{1/n} = 1$$

with equality if and only if the densities $f_1(x^n)$ and $f_2(x^n)$ are identical almost everywhere in R^n. In the latter case, the Bhattacharyya coefficient is useless, since the two densities are then indistinguishable. Otherwise, the Bhattacharyya coefficient is strictly less than one, and it decreases as the two densities are drawn further apart. For Bhattacharyya coefficient strictly less than one, the two bounds in (11.3.3) both decrease monotonically to zero as n increases to asymptotically large values. The probability of error $P_e(n)$ then decreases to zero as well, at an exponential rate, as clearly indicated by the bounds in (11.3.3). Therefore, for cases where the probability of error cannot be explicitly computed, the Bhattacharyya coefficient provides its convergence rate in a specific fashion.

From the above discussion, it is clear that the Bhattacharyya coefficient is a measure of distinguishability between two density functions. Another such measure is the Kullback-Leibler distance, whose properties we initially saw in Chapter 6, with Huber's robust detection model. The Kullback-Leibler distance between two densities f_1 and f_2 is generally given by

$$I(f_1, f_2) = \lim_{n \to \infty} n^{-1} \int_{R^n} f_1(x^n) \log \frac{f_1(x^n)}{f_2(x^n)} dx^n$$

$$= \left\{ -\lim_{n \to \infty} n^{-1} \int_{R^n} dx^n f_1(x^n) \log f_2(x^n) \right\} \quad (11.3.5)$$

$$- \left\{ -\lim_{n \to \infty} n^{-1} \int_{R^n} dx^n f_1(x^n) \log f_1(x^n) \right\}$$

The second bracket in (11.3.5) is the entropy of the process whose arbitrary dimensionality density is f_1. The first bracket in (11.3.5) is the mismatch entropy, when the process with density f_2 is assumed and the process with density f_1 is actually acting. The Kullback-Leibler distance is thus the difference between the entropy of one process and a mismatch entropy. The smaller this distance, the closer the two entropies, therefore the less dramatic the mismatch between the two processes, and thus the less distinguishable they are. The J-divergence distance has similar characteristics; there the difference between the two symmetric mismatch entropies and the entropies of the two processes is measured. As exhibited by Lemma 11.2.1, the Kullback-Leibler and J-divergence distances are stronger measures of distinguishability between two processes than is the Bhattacharyya distance. Thus, since (for example) the least favorable pair of densities in Huber's robust detection game (Chapter 6) is the closest in Kullback-Leibler and J-divergence distances, it is also close in Bhattacharyya distance.

Let us now focus on the Prohorov distance. As we discussed in Section 11.1, the Prohorov distance represents weak closeness of density functions. If f_0 is some well known nominal density function and ε is some given number in (0, 1), then the n-dimensional densities f included in the ball represented by $\Pi_{n, \rho_n}(f_0, f) \leq \varepsilon$ generate data sequences that with probability higher than $1 - \varepsilon$ are closer than ε to data sequences generated by f_0 (in the ρ_n distortion sense). As an example, let us consider the class of density functions that represent the occurrence of independent outliers on data generated by a stationary and memoryless process whose first order density is f_0. Let the probability with which an observed datum y_n is an outlier be ε, and let us then denote the same datum by x_n if it is instead generated by f_0. Then, clearly, $\Pr\{|Y_n - X_n| > \varepsilon\} \leq \varepsilon$. That means that the density functions which represent with probability ε the occurrence of independent outliers on data sequences from the stationary and memoryless density f_0 are included in the Prohorov ball $\Pi_{1, \rho}(f_0, f) \leq \varepsilon$, where $\rho(x, y) = |x - y|$. Therefore, the Prohorov distance best represents the outlier models that we discussed in Chapters 6 and 10, and it is a significant distance for statistical robustness, as we will further discuss in Chapter 12. Let us now consider the case where independent outliers may occur on data generated by a stationary process with memory, whose arbitrary dimensionality density function is denoted f_0. Let the outliers occur with probability ε, let X_n denote the nth random datum generated by f_0, let W_n denote the nth random datum generated by an outlier, and let Y_n denote the nth random observed datum.

11.3 Relation to Statistical Procedures

Then

$$Y_n = (1 - \varepsilon)X_n + \varepsilon W_n \qquad (11.3.6)$$

From the data model in (11.3.6), we conclude $\Pr\{n^{-1}\sum_{i=1}^n |Y_i - X_i| > \varepsilon\} \leq \varepsilon^n < \varepsilon$ for any given n. Therefore, given n, the density functions representing the above outlier model are included in the Prohorov ball $\Pi_{n,\rho_n}(f_0, f) \leq \varepsilon$, where $\rho_n(x^n, y^n)$ is the single-letter distortion measure $n^{-1}\sum_{i=1}^n |x_i - y_i|$. The latter ball is pessimistically large, however, for large values of n. For large n values, a physically meaningful alternative is then the Prohorov distance with distortion measure $\rho_n^*(x^n, y^n)$ as below.

$$\rho_n^*(x^n, y^n) = \inf\left\{\alpha: \frac{[\# i: |x_i - y_i| > \alpha]}{n} \leq \alpha\right\} \qquad (11.3.7)$$

For the model in (11.3.6), and for mutually ergodic nominal and outlier processes, the ratio $n^{-1}[\# i: |x_i - y_i| > \alpha]$ converges asymptotically ($n \to \infty$) to the probability $\Pr\{|X_i - Y_i| > \alpha\}$. Since the model in (11.3.6) implies $\Pr\{|X_i - Y_i| > \varepsilon\} \leq \varepsilon$, the infimum α in (11.3.7) is then asymptotically less than or equal to ε. In addition, the limit $\lim_{n \to \infty} \Pr\{\rho_n^*(X^n, Y^n) > \varepsilon\}$ is then definitely less than ε, thus the densities representing the model in (11.3.6) are asymptotically included in the Prohorov ball $\Pi_{n,\rho_n^*}(f_0, f) \leq \varepsilon$, where $\rho_n^*(\cdot, \cdot)$ is as in (11.3.7). For the class of stationary and ergodic processes, the Prohorov ball $\Pi_{n,\rho_n^*}(f_0, f) \leq \varepsilon$ is asymptotically ($n \to \infty$) tight, as compared to the Prohorov ball $\Pi_{1,\rho}(f_0, f) \leq \varepsilon$ with $\rho(x, y) = |x - y|$. In fact, it can be shown that the above two Prohorov balls are then asymptotically identical; that is, an ergodic and stationary process is contained in $\Pi_{1,\rho}(f_0, f) \leq \varepsilon$ if and only if it is contained in $\Pi_{n,\rho_n^*}(f_0, f) \leq \varepsilon$, for $n \to \infty$. We point out that the distortion measure in (11.3.7) is a metric (satisfies the triangular property). Thus, the Prohorov distance $\Pi_{n,\rho_n^*}(f_1, f_2)$ is also a metric.

The main conclusion from the above discussion is that the Prohorov distance best represents the outliers model on nominal data generated by both memoryless and nonmemoryless stochastic processes. In fact, the representation of the latter model is the main significant contribution of the Prohorov distance to statistical procedures, and to robust such procedures in particular. As we will see in Chapter 12, the use of this distance is indispensable in the derivation of sufficient conditions for qualitative robustness. We emphasize that the value of the Prohorov distance is qualitative rather than quantitative, since its computation (for given density functions) is generally impossible.

Let us now discuss the Vasershtein and the Rho-Bar distances. An interesting application of the Vasershtein distance corresponds to parameter estimation. Without lack in generality, let us consider an estimate $\hat{\theta}(x^n)$ of a scalar and real parameter θ, from an observation vector x^n. If the sequence x^n is generated by some process whose density function is f_1, then given the true parameter value θ, the performance of the estimate $\hat{\theta}(x^n)$ at f_1 can be measured by the expected value $E\{\rho[\theta, \hat{\theta}(X^n)] | f_1\}$, of some distortion function $\rho[\theta, \hat{\theta}(x^n)]$ between the estimate and the true parameter value. If f_2 is the density function of some other data generating stochastic process, and if $\rho(\cdot,\cdot)$ is a metric, we then obtain,

$$E\{\rho[\theta,\hat{\theta}(X^n)]|f_1\} \leq E\{\rho[\theta,\hat{\theta}(Y^n)]|f_2\} + E\{\rho[\hat{\theta}(X^n),\hat{\theta}(Y^n)]|p\}$$
$$E\{\rho[\theta,\hat{\theta}(Y^n)]|f_2\} \leq E\{\rho[\theta,\hat{\theta}(X^n)]|f_1\} + E\{\rho[\hat{\theta}(X^n),\hat{\theta}(Y^n)]|p\} \quad (11.3.8)$$

where p is any $2n$-dimensional density function with marginals f_1 and f_2. The inequalities in (11.3.8) are then also valid for the joint $2n$-dimensional density function that attains the Vasershtein distance between the random variables $\hat{\theta}(X^n)$ and $\hat{\theta}(Y^n)$, where the density function g_1 of the random variable $\hat{\theta}(X^n)$ is induced by f_1 and the function $\hat{\theta}(\cdot)$, while the density function g_2 of the random variable $\hat{\theta}(Y^n)$ is induced by f_2 and the function $\hat{\theta}(\cdot)$. Denoting the Vasershtein distance between g_1 and g_2 by $V(g_1,g_2)$, and due to the symmetry of the two expressions in (11.3.8), we obtain

$$|E\{\rho[\theta,\hat{\theta}(X^n)]|f_1\} - E\{\rho[\theta,\hat{\theta}(Y^n)]|f_2\}| \leq V_\rho(g_1,g_2) \quad (11.3.9)$$

From (11.3.9) we conclude that if the Vasershtein distance is small, then the performance of the estimate at f_1 and the same performance at f_2 are close to each other. In general, the Vasershtein distance is valuable as a measure of performance closeness for various statistical operations under varying data generating processes.

While the Vasershtein distance is useful for performance evaluation in parameter estimation, the Rho-Bar distance measures closeness between stochastic processes with memory. If f_1, f_2, and f_3 denote the arbitrary dimensionality density functions of three such processes, and if the scalar $\rho(\cdot,\cdot)$ in the single-letter distortion measure in (11.1.8) is a metric, then

$$\bar{\rho}(f_1,f_2) \leq \bar{\rho}(f_1,f_3) + \bar{\rho}(f_3,f_2)$$
$$\bar{\rho}(f_1,f_3) \leq \bar{\rho}(f_1,f_2) + \bar{\rho}(f_3,f_2) \quad (11.3.10)$$
$$|\bar{\rho}(f_1,f_2) - \bar{\rho}(f_1,f_3)| \leq \bar{\rho}(f_3,f_2)$$

From (11.3.10) we conclude that if f_3 and f_2 are Rho-Bar close, then the difference of the Rho-Bar closeness of each one of them to a third density function f_1 is close as well. Therefore, if f_3 and f_2 are Rho-Bar close, and if f_1 "estimates" either one of the processes represented by those two densities, then the Rho-Bar performance of this estimate at f_2 is close to its Rho-Bar performance at f_3. This latter property finds applications in problems such as filtering, where the data of some signal process are estimated from noisy observations. We will discuss this problem further in Chapter 12.

A popular distortion measure $\rho(\cdot,\cdot)$ in parameter estimation, as well as in problems such as filtering, is the difference squared—$\rho(x,y) = (x-y)^2$—which does not satisfy the triangular property. It is then interesting to see if, when the above data distortion measure is used, the Vasershtein and the Rho-Bar distances still measure difference in performance at different processes. Let us consider three stationary processes with first order density functions f_1, f_2, and

11.3 Relation to Statistical Procedures

f_3. Then, as we saw in Section 11.1, we have

$$\bar{\rho}(f_1, f_2) = V_{1,\rho}(f_1, f_2) = \inf_{g \in \mathscr{P}^1_{12}} \{\rho(X, Y)|g\} \qquad (11.3.11)$$

$$\bar{\rho}(f_1, f_3) = V_{1,\rho}(f_1, f_3) = \inf_{g \in \mathscr{P}^1_{13}} \{\rho(X, W)|g\} \qquad (11.3.12)$$

$$\bar{\rho}(f_2, f_3) = V_{1,\rho}(f_3, f_2) = \inf_{g \in \mathscr{P}^1_{23}} \{\rho(Y, W)|g\} \qquad (11.3.13)$$

where X, Y, and W, are scalar random variables respectively corresponding to densities f_1, f_2, and f_3, and where $\mathscr{P}^1_{ij}$ is the class of all the two-dimensional densities with marginals f_i and f_j. Let g_{12}, g_{13}, and g_{23} be respectively the densities that satisfy the infima in (11.3.11), (11.3.12), and (11.3.13), and let g be some four-dimensional density with marginals g_{13} and g_{23}. Then,

$$\begin{aligned}E\{(X-Y)^2|g\} &= E\{[(X-W)+(W-Y)]^2|g\} \\ &= E\{(X-W)^2|g\} + E\{(W-Y)^2|g\} \\ &\quad + 2E\{(X-W)(W-Y)|g\} \\ &\leq E\{(X-W)^2|g_{13}\} + E\{(W-Y)^2|g_{23}\} \\ &\quad + 2E\{|X-W||W-Y||g\} \end{aligned} \qquad (11.3.14)$$

Applying the Schwartz inequality, we obtain

$$\begin{aligned}E\{|X-W||W-Y||g\} &\leq [E\{(X-W)^2|g\}E\{(W-Y)^2|g\}]^{1/2} \\ &= E^{1/2}\{(X-W)^2|g_{13}\}E^{1/2}\{(W-Y)^2|g_{23}\} \end{aligned} \qquad (11.3.15)$$

We now substitute (11.3.15) in (11.3.14), to conclude

$$\begin{aligned}\bar{\rho}(f_1, f_2) &\leq E\{(X-Y)^2|g\} \\ &\leq E\{(X-W)^2|g_{13}\} + E\{(W-Y)^2|g_{23}\} \\ &\quad + 2E^{1/2}\{(X-W)^2|g_{13}\}E^{1/2}\{(W-Y)^2|g_{23}\} \\ &= [E^{1/2}\{(X-W)^2|g_{13}\} + E^{1/2}\{(W-Y)^2|g_{23}\}]^2 \\ &= \{[\bar{\rho}(f_1, f_3)]^{1/2} + [\bar{\rho}(f_2, f_3)]^{1/2}\}^2 \end{aligned}$$

or

$$[\bar{\rho}(f_1, f_2)]^{1/2} \leq [\bar{\rho}(f_1, f_3)]^{1/2} + [\bar{\rho}(f_2, f_3)]^{1/2} \qquad (11.3.16)$$

where $\rho(x, y) = (x - y)^2$. We similarly obtain the symmetric inequality

$$[\bar{\rho}(f_1, f_3)]^{1/2} \leq [\bar{\rho}(f_1, f_2)]^{1/2} + [\bar{\rho}(f_2, f_3)]^{1/2} \qquad (11.3.17)$$

From (11.3.16) and (11.3.17), we thus conclude that if $\rho(x, y) = (x - y)^2$, then

$$\left|[\bar{\rho}(f_1, f_2)]^{1/2} - [\bar{\rho}(f_1, f_3)]^{1/2}\right| \leq [\bar{\rho}(f_2, f_3)]^{1/2} \qquad (11.3.18)$$

Expression (11.3.18) provides an inequality similar to that in (11.3.10), where the square roots of the Rho-Bar distances are now involved. A parallel result evolves trivially for the Vasershtein distance. Therefore, although the difference square distortion measure does not satisfy the triangular property, the Vasershtein and the Rho-Bar distances that use it for data distortion measure still act as bounds on performance differences at the corresponding processes.

11.4 EXPLICIT EXPRESSIONS FOR GAUSSIAN PROCESSES

Some of the stochastic distance measures we discussed can be explicitly computed in certain special cases. Such an interesting case arises when Gaussian processes are involved. In this section we will consider this latter case when the Gaussian processes are scalar but not necessarily stationary. In particular, we consider discrete-time, real Gaussian processes of both finite and infinite dimensionality. For given dimensionality n, we denote by $M_n(l)$ the mean column vector of the Gaussian process indexed by l, and we denote by $R_n(l)$ its autocovariance matrix. If $|R_n(l)|$ denotes the determinant of the matrix $R_n(l)$, if $f_l(x^n)$ denotes the n-dimensional density function of the above Gaussian process at the vector point x^n, and if $R_n^{-1}(l)$ denotes the inverse of the matrix $R_n(l)$, then

$$f_l(x^n) = (2\pi)^{-n/2}|R_n(l)|^{-1/2} \exp\left\{-\frac{1}{2}[x^n - M_n(l)]^T R_n^{-1}(l)[x^n - M_n(l)]\right\} \quad (11.4.1)$$

For infinite dimensionalities, we will consider wide-sense stationary Gaussian processes. Considering an infinite random data sequence $X_1, X_2, \ldots$ generated by the Gaussian process indexed by l, we will then denote $m_k(l) = E\{X_k|f_l\}$ and $r_k(l) = E\{[X_i - m_i(l)][X_{i+k} - m_{i+k}(l)]|f_l\}$, and we will then assume that the following spectral densities exist for all λ in $[-\pi, \pi]$.

$$s_l(\lambda) = \sum_k r_k(l)e^{jk\lambda} \quad \lambda \in [-\pi, \pi]$$
$$m_l(\lambda) = \sum_k m_k(l)e^{jk\lambda} \quad \lambda \in [-\pi, \pi] \quad (11.4.2)$$

We start with n-dimensional Gaussian processes whose density functions are as in (11.4.1), and with the Chernoff coefficient in (11.1.2), the Bhattacharyya coefficient and distance in (11.1.6) and (11.1.5), the Kullback-Leibler distance in (11.1.3), and the J-divergence distance in (11.1.4). We express our results in Lemma 11.4.1 below, where we consider two Gaussian densities $f_1(x^n)$ and $f_2(x^n)$ as in (11.4.1), with respective mean vectors $M_n(1)$ and $M_n(2)$, and respective autocovariance matrices $R_n(1)$ and $R_n(2)$, where I_n is the n-dimensional identity matrix, and where we define

$$d_{n,t}(f_1, f_2) = \left[\int_{R^n} [f_1(x^n)]^t [f_2(x^n)]^{1-t} dx^n\right]^{1/n} \quad 0 \le t \le 1 \quad (11.4.3)$$

LEMMA 11.4.1: For the Gaussian real densities $f_1(x^n)$ and $f_2(x^n)$, the Chernoff coefficient in (11.4.3), the Bhattacharyya coefficient $c_n(f_1, f_2)$ in (11.1.6), the Kullback-Leibler distance $I_n(f_1, f_2)$ in (11.1.3), and the J-divergence distance $J_n(f_1, f_2)$ in (11.1.4) are respectively given by the following expressions, where the logarithm in the latter two distances is the natural logarithm.

11.4 Explicit Expressions for Gaussian Processes

$$d_{n,t}(f_1, f_2) = |R_n(2)|^{t/2n}|R_n(1)|^{(1-t)/2n}|tR_n(2) + (1-t)R_n(1)|^{-1/2n}$$

$$\cdot \exp\left\{-\frac{t(1-t)}{2n}[M_n(2) - M_n(1)]^T\right.$$

$$\left. \cdot [tR_n(2) + (1-t)R_n(1)]^{-1}[M_n(2) - M_n(1)]\right\} \quad (11.4.4)$$

$$c_n(f_1, f_2) = \sqrt{2}|R_n(1)R_n(2)|^{1/4n}|R_n(1) + R_n(2)|^{-1/2n}$$

$$\cdot \exp\left\{-\frac{1}{4n}[M_n(2) - M_n(1)]^T[R_n(1) + R_n(2)]^{-1}[M_n(2) - M_n(1)]\right\} \quad (11.4.5)$$

$$I_n(f_1, f_2) = -\frac{1}{2} + \frac{1}{2n}\ln|R_n(2)| \, |R_n^{-1}(1)| + \frac{1}{2n}\text{tr } R_n^{-1}(2)R_n(1)$$

$$+ \frac{1}{2n}[M_n(2) - M_n(1)]^T R_n^{-1}(2)[M_n(2) - M_n(1)] \quad (11.4.6)$$

$$J_n(f_1, f_2) = -1 + (2n)^{-1}\{\text{tr}[R_n^{-1}(2)R_n(1) + R_n^{-1}(1)R_n(2)]$$

$$+ [M_n(2) - M_n(1)]^T[R_n^{-1}(1) + R_n^{-1}(2)][M_n(2) - M_n(1)]\} \quad (11.4.7)$$

$$J_n(f_1, f_2) \geq -8 \ln c_n(f_1, f_2) = 8B_n(f_1, f_2)$$

Proof: For the density functions considered, we obtain

$$\int_{R^n}[f_1(x^n)]^t[f_2(x^n)]^{1-t}dx^n$$

$$= (2\pi)^{-n/2}|R_n(1)|^{-t/2}|R_n(2)|^{-(1-t)/2}\int_{R^n}dx^n \exp\left\{-\frac{t}{2}[x^n - M_n(1)]^T\right.$$

$$\left. \cdot R_n^{-1}(1)[x^n - M_n(1)] - \frac{1-t}{2}[x^n - M_n(2)]^T R_n^{-1}(2)[x^n - M_n(2)]\right\}$$

Expanding the expressions in the exponent and re-forming, we obtain

$$\int_{R^n}[f_1(x^n)]^t[f_2(x^n)]^{1-t}dx^n$$

$$= |R_n(1)|^{-t/2}|R_n(2)|^{-(1-t)/2}|[tR_n^{-1}(1) + (1-t)R_n^{-1}(2)]^{-1}|^{1/2}$$

$$\cdot \exp\left\{-\frac{1}{2}[tM_n^T(1)R_n^{-1}(1)M_n(1)\right.$$

$$\left. + (1-t)M_n^T(2)R_n^{-1}(2)M_n(2) - M_n^T\Lambda_n^{-1}M_n]\right\} \quad (11.4.8)$$

where

$$M_n = tR_n^{-1}(1)M_n(1) + (1-t)R_n^{-1}(2)M_n(2) \quad (11.4.9)$$

$$\Lambda_n = tR_n^{-1}(1) + (1-t)R_n^{-1}(2) = R_n^{-1}(1)[tR_n(2) + (1-t)R_n(1)]R_n^{-1}(2)$$

From (11.4.8) and (11.4.9), and via some straightforward operations, we obtain the expression in (11.4.4). The expression in (11.4.5) is obtained from (11.4.4) if t is replaced by 0.5.

To prove (11.4.6), we write

$$nI_n(f_1, f_2) = E\left\{\ln\frac{f_1(X^n)}{f_2(X^n)}\bigg|f_1\right\}$$

$$= \frac{1}{2}\ln|R_n(2)| - \frac{1}{2}\ln|R_n(1)|$$

$$- \frac{1}{2}E\{[X^n - M_n(1)]^T R_n^{-1}(1)[X^n - M_n(1)]\,|\,f_1\}$$

$$+ \frac{1}{2}E\{[X^n - M_n(2)]^T R_n^{-1}(2)[X^n - M_n(2)]\,|\,f_1\}$$

$$= \frac{1}{2}\ln|R_n(2)| - \frac{1}{2}\ln|R_n(1)|$$

$$- \frac{1}{2}\operatorname{tr} R_n^{-1}(1)E\{[X^n - M_n(1)][X^n - M_n(1)]^T\,|\,f_1\}$$

$$+ \frac{1}{2}\operatorname{tr} R_n^{-1}(2)E\{[X^n - M_n(2)][X^n - M_n(2)]^T\,|\,f_1\}$$

$$= \frac{1}{2}\ln|R_n(2)| - \frac{1}{2}\ln|R_n(1)| - \frac{1}{2}\operatorname{tr} R_n^{-1}(1)R_n(1)$$

$$+ \frac{1}{2}\operatorname{tr} R_n^{-1}(2)E\{[X^n - M_n(1)][X^n - M_n(1)]^T$$
$$- [M_n(2) - M_n(1)][X^n - M_n(1)]^T$$
$$- [X^n - M_n(1)][M_n(2) - M_n(1)]^T$$
$$+ [M_n(2) - M_n(1)][M_n(2) - M_n(1)]^T\,|\,f_1\}$$

$$= \frac{1}{2}\ln|R_n(2)| - \frac{1}{2}\ln|R_n(1)| - \frac{n}{2}$$

$$+ \frac{1}{2}\operatorname{tr} R_n^{-1}(2)\{R_n(1) + [M_n(2) - M_n(1)][M_n(2) - M_n(1)]^T\}$$

$$= \frac{1}{2}\ln|R_n(2)R_n^{-1}(1)| - \frac{n}{2} + \frac{1}{2}\operatorname{tr} R_n^{-1}(2)R_n(1)$$

$$+ \frac{1}{2}[M_n(2) - M_n(1)]^T R_n^{-1}(2)[M_n(2) - M_n(1)]$$

where tr denotes trace.

From the final expression we conclude (11.4.6). We obtain (11.4.7) by adding (11.4.6) to its symmetric expression $I_n(f_2, f_1)$. The inequality between the J-

11.4 Explicit Expressions for Gaussian Processes

divergence and the Bhattacharyya distances follows easily from the corresponding expressions. ∎

From the expressions in Lemma 11.4.1, we observe that if the two autocovariance matrices are equal to each other and equal to R_n, then

$$d_{n,t}(f_1, f_2) = \exp\left\{-\frac{t(1-t)}{2n}[M_n(2) - M_n(1)]^T R_n^{-1}[M_n(2) - M_n(1)]\right\} \quad (11.4.10)$$

$$c_n(f_1, f_2) = \exp\left\{-\frac{1}{8n}[M_n(2) - M_n(1)]^T R_n^{-1}[M_n(2) - M_n(1)]\right\} \quad (11.4.11)$$

$$B_n(f_1, f_2) = -\ln c_n(f_1, f_2)$$
$$= \frac{1}{8n}[M_n(2) - M_n(1)]^T R_n^{-1}[M_n(2) - M_n(1)] \quad (11.4.12)$$

$$I_n(f_1, f_2) = \frac{1}{2n}[M_n(2) - M_n(1)]^T R_n^{-1}[M_n(2) - M_n(1)] \quad (11.4.13)$$

$$J_n(f_1, f_2) = \frac{1}{n}[M_n(2) - M_n(1)]^T R_n^{-1}[M_n(2) - M_n(1)] \quad (11.4.14)$$

$$J_n(f_1, f_2) = 8B_n(f_1, f_2) \quad (11.4.15)$$

From the above expressions, we conclude that when the two autocovariance matrices are identical, then the Chernoff, Bhattacharyya, Kullback-Leibler, and J-divergence distances are all strictly functions of the quadratic form $[M_n(2) - M_n(1)]^T R_n^{-1}[M_n(2) - M_n(1)]$, so the larger the weighted difference of the mean vectors, the larger all the above distances as well. From (11.4.15) we also conclude that the J-divergence distance is then eight times the Bhattacharyya distance.

Let us now assume that the two Gaussian processes in Lemma 11.4.1 are both stationary and white, with respective means m_1 and m_2 and respective powers σ_1^2 and σ_2^2. From the expressions in the lemma, and by substitution, we then obtain,

$$d_{n,t}(f_1, f_2) = \sigma_2^t \sigma_1^{1-t}[t\sigma_2^2 + (1-t)\sigma_1^2]^{-1/2} \exp\left\{-\frac{t(1-t)(m_2 - m_1)^2}{2[t\sigma_2^2 + (1-t)\sigma_1^2]}\right\}$$
$$(11.4.16)$$

$$c_n(f_1, f_2) = [2\sigma_1\sigma_2]^{1/2}[\sigma_1^2 + \sigma_2^2]^{-1/2} \exp\left\{-\frac{(m_2 - m_1)^2}{4[\sigma_1^2 + \sigma_2^2]}\right\} \quad (11.4.17)$$

$$I_n(f_1, f_2) = \frac{1}{2}\left[-1 + \ln\frac{\sigma_2}{\sigma_1} + \left(\frac{\sigma_1}{\sigma_2}\right)^2 + \frac{(m_2 - m_1)^2}{\sigma_2^2}\right] \quad (11.4.18)$$

$$J_n(f_1, f_2) = -1 + \frac{1}{2\sigma_1^2\sigma_2^2}\{\sigma_1^4 + \sigma_2^4 + (m_2 - m_1)^2(\sigma_1^2 + \sigma_2^2)\} \quad (11.4.19)$$

$$B_n(f_1, f_2) = -\ln c_n(f_1, f_2) = \frac{1}{2}\ln\frac{\sigma_1^2 + \sigma_2^2}{2\sigma_1\sigma_2} + \frac{(m_2 - m_1)^2}{4(\sigma_1^2 + \sigma_2^2)} \quad (11.4.20)$$

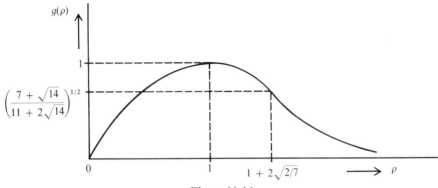

Figure 11.4.1

Let us consider the case where the two stationary and white Gaussian processes have equal means, and let us then denote by $\rho = \sigma_2 \sigma_1^{-1}$ the ratio of the square root of their powers. From (11.4.17) we then obtain

$$c_n(f_1, f_2) = \left[\frac{2\rho}{1+\rho^2}\right]^{1/2} = g(\rho) \qquad (11.4.21)$$

The Bhattacharyya coefficient in (11.4.21) is such that $g(\rho) = g(\rho^{-1})$ and it attains its maximum value one at $\rho = 1$. In the latter case, the two white and stationary processes are identical; they are thus indistinguishable. We plot the function $g(\rho)$ in Figure 11.4.1 below. For ρ values such that $g(\rho)$ is small, the bounds to the probability of error in (11.3.3), Lemma 11.3.1, both converge rapidly to zero with increasing n.

To this point, we have discussed explicit expressions of various distances for data sequences generated by Gaussian processes of arbitrary dimensionality n. It is now important to seek such expressions for asymptotically large data sequences. The asymptotic expressions exhibit the ultimate distinguishability between two Gaussian processes. We will consider in particular the limits of the Chernoff coefficient in (11.4.4), the Bhattacharyya coefficient in (11.4.5), the Kullback-Leibler distance in (11.4.6), and the J-divergence distance in (11.4.7). We will use results from Kazakos et al (1980), without proof, since the latter is quite involved. We express those results in Lemma 11.4.2 below, where we use the spectral densities in (11.4.2) for $l = 1, 2$ and where we denote

$$d_t(f_1, f_2) = \lim_{n \to \infty} d_{n,t}(f_1, f_2) \qquad c(f_1, f_2) = \lim_{n \to \infty} c_n(f_1, f_2)$$
$$\qquad (11.4.22)$$
$$I(f_1, f_2) = \lim_{n \to \infty} I_n(f_1, f_2) \qquad J(f_1, f_2) = \lim_{n \to \infty} J_n(f_1, f_2)$$

LEMMA 11.4.2: Consider two real, scalar, discrete-time, and wide-sense stationary processes, whose density functions, of arbitrary dimensionality, are denoted f_1 and f_2, and whose spectral densities in (11.4.2) exist and are respectively denoted $m_1(\lambda)$ versus $m_2(\lambda)$ and $s_1(\lambda)$ versus $s_2(\lambda)$; $\lambda \in [-\pi, \pi]$. The asymptotic Chernoff and Bhattacharyya coefficients $d_t(f_1, f_2)$ and $c(f_1, f_2)$ and the asymp-

11.4 Explicit Expressions for Gaussian Processes

totic Kullback-Leibler and J-divergence distances $I(f_1, f_2)$ and $J(f_1, f_2)$ are then given by

$$-2 \log d_t(f_1, f_2) = (2\pi)^{-1} \int_{-\pi}^{\pi} d\lambda \log\{[(1-t)s_1(\lambda) + ts_2(\lambda)]s_1^{t-1}(\lambda)s_2^{-t}(\lambda)\}$$
$$+ t(1-t)(2\pi)^{-1} \int_{-\pi}^{\pi} d\lambda [m_2(\lambda) - m_1(\lambda)]^2$$
$$\times [(1-t)s_1(\lambda) + ts_2(\lambda)]^{-2} \qquad (11.4.23)$$

$$2B(f_1, f_2) = -2 \log c(f_1, f_2)$$
$$= (2\pi)^{-1} \int_{-\pi}^{\pi} d\lambda \log\{2^{-1}[s_1(\lambda) + s_2(\lambda)]s_1^{-1/2}(\lambda)s_2^{-1/2}(\lambda)\}$$
$$+ (2\pi)^{-1} \int_{-\pi}^{\pi} d\lambda [m_2(\lambda) - m_1(\lambda)]^2 [s_1(\lambda) + s_2(\lambda)]^{-2} \qquad (11.4.24)$$

$$2I(f_1, f_2) = -1 + (2\pi)^{-1} \int_{-\pi}^{\pi} d\lambda \{s_1^{-1}(\lambda)s_2(\lambda) - \log[s_1^{-1}(\lambda)s_2(\lambda)]\}$$
$$+ (2\pi)^{-1} \int_{-\pi}^{\pi} d\lambda [m_2(\lambda) - m_1(\lambda)]^2 s_1^{-2}(\lambda) \qquad (11.4.25)$$

$$2J(f_1, f_2) = -2 + (2\pi)^{-1} \int_{-\pi}^{\pi} d\lambda [s_1^{-2}(\lambda) + s_2^{-2}(\lambda)]\{s_1(\lambda)s_2(\lambda)$$
$$+ [m_2(\lambda) - m_1(\lambda)]^2\} \qquad (11.4.26)$$

∎

When the two Gaussian processes in Lemma 11.4.2 are both stationary, with respective scalar constant means m_1 and m_2, we easily conclude that

$$m_1(\lambda) = m_1 \delta(\lambda) \qquad m_2(\lambda) = m_2 \delta(\lambda) \qquad (11.4.27)$$

where $\delta(\lambda)$ is the delta function at the point zero.

In the latter case, the expressions in Lemma 11.4.2 take the forms

$$-2 \log d_t(f_1, f_2) = (2\pi)^{-1} \int_{-\pi}^{\pi} d\lambda \log\{[(1-t)s_1(\lambda) + ts_2(\lambda)]s_1^{t-1}(\lambda)s_2^{-t}(\lambda)\}$$
$$+ t(1-t)(2\pi)^{-1}(m_2 - m_1)^2[(1-t)s_1(0) + ts_2(0)]^{-2} \qquad (11.4.28)$$

$$2B(f_1, f_2) = -2 \log c(f_1, f_2)$$
$$= (2\pi)^{-1} \int_{-\pi}^{\pi} d\lambda \log\{2^{-1}[s_1(\lambda) + s_2(\lambda)]s_1^{-1/2}(\lambda)s_2^{-1/2}(\lambda)\}$$
$$+ (2\pi)^{-1}(m_2 - m_1)^2[s_1(0) + s_2(0)]^{-2} \qquad (11.4.29)$$

$$2I(f_1, f_2) = -1 + (2\pi)^{-1}(m_2 - m_1)^2 s_1^{-2}(0)$$
$$+ (2\pi)^{-1} \int_{-\pi}^{\pi} d\lambda \{s_1^{-1}(\lambda)s_2(\lambda) - \log[s_1^{-1}(\lambda)s_2(\lambda)]\} \qquad (11.4.30)$$

$$2J(f_1, f_2) = -2 + (2\pi)^{-1}(m_2 - m_1)^2[s_1^{-2}(0) + s_2^{-2}(0)]$$
$$+ (2\pi)^{-1} \int_{-\pi}^{\pi} d\lambda s_1(\lambda)s_2(\lambda)[s_1^{-2}(\lambda) + s_2^{-2}(\lambda)] \qquad (11.4.31)$$

where $s_1(0)$ and $s_2(0)$ are the respective powers of the two processes; that is,

$$s_1(0) = \sum_k r_k(1) \qquad s_2(0) = \sum_k r_k(2) \qquad (11.4.32)$$

We will now develop an inequality between the Bhattacharyya distance in (11.4.29) and the J-divergence distance in (11.4.31), when the two Gaussian processes are both stationary and have identical means. Without lack in generality, we will assume that the common mean value is zero.

LEMMA 11.4.3: Consider two zero mean, stationary, real, discrete-time and scalar Gaussian stationary processes with power spectral densities $s_1(\lambda)$ and $s_2(\lambda)$ defined on $[-\pi, \pi]$. The asymptotic Bhattacharyya and J-divergence distances between those two processes then satisfy the inequality

$$J(f_1, f_2) \geq \pi^{-2} 2^{4\pi - 3} \exp\{8\pi B(f_1, f_2)\} - 2 \qquad (11.4.33)$$

where

$$B(f_1, f_2) = -2^{-1} \ln 2 + (4\pi)^{-1} \int_{-\pi}^{\pi} d\lambda \ln\{[s_1(\lambda)s_2(\lambda)]^{-1/2}[s_1(\lambda) + s_2(\lambda)]\} \qquad (11.4.34)$$

$$J(f_1, f_2) = -2 + (4\pi)^{-1} \int_{-\pi}^{\pi} d\lambda [s_1(\lambda)s_2(\lambda)]^{-1}[s_1(\lambda) + s_2(\lambda)]^2 \qquad (11.4.35)$$

Proof: The expression in (11.4.34) is derived from (11.4.29), by setting $m_1 = m_2$. Setting $m_1 = m_2$ in (11.4.31), we obtain

$$2J(f_1, f_2) = -2 + (2\pi)^{-1} \int_{-\pi}^{\pi} d\lambda [s_1(\lambda)s_2(\lambda)]^{-1}[s_1^2(\lambda) + s_2^2(\lambda)]$$

$$= 2 + (2\pi)^{-1} \int_{-\pi}^{\pi} d\lambda [s_1(\lambda)s_2(\lambda)]^{-1}\{[s_1(\lambda) + s_2(\lambda)]^2 - 2s_1(\lambda)s_2(\lambda)\}$$

$$= -4 + (2\pi)^{-1} \int_{-\pi}^{\pi} d\lambda [s_1(\lambda)s_2(\lambda)]^{-1}[s_1(\lambda) + s_2(\lambda)]^2$$

thus (11.4.35).

Applying Jensen's inequality, we now have

$$\int_{-\pi}^{\pi} d\lambda \ln\{[s_1(\lambda)s_2(\lambda)]^{-1/2}[s_1(\lambda) + s_2(\lambda)]\}$$

$$\leq \ln\left\{\int_{-\pi}^{\pi} d\lambda [s_1(\lambda)s_2(\lambda)]^{-1/2}[s_1(\lambda) + s_2(\lambda)]\right\}$$

$$= \frac{1}{2} \ln\left\{\int_{-\pi}^{\pi} d\lambda [s_1(\lambda)s_2(\lambda)]^{-1/2}[s_1(\lambda) + s_2(\lambda)]\right\}^2 \qquad (11.4.36)$$

11.4 Explicit Expressions for Gaussian Processes

From the Schwartz inequality, we have

$$\left\{\int_{-\pi}^{\pi} d\lambda [s_1(\lambda)s_2(\lambda)]^{-1/2}[s_1(\lambda) + s_2(\lambda)]\right\}^2$$

$$\leq 2\pi \int_{-\pi}^{\pi} d\lambda [s_1(\lambda)s_2(\lambda)]^{-1}[s_1(\lambda) + s_2(\lambda)]^2 \qquad (11.4.37)$$

From expressions (11.4.34), (11.4.36), and (11.4.37), we thus obtain

$$B(f_1, f_2) + 2^{-1} \ln 2 = (4\pi)^{-1} \int_{-\pi}^{\pi} d\lambda \ln\{[s_1(\lambda)s_2(\lambda)]^{-1/2}[s_1(\lambda) + s_2(\lambda)]\}$$

$$\leq (8\pi)^{-1} \ln\left\{2\pi \int_{-\pi}^{\pi} d\lambda [s_1(\lambda)s_2(\lambda)]^{-1}[s_1(\lambda) + s_2(\lambda)]^2\right\}$$

(11.4.38)

From (11.4.35), we conclude that the argument in the logarithm of the right part of (11.4.38) equals $8\pi^2\{J(f_1, f_2) + 2\}$. Substituting the latter expression in (11.4.38), we obtain

$$B(f_1, f_2) + 2^{-1} \ln 2 \leq (8\pi)^{-1} \ln\{8\pi^2 J(f_1, f_2) + 16\pi^2\} \qquad (11.4.39)$$

or

$$8\pi B(f_1, f_2) + \ln 2^{4\pi} \leq \ln\{8\pi^2 J(f_1, f_2) + 16\pi^2\}$$

or

$$2^{4\pi} \exp\{8\pi B(f_1, f_2)\} \leq 8\pi^2 J(f_1, f_2) + 16\pi^2$$

The final expression trivially gives (11.4.33). ∎

We point out that the left part of (11.4.33) is larger than the bound $8B(f_1, f_2)$ in Lemma 11.4.1. In Lemma 11.2.1 we found that $J_n(f_1, f_2) \geq 4B_n(f_1, f_2)$ for arbitrary processes and arbitrary dimensionalities n. The bound in (11.4.33) is significantly stronger, dramatizing the sensitivity of the asymptotic J-divergence distance when stationary zero mean Gaussian processes are involved. Qualitatively speaking, the Bhattacharyya distance between two such processes may be relatively small although their J-divergence distance is at the same time relatively large. Thus, the J-divergence distance picks up fine differences between the two processes which the Bhattacharyya distance misses. This is due to the fact that the J-divergence distance measures relative information.

In the previous section, we discussed the use of the Rho-Bar distance as a measure of closeness between stochastic processes with memory. We also discussed the relationship of this distance to the widely used mean-squared performance criterion. The natural questions arising at this point are: Can the Rho-Bar distance between two stationary Gaussian processes be computed for the mean-squared data distortion criterion? If so, what is it? To answer these questions, let us consider two stationary, discrete-time, real and scalar stationary processes with means m_1 versus m_2 and power spectral densities $s_1(\lambda)$ versus $s_2(\lambda)$, defined in $[-\pi, \pi]$. Let us consider the single-letter distortion measure $\rho_n^*(x^n, y^n) = n^{-1} \sum_{i=1}^n (x_i - y_i)^2$, and let us then denote the corresponding Rho-Bar distance between the two processes by $\bar{\rho}^*(f_1, f_2)$. We can then express the following lemma.

LEMMA 11.4.4: For arbitrary stationary processes with means m_1 and m_2 and power spectral densities $s_1(\lambda)$ and $s_2(\lambda)$ we have

$$\bar{\rho}^*(f_1, f_2) \geq (2\pi)^{-1} \int_{-\pi}^{\pi} [s_1^{1/2}(\lambda) - s_2^{1/2}(\lambda)]^2 \, d\lambda + (m_1 - m_2)^2 \quad (11.4.40)$$

If the processes are Gaussian, then (11.4.40) is satisfied with equality.

Proof: Let X denote a single random datum from the process f_1. Let Y denote a single random datum from the process f_2. Let p denote the two-dimensional density function whose marginals are the first order density functions, and which attains the first order Vasershtein distance between f_1 and f_2 for distortion measure $\rho(x, y) = (x - y)^2$. Then, as we have already discussed, since the two processes are stationary we have

$$\begin{aligned}
\bar{\rho}^*(f_1, f_2) &= V_{1,\rho}(f_1, f_2) = E\{(X - Y)^2 | p\} \\
&= E\{[(X - m_1) - (Y - m_2) + (m_1 - m_2)]^2 | p\} \\
&= E\{[(X - m_1) - (Y - m_2)]^2 | p\} + (m_1 - m_2)^2 \\
&= (m_1 - m_2)^2 + E\{(X - m_1)^2 | p\} + E\{(Y - m_2)^2 | p\} \\
&\quad - 2E\{(X - m_1)(Y - m_2) | p\} \\
&= (m_1 - m_2)^2 + E\{(X - m_1)^2 | f_1\} + E\{(Y - m_2)^2 | f_2\} \\
&\quad - 2E\{(X - m_1)(Y - m_2) | p\} \quad (11.4.41)
\end{aligned}$$

Let us now consider the power spectral densitites $s_1(\lambda)$ and $s_2(\lambda)$, and let $s_p(\lambda)$ denote the cross power spectral density, as induced by the generalization in all dimensions of the joint density function p. Then, it is commonly known that

$$\begin{aligned}
E\{(X - m_1)^2 | f_1\} &= (2\pi)^{-1} \int_{-\pi}^{\pi} s_1(\lambda) \, d\lambda \\
E\{(Y - m_2)^2 | f_2\} &= (2\pi)^{-1} \int_{-\pi}^{\pi} s_2(\lambda) \, d\lambda \quad (11.4.42) \\
E\{(X - m_1)(Y - m_2) | p\} &= (2\pi)^{-1} \int_{-\pi}^{\pi} s_p(\lambda) \, d\lambda
\end{aligned}$$

But from Rozanov (1967) and Gray et al (1975), we have $|s_p(\lambda)|^2 \leq s_1(\lambda) s_2(\lambda)$ $\forall \lambda \in [-\pi, \pi]$. Substituting (11.4.42) in (11.4.41) and using the inequality we

11.4 Explicit Expressions for Gaussian Processes

obtain

$$\bar{\rho}^*(f_1, f_2) = (m_1 - m_2)^2 + (2\pi)^{-1} \int_{-\pi}^{\pi} [s_1(\lambda) + s_2(\lambda) - 2s_p(\lambda)] \, d\lambda$$

$$\geq (m_1 - m_2)^2 + (2\pi)^{-1} \int_{-\pi}^{\pi} [s_1(\lambda) + s_2(\lambda) - 2s_1^{1/2}(\lambda)s_2^{1/2}(\lambda)] \, d\lambda$$

$$= (m_1 - m_2)^2 + (2\pi)^{-1} \int_{-\pi}^{\pi} [s_1^{1/2}(\lambda) - s_2^{1/2}(\lambda)]^2 \, d\lambda \quad (11.4.43)$$

We now observe that the matrix

$$\begin{bmatrix} s_1(\lambda) & s_1^{1/2}(\lambda)s_2^{1/2}(\lambda) \\ s_1^{1/2}(\lambda)s_2^{1/2}(\lambda) & s_2(\lambda) \end{bmatrix}$$

is nonnegative definite. Therefore, there exists a jointly Gaussian and stationary pair of processes whose power spectral densities are $s_1(\lambda)$ and $s_2(\lambda)$, and whose cross power spectral density is $s_1^{1/2}(\lambda)s_2^{1/2}(\lambda)$ [see Rozanov (1967)]. As a result, if the two processes are Gaussian, an appropriate jointly Gaussian process can be found, so that (11.4.43) is attained with equality. The proof of the lemma is now complete. ∎

Due to Lemma 11.4.4, we conclude that among all the pairs of stationary processes with fixed means and power spectral densities, the Gaussian pair attains the minimum Rho-Bar distance whose distortion measure is the mean squared. So, if we consider the class $\mathscr{C}_1$ of all stationary processes with mean m_1 and power spectral density $s_1(\lambda)$ and the class $\mathscr{C}_2$ of such processes with mean m_2 and power spectral density $s_2(\lambda)$, then the processes that are closest across the two classes (or least favorable, since least distinguishable) in $\bar{\rho}^*$ are the two Gaussian processes. Let us now consider the class $\mathscr{D}_1$ of all zero mean stationary, scalar, real, and discrete-time processes whose power spectral densities are described by some power spectral class $\mathscr{F}_1$. Let $\mathscr{D}_2$ be the class of all zero mean stationary, scalar, real, and discrete-time processes whose power spectral densities are described by another spectral class $\mathscr{F}_2$ which is disjoint from $\mathscr{F}_1$. Then, the processes in $\bar{\rho}^*$ that are closest across $\mathscr{D}_1$ and $\mathscr{D}_2$ are the two Gaussian processes with spectral densities $s_1^*(\lambda) \in \mathscr{F}_1$ and $s_2^*(\lambda) \in \mathscr{F}_2$ such that

$$(2\pi)^{-1} \int_{-\pi}^{\pi} [\sqrt{s_1^*(\lambda)} - \sqrt{s_2^*(\lambda)}]^2 \, d\lambda = \inf_{\substack{s_1(\lambda) \in \mathscr{F}_1 \\ s_2(\lambda) \in \mathscr{F}_2}} (2\pi)^{-1} \int_{-\pi}^{\pi} [s_1^{1/2}(\lambda) - s_2^{1/2}(\lambda)]^2 \, d\lambda$$

(11.4.44)

If both $\mathscr{F}_1$ and $\mathscr{F}_2$ contain power spectral densities such that $(2\pi)^{-1} \cdot \int_{-\pi}^{\pi} s(\lambda) \, d\lambda = 1 \;\forall\; s(\lambda) \in \mathscr{F}_1$ and $\forall\; s(\lambda) \in \mathscr{F}_2$ and since

$$(2\pi)^{-1} \int_{-\pi}^{\pi} [s_1^{1/2}(\lambda) - s_2^{1/2}(\lambda)]^2 \, d\lambda$$

$$= (2\pi)^{-1} \int_{-\pi}^{\pi} s_1(\lambda) \, d\lambda + (2\pi)^{-1} \int_{-\pi}^{\pi} s_2(\lambda) \, d\lambda - 2(2\pi)^{-1} \int_{-\pi}^{\pi} [s_1(\lambda)s_2(\lambda)]^{1/2} \, d\lambda$$

$$= 2\left\{1 - (2\pi)^{-1} \int_{-\pi}^{\pi} [s_1(\lambda)s_2(\lambda)]^{1/2} \, d\lambda\right\}$$

then the infimum in (11.4.44) is clearly equivalent to the supremum,

$$\sup_{\substack{s_1(\lambda) \in \mathscr{F}_1 \\ s_2(\lambda) \in \mathscr{F}_2}} (2\pi)^{-1} \int_{-\pi}^{\pi} [s_1(\lambda)s_2(\lambda)]^{1/2} \, d\lambda \qquad (11.4.45)$$

The integral in (11.4.45) is basically the Bhattacharyya coefficient of the power spectral densities $s_1(\lambda)$ and $s_2(\lambda)$. Therefore, denoting $C(s_1, s_2) = (2\pi)^{-1} \int_{-\pi}^{\pi} [s_1(\lambda)s_2(\lambda)]^{1/2} \, d\lambda$ and $B(s_1, s_2) = -\ln C(s_1, s_2)$, we conclude that if the spectral classes $\mathscr{F}_1$ and $\mathscr{F}_2$ are as above, then the least favorable spectral densities $s_1^*(\lambda)$ and $s_2^*(\lambda)$ in (11.4.45) attain the smallest spectral Bhattacharyya distance in $\mathscr{F}_1 \times \mathscr{F}_2$. That is

$$B(s_1^*, s_2^*) = \inf_{\substack{s_1 \in \mathscr{F}_1 \\ s_2 \in \mathscr{F}_2}} B(s_1, s_2) \qquad (11.4.46)$$

where

$$B(s_1, s_2) = -\ln C(s_1, s_2)$$
$$C(s_1, s_2) = (2\pi)^{-1} \int_{-\pi}^{\pi} [s_1(\lambda)s_2(\lambda)]^{1/2} \, d\lambda \qquad (11.4.47)$$

We will complete this section with two examples, which represent two interesting cases of spectral classes $\mathscr{F}_1$ and $\mathscr{F}_2$ in the above formalization.

Example 11.4.1

Consider two classes $\mathscr{D}_1$ and $\mathscr{D}_2$ of zero mean, stationary, scalar, real, and discrete-time processes. Class $\mathscr{D}_1$ contains all such processes whose power spectral densities are identical and equal to $s_1(\lambda)$, where $(2\pi)^{-1} \int_{-\pi}^{\pi} s_1(\lambda) \, d\lambda = 1$. Class $\mathscr{D}_2$ contains processes with power spectral densities in the class $\mathscr{F}_2$ described below, where $s_0(\lambda)$ is known and given, and such that $(2\pi)^{-1} \int_{-\pi}^{\pi} s_0(\lambda) \, d\lambda = 1$, and where ε is a given number in $(0, 1)$.

$$\mathscr{F}_2 = \{s(\lambda): s(\lambda) = (1 - \varepsilon)s_0(\lambda) + \varepsilon h(\lambda);$$
$$\text{where } (2\pi)^{-1} \int_{-\pi}^{\pi} h(\lambda) \, d\lambda = 1, h(\lambda) \geq 0 \quad \forall \lambda \in [-\pi, \pi]\} \qquad (11.4.48)$$

The spectral class in (11.4.48) can be considered as a spectral outlier model where spectral outliers may occur with probability ε on a nominal spectral density $s_0(\lambda)$. We have already established that the closest across the classes $\mathscr{D}_1$ and $\mathscr{D}_2$ processes in $\bar{\rho}^*$ are the Gaussian processes with spectral densities $s_1^*(\lambda)$ and $s_2^*(\lambda)$, such that,

$$(2\pi)^{-1} \int_{-\pi}^{\pi} [s_1^*(\lambda)s_2^*(\lambda)]^{1/2} \, d\lambda = \sup_{\substack{s_1 \in \mathscr{F}_1 \\ s_2 \in \mathscr{F}_2}} (2\pi)^{-1} \int_{-\pi}^{\pi} [s_1(\lambda)s_2(\lambda)]^{1/2} \, d\lambda \qquad (11.4.49)$$

where here $s_1^*(\lambda) = s_1(\lambda) \; \forall \lambda \in [-\pi, \pi]$. Thus, we only need to find $s_2^*(\lambda)$. The expression $(2\pi)^{-1} \int_{-\pi}^{\pi} [s_1(\lambda)s_2(\lambda)]^{1/2} \, d\lambda$ is strictly concave with respect to $s_2(\lambda)$. To find its supremum in $\mathscr{F}_2$, we can thus apply the Kuhn-Tucker conditions.

11.4 Explicit Expressions for Gaussian Processes

Each spectral density $s(\lambda)$ in $\mathscr{F}_2$ satisfies the constraints

$$(2\pi)^{-1} \int_{-\pi}^{\pi} s(\lambda)\, d\lambda = 1$$
$$s(\lambda) - (1-\varepsilon)s_0(\lambda) \geq 0 \qquad \forall\, \lambda \in [-\pi, \pi] \tag{11.4.50}$$

If $s^*(\lambda)$ satisfies the supremum with respect to $s_2(\lambda)$ in (11.4.49), then the variational equation that includes the function to be optimized as well as the constraints in (11.4.50), takes the form

$$\frac{\partial}{\partial \delta}\Bigg\{ (2\pi)^{-1} \int_{-\pi}^{\pi} [s_1(\lambda)]^{1/2}[(1-\delta)s^*(\lambda) + \delta h(\lambda)]^{1/2}\, d\lambda$$
$$- \rho(2\pi)^{-1} \int_{-\pi}^{\pi} [(1-\delta)s^*(\lambda) + \delta h(\lambda)]\, d\lambda$$
$$- (2\pi)^{-1} \int_{-\pi}^{\pi} f(\lambda)[(1-\delta)s^*(\lambda) + \delta h(\lambda) - (1-\varepsilon)s_0(\lambda)]\, d\lambda \Bigg\}\Bigg|_{\delta=0} = 0$$
$$\tag{11.4.51}$$

where ρ is a Langrange multiplier, and $f(\lambda): f(\lambda) > 0\ \forall\, \lambda$. Performing the differentation in (11.4.51) and setting $\delta = 0$, we obtain

$$(4\pi)^{-1} \int_{-\pi}^{\pi} d\lambda\, [h(\lambda) - s^*(\lambda)]\{[s_1(\lambda)]^{1/2}[s^*(\lambda)]^{-1/2} - 2\rho - 2f(\lambda)\} = 0 \qquad \forall\, h(\lambda) \tag{11.4.52}$$

or

$$[s_1(\lambda)]^{1/2}[s^*(\lambda)]^{-1/2} - 2\rho = 2f(\lambda) \qquad \forall\, \lambda \in [-\pi, \pi] \tag{11.4.53}$$

where the Kuhn-Tucker conditions impose

$$f(\lambda)[s^*(\lambda) - (1-\varepsilon)s_0(\lambda)] = 0 \qquad \forall\, \lambda \in [-\pi, \pi] \tag{11.4.54}$$

From (11.4.53) and (11.4.54), we conclude

$$s^*(\lambda): \{[s_1(\lambda)]^{1/2}[s^*(\lambda)]^{-1/2} - 2\rho\}\{s^*(\lambda) - (1-\varepsilon)s_0(\lambda)\} = 0 \qquad \forall\, \lambda \in [-\pi, \pi] \tag{11.4.55}$$

Since $s^*(\lambda) - (1-\varepsilon)s_0(\lambda) \geq 0\ \forall\, \lambda \in [-\pi, \pi]$, (11.4.55) finally gives

$$\text{Either} \qquad s^*(\lambda) = (1-\varepsilon)s_0(\lambda)$$
$$\text{Or} \qquad [s_1(\lambda)]^{1/2}[s^*(\lambda)]^{-1/2} = 2\rho$$

or equivalently

$$\left.\begin{array}{ll}\text{Either} & s^*(\lambda) = (1-\varepsilon)s_0(\lambda) \\ \text{Or} & s^*(\lambda) = (2\rho)^{-2}s_1(\lambda)\end{array}\right\} \text{depending on } \max[(1-\varepsilon)s_0(\lambda), (2\rho)^{-2}s_1(\lambda)]$$

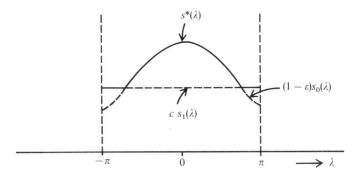

Figure 11.4.2

That is, given λ in $[-\pi, \pi]$, we have

$$s^*(\lambda) = \max[(1 - \varepsilon)s_0(\lambda), cs_1(\lambda)] \tag{11.4.56}$$

where $c = (2\rho)^{-2}$.

The constant c in (11.4.56) is unique and such that

$$(2\pi)^{-1} \int_{-\pi}^{\pi} d\lambda \max[(1 - \varepsilon)s_0(\lambda), cs_1(\lambda)] = 1 \tag{11.4.57}$$

In Figure 11.4.2, we plot the least favorable power spectral density in (11.4.56) when $s_1(\lambda) = 1 \ \forall \ \lambda \in [-\pi, \pi]$.

We observe that the power spectral density $s^*(\lambda)$ is "flattest" in $\mathscr{F}_2$, having at the same time the highest entropy in $\mathscr{F}_2$. We observe that if $(2\pi)^{-1} \cdot \int_{-\pi}^{\pi} \max[(1 - \varepsilon)s_0(\lambda), s_1(\lambda)] d\lambda \leq 1$, we can select $s^*(\lambda) = s_1(\lambda) \ \forall \ \lambda \in [-\pi, \pi]$, which attains the supremum in (11.4.49). Then the least favorable power spectral density is not a function of the outlier frequency ε in class $\mathscr{F}_2$. As a conclusion, the maximum such ε^* that provides a least favorable power spectral density that still depends on the outlier frequency is given by

$$(2\pi)^{-1} \int_{-\pi}^{\pi} \max[(1 - \varepsilon^*)s_0(\lambda), s_1(\lambda)] d\lambda = 1 \tag{11.4.58}$$

The ε^* in (11.4.58) is called the *breakdown point* of the problem.

Example 11.4.2

Consider the two classes of stochastic processes in Example 11.4.1. Let the spectral class $\mathscr{F}_2$ be exactly as in the latter example, but let the power spectral densities of class $\mathscr{D}_1$ belong to some class $\mathscr{F}_1$, described by

$$\mathscr{F}_1 = \left\{ s(\lambda) \colon s(\lambda) = (1 - \delta)s_{10}(\lambda) + \varepsilon h(\lambda); \right. \\ \left. \text{where } (2\pi)^{-1} \int_{-\pi}^{\pi} h(\lambda) d\lambda = 1, h(\lambda) \geq 0 \ \forall \ \lambda \in [-\pi, \pi] \right\} \tag{11.4.59}$$

11.4 Explicit Expressions for Gaussian Processes

where δ is a given number in $(0, 1)$ and $s_{10}(\lambda)$ is a given power spectral density such that $(2\pi)^{-1} \int_{-\pi}^{\pi} s_{10}(\lambda) \, d\lambda = 1$, and such that it is nonidentical to the nominal power spectral density $s_0(\lambda)$ in class $\mathscr{F}_2$. Class $\mathscr{F}_1$ represents thus a spectral outlier model.

In seeking the processes across the classes $\mathscr{D}_1$ and $\mathscr{D}_2$ that are closest in $\bar{\rho}^*$, we are searching for the two Gaussian processes with power spectral densities $s_1^*(\lambda)$ in $\mathscr{F}_1$ and $s_2^*(\lambda)$ in $\mathscr{F}_2$ such that

$$(2\pi)^{-1} \int_{-\pi}^{\pi} [s_1^*(\lambda) s_2^*(\lambda)]^{1/2} \, d\lambda = \sup_{\substack{s_1 \in \mathscr{F}_1 \\ s_2 \in \mathscr{F}_2}} (2\pi)^{-1} \int_{-\pi}^{\pi} [s_1(\lambda) s_2(\lambda)]^{1/2} \, d\lambda \quad (11.4.60)$$

If we temporarily fix the power spectral density $s_1(\lambda)$, the supremum in (11.4.60) will be attained by a spectral density $s_2'(\lambda)$ in $\mathscr{F}_2$ which is as in (11.4.56), Example 11.4.1. A symmetric conclusion is drawn if $s_2(\lambda)$ is temporarily fixed instead. Combining the above two expressions, we finally conclude that the pair $[s_1^*(\lambda), s_2^*(\lambda)]$ in (11.4.60) is such that,

$$\begin{aligned} s_1^*(\lambda) &= \max[(1 - \delta) s_{10}(\lambda), c_1(1 - \varepsilon) s_0(\lambda)] & \lambda \in [-\pi, \pi] \\ s_2^*(\lambda) &= \max[(1 - \varepsilon) s_0(\lambda), c_2(1 - \delta) s_{10}(\lambda)] & \lambda \in [-\pi, \pi] \end{aligned} \quad (11.4.61)$$

where the constants c_1 and c_2 are unique and such that

$$\begin{aligned} c_1 &: (2\pi)^{-1} \int_{-\pi}^{\pi} \max[(1 - \delta) s_{10}(\lambda), c_1(1 - \varepsilon) s_0(\lambda)] \, d\lambda = 1 \\ c_2 &: (2\pi)^{-1} \int_{-\pi}^{\pi} \max[(1 - \varepsilon) s_0(\lambda), c_2(1 - \delta) s_{10}(\lambda)] \, d\lambda = 1 \end{aligned} \quad (11.4.62)$$

As in Example 11.4.1, we observe that if the constants ε and δ are such that

$$C(\varepsilon, \delta) = (2\pi)^{-1} \int_{-\pi}^{\pi} \max[(1 - \delta) s_{10}(\lambda), (1 - \varepsilon) s_0(\lambda)] \, d\lambda \leq 1 \quad (11.4.63)$$

then $s_1^*(\lambda) = s_2^*(\lambda)$ can be selected, which makes the supremum in (11.4.60) equal to one. In this latter case, the least favorable pair $[s_1^*(\lambda), s_2^*(\lambda)]$ of power spectral densities does not depend on the outlier frequencies ε and δ. Such dependency is guaranteed if instead $C(\varepsilon, \delta) > 1$. Then, the resulting $\bar{\rho}^*$ distance between the least favorable pair of Gaussian processes is also a function of the outlier frequencies ε and δ. The boundary between such dependency and nondependency is represented by the pairs (ε, δ) that satisfy

$$C(\varepsilon, \delta) = (2\pi)^{-1} \int_{-\pi}^{\pi} \max[(1 - \delta) s_{10}(\lambda), (1 - \varepsilon) s_0(\lambda)] \, d\lambda = 1 \quad (11.4.64)$$

It can be easily seen that the curve represented by (11.4.64) is generally convex and becomes a straight line if the two nominal spectral densities $s_{10}(\lambda)$ and $s_0(\lambda)$ do not overlap. The latter curve, called the *breakdown curve* of the problem,

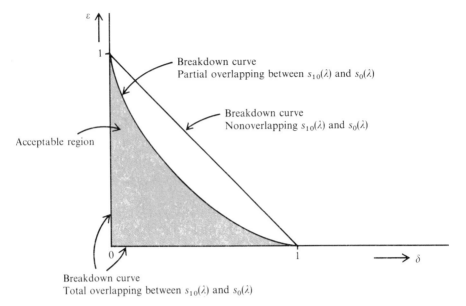

Figure 11.4.3

is plotted in Figure 11.4.3. The (ε, δ) region where the $\bar{\rho}^*$ distance between the least favorable pair of stochastic processes is still a function of the two outlier frequencies and has not yet reached its absolutely maximum value is called the *acceptable region*. Total overlapping of the nominal power spectral densities means that they are identical. The breakdown curve falls then onto the ε and δ axes. In figure 11.4.4, we plot the least favorable power spectral densities in (11.4.61) for partially overlapping nominal power spectral densities $s_0(\lambda)$ and $s_{10}(\lambda)$ and within the acceptable (ε, δ) region.

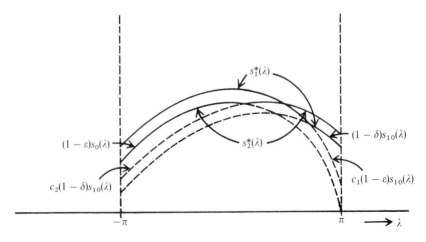

Figure 11.4.4

11.5 APPLICATIONS AND EXAMPLES

The stochastic distance measures have numerous applications in a variety of problems. We have already discussed some of those applications in this chapter, while in Chapter 12 we will exhibit the use of the Prohorov and Rho-Bar distances in qualitative robustness. In this section, we will demonstrate the use of some of those distances in two classical problems.

Example 11.5.1

A classical problem corresponds to the discrimination between two distinct Gaussian processes. In particular, let us suppose that a data vector x^n is observed, and let it be known that it has been generated by one of two known nonstationary Gaussian processes. Let the mean vector and the autocovariance matrix of one of those processes be respectively $M_n(1)$ and $R_n(1)$, and let the other process have mean vector $M_n(2)$ and autocovariance matrix $R_n(2)$. If either a priori probabilities on the two processes or a false alarm rate were available, then the Bayesian and Neyman-Pearson detection schemes respectively could be applied for discrimination. As we saw in Chapters 4 and 5, the resulting tests would then involve nonlinear transformations on the data [for nonidentical $R_n(1)$ and $R_n(2)$]. Let us now assume either that a priori probabilities and false alarm rate are not available, or that even if they are, we wish to apply a linear data transformation for initial discrimination or classification. In this latter case, we wish to find the "best" such linear transformation, in some optimality sense. The first such optimality measure that comes to mind is the Bhattacharyya distance, due to its direct relationship to the probability of error in detection or discrimination. The other distance that may be used is the J-divergence. We now pose the problem.

Find the row constant vector $A_n = [a_1, \ldots, a_n]$, that maximizes either the Bhattacharyya or the J-divergence distance between the density functions f_1 and f_2, where f_1 is the density function of the random variable $A_n X^n = \sum_{i=1}^{n} a_i X_i$ when X^n is generated by the Gaussian process with $M_n(1)$ and $R_n(1)$, and f_2 is the density function of $A_n X^n$ when X^n is generated by the Gaussian process with $M_n(2)$ and $R_n(2)$.

Given row vector A_n, the random variable $A_n X^n$ is Gaussian under both Gaussian processes (1) and (2). Under process (1), the mean m_1 and the variance σ_1^2 are given respectively by

$$m_1 = A_n M_n(1) \tag{11.5.1}$$

$$\sigma_1^2 = E_{(1)}\{A_n[X^n - M_n(1)][X^n - M_n(1)]^T A_n^T\}$$
$$= A_n E_{(1)}\{[X^n - M_n(1)][X^n - M_n(1)]^T\} A_n^T = A_n R_n(1) A_n^T \tag{11.5.2}$$

Under process (2), the mean m_2 and the variance σ_2^2 of the random variable $A_n X^n$ are clearly

$$m_2 = A_n M_n(2) \tag{11.5.3}$$

$$\sigma_2^2 = A_n R_n(2) A_n^T \tag{11.5.4}$$

The density functions $f_1(y)$ and $f_2(y)$ at the scalar point y are then

$$f_1(y) = (2\pi)^{-1/2}\sigma_1^{-1}\exp\left\{-\frac{(y-m_1)^2}{2\sigma_1^2}\right\} \quad (11.5.5)$$

$$f_2(y) = (2\pi)^{-1/2}\sigma_2^{-1}\exp\left\{-\frac{(y-m_2)^2}{2\sigma_2^2}\right\} \quad (11.5.6)$$

Directly from (11.4.19) and (11.4.20) in Section 11.4, we have the following expressions for the Bhattacharyya distance $B(f_1, f_2)$ and the J-divergence distance $J(f_1, f_2)$ between the densities in (11.5.5) and (11.5.6).

$$B(f_1, f_2) = \frac{1}{2}\ln\frac{\sigma_1^2 + \sigma_2^2}{2\sigma_1\sigma_2} + \frac{(m_2-m_1)^2}{4(\sigma_1^2+\sigma_2^2)} \quad (11.5.7)$$

$$J(f_1, f_2) = -1 + \frac{\sigma_1^4+\sigma_2^4}{2\sigma_1^2\sigma_2^2} + (m_2-m_1)^2\frac{\sigma_1^2+\sigma_2^2}{2\sigma_1^2\sigma_2^2}$$

$$= \frac{[\sigma_1^2-\sigma_2^2]^2}{2\sigma_1^2\sigma_2^2} + (m_2-m_1)^2\frac{\sigma_1^2+\sigma_2^2}{2\sigma_1^2\sigma_2^2} \quad (11.5.8)$$

Let us now define,

$$x = [A_n R_n(2) A_n^T]^{1/2}[A_n R_n(1) A_n^T]^{-1/2} \quad (11.5.9)$$

By substitution in (11.5.7) and (11.5.8), due to (11.5.2) and (11.5.4), and via some straightforward transformations, we find

$$B(f_1, f_2) = \frac{1}{2}\ln(x + x^{-1}) + (m_2 - m_1)^2 \frac{1}{4\sigma_1^2 x(x+x^{-1})} \quad (11.5.10)$$

$$J(f_1, f_2) = \frac{1}{2}(x - x^{-1})^2 + (m_2 - m_1)^2\frac{x + x^{-1}}{2\sigma_1^2 x} \quad (11.5.11)$$

We will now distinguish between two cases: the case of equal mean vectors $M_n(1)$ and $M_n(2)$, and the case of equal covariance matrices $R_n(1)$ and $R_n(2)$.

CASE 1 $M_n(1) = M_n(2)$: Then $m_1 = m_2$ for all vectors A_n and (11.5.10) and (11.5.11) give, where x is as in (11.5.9),

$$B(f_1, f_2) = \frac{1}{2}\ln(x + x^{-1}) = g(x) \quad (11.5.12)$$

$$J(f_1, f_2) = \frac{1}{2}(x - x^{-1})^2 = h(x) \quad (11.5.13)$$

11.5 Applications and Examples

As functions of x, both (11.5.12) and (11.5.13) are such that $g(x) = g(x^{-1})$, $h(x) = h(x^{-1})$, they attain minima at $x = 1$, and if x is such that $1 < x < c$, then they both attain their maxima at $x = c$ and $x = c^{-1}$. Now let W_n denote the matrix of relative eigenvectors of $R_n(2)$ with respect to $R_n(1)$, and let Λ_n denote the diagonal matrix of the respective relative eigenvalues. We can then write

$$R_n(1) = W_n W_n^T \qquad R_n(2) = W_n \Lambda_n W_n^T \qquad (11.5.14)$$

∎

Let us denote by $\lambda_{\min}$ and $\lambda_{\max}$ the minimum versus the maximum eigenvalue in the diagonal of Λ_n, and let us then define the following two row vectors.

$$V_n(\max) = [0, 0, \ldots, 1, 0, \ldots, 0]$$
$$\uparrow$$
position corresponding to $\lambda_{\max}$ in the Λ_n diagonal

$$V_n(\min) = [0, 0, \ldots, 1, 0, \ldots, 0]$$
$$\uparrow$$
position corresponding to $\lambda_{\min}$ in the Λ_n diagonal

Let us then define the row vectors $T_n(\max)$ and $T_n(\min)$ as follows, where c is some arbitrary scalar constant.

$$T_n(\max) = cV_n(\max)W_n^{-1} \qquad (11.5.15)$$

$$T_n(\min) = cV_n(\min)W_n^{-1} \qquad (11.5.16)$$

From (11.5.14), (11.5.15), and (11.5.16), we obtain

$$x_{\max} = [T_n(\max)R_n(2)T_n^T(\max)]^{1/2}[T_n(\max)R_n(1)T_n(\max)]^{-1/2}$$
$$= [V_n(\max)\Lambda_n V_n^T(\max)]^{1/2}[V_n(\max)V_n^T(\max)]^{-1/2}$$
$$= \lambda_{\max}^{1/2} \qquad (11.5.17)$$

and similarly

$$x_{\min} = \lambda_{\min}^{1/2} \qquad (11.5.18)$$

One of the values in (11.5.17) and (11.5.18) satisfies simultaneously the maxima in (11.5.12) and (11.5.13). If $\lambda_{\max} > \lambda_{\min}^{-1}$ the maxima occur at $x_{\max}$. If $\lambda_{\max} < \lambda_{\min}^{-1}$ the maxima occur at $x_{\min}$. In the first case, the "best" linear transformation on the data (that maximizes the Bhattacharrya and the J-divergence distances) is the one in (11.5.15). In the second case, the best linear transformation is given by (11.5.16). We note that in the case of unequal covariance matrices, both the Bayesian and the Neyman-Pearson detection schemes induce quadratic data transformations.

CASE 2 $R_n(1) = R_n(2) = R_n$: Then, for every A_n, $\sigma_1^2 = \sigma_2^2$ and $x = 1$ in (11.5.9). The distances in (11.5.10) and (11.5.11) then reduce to

$$B(f_1, f_2) = \frac{1}{2} \ln 2 + \frac{\{A_n[M_n(2) - M_n(1)]\}^2}{8 A_n R_n A_n^T}$$

$$= \frac{1}{2} \ln 2 + \frac{A_n \{[M_n(2) - M_n(1)][M_n(2) - M_n(1)]^T\} A_n^T}{8 A_n R_n A_n^T} \quad (11.5.19)$$

$$J(f_1, f_2) = \frac{A_n \{[M_n(2) - M_n(1)][M_n(2) - M_n(1)]^T\} A_n^T}{A_n R_n A_n^T} \quad (11.5.20)$$

It is easily verified that both distances are maximized for

$$A_n = c[M_n(2) - M_n(1)]^T R_n \quad (11.5.21)$$

where c is some arbitrary scalar constant. ∎

We note that the linear data transformation represented by (11.5.21) is in this case identical to that induced by the Bayesian and Neyman-Pearson detection schemes. This linear transformation is then the optimal discriminant transformation in the above optimal detection schemes as well.

Completing the present example, we point out that if we use the Vasershtein distance as a discrimination criterion, we obtain identical results as with the Bhattacharyya and J-divergence distances, in both cases 1 and 2. Indeed, the Vasershtein distance between the densities f_1 and f_2, with distortion measure the mean squared, is given by the expression, established in parallel to (11.4.40),

$$V(f_1, f_2) = (m_2 - m_1)^2 + (\sigma_2 - \sigma_1)^2 \quad (11.5.22)$$

where m_1, σ_1, m_2, σ_2 are given by (11.5.1), (11.5.2), (11.5.3) and (11.5.4). The maximization of $V(f_1, f_2)$ in cases 1 and 2 follows the same procedures and provides identical results.

Example 11.5.2

Let us now consider a data smoothing problem. Suppose that data sequences y^n are initially generated by some zero mean, stationary, scalar, real, and discrete-time process, which is also first order Markov with power spectral density $s_0(\lambda)$ given below, for $\alpha: 0 < \alpha < 1$.

$$s_0(\lambda) = [1 - e^{-2\alpha}][1 + e^{-2\alpha} - 2e^{-\alpha} \cos \lambda]^{-1} \quad \lambda \in [-\pi, \pi] \quad (11.5.23)$$

The power spectral density $s_0(\lambda)$ induces autocovariance coefficients $\{r_k\}$ such that $r_k = e^{-\alpha|k|}$. Let us now assume that, with frequency $\varepsilon/2$, outliers with amplitudes y and $-y$ may occur per datum, where these outliers are independent from the nominal process and independent from each other. If x^n denotes then

11.5 Applications and Examples

the observed data sequence, its mean, autocovariance coefficients μ_k, and power spectral density $s_{\varepsilon,y}(\lambda)$ are then

$$E\{X_n\} = (1-\varepsilon)E\{Y_n\} + \frac{\varepsilon}{2}[y - y] = 0 \tag{11.5.24}$$

$$\mu_0 = E\{X_n^2\} = (1-\varepsilon)E\{Y_n^2\} + \frac{\varepsilon}{2}\int_R u^2 \delta(u-y)\,du + \frac{\varepsilon}{2}\int_R u^2 \delta(u+y)\,du$$

$$= (1-\varepsilon)r_0 + \varepsilon y^2 = 1 - \varepsilon + \varepsilon y^2 \tag{11.5.25}$$

$$\mu_k = E\{X_0 X_k\} = (1-\varepsilon)^2 E\{Y_0 Y_k\} = (1-\varepsilon)^2 r_k = (1-\varepsilon)^2 e^{-\alpha|k|} \quad |k| \geq 1 \tag{11.5.26}$$

$$s_{\varepsilon,y}(\lambda) = \sum_k \mu_k e^{jk\lambda} = 1 - \varepsilon + \varepsilon y^2 + \sum_{k \neq 0}(1-\varepsilon)^2 r_k e^{jk\lambda}$$

$$= 1 - \varepsilon + \varepsilon y^2 - (1-\varepsilon)^2 r_0 + (1-\varepsilon)^2 \sum_k r_k e^{jk\lambda}$$

$$= \varepsilon[1 - \varepsilon + y^2] + (1-\varepsilon)^2 s_0(\lambda) \tag{11.5.27}$$

where $s_0(\lambda)$ is the nominal power spectral density in (11.5.23).

Let us denote by $\bar{\rho}^*(s_0, s_{\varepsilon,y})$ the Rho-Bar distance between the nominal process with power spectral density s_0 and the outlier process with power spectral density $s_{\varepsilon,y}$, when $\rho(u, v) = (u-v)^2$. From Lemma 11.4.4 we then have

$$\bar{\rho}^*(s_0, s_{\varepsilon,y}) \geq (2\pi)^{-1} \int_{-\pi}^{\pi} [s_0^{1/2}(\lambda) - s_{\varepsilon,y}^{1/2}(\lambda)]^2\, d\lambda$$

$$= (2\pi)^{-1}\int_{-\pi}^{\pi} s_0(\lambda)\,d\lambda + (2\pi)^{-1}\int_{-\pi}^{\pi} s_{\varepsilon,y}(\lambda)\,d\lambda$$

$$- 2(2\pi)^{-1}\int_{-\pi}^{\pi}[s_0(\lambda)s_{\varepsilon,y}(\lambda)]^{1/2}\, d\lambda \tag{11.5.28}$$

From (11.5.23) and (11.5.27), we have

$$(2\pi)^{-1}\int_{-\pi}^{\pi} s_0(\lambda)\,d\lambda = 1$$

$$(2\pi)^{-1}\int_{-\pi}^{\pi} s_{\varepsilon,y}(\lambda)\,d\lambda = \varepsilon[1 - \varepsilon + y^2] + (1-\varepsilon)^2 = 1 - \varepsilon + \varepsilon y^2 \tag{11.5.29}$$

$$(2\pi)^{-1}\int_{-\pi}^{\pi}[s_0(\lambda)s_{\varepsilon,y}(\lambda)]^{1/2}\, d\lambda$$

$$= (1-\varepsilon)(2\pi)^{-1}\int_{-\pi}^{\pi} d\lambda\, s_0(\lambda)\left[1 + \frac{\varepsilon}{(1-\varepsilon)^2}\frac{1-\varepsilon+y^2}{s_0(\lambda)}\right]^{1/2}$$

Substituting expressions (11.5.29) in (11.5.28), we obtain,

$$\bar{\rho}^*(s_0, s_{\varepsilon,y}) \geq 2 + \varepsilon(y^2 - 1) - 2(1-\varepsilon)(2\pi)^{-1}$$

$$\times \int_{-\pi}^{\pi} d\lambda\, s_0(\lambda)\left[1 + \frac{\varepsilon}{(1-\varepsilon)^2}\frac{1-\varepsilon+y^2}{s_0(\lambda)}\right]^{1/2} \tag{11.5.30}$$

The power spectral density $s_0(\lambda)$ in (11.5.23) is bounded from below by $(1 - e^{-\alpha})(1 + e^{-\alpha})$ for every λ in $[-\pi, \pi]$. Therefore,

$$-\left[1 + \frac{\varepsilon}{(1-\varepsilon)^2} \frac{1-\varepsilon+y^2}{s_0(\lambda)}\right]^{1/2}$$

$$\geq -\left[1 + \frac{\varepsilon}{(1-\varepsilon)^2} \frac{1+e^{-\alpha}}{1-e^{-\alpha}}(1-\varepsilon+y^2)\right]^{1/2} \quad \forall \lambda \in [-\pi, \pi]$$

and from (11.5.30)

$$\bar{\rho}^*(s_0, s_{\varepsilon,y}) \geq 2 + \varepsilon(y^2 - 1) - 2(1-\varepsilon)\left[1 + \frac{\varepsilon}{(1-\varepsilon)^2} \frac{1+e^{-\alpha}}{1-e^{-\alpha}}(1-\varepsilon+y^2)\right]^{1/2}$$

(11.5.31)

From (11.5.31) we conclude that the Rho-Bar distance between the nominal and the outlier processes can be arbitrarily large, if the outlier amplitude is large, for any nonzero outlier frequency ε. We thus wish to "smooth" the data sequences to eliminate the undesirable extreme outliers and to achieve data that closely represent the nominal process. A good measure of such closeness is the Rho-Bar distance with $\rho(u, v) = (u - v)^2$, so we propose the following smoothing operation. If $\Phi(u)$ denotes the cumulative distribution of the zero mean, unit variance Gaussian random variable at the point u, then we first transform the observed data $\{y_i\}$ into the sequence $\{w_i = \Phi(y_i) - \frac{1}{2}\}$. Then we form the sequence $\{z_i\}$ by

$$z_1 = m^{-1} \sum_{j=1}^{m} \left[\Phi(y_j) - \frac{1}{2}\right] \quad z_{k+1} = m^{-1} \sum_{j=k+1}^{k+m} \left[\Phi(y_j) - \frac{1}{2}\right] \quad k \geq 1$$

(11.5.32)

Under both the nominal and the outlier models, the random sequence $\{Z_k\}$ is stationary. Due to the Markovian structure of both models, and due to the Markovian theorem, the process determined by the sequence $\{Z_k\}$ is also asymptotically ($m \to \infty$) Gaussian for both the nominal and the outlier models. It is also zero mean under both models, if the first order density function of the nominal process is symmetric around zero, since the operation $g(u) = \Phi(u) - \frac{1}{2}$ is such that, $g(u) = -g(-u)$. Now let the nominal data process whose power spectral density is as in (11.5.23) be also Gaussian. Then let $f_0(\lambda)$ denote the power spectral density of the sequence $\{W_k\}$ if the data sequence $\{y_i\}$ is generated by the nominal Gaussian process, and let $f_{\varepsilon,y}(\lambda)$ denote the power spectral density of $\{W_k\}$ if the nominal process is Gaussian, and for the outlier model explained earlier. Let $\{\rho_k\}$ be the autocovariance coefficients induced by $f_0(\lambda)$, and let $\{v_k\}$ be the autocovariance coefficients induced by $f_{\varepsilon,y}(\lambda)$. Then, as in (11.5.25)

11.5 Applications and Examples

and (11.5.26), we have

$$v_0 = (1-\varepsilon)\rho_0 + \frac{\varepsilon}{2}\int_R \left[\Phi(u) - \frac{1}{2}\right]^2 \delta(u-y)\,du + \frac{\varepsilon}{2}\int_R \left[\Phi(u) - \frac{1}{2}\right]^2 \delta(u+y)\,du$$

$$= (1-\varepsilon)\rho_0 + \varepsilon\left[\Phi(y) - \frac{1}{2}\right]^2 = (1-\varepsilon)^2\rho_0 + \varepsilon(1-\varepsilon)\rho_0 + \varepsilon\left[\Phi(y) - \frac{1}{2}\right]^2$$

(11.5.33)

$$v_k = (1-\varepsilon)^2 \rho_k \qquad |k| \geq 1 \tag{11.5.34}$$

Thus

$$f_{\varepsilon,y}(\lambda) = \sum_k v_k e^{jk\lambda} = (1-\varepsilon)^2 \sum_k \rho_k e^{jk\lambda} + \varepsilon(1-\varepsilon)\rho_0 + \varepsilon\left[\Phi(y) - \frac{1}{2}\right]^2$$

$$= \varepsilon(1-\varepsilon)\rho_0 + \varepsilon\left[\Phi(y) - \frac{1}{2}\right]^2 + (1-\varepsilon)^2 f_0(\lambda) \qquad \lambda \in [-\pi, \pi] \quad (11.5.35)$$

To find the power spectral density $f_0(\lambda)$, we need to compute the autocovariance coefficients $\{\rho_k\}$, where

$$\rho_0 = \int_R \left[\Phi(u) - \frac{1}{2}\right]^2 \phi(u)\,du = \int_R \Phi^2(u)\phi(u)\,du - \frac{1}{4} \tag{11.5.36}$$

$$\rho_k = \iint_R du\,dv \left[\Phi(u) - \frac{1}{2}\right]\left[\Phi(v) - \frac{1}{2}\right] \frac{\exp\left\{-\frac{1}{2}[u,v]\begin{bmatrix}1 & r_k \\ r_k & 1\end{bmatrix}^{-1}\begin{bmatrix}u\\v\end{bmatrix}\right\}}{2\pi[1-r_k^2]^{1/2}}\,du\,dv$$

$$= \int_R \phi(u)\Phi(u)\Phi\left(\frac{r_k u}{\sqrt{2-r_k^2}}\right) du - \frac{1}{4} \qquad r_k = e^{-\alpha|k|} \quad k \geq 1 \tag{11.5.37}$$

where $\phi(u)$ denotes the zero mean, unit variance Gaussian density function at the point u, and where (11.5.37) evolved from some straightforward operations, and the easily proven equation

$$\int_R \phi(u)\Phi(au+b)\,du = \Phi\left(\frac{b}{(1+a^2)^{1/2}}\right) \tag{11.5.38}$$

To compute ρ_0 and ρ_k, we use the easily proven expressions,

$$\int_R \phi(u)\Phi^2(au)\,du = \pi^{-1}\tan^{-1}(\sqrt{1+2a^2}) \tag{11.5.39}$$

$$\int_R \phi(u)\Phi(u)\Phi\left(\frac{u}{a}\right)du = \frac{1}{2}\left[1 - \pi^{-1}\tan^{-1}(\sqrt{1+2a^2})\right] \tag{11.5.40}$$

Substituting (11.5.39) and (11.5.40) in (11.5.36) and (11.5.39) finally gives

$$\rho_0 = \pi^{-1} \tan^{-1}(\sqrt{3}) - \frac{1}{4} = \frac{1}{3} - \frac{1}{4} = \frac{1}{12} \qquad (11.5.41)$$

$$\rho_k = \frac{1}{4} - \frac{1}{2\pi} \tan^{-1}(r_k^{-1}[4 - r_k^2]^{1/2}) \qquad |k| \geq 1 \qquad (11.5.42)$$

$$f_0(\lambda) = \sum_k \rho_k e^{jk\lambda} = \frac{1}{4}\delta(\lambda) - \frac{1}{2\pi} \sum_k e^{jk\lambda} \tan^{-1}(r_k^{-1}[4 - r_k^2]^{1/2})$$

$$r_k = e^{-\alpha|k|} \qquad 0 \leq \rho_k \leq \frac{1}{12} \qquad \forall k \qquad (11.5.43)$$

If now $g_0(\lambda)$ and $g_{\varepsilon,y}(\lambda)$ denote the power spectral densities of the sequence $\{Z_k\}$ in (11.5.32), under the Gaussian nominal data process and the outlier model respectively, then

$$g_0(\lambda) = \frac{1 - \cos m\lambda}{m^2(1 - \cos \lambda)} f_0(\lambda) \qquad \lambda \in [-\pi, \pi]$$

$$g_{\varepsilon,y}(\lambda) = \frac{1 - \cos m\lambda}{m^2(1 - \cos \lambda)} f_{\varepsilon,y}(\lambda) \qquad \lambda \in [-\pi, \pi] \qquad (11.5.44)$$

where $f_0(\lambda)$ and $f_{\varepsilon,y}(\lambda)$ are given respectively by (11.5.43) and (11.5.35), and where from the latter expressions and (11.5.44), we can easily obtain

$$(2\pi)^{-1} \int_{-\pi}^{\pi} g_0(\lambda) d\lambda < \frac{1}{12} + \rho_0 = \frac{1}{6} \qquad \forall m \text{ in } (11.5.32)$$

$$(2\pi)^{-1} \int_{-\pi}^{\pi} g_{\varepsilon,y}(\lambda) d\lambda$$

$$= (1 - \varepsilon)^2 (2\pi)^{-1} \int_{-\pi}^{\pi} g_0(\lambda) d\lambda$$

$$+ \left\{\varepsilon(1 - \varepsilon)\rho_0 + \varepsilon\left[\Phi(y) - \frac{1}{2}\right]^2\right\}(2\pi)^{-1} \int_{-\pi}^{\pi} \frac{1 - \cos m\lambda}{m^2(1 - \cos \lambda)} d\lambda$$

$$< \frac{(1 - \varepsilon)^2}{6} + \frac{\varepsilon(1 - \varepsilon)}{12} + \varepsilon\left[\Phi(y) - \frac{1}{2}\right]^2 \qquad \forall m \text{ in } (11.5.33)$$

Let us now consider the Rho-Bar distance $\bar{\rho}^*(g_0, g_{\varepsilon,y})$, with $\rho(u, v) = (u - v)^2$, between the transformed sequence $\{Z_k\}$ in (11.5.32) when induced by the Gaussian data process, and the same sequence when induced by the outlier model.

Then, due to the above inequalities, we have

$$\bar{\rho}^*(g_0, g_{\varepsilon,y}) \le (2\pi)^{-1} \int_{-\pi}^{\pi} g_0(\lambda) \, d\lambda + (2\pi)^{-1} \int_{-\pi}^{\pi} g_{\varepsilon,y}(\lambda) \, d\lambda$$

$$< \frac{1}{6} + \frac{(1-\varepsilon)^2}{6} + \frac{\varepsilon(1-\varepsilon)}{12} + \varepsilon \left[\Phi(y) - \frac{1}{2} \right]^2$$

$$= \frac{1}{6} + \frac{(1-\varepsilon)(2-\varepsilon)}{12} + \varepsilon \left[\Phi(y) - \frac{1}{2} \right]^2$$

$$\le \frac{1}{6} + \frac{(1-\varepsilon)(2-\varepsilon)}{12} + \frac{\varepsilon}{4} < \frac{7}{12} \quad \forall \ m \ \text{in (11.5.32)} \quad (11.5.45)$$

The Rho-Bar distance $\bar{\rho}^*(g_0, g_{\varepsilon,y})$ is thus less than 7/12 for every outlier amplitude and frequency, while the Rho-Bar distance in (11.5.31) can be arbitrarily large for large outlier amplitudes and any nonzero outlier frequency. In fact, a stronger property holds. That is, the Rho-Bar distance $\bar{\rho}^*(g_0, g_{\varepsilon,y})$ is continuous with respect to ε for every value y, since the spectral density $g_{\varepsilon,y}(\lambda)$ is also continuous with respect to ε and bounded. The operation in (11.5.32) thus smooths the data, compressing them to a transformed sequence that closely represents a transformed form of the nominal Gaussian process. Estimates $\{\hat{x}_i\}$ of the Gaussian data can then be obtained by the inverse transformation

$$\hat{x}_{k+1} = \Phi^{-1}(z_{k+1}) = \Phi^{-1}\left(m^{-1} \sum_{j=k+1}^{m+k} \left[\Phi(y_j) - \frac{1}{2} \right] \right)$$

11.6 PROBLEMS

11.6.1 Show that the Bhattacharyya and the J-divergence distances are strictly convex with respect to both the involved densities.

11.6.2 Prove that the Prohorov, Vasershtein, and Rho-Bar distances are metrics, if their distortion measure is a metric.

11.6.3 For the problem in Example 11.5.1, show that the Vasershtein distance in (11.5.22) is maximized either for the linear data transformations in (11.5.15) and (11.5.16), if $m_1 = m_2$, or for the linear data transformation in (11.5.21), if $R_n(1) = R_n(2)$.

11.6.4 Consider the statistics,

$$T_1(x^n) = n^{-1} \sum_{i=1}^{n} x_i \quad \text{and} \quad T_2(x^n) = n^{-1} \sum_{i=1}^{n} \operatorname{sgn} x_i$$

where $\operatorname{sgn} x = \{0; x > 0 \text{ and } 1; x \le 0\}$.

Assume first that x^n is generated by a zero mean, unit variance Gaussian memoryless process. Then assume that x^n is generated by a stationary memoryless process such that,

$$f(x_i) = (1 - \varepsilon)\phi(x_i) + \varepsilon\delta(x_i - y) \qquad 0 < \varepsilon < 1$$

Denote by f_{11} and f_{12} the asymptotic $(n \to \infty)$ density functions of $T_1(x^n)$ when x^n is respectively generated by the Gaussian process, and the memoryless process with density $f(x_i)$ as above. Denote by f_{21} and f_{22} the asymptotic densities of $T_2(x^n)$ when x^n is respectively generated by the Gaussian process, and the non-Gaussian memoryless process. For $\rho(u, v) = (u - v)^2$, find the Vasershtein distances $V_\rho(f_{11}, f_{12})$ and $V_\rho(f_{21}, f_{22})$. Compare those two distances. Are they continuous with respect to ε? What happens when $y \to \infty$?

11.6.5 Consider a Gaussian density $f_1(x^n)$ with mean vector M_n and autocovariance matrix R_n. Consider then the class $\mathscr{F}$ of all n-dimensional density functions with common mean vectors $M'_n \neq M_n$, and common autocovariance matrices $R'_n \neq R_n$. Find the closest to $f_1(x^n)$ density function in $\mathscr{F}$, in Kullback-Leibler and in J-divergence distances.

11.6.6 For $\rho(u, v) = |u - v|$, and for zero mean, discrete-time stationary processes with power spectral densities $s_1(\lambda)$ and $s_2(\lambda)$; $\lambda \in [-\pi, \pi]$, and $\sigma_i^2 = (2\pi)^{-1}\int_{-\pi}^{\pi} s_i(\lambda)\, d\lambda$; $i = 1, 2$, show that the Rho-Bar distance is bounded from above by $\pi^{-1}\{\int_{-\pi}^{\pi}[s_1^{1/2}(\lambda) - s_2^{1/2}(\lambda)]^2\, d\lambda\}^{1/2}$ and from below by $(2\pi)^{1/2}|\sigma_1 - \sigma_2|$ [(Gray et al (1975)].

11.6.7 Consider some time interval Δ. Consider two Poisson processes with respective intensities λ_1 and λ_2 in Δ. We wish to distinguish between the two processes, from observations in n time intervals of length Δ. Find the Bhattacharyya and the J-divergence distances between the two processes in n length-Δ time intervals, and compare them with the probability of error when the two processes occur with equal a priori probabilities.

11.6.8 Consider two deterministic, real, and scalar signal waveforms $s_1(t)$ and $s_2(t)$; $t \in [0, T]$ with equal energy $E = \int_0^T s_1^2(t)\, dt = \int_0^T s_2^2(t)\, dt$. Let us also assume that the two signals are orthogonal in $[0, T]$; that is, $\int_0^T s_1(t)s_2(t)\, dt = 0$. Consider L channels, inducing both multiplicative and additive noise. If some waveform $m(t)$ is transmitted through the kth such channel, then $x_k(t) = a_k m(t) + n_k(t)$; $t \in [0, T]$ is observed, where a_k is the realization of a zero mean, variance σ^2 Gaussian random variable A_k; where $n_k(t)$ is the realization of zero mean, white, power $N_0/2$ Gaussian noise $N_k(t)$; and where $E\{A_k A_j\} = 0$; $k \neq j$ $E\{N_k(t)N_j(t)\} = 0 \; \forall\, t \in [0, T]$, for $k \neq j$. Given the signal waveform $s_i(t)$; $i = 1, 2$, as above, let $L^{-1}s_i(t)$ be transmitted through each of the L above channels. Use the Bhattacharyya distance to show that for best distinction among the two signals $s_1(t)$ and $s_2(t)$, the number of channels needed is $L = \lfloor R/3.07 \rfloor$, where $\lfloor\ \rfloor$ denotes integer part, and where $R = 2E\sigma^2/N_0$ [Kailath (1967)].

REFERENCES

Bhattacharyya, A. (1943), "On a measure of divergence between two statistical populations defined by their probability distributions," *Bull. Calcutta Math. Soc.* 35, 99–109.

Cambanis, S., G. Sinous, and W. Strout (1976), "Inequalities for EK (X,Y) when the Marginals are Fixed," *Institute of Statistics Mimeo Series No. 1053*, Univ. of North Carolina, Chapel Hill.

Chernoff, H. (1952), "A Measure of Asymptotic Efficiency for Tests of a Hypothesis Based on a Sum of Observations," *Ann. Math. Stat.* 23, 493–507.

Csiszar, I. (1975), "I-divergence Geometry of Probability Distributions and Minimization Problems," *Ann. Math. Prob.* 3, No. 1, 146–159.

Dobrushin, R. L. (1970) "Prescribing a System of Random Variables by Conditional Distributions," *Theory of Prob. and Appl.* XV, No. 3, 458–486.

Dudley, R. M. (1968), "Distances of Probability Measures and Random Variables," *Ann. Math. Statist.* 39, 1563–1572.

Gray, R. M., D. Neuhoff, and P. C. Shields (1975), "A generalization of Orstein's distance with applications to Information Theory," *Ann. Prob.* 3, No. 2, 315–328.

Hájek, J. (1958), "On a property of normal distribution of any stochastic process" (in Russian), *Czech. Math. J.* 83, 610–18, (a translation appears in "Selected Translations" in *Math. Statist. and Prob.* 1, 245–252.

Hawkes, R. M. and J. B. Moore (1976), "Performance bounds for adaptive estimation," *IEEE Proc.* 64, 1143–1150.

Kadota, T. and L. A. Shepp (1967), "On the Best Finite Set of Linear Observables for Discriminating Two Gaussian Signals," *IEEE Trans. Inform. Theory* 1T-13, 278–284.

Kailath, T. (1967), "The Divergence and Bhattacharyya Distance Measures in Signal Selection," *IEEE Trans. Comm. Tech.* 15, No. 1, 52–60.

Kakutani, S. (1948), "On equivalence of infinite product measures," *Am. Math. Stat.* 49, 214–224.

Kazakos, D. (1977), "Maximin Linear Discrimination I," *IEEE Trans.* on Systems, Man and Cybermetrics, to appear in September 1977.

——— (1980), "On Resolution and Exponential Discrimination between Gaussian Stationary Vector Processes and Dynamic Models," *IEEE Trans. Automat. Contr.* AC-25.

Kazakos, D. and P. Papantoni-Kazakos (1980), "Spectral Distance Measures Between Gaussian Processes," *IEEE Trans. Automat. Contr.* AC-25, 950–959.

Kobayashi, H. (1970), "Distance Measures and Asymptotic Relative Efficiency," *IEEE Trans. Inform. Theory* IT-16, 288–291.

Kullback, S. (1959), *Information Theory and Statistics*, Wiley, New York.

——— (1967), "A Lower Bound for Discrimination Information in Terms of Variation," *IEEE Trans. Inform. The.* Vol. IT-13, 126–127.

Kullback, S. and R. A. Leibler (1951) "On Information and Sufficiency," *Am. Math. Stat.* 22, 79–86.

Lainiotis, D. G. and S. K. Park (1971) "Probability of Error Bounds," *IEEE Trans. Syst. Man. and Cyber.* SMC-1, No. 2, 175–178.

Parthasarathy, K. R. (1968), Probability Measures on Metric Spaces, Academic Press, New York.

Pierce, J. N. (1961), "Theoretical Limitations of Frequency and Time Diversity for Fading Binary Transmissions, *IRE Trans. Comm. Sys.* (Corres.), CS-9, 186–187.

Pinsker, M. S. (1964), *Information and Information Stability of Random Variables and Processes*, Holden-Day, San Francisco, CA.

Prohorov, Yu. V. (1956), "Convergence of Random Processes and Limit Theorems in Probability Theory," *Theor. Prob. Appl.* 1, 157–214.

Rozanov, Y. (1967), *Stationary Random Processes*, Holden-Day, San Francisco, CA.

Schweppe, F. (1967), "On the Distance Between Gaussian Processes: The State Space Approach," *Information and Control*.

Shepp, L. A. (1965), "Distinguishing a Sequence of Random Variables From a Translate of Itself," *Ann. Math. Stat.* 36, 1107–1112.

Shumway, R. H. and A. N. Unger, (1974) "Linear Discriminant Functions of Stationary Time Series," *J. Amer. Statist. Ass.* 69, 948–956.

Stein, C. (1964) "Approximation of Improper Prior Measures by Prior Probability Measures," *Tech. Rep. 12*. Department of Statistics, Stanford Univ., Stanford, CA.

Strassen, V. (1965) "The Existence of Probability Measures With Given Marginals," *Am. Math. Statist.* 36, 423–439.

Tou, J. T. and R. P. Heydorn, (1967) "Some Approaches to Optimum Feature Extraction," in *Computer and Information Sciences*, Vol. 2, J. T. Tou, ed., Academic Press, New York.

Vasershtein, L. N. (1969), "Markov Processes on a Countable Product Space, Describing Large Systems of Automata," *Problemy Peredachi Informatsii* 5, No. 3, 64–73 (in Russian).

11.7 References

Vasershtein, L. N. and A. M. Leonfovich, (1970) "On Invarient Measures of Certain Markov Operators, Describing a Random Medium," *Problemy Peredachi Informatsii*, 6, No. 1, 71–80 (in Russian).

Wyner, A. D. (1975), "The Common Information of Two Dependent Random Variables," *IEEE Trans. Inform. Th.*, IT-21, 163–179.

Zolotarev, V. M. and Senatov, V. V. (1975) "Two-Sided Estimates of Levy's Metric," *Theory Prob. and Appl.* XX, No. 2, 234–245.

CHAPTER 12

QUALITATIVE ROBUSTNESS

12.1 INTRODUCTION

Until now, statistical robustness has been presented as the solution of some saddle-point game formalization on a class $\mathscr{C}$ of stochastic processes and a class $\mathcal{O}$ of data operations. In fact, the robust statistical procedures presented in Chapters 6 and 10 were limited to the outliers model on memoryless stochastic processes, and to either single hypothesis testing or location parameter estimation, and only asymptotic performance was considered. The natural question arising at this point is: Can statistical robustness be defined in terms such that general classes $\mathscr{C}$ and $\mathcal{O}$ of stochastic processes and data operations, as well as both asymptotic and nonasymptotic performance, are included? The answer to this question is provided by *qualitative robustness*.

The analytical reader will naturally conclude that the fundamental common characteristic of the robust game formalizations in Chapters 6 and 10, and the breakdown point and influence curve performance criteria in Chapter 10, is performance stability. In loose terms, a robust statistical procedure guarantees small performance deviations for small perturbations in the data generating stochastic process. Thus, robustness can be qualitatively defined along those lines, where for an analytical definition, the use of stochastic distance measures is essential. In particular, let x^n and y^n be n-dimensional data sequences, generated respectively by two nonidentical n-dimensional density functions f_0^n and f^n. Let $g(\cdot)$ denote some function or operation on n-dimensional data sequences, where $g(\cdot)$ could be, for example, a test function in hypothesis testing or a parameter estimate. Let h_{0g} and h_g denote respectively the density functions of the random variables $g(X^n)$ and $g(Y^n)$ (where X^n is generated by f_0^n, and where Y^n is generated by f^n), and let $d_1(f_0^n, f^n)$ and $d_2(h_{0g}, h_g)$ be two stochastic distance measures respectively between the densities f_0^n and f^n, and the densities h_{0g} and h_g. Then we can present the following definition.

12.1 Introduction

DEFINITION 12.1.1: The operation $g(\cdot)$ is qualitatively robust at the density function f_0^n, in stochastic distance measures $d_1(\cdot, \cdot)$ and $d_2(\cdot, \cdot)$, iff

Given $\varepsilon > 0$, there exists $\delta > 0$ such that
if f^n is such that $d_1(f_0^n, f^n) < \delta$, then
h_g is such that $d_2(h_{0g}, h_g) < \varepsilon$. ∎

From Definition 12.1.1, we conclude that qualitative robustness is a local (around f_0^n) stability property, parallel to the continuity property of real functions. The specific analytical properties of a qualitatively robust data operation $g(\cdot)$ depend on the choice of the stochastic distance measures $d_1(\cdot, \cdot)$ and $d_2(\cdot, \cdot)$. The latter stochastic distances are initially selected to best reflect the desired stability properties of the qualitatively robust data operation, where the weaker the distance $d_1(\cdot, \cdot)$ and the stronger the distance $d_2(\cdot, \cdot)$, then the stronger the qualitative robustness property. The main issue arising here is the relationship of qualitative robustness to the robust saddle-point game-theory formalizations, and the choice of the stochastic distance measures $d_1(\cdot, \cdot)$. We will first address the relationship to the robust saddle-point game-theory formalizations. Our discussion will be qualitative, drawing from the relationship with real functions. The complete analytical analogy can be found in Papantoni-Kazakos (1984a, 1984b, 1983).

Let us consider a saddle-point game with payoff function $f(x, y)$, where the function $f(\cdot, \cdot)$ and its arguments x and y are all real and scalar, and where x and y take values respectively in the subsets A and B of the real line R. Consider the metric $d(u, v) = |u - v|$ on the real line, and let the subsets A and B both be convex with respect to that metric. Let at least one of those two subsets also be compact with respect to the metric $d(\cdot, \cdot)$, and let the payoff function $f(x, y)$ be convex in x, concave in y, and continuous in x and y, with respect to the same metric. Then, as we saw in Chapter 2, the existence of a saddle-point solution (x^*, y^*) such that $f(x^*, y) \leq f(x^*, y^*) \leq f(x, y^*); \forall\, x \in A$ and $\forall\, y \in B$ is guaranteed and it is unique. If, on the other hand, the function $f(x, y)$ is not continuous in x and y, then the existence of a saddle-point solution is not generally guaranteed. The continuity of the payoff function is thus an essential property for the guaranteed existence of a saddle-point solution. The same is true when instead of x and y, we have density functions f^n and h_g as in Definition 12.1.1. In the latter case, the metric $|u - v|$ on the real line is replaced by the stochastic distance measure $d_1(\cdot, \cdot)$ for data generating densities f^n, and by the stochastic distance measure $d_2(\cdot, \cdot)$, for densities h_g induced by some f^n and some data operation g. Therefore, qualitative robustness is essential for the guaranteed solutions of robust saddle-point game-theory formalizations.

Let us now turn to the choice of the distances $d_1(\cdot, \cdot)$ and $d_2(\cdot, \cdot)$ in Definition 12.1.1. As we already pointed out, to make the qualitative robustness property strong, we need a weak distance $d_1(\cdot, \cdot)$ and a strong distance $d_2(\cdot, \cdot)$. As we saw in Chapter 11, a weak distance that also represents closeness in data sequences and best reflects the outlier model as well is the Prohorov distance, with data distortion measure $\rho_n(x^n, y^n)$, as follows.

$$\rho_n(x^n, y^n) = \begin{cases} n^{-1} \sum_{i=1}^{n} |x_i - y_i| = \gamma_n(x_1^n, y_1^n) & \text{if } n \text{ given and finite} \\ \inf\{\alpha: n^{-1}[\#i: \gamma_m(x_{i+1}^{i+m}, y_{i+1}^{i+m}) > \alpha] \leq \alpha\} \\ \text{if } n > n_0, \text{ where } m \text{ and } n_0 \text{ are positive integers} \end{cases}$$
(12.1.1)

The Prohorov distance with data distortion measure as in (12.1.1) is a metric; that is, it satisfies the triangular property. For classes of memoryless processes, this distance is identical to the Prohorov distance with data distortion measure $\rho_1(x, y) = |x - y|$. Regarding the choice of the distance $d_2(\cdot, \cdot)$, the Vasershtein or Rho-Bar distances are appropriate. Indeed, as we saw in Chapter 11, those two distances are strong and they both bound difference in expected error performance. The choice of the data distortion measure within the latter distances depends on the particular application, where a popular and useful such choice is the difference squared distortion measure $\rho^*(x, y) = (x - y)^2$. As we saw in Chapter 11, the Rho-Bar distance is used for closeness in stochastic processes. Given some data sequence $y_1^{N+n} = \{y_1, \ldots, y_{N+n}\}$ and some scalar data operation $g(\cdot)$, let $g(y_i^{i+n})$ estimate the datum x_k of some process whose arbitrary dimensionality density function is f_2 and whose data sequences are $\ldots, x_{-1}, x_0, x_1, \ldots$. If the sequence y_1^{N+n} is generated by the density function f_0^{N+n}, let h_{0g} denote the arbitrary dimensionality density induced by f_0^{N+n} and the data operation $g(\cdot)$. Let h_g denote the arbitrary dimensionality density induced by $g(\cdot)$ and some other data density function f^{N+n}. Then, h_{0g} and h_g both estimate f_2. Given some data distortion measure $\rho(\cdot, \cdot)$, the goodness of those two estimates is respectively measured by the Rho-Bar distances $\bar{\rho}(f_2, h_{0g})$ and $\bar{\rho}(f_2, h_g)$. As we saw in Chapter 11, if $\rho(u, v) = |u - v|$, then $|\bar{\rho}(f_2, h_{0g}) - \bar{\rho}(f_2, h_g)| \leq \bar{\rho}(h_{0g}, h_g)$; thus, the Rho-Bar distance $\bar{\rho}(h_{0g}, h_g)$ measures how closely h_{0g} fits f_2, as compared to the fitness of h_g to f_2. A similar conclusion is drawn, when the data distortion measure is the difference squared, $\rho^*(u, v) = (u - v)^2$, where then $|[\bar{\rho}^*(f_2, h_{0g})]^{1/2} - [\bar{\rho}^*(f_2, h_g)]^{1/2}| \leq [\bar{\rho}^*(h_{0g}, h_g)]^{1/2}$ [see (11.3.18)].

Definition 12.1.1 of qualitative robustness was first given by Hampel (1971), who considered only memoryless processes and used the Prohorov distance in place of the $d_2(\cdot, \cdot)$ distance. The definition was extended to processes with memory, first by Papantoni-Kazakos and Gray (1979), and then by Cox (1978), Bustos et al (1984), and Papantoni-Kazakos (1983, 1984a, 1984b, 1981a). It turns out that the above definition provides sufficient conditions that data operations should satisfy, to guarantee the local stability exhibited by Definition 12.1.1. Those conditions are constructive, and their derivation can be found in the above references. Here, we only state them in Theorem 12.1.1 below, whose proof is beyond the scope of this book.

THEOREM 12.1.1: Consider a scalar real operation $g(x^n)$ on data sequences x^n of length n. Let $g(x^n)$ be bounded, and such that

i. If n is finite, then $g(x^n)$ is pointwise continuous as a function on the data. That is, given $\varepsilon > 0$, there exists $\delta > 0$, such that $n^{-1} \sum_i |x_i - y_i| < \delta$ implies $|g(x^n) - g(y^n)| < \varepsilon$

12.1 Introduction

ii. If n is asymptotically large, and given some data generating density function f_0, then $g(x^n)$ is pointwise asymptotically continuous at f_0. That is, given $\varepsilon > 0$ and $\eta > 0$, there exist $\delta > 0$, positive integers m and n_0, and for each $n > n_0$ some set $A^n \in R^n$, such that $\Pr(x^n \in A^n | f_0^n) > 1 - \eta$ and $x^n \in A^n$ and $\inf\{\alpha: n^{-1}[\#i: \gamma_m(x_{i+1}^{i+m}, y_{i+1}^{i+m}) > \alpha] \leq \alpha\} < \delta$ implies $|g(x^n) - g(y^n)| < \varepsilon \ \forall \ n > n_0$, where $\gamma_m(x_{i+1}^{i+m}, y_{i+1}^{i+m}) = m^{-1} \sum_{j=i+1}^{i+m} |x_j - y_j|$

Then the operation $g(\cdot)$ is qualitatively robust at the density function f_0^n, where in Definition 12.1.1, $d_1(\cdot, \cdot)$ is replaced by the Prohorov distance with data distortion measure as in (12.1.1), and $d_2(\cdot, \cdot)$ is replaced by either the Vasershtein or the Rho-Bar distances with distortion measure $\rho(u, v)$, equal either to $|u - v|$ or some continuous function of $|u - v|$. ∎

From Theorem 12.1.1, we conclude that to be qualitatively robust, it suffices that a data operation be bounded and continuous. For data sequences of finite length continuity is defined in the usual functional sense. For asymptotically large data sequences, continuity is defined as follows at some data generating density function: If some sequence x^n is representative of the latter density function, in the sense that it belongs to a high-probability set A^n, and if the majority of the elements of another sequence y^n are close to the corresponding elements of the sequence x^n, then the values $g(x^n)$ and $g(y^n)$ of the data operation are close as well. Due to the above results, we conclude that linear data operations are not qualitatively robust. This is so because such operations are not bounded, and because closeness between the majority of corresponding elements of two sequences does not guarantee closeness in the values of those operations. Thus, the empirical mean location parameter estimate in Chapter 10 is not qualitatively robust. Also,

i. The median location parameter estimate in Chapter 10 is qualitatively robust for asymptotically large data sequences, but not for those of finite length. Indeed, this estimate is bounded and asymptotically continuous [condition ii., Theorem 12.1.1) at every ergodic and stationary process, but is not continuous as a function on data sequences, since the sgn x is not. For the same reasons, the sign and the Wilcoxon nonparametric tests in Chapter 7 are asymptotically qualitatively robust, but not nonasymptotically.
ii. The M-estimates in Chapter 10, with bounded and continuous function $\psi(\cdot)$, are qualitatively robust at every stationary and memoryless or Markovian process for data sequences both of finite length and asymptotically large. Indeed, the operation $n^{-1}\sum_i \psi(x_i - \alpha)$ then satisfies both the continuity conditions i. and ii. in Theorem 12.1.1, and it is bounded.
iii. Huber's robust detector in Chapter 6 is qualitatively robust at every stationary and memoryless process, both asymptotically and nonasymptotically. Indeed, the latter detector induces bounded and continuous operations on the data, and satisfies both conditions i. and ii. in Theorem 12.1.1.

Qualitative robustness is a property that does not induce uniqueness. That is, given a specific problem, and some data generating density function f_0, there generally exists a whole class $\mathcal{O}$ of data operations that are qualitatively robust at f_0. Additional performance criteria are thus needed, to evaluate and compare different data operations in class $\mathcal{O}$. Such performance criteria are the breakdown point and the sensitivity, both defined asymptotically ($n \to \infty$) and at the density function f_0. Given f_0 and given some operation $g(\cdot)$ in class $\mathcal{O}$, consider the density functions f, that are included in the Prohorov ball $\Pi_{n,\rho_n}(f_0,f) \le \varepsilon$, where ρ_n is as in (12.1.1). Let h_{0g} and h_g be the density functions induced by the data operation $g(\cdot)$ and the densities f_0 and f respectively. Given some scalar data distortion measure $\rho(\cdot,\cdot)$, consider the Rho-Bar distance $\bar{\rho}(h_{0g}, h_g)$. Then, the *breakdown point*, ε^*, of the operation $g(\cdot)$ at f_0, is the largest value ε, such that, if f is some density in the ball $\lim_{n\to\infty} \Pi_{n,\rho_n}(f_0,f) \le \varepsilon$, then the distance $\bar{\rho}(h_{0g}, h_g)$ is a function of ε. The *sensitivity* of the operation $g(\cdot)$ at the density f_0 is defined as

$$\lim_{\substack{n\to\infty \\ \varepsilon \to \infty}} \frac{\bar{\rho}(h_{0g}, h_g)}{\Pi_{n,\rho_n}(f_0,f)}$$

We note that the breakdown point defined here is a generalization of the parallel definition in Chapter 10. We also note that the sensitivity is parallel to the derivative of real functions. It can be found that if bounded sensitivity at f_0 is required (parallel to bound derivative) then the qualitatively robust operation $g(\cdot)$ should also be differentiable almost everywhere as a real function of the data, and for asymptotically large sequences it should be such that

$$|g(x^n) - g(y^n)| \le c \inf\{\alpha: n^{-1}[\#i: \gamma_m(x_{i+1}^{i+m}, y_{i+1}^{i+m}) > \alpha] \le \alpha\}$$

where c is some bounded constant, and where $x^n \in A^n$ for A^n as in part ii of Theorem 12.1.1 [see Papantoni-Kazakos (1983, 1984b)].

Completing this section, we point out that qualitative robustness is a performance stability property, which extends beyond the hypothesis testing and parameter estimation problems. Its applications include source coding, quantization, prediction, interpolation, and filtering. For qualitatively robust source coding and quantization, see Papantoni-Kazakos (1981a, 1981b). Prediction, interpolation, and filtering will be discussed in the remainder of this chapter.

12.2 PARAMETRIC PREDICTION, INTERPOLATION, AND FILTERING

Prediction, interpolation, and filtering all refer to the estimation of unobserved data values from a sequence of observed data. In all three cases, the optimality criterion is the minimization of an expected penalty function; the optimization problems are thus Bayesian. In prediction and interpolation, it is assumed that the data are generated by a single stochastic process, and are observed noiselessly, while in filtering the data are corrupted by noise. The specific parametric formalization of those problems is as follows.

12.2 Parametric Prediction, Interpolation, and Filtering

Prediction

Consider a well known, real, discrete-time stochastic process, whose arbitrary dimensionality density function is denoted f. Let the sequence $x_{-n}^{-1} = \{x_{-n}, \ldots, x_{-1}\}$ of data generated by the process be observed, and let some scalar penalty function $c(u, v)$ be available. The objective is the estimation $\hat{x}_k = \hat{x}_k(x_{-n}^{-1})$ of the unobserved datum x_k; $k \geq 0$, from the process, so that the following expected value is minimized.

$$E\{c(\hat{x}_k(X_{-n}^{-1}), X_k)|f\} \tag{12.2.1}$$

Interpolation

Given a stochastic process, and a penalty function, as in prediction, let the data sequence $x_{-n}^{-1} \cup x_1^m = \{x_{-n}, \ldots, x_{-1}, x_1, \ldots, x_m\}$ generated by the process be observed. The objective is then the estimation $\hat{x}_0 = \hat{x}_0(x_{-n}^{-1} \cup x_1^m)$ of the unobserved datum x_0 from the process, so that the following expected value is minimized.

$$E\{c(\hat{x}_0(X_{-n}^{-1} \cup X_1^m), X_0)|f\} \tag{12.2.2}$$

Filtering

Consider a well known, discrete-time, information process whose data sequences are denoted $\ldots, x_{-1}, x_0, x_1, \ldots$ and a well known, discrete-time noise process whose data sequences are denoted $\ldots, w_{-1}, w_0, w_1, \ldots$. Let the mutual statistics between those two processes be known and let f denote the arbitrary dimensionality density function of the well known discrete-time joint process whose marginals are the above information and noise processes. Let y_{-n}^n be an observed data sequence, generated by the joint process, and let some scalar penalty function $c(u, v)$ be available. The objective is then to derive an estimate $\hat{x}_k^m = \hat{x}_k^m(y_{-n}^n) = \{\hat{x}_k(y_{-n}^n), \ldots, \hat{x}_m(y_{-n}^n)\}$ of the information data sequence x_k^m so that the following expected value is minimized.

$$E\{(m - k + 1)^{-1} \sum_{i=k}^{m} c(\hat{x}_i(Y_{-n}^n), X_i)|f\} \tag{12.2.3}$$

The expected values in (12.2.1), (12.2.2), and (12.2.3) respectively can be written as

$$E\{c(\hat{x}_k(X_{-n}^{-1}), X_k)|f\} = \int_{R^n} dx^n f_{X_{-n}^{-1}}(x^n) \int_R du\, c(\hat{x}_k(x^n), u) f_{X_k|X_{-n}^{-1}}(u|x^n) \tag{12.2.4}$$

$$E\{c(\hat{x}_0(X_{-n}^{-1} \cup X_1^m), X_0)|f\} = \int_{R^{n+m}} dx^n dx^m f_{X_{-n}^{-1} \cup X_1^m}(x^n, x^m) \int_R du\, c(\hat{x}_0(x^n, x^m), u) \tag{12.2.5}$$

$$E\{(m - k + 1)^{-1} \sum_{i=k}^{m} c(\hat{x}_i(Y_{-n}^n), X_i)|f\}$$

$$= (m - k - 1)^{-1} \sum_{i=k}^{m} \int_{R^{2n}} dy^{2n} f_{Y_{-n}^n}(y^{2n}) \int_R du\, c(\hat{x}_i(y^{2n}), u) f_{X_i|Y_{-n}^n}(u|y^{2n}) \tag{12.2.6}$$

where $f_{W_i^\rho}(w^{\rho-l+1})$ denotes the density function of the random sequence W_i^ρ at the vector point $w^{\rho-l+1}$, and where $f_{W_i|W_i^\rho}(u|w^{\rho-l+1})$ denotes the conditional density function of the random variable W_i at the scalar point u, conditioned on the vector value $w^{\rho-l+1}$ of the random sequence W_i^ρ. Given the corresponding sequence values, the expected values in (12.2.4), (12.2.5), and (12.2.6) are minimized when the respective conditional expected values below are minimized, assuming that such minima exist.

$$E\{c(\hat{x}_k(x^n), X_k)|X_{-n}^{-1} = x^n, f\} = \int_R du\, c(\hat{x}_k(x^n), u) f_{X_k|X_{-n}^{-1}}(u|x^n) \quad (12.2.7)$$

$$E\{c(\hat{x}_0(x^n, x^m), X_0)|X_{-n}^{-1} \cup X_1^m = x^n \cup x^m, f\}$$
$$= \int_R du\, c(\hat{x}_0(x^n, x^m), u) f_{X_0|X_{-n}^{-1} \cup X_1^m}(u|x^n, x^m) \quad (12.2.8)$$

$$E\{c(\hat{x}_i(y^{2n}), X_i)|Y_{-n}^n = y^{2n}, f\} = \int_R du\, c(\hat{x}_i(y^{2n}), u) f_{X_i|Y_{-n}^n}(u|y^{2n}) \quad k \le i \le m$$
$$(12.2.9)$$

For any strictly convex penalty function $c(u, v)$, the minima of the expected values in (12.2.7), (12.2.8), and (12.2.9) with respect to the corresponding estimates exist and are unique. For the popular and useful mean-squared penalty function $c(u, v) = (u - v)^2$ those minima are all easily found to be satisfied by the respective conditional expected values. That is, the optimal mean-squared predictor $\hat{x}_k^*(x^n)$ in (12.2.7), the optimal mean-squared interpolator, $\hat{x}_0^*(x^n, x^m)$ in (12.2.8), and the optimal mean-squared filter $\hat{x}_i^*(y^{2n})$; $k \le i \le m$ in (12.2.9), respectively, are

$$\hat{x}_k^*(x^n) = E\{X_k|X_{-n}^{-1} = x^n, f\} \qquad k \ge 0 \quad (12.2.10)$$

$$\hat{x}_0^*(x^n, x^m) = E\{X_0|X_{-n}^{-1} = x^n, X_1^m = x^m, f\} \quad (12.2.11)$$

$$\hat{x}_i^*(y^{2n}) = E\{X_i|Y_{-n}^n = y^{2n}, f\} \qquad k \le i \le m \quad (12.2.12)$$

If the indices i in (12.2.12) are all larger than n, the filter is called *causal*, and it estimates information data from past observed data. Otherwise, the filter is called *noncausal*, or *smoother*. For arbitrary density functions, the conditional expectations in (12.2.10), (12.2.11), and (12.2.12) may be complex and intractable functions of the data sequences they are conditioned on. A popular simplification has been linearization. That is, given the mean-squared penalty function, the optimal linear mean-squared predictor $\hat{x}_{Lk}^*(x^n) = \sum_{i=-n}^{-1} a_i^* x_i$, interpolator $\hat{x}_{L0}^*(x^n, x^m) = \sum_{i=-n}^{-1} b_i^* x_i + \sum_{i=1}^{m} b_i^* x_i$, and filter $\hat{x}_{Li}^*(y^{2n}) = \sum_{j=-n}^{n} d_{ij}^* y_j$; $k \le i \le m$ are sought. Due to the strict convexity of the difference squared function, those linear data operations are found directly from differentiation of the expected values in (12.2.4), (12.2.5), and (12.2.6), with respect to the corresponding linear coefficients. In particular,

$$\hat{x}_{Lk}^*(x^n) = \sum_{i=-n}^{-1} a_i^* x_i$$

$$\{a_i^*\}: \frac{\partial}{\partial a_l} E\left\{\left[X_k - \sum_{i=-n}^{-1} a_i X_i\right]^2 \bigg| f\right\}\bigg|_{\{a_i\} = \{a_i^*\}} = 0 \qquad -n \le l \le -1$$

12.2 Parametric Prediction, Interpolation, and Filtering

or

$$\{a_i^*\}: E\left\{\left[X_k - \sum_{i=-n}^{-1} a_i^* X_i\right] X_l \bigg| f\right\} = 0 \qquad -n \le l \le -1 \quad (12.2.13)$$

$$\hat{x}_{L0}^*(x^n, x^m) = \sum_{i=-n}^{-1} b_i^* x_i + \sum_{i=1}^{m} b_i^* x_i$$

$$\{b_i^*\}: \frac{\partial}{\partial b_l} E\left\{\left[X_0 - \sum_{i=-n}^{-1} b_i X_i - \sum_{i=1}^{m} b_i X_i\right]^2 \bigg| f\right\}\bigg|_{\{b_i\}=\{b_i^*\}} = 0 \qquad \begin{cases} -n \le l \le -1 \\ 1 \le l \le m \end{cases}$$

or

$$\{b_i^*\}: E\left\{\left[X_0 - \sum_{i=-n}^{-1} b_i^* X_i - \sum_{i=1}^{m} b_i^* X_i\right] X_l \bigg| f\right\} = 0 \qquad \begin{cases} -n \le l \le -1 \\ 1 \le l \le m \end{cases}$$

(12.2.14)

$$\hat{x}_{Li}^*(y^{2n}) = \sum_{j=-n}^{n} d_{ij}^* y_j$$

$$\{d_{ij}^*\}: \frac{\partial}{\partial d_{il}} E\left\{\left[X_i - \sum_{j=-n}^{n} d_{ij} Y_j\right]^2 \bigg| f\right\}\bigg|_{\{d_{ij}\}=\{d_{ij}^*\}} = 0 \qquad \begin{cases} -n \le l \le n \\ k \le i \le m \end{cases}$$

or

$$\{d_{ij}^*\}: E\left\{\left[X_i - \sum_{j=-n}^{n} d_{ij}^* Y_j\right] Y_l \bigg| f\right\} = 0 \qquad \begin{cases} -n \le l \le n \\ k \le i \le m \end{cases} \quad (12.2.15)$$

The equations in (12.2.13), (12.2.14), and (12.2.15) represent orthogonality principles, and they provide the optimal linear mean-squared predictor, interpolator, and filter respectively. These optimal linear operations are identical to the respective conditional expectations in (12.2.10), (12.2.11), and (12.2.12) if the data generating process in the first two is Gaussian, and if the information and noise processes in (12.2.12) are jointly Gaussian. In general, let us now assume that in the prediction and interpolation problems the data generating process is real, zero mean, and stationary, with power spectral density $f_1(\lambda)$; $\lambda \in [-\pi, \pi]$. Let us assume that in the filtering problem, the information and noise processes are both zero mean and stationary, with respective power spectral densities $f_1(\lambda)$ and $f_2(\lambda)$; $\lambda \in [-\pi, \pi]$, they are mutually independent, and the noise process is additive to the information process; that is, $Y_l = X_l + W_l$. Denoting,

$$r_k = (2\pi)^{-1} \int_{-\pi}^{\pi} f_1(\lambda) e^{-jk\lambda} d\lambda \qquad f_1(\lambda) = \sum_k r_k e^{jk\lambda} \qquad \lambda \in [-\pi, \pi]$$

$$\rho_k = (2\pi)^{-1} \int_{-\pi}^{\pi} f_2(\lambda) e^{-jk\lambda} d\lambda \qquad f_2(\lambda) = \sum_k \rho_k e^{jk\lambda} \qquad \lambda \in [-\pi, \pi]$$

(12.2.16)

we then have in (12.2.13) and (12.2.14)

$$E\{X_k X_l | f\} = r_{|k-l|} \qquad E\{X_0 X_l | f\} = r_l \qquad (12.2.17)$$

in (12.2.15)

$$E\{X_i Y_l | f\} = E\{X_i(X_l + W_l) | f\}$$
$$= E\{X_i X_l | f\} + E\{X_i | f\} E\{W_l | f\}$$
$$= E\{X_i X_l | f\} = r_{|i-l|} \qquad (12.2.18)$$

$$E\{Y_j Y_l | f\} = E\{(X_j + W_j)(X_l + W_l) | f\}$$
$$= E\{X_j X_l | f\} + E\{W_j W_l | f\}$$
$$= r_{|j-l|} + \rho_{|j-l|} \qquad (12.2.19)$$

Substituting (12.2.17), (12.2.18), and (12.2.19) in (12.2.13), (12.2.14), and (12.2.15), we conclude that for zero mean, stationary, discrete-time processes, and an additive independent noise process in filtering, the optimal linear mean-squared predictor, interpolator, and filter are respectively given by

$$\hat{x}_{Lk}^*(x^n) = \sum_{i=-n}^{-1} a_i^* x_i$$

$$\{a_i^*\}: r_{|k-l|} = \sum_{i=-n}^{-1} a_i^* r_{|i-l|} \qquad -n \le l \le -1 \qquad (12.2.20)$$

$$\hat{x}_{LO}^*(x^n, x^m) = \sum_{i=-n}^{-1} b_i^* x_i + \sum_{i=1}^{m} b_i^* x_i$$

$$\{b_i^*\}: r_l = \sum_{i=-n}^{-1} b_i^* r_{|i-l|} + \sum_{i=1}^{m} b_i^* r_{|i-l|} \qquad \begin{cases} -n \le l \le -1 \\ 1 \le l \le m \end{cases} \qquad (12.2.21)$$

$$\hat{x}_{Li}^*(y^{2n}) = \sum_{j=-n}^{n} d_{ij}^* y_j$$

$$\{d_{ij}^*\}: r_{|j-l|} = \sum_{j=-n}^{n} d_{ij}^* [r_{|j-l|} + \rho_{|j-l|}] \qquad \begin{cases} -n \le l \le n \\ k \le i \le m \end{cases} \qquad (12.2.22)$$

The functions $a(\lambda) = \sum_{i=-n}^{-1} a_i^* e^{ji\lambda}$; $\lambda \in [-\pi, \pi]$, $b(\lambda) = \sum_{i=-n}^{-1} b_i^* e^{ji\lambda} + \sum_{i=1}^{m} b_i^* e^{ji\lambda}$; $\lambda \in [\pi, \pi]$, and $d_i(\lambda) = \sum_{m=-n}^{n} d_{im}^* e^{jm\lambda}$; $\lambda \in [-\pi, \pi]$, $k \le i \le m$ are then called the *transfer functions* of the optimal linear mean-squared predictor, interpolator, and filter respectively; they are strictly functions of the power spectral densities (or the second order statistics) of the involved processes. Asymptotically [$n \to \infty$ in (12.2.20), and $n, m \to \infty$ in (12.2.21)] the mean-squared error $e_p(f_1)$ induced by the optimal linear mean-squared predictor in (12.2.20) and the mean-squared error $e_I(f_1)$ induced by the optimal linear mean-squared interpolator in (12.2.21) are given below, as functions of the power spectral density of the data process [see Kolmogorov (1941) and Hannan (1970)].

$$e_p(f_1) = \exp\left\{(2\pi)^{-1} \int_{-\pi}^{\pi} \log[2\pi f_1(\lambda)] \, d\lambda\right\}$$

$$e_I(f_1) = 4\pi^2 \left[\int_{-\pi}^{\pi} f_1^{-1}(\lambda) \, d\lambda\right]^{-1} \qquad (12.2.23)$$

12.2 Parametric Prediction, Interpolation, and Filtering

The transfer functions $a(\lambda)$ and $b(\lambda)$ of respectively the optimal linear mean-squared predictor and interpolator that attain the errors in (12.2.23) are as follows, where * denotes conjugate.

$$a(\lambda)a^*(\lambda) = \exp\left\{(2\pi)^{-1}\int_{-\pi}^{\pi} d\omega \log f_1(\omega) - \log f_1(\lambda)\right\} \quad \text{a.e. in } [-\pi, \pi] \quad (12.2.24)$$

$$b(\lambda) = 2\pi \left[\int_{-\pi}^{\pi} f_1^{-1}(\omega)\,d\omega\right]^{-1} f_1^{-1}(\lambda) \quad \text{a.e. in } [-\pi, \pi]$$

Regarding optimal linear mean-squared filtering, let us without lack in generality assume that the datum x_0 from the information process is estimated. In the case of asymptotic [$n \to \infty$ in (12.2.22)] and noncausal filtering or smoothing, both the transfer function $d_0(\lambda)$ of the optimal linear mean-sqaured filter, and the induced mean-squared error $e_F(f_1, f_2)$ can be found easily as functions of the power spectral densities f_1 and f_2 of the mutually independent and additive information and noise processes [see Hannan (1970)], and are

$$d_0(\lambda) = f_1(\lambda)[f_1(\lambda) + f_2(\lambda)]^{-1} \quad \lambda \in [-\pi, \pi] \quad (12.2.25)$$

$$e_F(f_1, f_2) = (2\pi)^{-1} \int_{-\pi}^{\pi} f_1(\lambda)[f_1(\lambda) + f_2(\lambda)]^{-1} f_2(\lambda)\,d\lambda \quad (12.2.26)$$

Let us now consider optimal linear mean-squared causal filtering. In particular, let the datum x_n from the information process be estimated from the $y_1^n = \{y_1, \ldots, y_n\}$ observation sequence, where we denote $\hat{x}_n^*(y_1^n) = \sum_{j=1}^n d_{nj}^* y_j$. Let the information and noise processes be zero mean, real, additive, and mutually independent. In addition, let the noise process be white and generally nonstationary; that is, $E\{W_j W_l\} = \sigma_j^2 \Delta(j-l)$, where $\Delta(\cdot)$ denotes the Kronecker delta. Let the information process be generally nonstationary, and first order wide-sense Markov; that is, $E\{X_n | X_1^{n-1} = x_1^n\} = E\{X_n | X_{n-1} = x_{n-1}\}$. Then, a trivially found recursive form of the optimal linear mean-squared filter exists. In particular, denoting

$$r_{kl} = E\{X_k X_l\} \qquad d_n = r_{n-1,n} r_{n-1,n-1}^{-1} \quad (12.2.27)$$

we have

$$\hat{x}_1^*(y_1) = r_{11}[\sigma_1^2 + r_{11}]^{-1} y_1 \qquad \hat{x}_n^*(y_1^n) = \alpha_n \hat{x}_{n-1}^*(y_1^{n-1}) + \beta_n y_n \quad (12.2.28)$$

where if

$$e_n = E\{[\hat{x}_n^*(Y_1^n) - X_n]^2\} \quad (12.2.29)$$

Then

$$e_1 = r_{11}\sigma_1^2[\sigma_1^2 + r_{11}]^{-1} \quad (12.2.30)$$

$$e_n = [r_{nn} - d_n^2 r_{n-1,n-1} + d_n^2 e_{n-1}]$$
$$\times [r_{nn} - d_n^2 r_{n-1,n-1} + d_n^2 e_{n-1} + \sigma_n^2]^{-1}\sigma_n^2 \quad (12.2.31)$$

and

$$\beta_n = e_n \sigma_n^{-2} \qquad \alpha_n = d_n(1 - e_n \sigma_n^{-2}) \qquad (12.2.32)$$

The above special case of recursive linear filtering is called *Kalman filtering* [see Kalman (1960), (1963)], it corresponds to estimating information data from the whole past and present observed data sequence. If, in addition to being wide-sense Markov, the information process is also Gaussian and if in addition to being white, additive, and independent of the information process, the noise is also Gaussian, then the Kalman filter is also the absolutely optimal mean-squared filter. We point out that the results for discrete-time scalar processes can be extended to discrete-time vector processes, and to continuous-time processes as well. For the former, see Zasuhin (1941) and Hannan (1970). The latter is attained via the Karhunen-Loève expansion, as in Chapter 4, and for particular applications see, for example, Gelb (1975).

Let us now consider asymptotic prediction and interpolation for zero mean Gaussian processes whose power spectral densities are only known to belong to some given class. In particular, let $f_0(\lambda); \lambda \in [-\pi, \pi]$ be a given power spectral density, let ε be a given number in $(0, 1)$, and let the class $\mathscr{F}$ of power spectral densities be defined as

$$\mathscr{F} = \left\{ f : f(\lambda) = (1 - \varepsilon) f_0(\lambda) + \varepsilon h(\lambda); \lambda \in [-\pi, \pi], \right.$$

$$\left. \text{where } h(\lambda) \geq 0 \ \forall \ \lambda \text{ and } (2\pi)^{-1} \int_{-\pi}^{\pi} h(\lambda) \, d\lambda = (2\pi)^{-1} \int_{-\pi}^{\pi} f_0(\lambda) \, d\lambda = 1 \right\}$$

(12.2.33)

Given some spectral density f, the optimal at the Gaussian process mean-squared asymptotic predictor and interpolator are both linear, with transfer functions as in (12.2.24). We may thus restrict ourselves to the classes $\mathscr{C}_p$ and $\mathscr{C}_I$ of linear predictors and interpolators, denoting their members respectively g_p and g_I. Let us denote by $e_p(f, g_p)$ the asymptotic mean-squared prediction error induced by the power spectral density f and the linear predictor g_p. Let $e_I(f, g_I)$ denote the asymptotic mean-squared interpolation error induced by the same power spectral density and the linear interpolator g_I. We may search then for pairs (f_p^*, g_p^*) and (f_I^*, g_I^*) such that $f_p^* \in \mathscr{F}$, $f_I^* \in \mathscr{F}$, $g_p^* \in \mathscr{C}_p$ and $g_I^* \in \mathscr{C}_I$, which satisfy the following saddle-point games.

$$\forall f \in \mathscr{F} \qquad e_p(f, g_p^*) \leq e_p(f_p^*, g_p^*) \leq e_p(f_p^*, g_p) \qquad \forall g_p \in \mathscr{C}_p \quad (12.2.34)$$

$$\forall f \in \mathscr{F} \qquad e_I(f, g_I^*) \leq e_I(f_I^*, g_I^*) \leq e_I(f_I^*, g_I) \qquad \forall g_I \in \mathscr{C}_I \quad (12.2.35)$$

Both the above games have solutions. The solution of the game in (12.2.34) exists in Hosoya (1978), and in Tsaknakis et al (1982). In the latter reference, the solution of the game in (12.2.35) is also found, and vector processes are generally considered. Due to (12.2.23), the solutions of the games in (12.2.34)

12.2 Parametric Prediction, Interpolation, and Filtering

and (12.2.35) are such that

$$e_p(f_p^*, g_p^*) = \exp\left\{(2\pi)^{-1} \int_{-\pi}^{\pi} \log[2\pi f_p^*(\lambda)] \, d\lambda\right\} \quad (12.2.36)$$

$$e_I(f_I^*, g_I^*) = 4\pi\left\{\int_{-\pi}^{\pi} [f_I^*(\lambda)]^{-1} \, d\lambda\right\}^{-1} \quad (12.2.37)$$

where the asymptotic linear operations g_p^* and g_I^* are given by the expressions in (12.2.24), with respective substitution of $f_p^*(\omega)$ and $f_I^*(\omega)$ for $f_1(\omega)$. The search for the pairs (f_p^*, g_p^*) and (f_I^*, g_I^*) reduces then to the search for those power spectral densities $f_p^*(\lambda)$ and $f_I^*(\lambda)$ in $\mathscr{F}$ that maximize the expressions in (12.2.36) and (12.2.37) respectively. It is easily found that the latter expressions are both strictly concave with respect to the power spectral densities they involve. Applying calculus of variations, with the use of some additional inequalities for (12.2.37), then gives (for proof see Section 12.6).

$$f_p^*(\lambda) = \max[(1-\varepsilon)f_0(\lambda), c] \quad \lambda \in [-\pi, \pi]$$
$$c: (2\pi)^{-1} \int_{-\pi}^{\pi} \max[(1-\varepsilon)f_0(\lambda), c] \, d\lambda = 1 \quad (12.2.38)$$

$$f_I^*(\lambda) = \max[T^{-1}(d), (1-\varepsilon)f_0(\lambda)] \quad \lambda \in [-\pi, \pi]$$
$$T(\gamma) = \gamma \int_{-\pi}^{\pi} (\max[\gamma, (1-\varepsilon)f_0(\lambda)])^{-1} \, d\lambda$$
$$d: (2\pi)^{-1} \int_{-\pi}^{\pi} \max[T^{-1}(d), (1-\varepsilon)f_0(\lambda)] \, d\lambda = 1 \quad (12.2.39)$$

where $T^{-1}(x) = y$, means $x = T(y)$.

We observe that the least favorable power spectral densities in (12.2.38) and (12.2.39) are both the "flattest" such densities in class $\mathscr{F}$ in (12.2.33), and they have the same general form. In Figure 12.2.1, we plot the power spectral density $f_p^*(\lambda)$ in (12.2.38).

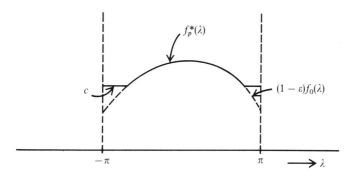

Figure 12.2.1

When noncausal asymptotic mean-squared filtering for zero mean, scalar, stationary Gaussian processes is considered, in the presence of zero mean, scalar, additive Gaussian noise, and when the power spectral densities of the information and noise processes belong to two disjoint given spectral classes $\mathscr{F}_1$ and $\mathscr{F}_2$, then a game on $\mathscr{F}_1$, $\mathscr{F}_2$, and the class $\mathscr{C}$ of asymptotic linear filters can be formalized. Let $f_1 \in \mathscr{F}_1$, $f_2 \in \mathscr{F}_2$, and $g \in \mathscr{C}$, and let $e_F(f_1, f_2, g)$ be the mean squared filtering error induced by f_1, f_2, and g. We then search for a triple (f_1^*, f_2^*, g^*), such that $f_1^* \in \mathscr{F}_1$, $f_2^* \in \mathscr{F}_2$, and $g^* \in \mathscr{C}$, and

$$\forall f_1 \in \mathscr{F}_1 \quad \forall f_2 \in \mathscr{F}_2 \quad e_F(f_1, f_2, g^*) \le e_F(f_1^*, f_2^*, g^*)$$
$$\le e_F(f_1^*, f_2^*, g) \quad \forall g \in \mathscr{C} \quad (12.2.40)$$

Due to (12.2.25) and (12.2.26), the right part of (12.2.40) is satisfied for a g^* whose transfer function $g^*(\lambda)$ is

$$g^*(\lambda) = f_1^*(\lambda)[f_1^*(\lambda) + f_2^*(\lambda)]^{-1} \quad \lambda \in [-\pi, \pi]$$

and then

$$e_F(f_1^*, f_2^*, g^*) = (2\pi)^{-1} \int_{-\pi}^{\pi} f_1^*(\lambda)[f_1^*(\lambda) + f_2^*(\lambda)]^{-1} f_2^*(\lambda) \, d\lambda \quad (12.2.41)$$

To satisfy the left part of (12.2.40), we must search for the pair (f_1^*, f_2^*) that maximizes the expression in (12.2.41), where the latter expression is strictly concave with respect to both f_1^* and f_2^*. Now let $f_{01}(\lambda); \lambda \in [-\pi, \pi]$ and $f_{02}(\lambda); \lambda \in [-\pi, \pi]$ be two given nominal power spectral densities, and let the classes $\mathscr{F}_1$ and $\mathscr{F}_2$ be as in (12.2.33), with ε and f_0 respectively replaced by ε_1 and f_{01} for class $\mathscr{F}_1$, and by ε_2 and f_{02} for class $\mathscr{F}_2$. Then, applying calculus of variation [see Tsaknakis et al (1984)], we find that the pair (f_1^*, f_2^*) in (12.2.40) is

$$f_1^*(\lambda) = \max[(1 - \varepsilon_1)f_{01}(\lambda), c_1(1 - \varepsilon_2)f_{02}(\lambda)] \quad \lambda \in [-\pi, \pi]$$
$$c_1: (2\pi)^{-1} \int_{-\pi}^{\pi} \max[(1 - \varepsilon_1)f_{01}(\lambda), c_1(1 - \varepsilon_2)f_{02}(\lambda)] \, d\lambda = 1 \quad (12.2.42)$$

$$f_2^*(\lambda) = \max[(1 - \varepsilon_2)f_{02}(\lambda), c_2(1 - \varepsilon_1)f_{01}(\lambda)] \quad \lambda \in [-\pi, \pi]$$
$$c_2: (2\pi)^{-1} \int_{-\pi}^{\pi} \max[(1 - \varepsilon_2)f_{02}(\lambda), c_2(1 - \varepsilon_1)f_{01}(\lambda)] \, d\lambda = 1 \quad (12.2.43)$$

The analysis that led to (12.2.42) and (12.2.43) is exactly as in Example 11.4.2, and so are the results. There is thus also a breakdown curve, $\mathscr{C}(\varepsilon_1, \varepsilon_2)$, defined exactly as in (11.4.64) in the example, whose expression here is

$$C(\varepsilon_1, \varepsilon_2) = (2\pi)^{-1} \int_{-\pi}^{\pi} \max[(1 - \varepsilon_1)f_{01}(\lambda), (1 - \varepsilon_2)f_{02}(\lambda)] \, d\lambda = 1 \quad (12.2.44)$$

The breakdown curve in (12.2.44) is exactly as in Figure 11.4.3 in Example 11.4.2, when ε, δ, $s_0(\lambda)$, and $s_{10}(\lambda)$ respectively are replaced by ε_1, ε_2, $f_{01}(\lambda)$, and $f_{02}(\lambda)$. We complete this section by emphasizing that the solutions in (12.2.38), (12.2.39), (12.2.42), and (12.2.43) provide protection against spectral contaminations within the class of Gaussian processes, and they are parametric. They do not, however, provide protection against data outliers, and they are grossly nonrobust, since they induce linear and unbounded operations.

12.3 ROBUST PREDICTION AND INTERPOLATION

As we saw in Section 12.2, when the data generating process is well known, and the mean-squared performance criterion is used, then the optimal mean-squared predictor and interpolator are the conditional expectations in (12.2.10) and (12.2.11) respectively. If the data generating process is Gaussian, then these conditional expectations are identical to the optimal linear mean-squared operations in (12.2.13) and (12.2.14) respectively. Unfortunately, those linear operations are grossly nonrobust. Indeed, they are unbounded, and they do not satisfy the asymptotic continuity condition in part ii. of Theorem 12.1.1. In fact, the linear operations in (12.2.13) and (12.2.14) are identical for all data generating processes with identical power spectral densities; they are thus totally insensitive to non–second order statistical characteristics. In this section, we will focus on the discussion and analysis of qualitatively robust prediction and interpolation operations.

As we discussed in Section 11.3, the selection of the $d_1(\cdot, \cdot)$ distance in Theorem 12.1.1 as the Prohorov distance with data distortion measures as in (12.1.1) basically maps the outlier model in (11.3.6). To qualify our presentation, we will thus consider this model. In particular, let us consider data sequences ... $x_{-1}, x_0, x_1 \ldots$ generated by a well known nominal process whose arbitrary dimensionality density function is denoted f_0. Let us also consider an outlier process whose density function is unknown and whose data sequences are denoted ..., $w_{-1}, w_0, w_1 \ldots$. Let an outlier occur with probability ε per datum, and let the observation sequence be denoted ... $y_{-1}, y_0, y_1 \ldots$. Then if Y_n, X_n, and W_n denote the nth random elements of the sequences y^n, x^n, and w^n respectively, we have

$$Y_n = (1 - \varepsilon)X_n + \varepsilon W_n \qquad (12.3.1)$$

Let us initially assume that the probability ε is known, and that the nominal well known density function f_0 is Markovian. In particular, let us assume that there exists some finite integer m, such that

$$f_0(x_0 | x_{-n}^{-1}, x_1^n) = f_0(x_0 | x_{-m}^{-1}, x_1^m) \qquad \forall \, n > m$$
$$f_0(x_0 | x_{-n}^{-1}) = f_0(x_{-m}^{-1}) \qquad \forall \, n > m \qquad (12.3.2)$$

Let us consider an outlier model different than the one in (12.3.1). In particular, we will consider the class $\mathscr{F}$ of density functions f defined as follows, where h is any density.

$$f \in \mathscr{F} \Rightarrow f(x_1^n) = (1-\varepsilon)f_0(x_1^n) + \varepsilon h(x_1^n) \qquad n \leq 2m+1 \qquad (12.3.3)$$

Given the above assumptions, let us now consider saddle-point games for prediction and interpolation. If given y_{-n}^{-1} and $y_{-n}^{-1} \cup y_1^n$, $g_{np}(y_{-n}^{-1})$ and $g_{nI}(y_{-n}^{-1}, y_1^n)$ denote respectively a prediction and an interpolation operation for the random datum Y_0, then, if they exist, the solutions $g_{np}^*(y_{-n}^{-1})$ and $g_{nI}^*(y_{-n}^{-1}, y_1^n)$ of the corresponding games are defined as follows.

$$\forall f \in \mathscr{F} \qquad E\{[Y_0 - g_{np}^*(Y_{-n}^{-1})]^2 | f\} \leq E\{[Y_0 - g_{np}^*(Y_{-n}^{-1})]^2 | f^*\}$$
$$\leq E\{[Y_0 - g_{np}(Y_{-n}^{-1})]^2 | f^*\} \qquad \forall g_{np}$$
$$(12.3.4)$$

$$\forall f \in \mathscr{F} \qquad E\{[Y_0 - g_{nI}^*(Y_{-n}^{-1}, Y_1^n)]^2 | f\} \leq E\{[Y_0 - g_{nI}^*(Y_{-n}^{-1}, Y_1^n)]^2 | f^*\}$$
$$\leq E\{[Y_0 - g_{nI}(Y_{-n}^{-1}, Y_1^n)]^2 | f^*\} \qquad \forall g_{nI}$$
$$(12.3.5)$$

where $\mathscr{F}$ is the class in (12.3.3). Let us define the following conditional expected values, induced by the nominal density f.

$$g_{kp}^0(y_{-k}^{-1}) = E\{Y_0 | y_{-k}^{-1}, f_0\}$$
$$g_{kI}^0(y_{-k}^{-1}, y_1^k) = E\{Y_0 | y_{-k}^{-1}, y_1^k, f_0\} \qquad (12.3.6)$$

Then, applying variational analysis [see Tsaknakis (1986)], we find that the solutions of the games in (12.3.4) and (12.3.5) exist, and they are as follows, where $m_0 = E\{Y_0 | f_0\}$, and where $\operatorname{sgn} x = \begin{cases} 1; & x > 0 \\ 0; & x = 0. \\ -1; & x < 0 \end{cases}$

$$g_{np}^*(y_{-n}^{-1}) = \begin{cases} g_{np}^*(y_{-n}^{-1}) & n \leq m \\ g_{mp}^*(y_{-m}^{-1}) & n > m \end{cases} \qquad (12.3.7)$$

$$g_{kp}^*(y_{-k}^{-1}) = m_0 + \lambda_k^{-1} \operatorname{sgn}(g_{kp}^0(y_{-k}^{-1}) - m_0) \min[\lambda_k | g_{kp}^0(y_{-k}^{-1}) - m_0 |, (1-\varepsilon)] \qquad (12.3.8)$$

where λ_k is the unique positive number that satisfies the following equation, the integrand in which corresponds to the least favorable density f^* in (12.3.4).

$$\int_{R^k} dx_{-k}^{-1} f_0(x_{-k}^{-1}) \max(1-\varepsilon, \lambda_k | g_{kp}^0(x_{-k}^{-1}) - m_0 |) = 1 \qquad (12.3.9)$$

$$g_{nI}^*(y_{-n}^{-1}, y_1^n) = \begin{cases} g_{nI}^*(y_{-n}^{-1}, y_1^n) & n \leq m \\ g_{mI}^*(y_{-m}^{-1}, y_1^m) & n > m \end{cases} \qquad (12.3.10)$$

12.3 Robust Prediction and Interpolation

$$g^*_{kI}(y_{-k}^{-1}, y_1^k) = m_0 + \mu_k^{-1} \, \mathrm{sgn}(g^0_{kI}(y_{-k}^{-1}, y_1^k) - m_0) \min[\mu_k|g^0_{kI}(y_{-k}^{-1}, y_1^k) - m_0|, (1-\varepsilon)]$$
(12.3.11)

where μ_k is the unique positive number, that satisfies the following equation, the integrand in which corresponds to the least favorable density f^* in (12.3.5).

$$\int_{R^{2k}} dx_{-k}^{-1} dx_1^k f_0(x_{-k}^{-1}, x_1^k) \max(1 - \varepsilon, \mu_k|g^0_{kI}(x_{-k}^{-1}, x_1^k) - m_0|) = 1 \quad (12.3.12)$$

We observe that the operations in (12.3.8) and (12.3.11) are bounded. From (12.3.7) and (12.3.10), we also observe that the saddle-point solutions of the games in (12.3.4) and (12.3.5) are always operating on data sequences of finite length. If the conditional expected values in (12.3.6) are also continuous as functions of the data sequences they are conditioned on, then the latter saddle-point solutions are also qualitatively robust, since they then satisfy part i of Theorem 12.1.1. We will now proceed with a simple example.

Example 12.3.1

Consider the outlier model in (12.3.3), and let the nominal process be real, stationary, first order Markov Gaussian, with mean m_0 and autocovariance coefficients $r_k = \alpha^{-k}$. Then we easily obtain

$$g^0_{kp}(y_{-k}^{-1}) = E\{Y_0 | y_{-k}^{-1}, f_0\} = m_0 \frac{r_0 - r_1}{r_0} + \frac{r_1}{r_0} y_{-1}$$

$$= m_0(1 - \alpha^{-1}) + \alpha^{-1} y_{-1} \quad \forall k \geq 1 \quad (12.3.13)$$

$g^0_{kI}(y_{-k}^{-1}, y_1^k)$

$$= E\{Y_0 | y_{-k}^{-1}, y_1^k, f_0\} = m_0 \frac{r_0 - r_1}{r_0 + r_2} + \frac{r_1}{r_0 + r_2}(y_{-1} + y_1)$$

$$= [1 + \alpha^{-2}]^{-1}[m_0(1 - \alpha^{-1}) + \alpha^{-1}(y_{-1} + y_1)] \quad \forall k \geq 1 \quad (12.3.14)$$

$$g^0_{kp}(y_{-k}^{-1}) - m_0 = \alpha^{-1}(y_{-1} - m_0) \quad \forall k \geq 1 \quad (12.3.15)$$

$$g^0_{kI}(y_{-k}^{-1}, y_1^k) - m_0 = \alpha^{-1}[1 + \alpha^{-2}]^{-1}[y_{-1} + y_1 - m_0(1 + \alpha^{-1})] \quad \forall k \geq 1$$
(12.3.16)

Equations (12.3.9) and (12.3.12) take here the form

$$\lambda: E\{\max(1 - \varepsilon, \lambda\alpha^{-1}|Y - m_0|)|f_0\} = 1 \quad (12.3.17)$$

$$\mu: E\{\max(1 - \varepsilon, \mu\alpha^{-1}[1 + \alpha^{-2}]^{-1}|Y_{-1} + Y_1 - m_0(1 + \alpha^{-1})|)|f_0\} = 1 \quad (12.3.18)$$

Conditioned on f_0, the random variable $Y - m_0$ in (12.3.17) is zero mean, unit variance Gaussian; the variable $Y_{-1} + Y_1 - m_0(1 + \alpha^{-1})$ in (12.3.18) is then Gaussian, with mean $m_0(1 - \alpha^{-1})$ and variance $2(1 + \alpha^{-2})$. Equations (12.3.17) and (12.3.18) respectively can thus be written as

$$\lambda: G(\lambda) = (1-\varepsilon)\int_{-[(1-\varepsilon)/\lambda]\alpha}^{[(1-\varepsilon)/\lambda]\alpha} \phi(u)\,du + 2\lambda\alpha^{-1}\int_{[(1-\varepsilon)/\lambda]\alpha}^{\infty} u\phi(u)\,du$$

$$= (1-\varepsilon)\left[2\Phi\left(\frac{1-\varepsilon}{\lambda}\alpha\right) - 1\right] + 2\lambda\alpha^{-1}\phi\left(\frac{1-\varepsilon}{\lambda}\alpha\right) = 1 \quad (12.3.19)$$

$$\mu: H(\mu) = (1-\varepsilon)\int_{-[(1-\varepsilon)/\mu]\alpha(1+\alpha^{-2})}^{[(1-\varepsilon)/\mu]\alpha(1+\alpha^{-2})} du(2\pi)^{-1/2}[2(1+\alpha^{-2})]^{-1/2}$$

$$\times \exp\left\{-\frac{[u - m_0(1-\alpha^{-1})]^2}{4(1+\alpha^{-2})}\right\} + 2\mu\alpha^{-1}[1+\alpha^{-2}]^{-1}$$

$$\times \int_{[(1-\varepsilon)/\mu]\alpha(1+\alpha^{-2})}^{\infty} du\, u(2\pi)^{-1/2}[2(1+\alpha^{-2})]^{-1/2}$$

$$\times \exp\left\{-\frac{[u - m_0(1-\alpha^{-1})]^2}{4(1+\alpha^{-2})}\right\}$$

$$= (1-\varepsilon)\left[\Phi\left(\frac{\mu^{-1}(1-\varepsilon)\alpha(1+\alpha^{-2}) - m_0(1-\alpha^{-1})}{[2(1+\alpha^{-2})]^{1/2}}\right)\right.$$

$$\left. + \Phi\left(\frac{\mu^{-1}(1-\varepsilon)\alpha(1+\alpha^{-2}) + m_0(1-\alpha^{-1})}{[2(1+\alpha^{-2})]^{1/2}}\right) - 1\right]$$

$$- \mu 2 m_0 \frac{1-\alpha^{-1}}{1+\alpha^{-2}}\left[1 - \Phi\left(\frac{\mu^{-1}(1-\varepsilon)\alpha(1+\alpha^{-2}) - m_0(1-\alpha^{-1})}{[2(1+\alpha^{-2})]^{1/2}}\right)\right]$$

$$+ \mu 2\sqrt{2}\alpha^{-1}[1+\alpha^{-2}]^{-1/2}\phi\left(\frac{\mu^{-1}(1-\varepsilon)\alpha(1+\alpha^{-2}) - m_0(1-\alpha^{-1})}{[2(1+\alpha^{-2})]^{1/2}}\right) = 1$$

$$(12.3.20)$$

For $m_0 = 0$, we have, from (12.3.19), (12.3.20), (12.3.13), and (12.3.14),

$$\lambda: G(\lambda) = (1-\varepsilon)\left[2\Phi\left(\frac{1-\varepsilon}{\lambda}\alpha\right) - 1\right] + 2\lambda\alpha^{-1}\phi\left(\frac{1-\varepsilon}{\lambda}\alpha\right) = 1$$
$$(12.3.21)$$
$$\mu: H(\mu) = (1-\varepsilon)\left[2\Phi\left(\frac{1-\varepsilon}{\mu}\beta\right) - 1\right] + 2\mu\beta^{-1}\phi\left(\frac{1-\varepsilon}{\mu}\beta\right) = 1$$

where $\beta = \alpha 2^{-1/2}(1+\alpha^{-2})^{1/2} < 1$

$$g^*_{kp}(y^{-1}_{-k}) = \lambda^{-1}\,\text{sgn}(y_{-1})\min[\lambda\alpha^{-1}|y_{-1}|, (1-\varepsilon)] \quad \forall k \geq 1 \quad (12.3.22)$$

$$g^*_{kl}(y^{-1}_{-k}, y^k_1)$$
$$= \mu^{-1}\,\text{sgn}(y_{-1} + y_1)\min[\mu\alpha^{-1}[1+\alpha^{-2}]^{-1}|y_{-1} + y_1|, (1-\varepsilon)] \quad \forall k \geq 1$$
$$(12.3.23)$$

In Figure 12.3.1, we plot the function $G(\lambda)$ in (12.3.21), where $H(\mu)$ behaves similarly. We observe that the solution λ^* of the equation $G(\lambda) = 1$ increases monotonically, as α does, where α takes values in $(0, 1)$. Also, this solution de-

12.3 Robust Prediction and Interpolation

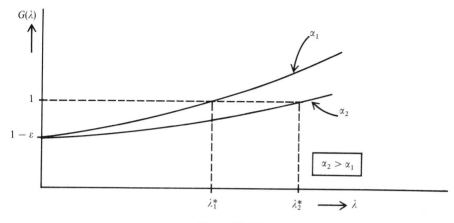

Figure 12.3.1

creases monotonically as ε does, and it equals zero for $\varepsilon = 0$. Let us now turn to the qualitatively robust predictor and interpolator, respectively given by (12.3.22) and (12.3.23). They both behave similarly; thus, we will only discuss the predictor in (12.3.22). In Figure 12.3.2, we plot this predictor $g_p^*(y)$ as a function of its argument y for two different values of ε. As ε decreases, the predictor tends more towards the optimal at the Guassian density mean-squared predictor, which is linear. Since the predictor $g_p^*(y)$ is a.e. differentiable, it has bounded sensitivity.

The predictor in (12.3.22) has been designed for the outlier model in (12.3.3). We wish to quantify its performance for the outlier model in (12.3.1), however.

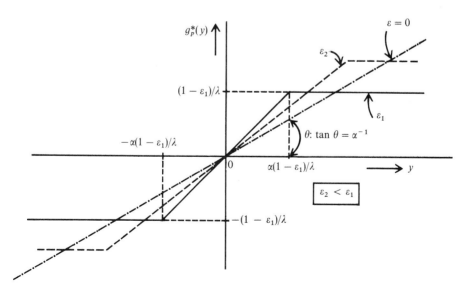

Figure 12.3.2

In particular, we will exhibit its mean-squared performance when it estimates the datum X_0 from the Gaussian nominal process, while the observed datum Y_{-1} has density function f defined as follows for some ζ in (0, 1) and some given y.

$$f_{Y_{-1}}(x) = (1 - \zeta)\phi(x) + \zeta\delta(x - y) \qquad (12.3.24)$$

where $\delta(w)$ is the delta function at zero, and where the model in (12.3.24) represents the occurrence of an extreme outlier. Let us denote by f_0 the nominal Gaussian density, and let us denote by f^* the density induced by the Gaussian density for the datum X_0, and the density in (12.3.24) for the observed datum y_{-1}. For the predictor $g_p^*(y)$ in (12.3.22) and for the optimal at the Gaussian density predictor $\alpha^{-1}y$ we then obtain

$$\begin{aligned} E\{[X_0 - g_p^*(Y_{-1})]^2 | f^*\} &= (1 - \zeta)E\{[X_0 - g_p^*(X_{-1})]^2 | f_0\} \\ &\quad + \zeta E\{[X_0 - g_p^*(y)]^2 | \phi\} \\ &= (1 - \zeta)\bigg\{1 + \alpha^{-2} + 2\lambda^{-2}(1 - \varepsilon)^2 \\ &\quad - 2\frac{1 - \varepsilon}{\lambda}\alpha^{-1}\phi\bigg(\frac{1 - \varepsilon}{\lambda}\alpha\bigg) \\ &\quad - 2[\alpha^{-2} + \lambda^{-2}(1 - \varepsilon)^2]\Phi\bigg(\frac{1 - \varepsilon}{\lambda}\alpha\bigg)\bigg\} \\ &\quad + \zeta\{1 + [g_p^*(y)]^2\} \qquad (12.3.25) \end{aligned}$$

$$\begin{aligned} E\{[X_0 - \alpha^{-1}Y_{-1}]^2 | f^*\} &= (1 - \zeta)E\{[X_0 - \alpha^{-1}X_{-1}]^2 | f_0\} \\ &\quad + \zeta E\{[X_0 - \alpha^{-1}y]^2 | \phi\} \\ &= 1 - \alpha^{-2} + \zeta\alpha^{-2}[1 + y^2] \qquad (12.3.26) \end{aligned}$$

Comparing (12.3.25) and (12.3.26), we conclude that the robust predictor always induces bounded mean-squared error, while the optimal at the Gaussian predictor does not. In fact, for outlier value y such that $|y| \to \infty$, the latter predictor induces infinite error for any nonzero value ζ; its breakdown point is thus zero. The mean-squared error induced by the robust predictor, on the other hand, remains then a bounded function of the statistic α^{-1} of the nominal process for every ζ less than one. For ζ values above $\zeta^* = [2\Phi(\lambda^{-1}(1 - \varepsilon)\alpha) - 1]$ $\{\lambda^{-2} + \alpha^{-2}[2\Phi(\lambda^{-1}(1 - \varepsilon)\alpha) - 1]\}^{-1}$ the latter predictor induces mean-squared error larger than $E\{X_0^2/f^*\} = 1$; its breakdown point is thus ζ^*. Since the robust predictor is bounded and a.e. differentiable, it also induces bounded sensitivity.

We emphasize that although the operations in (12.3.8) and (12.3.11) were found for the outlier model in (12.3.3), they also provide good protection against outliers as in the model (12.3.1), subject to the condition that the nominal data process is Markovian, as in (12.3.2). If the latter process is also Gaussian, the

12.3 Robust Prediction and Interpolation

operations in (12.3.8) and (12.3.11) are qualitatively robust, since they are then continuous, as functions of the data, and bounded. If the existence of outliers is suspected, but the probability of their occurrence is unknown, then a small ε should be selected ad hoc, and the resulting robust prediction and interpolation operations should be adopted.

Let us now consider the case where the nominal process in the model (12.3.1) is not Markovian, and let us initially consider the alternative outlier model in (12.3.3), for any data dimensionality n. Then, given n, saddle-point game formalizations as in (12.3.4) and (12.3.5) exist, whose respective solutions, for $g_{np}^0(y_{-n}^{-1})$ and $g_{nI}^0(y_{-n}^{-1}, y_1^n)$ as in (12.3.6), are

$$g_{np}^*(y_{-n}^{-1}) = m_0 + \lambda_n^{-1}\,\mathrm{sgn}(g_{np}^0(y_{-n}^{-1}) - m_0)\min(\lambda_n|g_{np}^0(y_{-n}^{-1}) - m_0|, 1-\varepsilon) \quad (12.3.27)$$

$$\lambda_n\colon \int_{R^n} dx_{-n}^{-1}\, f_0(x_{-n}^{-1})\max(1-\varepsilon,\, \lambda_n|g_{np}^0(x_{-n}^{-1}) - m_0|) = 1 \quad (12.3.28)$$

$$g_{nI}^*(y_{-n}^{-1}, y_1^n)$$
$$= m_0 + \mu_n^{-1}\,\mathrm{sgn}(g_{nI}^0(y_{-n}^{-1}, y_1^n) - m_0)\min(\mu_n|g_{nI}^0(y_{-n}^{-1}, y_1^n) - m_0|, 1-\varepsilon) \quad (12.3.29)$$

$$\mu_n\colon \int_{R^{2n}} dx_{-n}^{-1}\,dx_1^n\, f_0(x_{-n}^{-1}, x_1^n)\max(1-\varepsilon,\, \mu_n|g_{nI}^0(x_{-n}^{-1}, x_1^n) - m_0|) = 1 \quad (12.3.30)$$

The operations in (12.3.27) and (12.3.29) are clearly bounded for every n, and their bounds are monotonically nonincreasing with increasing n. Indeed, it can be shown [see Tsaknakis (1986)] that the positive constants λ_n and μ_n in (12.3.28) and (12.3.30) are monotonically nondecreasing with increasing n. If the nominal conditional expectations in (12.3.6) are also pointwise continuous as functions of the data they are conditioned on (as with the Gaussian nominal process), then the operations in (12.3.27) and (12.3.29) are also pointwise continuous, as functions of y_{-n}^{-1} and $y_{-n}^{-1} \cup y_1^n$ respectively. That does not, however, guarantee asymptotic continuity of those operations, as in part ii of Theorem 12.1.1. We must thus consider alternative asymptotic prediction and interpolation operations that satisfy the latter asymptotic continuity. Let us select a finite positive integer m, and let us consider observation sequences $y_1^n = \{y_1, \ldots, y_n\}$. Let us denote by $\hat{y}_{kp}$ the prediction estimate of the datum y_k from the past data y_1^{k-1}. Let us denote by $\hat{y}_{kI}$, the interpolation estimate from the data y_1^{k-1} and y_{k+1}^{2k-1}. Consider then, the following estimates $\hat{y}_{kp}$ and $\hat{y}_{kI}$, where the operations $g_{np}^*(\cdot)$ and $g_{nI}^*(\cdot)$ are respectively as in (12.3.27) and (12.3.29).

$$\hat{y}_{kp} = \begin{cases} g_{k-1,\,p}^*(y_1^{k-1}) & k \le m+1 \\ g_{k-1,\,p}^0(m_0,\, \hat{y}_{2,p}^{k-2},\, \hat{y}_{k-1,\,p}) \\ \quad + g_p[g_{k-1,\,p}^0(0,\ldots,0,\, y_{k-m}^{k-1} - \hat{y}_{k-m,\,p}^{k-1})] & k > m+1 \end{cases} \quad (12.3.31)$$

$$\hat{y}_{kI} = \begin{cases} g_{k-1,\,I}^*(y_1^{k-1},\, y_{k+1}^{2k-1}) & k \le m+1 \\ g_{k-1,\,I}^0(m_0,\, \hat{y}_{2,I}^{k-2},\, \hat{y}_{k-1,\,I}) \\ \quad + g_I[g_{k-1,\,I}^0(0,\ldots,0,\, y_{k-m}^{k-1} - \hat{y}_{k-m,\,I}^{k-1},\, y_{k+1}^{k+m} \\ \quad - \hat{y}_{k+1,\,I}^{k+m},\, 0,\ldots,0)] & k > m+1 \end{cases} \quad (12.3.32)$$

where

$$\hat{y}_{i,p}^j = [\hat{y}_{ip}, \ldots, \hat{y}_{jp}]^T;$$

$$j \geq i, \quad g(x) = m_0 + \lambda_m^{-1} \operatorname{sgn}(x - m_0) \min(\lambda_m |x - m_0|, 1 - \varepsilon)$$

$$\hat{y}_{i,I}^j = [\hat{y}_{iI}, \ldots, \hat{y}_{jI}]$$

$$j \geq i, \quad g(x) = m_0 + \mu_m^{-1} \operatorname{sgn}(x - m_0) \min(\mu_m |x - m_0|, 1 - \varepsilon)$$

The operations in (12.3.31) and (12.3.32) respectively utilize the closest m and $2m$ raw observation data, while in place of the remaining observation data, they utilize the corresponding estimates. Since the latter estimates are bounded, if they are pointwise continuous (as with the Gaussian nominal process) as functions of their data, then they are also asymptotically robust, in the sense of Definition 12.1.1, with respectively m and $2m$ in (12.1.1). In addition, the above operations provide good protection against outliers as in model (12.3.1), with, simultaneously, relatively minor loss in performance at the nominal process. The integer m in (12.3.31) and (12.3.32), and the constant ε in (12.3.27) and (12.3.29), can be both selected, to guarantee relatively minor performance loss at the nominal process, where ε represents a tradeoff. The larger the ε, the better the protection from high-frequency outliers, but the larger the performance loss at the nominal process. The full analysis and evaluation of the robust operations in (12.3.31) and (12.3.32) is provided by Tsaknakis (1986).

12.4 ROBUST FILTERING

In Section 12.2, we discussed parametric causal and noncausal filtering. In this section, we will study robust filtering, considering possible uncertainty in the information process and the presence of outliers in the noise process. In particular, let $\mathscr{F}_s$ and $\mathscr{F}_N$ be two disjoint classes of information and noise processes, respectively, whose arbitrary dimensionality density functions are respectively denoted f_s and f_N. Let f_{0s} and f_{0N} be two nominal well known, stationary density functions, such that $f_{0s} \in \mathscr{F}_s$ and $f_{0N} \in \mathscr{F}_N$. Let us assume that some density function f_s from class $\mathscr{F}_s$ is a priori selected by the system designer to represent the information process throughout the overall observation interval, and let us denote by $\ldots, X_{-1}, X_0, X_1, \ldots$ a random data sequence generated by f_s. We will initially assume that the class $\mathscr{F}_s$ consists of f_{0s} only. Then, we will consider a more general class $\mathscr{F}_s$, which we will define later.

Let us denote by $\ldots, W_{-1}, W_0, W_1, \ldots$ random noise data sequences, and let $\ldots, Z_{-1}, Z_0, Z_1, \ldots$ be data sequences from the nominal noise density function f_{0N}. Given some number ε_N in $(0, 1)$, let the class $\mathscr{F}_N$ of noise processes then be such that

$$W_n = (1 - \varepsilon_N)Z_n + \varepsilon_N V_n \qquad (12.4.1)$$

where $\ldots, V_{-1}, V_0, V_1, \ldots$ is a random sequence generated by any arbitrary dimensionality stationary density function.

The noise model in (12.4.1) represents the occurrence of outliers, with probability ε_N per datum. Given f_s in $\mathscr{F}_s$ and f_N in $\mathscr{F}_N$, we assume that the data

12.4 Robust Filtering

sequences from f_s and f_N are additive and that f_s and f_N are mutually independent. The, if ..., $Y_{-1}, Y_0, Y_1, \ldots$ denote random observation sequences, we have,

$$Y_n = X_n + W_n \quad \forall\, n \qquad (12.4.2)$$

where X_n is generated by f_s, W_n is generated by f_N [as in (12.4.1)], and the sequences $\ldots, X_{-1}, X_0, X_1, \ldots$ and $\ldots, W_{-1}, W_0, W_1, \ldots$ are mutually independent.

Let $g_{n+l,F}(y_{-n}^{l-1})$ denote a filtering operation, estimating the information datum X_0, via the observation sequence y_{-n}^{l-1}. Let $e_F(g_{n+l,F}, f_s, f_N)$ denote the mean-squared error induced by the operation $g_{n+l,F}(y_{-n}^{l-1})$ at the density functions $f_s \in \mathscr{F}_s$ and $f_N \in \mathscr{F}_N$. That is,

$$e_F(g_{n+l,F}, f_s, f_N) = E\{[X_0 - g_{n+l,F}(Y_{-n}^{l-1})]^2 \,|\, f_s, f_N\} \qquad (12.4.3)$$

Consider then the following saddle-point game. Search for the triple $(g_{n+l,F}^*, f_s^*, f_N^*)$ such that $f_s^* \in \mathscr{F}_s$, $f_N^* \in \mathscr{F}_N$, and

$$\forall\, f_s \in \mathscr{F}_s \quad \forall\, f_N \in \mathscr{F}_N \quad e_F(g_{n+l,F}^*, f_s, f_N)$$
$$\leq e_F(g_{n+l,F}^*, f_s^*, f_N^*) \leq e_F(g_{n+l,F}, f_s^*, f_N^*) \quad \forall\, g_{n+l,F} \qquad (12.4.4)$$

The right part of (12.4.4) is satisfied for $g_{n+l,F}^*(y_{-n}^{l-1})$ being the conditional expectation of X_0 at f_s^* and f_N^*. That is

$$g_{n+l,F}^*(y_{-n}^{l-1}) = E\{X_0 \,|\, y_{-n}^{l-1}, f_s^*, f_N^*\} \qquad (12.4.5)$$

The game in (12.4.4) reduces then to the following search. Find the pair (f_s^*, f_N^*), such that $f_s^* \in \mathscr{F}_s$, $f_N^* \in \mathscr{F}_N$, and

$$E\{[X_0 - E\{X_0 | y_{-n}^{l-1}, f_s^*, f_N^*\}]^2 \,|\, f_s^*, f_N^*\}$$
$$= \sup_{\substack{f_s \in \mathscr{F}_s \\ f_N \in \mathscr{F}_N}} E\{[X_0 - E\{X_0 | y_{-n}^{l-1}, f_s, f_N\}]^2 \,|\, f_s, f_N\} \qquad (12.4.6)$$

and select $g_{n+l,F}^*(y_{-n}^{l-1})$ as in (12.4.5).

Given f_s in $\mathscr{F}_s$ and f_N in $\mathscr{F}_N$, and due to their additivity and mutual independence, the induced observation density f equals the convolution, $f_s * f_N$, between the densities f_s and f_N. If μ_s and σ_s^2 denote respectively the mean and variance of the density f_s and defining then

$$\alpha(y_{-n}^{l-1}) = \int_{R^{n-l}} dx_{-n}^{l-1}\, x_0 f_s(x_{-n}^{l-1}) f_N(y_{-n}^{l-1} - x_{-n}^{l-1}) \qquad (12.4.7)$$

we easily find, for $f = f_s * f_N$,

$$E\{[X_0 - E\{X_0 | y_{-n}^{l-1}, f_s, f_N\}]^2 \,|\, f_s, f_N\} = \sigma_s^2 - \int_{R^{n+l}} dy_{-n}^{l-1}\, \frac{[\alpha(y_{-n}^{l-1}) - \mu_s f(y_{-n}^{l-1})]^2}{f(y_{-n}^{l-1})}$$

$$(12.4.8)$$

The expression in (12.4.8) is translation invariant, so without lack in generality we can assume that the mean μ_s is zero. Let now, $\{D_i; -n \le i \le l-1\}$, be a set of purely imaginary numbers, and let $\Phi_s(D_{-n}, \ldots, D_{l-1})$, $\Phi_N(D_{-n}, \ldots, D_{l-1})$, $\Phi(D_{-n}, \ldots, D_{l-1})$, and $A(D_{-n}, \ldots, D_{l-1})$ denote the characteristic functions (or Fourier transforms) at $\{D_i; -n \le i \le l-1\}$ of respectively the densities $f_s(x_{-n}^{l-1})$, $f_N(x_{-n}^{l-1})$, $f(x_{-n}^{l-1})$, and the function $\alpha(y_{-n}^{l-1})$ in (12.4.7), assuming that the former exist. Then, we have,

$$\Phi(D_{-n}, \ldots, D_{l-1}) = \Phi_s(D_{-n}, \ldots, D_{l-1}) \cdot \Phi_N(D_{-n}, \ldots, D_{l-1}) \qquad (12.4.9)$$

$$A(D_{-n}, \ldots, D_{l-1}) = \left[\frac{\partial}{\partial D_0} \Phi_s(D_{-n}, \ldots, D_{l-1}) \right] \Phi_N(D_{-n}, \ldots, D_{l-1})$$

$$= P(D_{-n}, \ldots, D_{l-1}) \cdot \Phi(D_{-n}, \ldots, D_{l-1}) \qquad (12.4.10)$$

where

$$P(D_{-n}, \ldots, D_{l-1}) = \frac{\frac{\partial}{\partial D_0} \Phi_s(D_{-n}, \ldots, D_{l-1})}{\Phi_s(D_{-n}, \ldots, D_{l-1})} \qquad (12.4.11)$$

If the characteristic function $\Phi_s(D_{-n}, \ldots, D_{l-1})$ is an analytic function, then the expression in (12.4.11) defines an integrodifferential operator. Via the latter operator, and through the use of inverse Fourier transforms, the function in (12.4.7) can be expressed, via the application of the operator $P(D_{-n}, \ldots, D_{l-1})$ on the observation density f, as

$$\alpha(y_{-n}^{l-1}) = P(D_{-n}, \ldots, D_{l-1})[f(y_{-n}^{l-1})] \qquad (12.4.12)$$

Assuming $\mu_s = 0$, the expectation in (12.4.8) can be then expressed as

$$E\{[X_0 - E\{X_0 | y_{-n}^{l-1}, f_s, f_N\}]^2 | f_s, f_N\}$$

$$= \sigma_s^2 - \int_{R^{n+l}} dy_{-n}^{l-1} \frac{\{P(D_{-n}, \ldots, D_{l-1})[f(y_{-n}^{l-1})]\}^2}{f(y_{-n}^{l-1})} \qquad (12.4.13)$$

We note that if the characteristic function $\Phi_s(D_{-n}, \ldots, D_{l-1})$ is nonzero and analytic, and admits a Taylor series expansion everywhere, then the operator $P(D_{-n}, \ldots, D_{l-1})$ can be written as a uniformly converging Taylor series of operators. That is, if D denotes the vector $[D_{-n}, \ldots, D_{l-1}]^T$, and if ∇ denotes the $(n+l)$-dimensional gradient, then

$$P(D) = \nabla P(0)D + 2^{-1} D^T \nabla^2 P(0) D + \cdots \qquad (12.4.14)$$

where $\nabla P(0)$ denotes the gradient of $P(D)$ at $D = 0$, where $\nabla^2 P(0)$ is the Hessian of $P(D)$ at $D = 0$, and where the constant term in (12.4.14) is zero since the zero mean assumption $\mu_s = 0$ implies $(\partial/\partial D_0)\Phi_s(D)|_{D=0} = 0$.

Let us now assume that the class $\mathscr{F}_s$ in (12.4.6) consists strictly of a single well known density f_{0s}, whose variance is σ_s^2, and whose characteristic function $\Phi_s(D_{-n}, \ldots, D_{l-1})$ is an analytic function. Then, due to (12.4.13), we observe that

12.4 Robust Filtering

the supremum in (12.4.6) reduces to the search for the infimum below, where $\mathscr{F}$ denotes the class induced by f_{0s} and f_N; that is, $\mathscr{F} = \{f = f_{0s} * f_N, f_N \in \mathscr{F}_N\}$.

$$\inf_{f \in \mathscr{F}} \int_{R^{n+l}} dy_{-n}^{l-1} \frac{\{P(D_{-n}, \ldots, D_{l-1})[f(y_{-n}^{l-1})]\}^2}{f(y_{-n}^{l-1})} \quad (12.4.15)$$

For the class $\mathscr{F}_N$ of noise processes as in (12.4.1), the infimum in (12.4.15) is very complex. We thus select initially an alternative class $\mathscr{F}'_N$ as we did in Section 12.3. In particular, let $\mathscr{F}'_N$ be such that

$$\mathscr{F}'_N = \{f_N : f_N = (1 - \varepsilon_N) f_{0N} + \varepsilon_N h,$$
$$h \text{ is any arbitrary dimensionality density function}\} \quad (12.4.16)$$

The class $\mathscr{F}$ of observation densities induced by $\mathscr{F}'_N$ and the density f_{0s} is clearly as follows, where $f_{0s} * f_N$ denotes convolution.

$$\mathscr{F} = \{f : f = (1 - \varepsilon_N) f_{0s} * f_{0N} + \varepsilon_N f_{0s} * h,$$
$$h \text{ is any arbitrary dimensionality density function}\} \quad (12.4.17)$$

The class in (12.4.17) is not compact under an appropriate metric on density functions. We thus consider instead the larger class $\mathscr{F}'$ and find the infimum in (12.4.15) within $\mathscr{F}'$. We define,

$$\mathscr{F}' = \{f : f = (1 - \varepsilon_N) f_{0s} * f_{0N} + \varepsilon_N h,$$
$$h \text{ is any arbitrary dimensionality density function}\} \quad (12.4.18)$$

We then express Theorem 12.4.1 below. This theorem and the subsequent Lemma 12.4.1 are due to Tsaknakis (1986). We include its proof in the appendix.

THEOREM 12.4.1: Let the density f_{0s} have a nonzero and analytic characteristic function $\Phi_s(D_{-n}, \ldots, D_{l-1}) = \Phi_s(\underline{D})$, that also admits a Taylor series expansion everywhere. Consider then the operator $P(\underline{D}) = P(D_{-n}, \ldots, D_{l-1})$ in (12.4.11), which also admits then a Taylor series expansion as in (12.4.14). Consider the class $\mathscr{F}'$ in (12.4.18), and denote

$$f_0 = f_{0s} * f_{0N} \quad (12.4.19)$$

Let $d(y_{-n}^{l-1})$ be a positive solution of the equation

$$|P(\underline{D}) d(y_{-n}^{l-1})| = \lambda d(y_{-n}^{l-1}) \quad \lambda > 0 \quad (12.4.20)$$

such that $d(y_{-n}^{l-1})$ is integrable over R^{n+l}, it is analytic for all nonzero vectors y_{-n}^{l-1}, and the quantity $P(\underline{D})[d^*(y_{-n}^{l-1})]$ exists for all y_{-n}^{l-1} in R^{n+l}, where

$$d^*(y_{-n}^{l-1}) = \begin{cases} (1 - \varepsilon_N) f_0(y_{-n}^{l-1}) & \text{for } y_{-n}^{l-1} \in A^{n+l} \\ \lambda d(y_{-n}^{l-1}) & \text{otherwise} \end{cases} \quad (12.4.21)$$

where, A^{n+l} includes all y_{-n}^{l-1}, such that $|P(\underline{D})[f_0(y_{-n}^{l-1})]/f_0(y_{-n}^{l-1})| \leq \lambda$.

Then, the infimum in (12.4.15), with substitution of $\mathscr{F}'$ for $\mathscr{F}$, exists and is attained by the following density f^* in $\mathscr{F}'$.

$$f^*(y_{-n}^{l-1}) = d^*(y_{-n}^{l-1}) \tag{12.4.22a}$$

with λ such that

$$\int_{R^{n+l}} f^*(y_{-n}^{l-1}) dy_{-n}^{l-1} = 1.$$

Furthermore, the filtering operation $g_{n+l, F}^*(y_{-n}^{l-1}) = E\{X_0 | y_{-n}^{l-1}, f^*\}$ that satisfies then the game in (12.4.4) on $\mathscr{F}'$ is

$$g_{n+l, F}^*(y_{-n}^{l-1}) = \begin{cases} \dfrac{P(\underline{D})[f_0(y_{-n}^{l-1})]}{f_0(y_{-n}^{l-1})} & \text{for } y_{-n}^{l-1} \in A^{n+l} \\ \pm \lambda & \text{for } y_{-n}^{l-1} \in [R^{n+l} - A^{n+l}] \end{cases} \tag{12.4.22b}$$ ∎

We note that the density f^* in (12.4.22a) belongs to the class $\mathscr{F}'$ in (12.4.18), but not to the class $\mathscr{F}$ in (12.4.17). It can be shown, however, that the density f^* is close to the class $\mathscr{F}$ in a precise sense [Tsaknakis (1986)]. We do not include this result here, since Theorem 12.4.1 represents only an intermediate step in the design of robust filtering operations. What are important here are the properties of the operation in (12.4.22b). For densities f_{0s} that satisfy the conditions of Theorem 12.4.1, but are arbitrary otherwise, the latter properties are not evident. An interesting and tractable special case arises, however, when f_{0s} is a Gaussian density function. Let the latter density have zero mean and covariance coefficients, $m_{ij} = E\{X_i X_j | f_{0s}\}$. Let M_{n+l} denote the autocovariance matrix, $E\{X_{-n}^{l-1}(X_{-n}^{l-1})^T | f_{0s}\}$. Then, the operator in (12.4.11) takes the form

$$P(\underline{D}) = P(D_{-n}, \ldots, D_{l-1}) = \sum_{j=-n}^{l-1} m_{0j} D_j \tag{12.4.23}$$

The quantity in the infimum of (12.4.15) is then a weighted Fisher information measure, and the equation in (12.4.20) becomes

$$\sum_{j=-n}^{l-1} m_{0j} \frac{\partial}{\partial y_j} d(y_{-n}^{l-1}) = \lambda d(y_{-n}^{l-1}) \qquad \lambda > 0 \tag{12.4.24}$$

Let us denote by a_{n+l}^T the row vector $[m_{0, l-1}, \ldots, m_{0, -n}]$ and now let the density function f_{0N} in Theorem 12.4.1 be zero mean Gaussian with autocovariance matrix N_{n+l}. Then, due to the above, and via relatively straightforward manipulations, Theorem 12.4.1 is simplified to the form given by Lemma 12.4.1 below.

LEMMA 12.4.1: Let the densities f_{0s} and f_{0N} in Theorem 12.4.1 be both zero mean Gaussian, with respective autocovariance matrices M_{n+l} and N_{n+l}, where the elements of M_{n+l} are denoted $\{m_{ij}\}$. Then, the density f_0 in (12.4.19) is zero mean Gaussian, with autocovariance matrix $\Lambda_{n+l} = M_{n+l} + N_{n+l}$, and the density f^* in (12.4.22a) and the filtering operator g^* in (12.4.22b) take then the following

12.4 Robust Filtering

special form, where $|\Lambda_{n+l}|$ means determinant, T means transpose and (-1) denotes inverse, where it is assumed that Λ_{n+l} is nonsingular, and where $a_{n+l}^T = [m_{0,\,l-1},\ldots, m_{0,\,-n}]$, $\operatorname{sgn} x = \{1;\, x \geq 0 \text{ and } -1;\, x < 0\}$.

$$f^*(y_{-n}^{l-1}) = \begin{cases} (1-\varepsilon_N)(2\pi)^{-(n-l)/2}|\Lambda_{n+l}|^{-1/2}\exp\{-2^{-1}(y_{-n}^{l-1})^T\Lambda_{n+l}^{-1}y_{-n}^{l-1}\} \\ \qquad\qquad\qquad\qquad \text{for } y_{-n}^{l-1}:|a_{n+l}^T\Lambda_{n+l}^{-1}y_{-n}^{l-1}| \leq \lambda \\[4pt] (1-\varepsilon_N)(2\pi)^{-(n+l)/2}|\Lambda_{n+l}|^{-1/2}\exp\Big\{-2^{-1}(y_{-n}^{l-1})^T\Lambda_{n+l}^{-1}y_{-n}^{l-1} \\[2pt] \qquad + \dfrac{[\lambda-|a_{n+l}^T\Lambda_{n+l}^{-1}y_{-n}^{l-1}|]^2}{2a_{n+l}^T\Lambda_{n+l}^{-1}a_{n+l}}\Big\} \quad y_{-n}^{l-1}:|a_{n+l}^T\Lambda_{n+l}^{-1}y_{-n}^{l-1}| > \lambda \end{cases}$$

(12.4.25)

$$g_{n+l,\,F}^*(y_{-n}^{l-1}) = \begin{cases} a_{n+l}^T\Lambda_{n+l}^{-1}y_{-n}^{l-1} & \text{for } y_{-n}^{l-1}:|a_{n+l}^T\Lambda_{n+l}^{-1}y_{-n}^{l-1}| \leq \lambda \\ \lambda\,\operatorname{sgn}(a_{n+l}^T\Lambda_{n+l}^{-1}y_{-n}^{l-1}) & \text{for } y_{-n}^{l-1}:|a_{n+l}^T\Lambda_{n+l}^{-1}y_{-n}^{l-1}| > \lambda \end{cases} \qquad (12.4.26)$$

where denoting $c = \lambda[a_{n+l}^T\Lambda_{n+l}^{-1}a_{n+l}]^{-1/2}$, and for $\phi(x)$ and $\Phi(x)$ denoting respectively the density at x and the cumulative distribution at x of the zero mean, unit variance Gaussian random variable, the constant λ is such that,

$$\Phi(c) + c^{-1}\phi(c) = 2^{-1}[1 + (1-\varepsilon_N)^{-1}] \qquad (12.4.27)$$

Given ε_N, n, and l, the constant λ is positive and unique. Given n and l, λ decreases monotonically with increasing ε_N. For $\varepsilon_N = 0$, λ equals infinity, and the filtering operation in (12.4.26) becomes then identical to the optimal at the Gaussian noise, linear, mean-squared filter.

Denoting, $I(f) = \int_{R^{n+l}} dy_{-n}^{l-1}\, f^{-1}(y_{-n}^{l-1})\{P(\underline{D})[f(y_{-n}^{l-1})]\}^2$, for the operator, $P(\underline{D})$, in (12.4.23), we also find $I(f^*)$ for the density f^* in (12.4.25), where c is as in (12.4.27).

$$I(f^*) = 2(1-\varepsilon_N)a_{n+l}^T\Lambda_{n+l}^{-1}a_{n+l}[\Phi(c) - 2^{-1}] \qquad (12.4.28)$$

∎

We observe that the filtering operation in (12.4.26) is a truncated linear function of the data; it is thus bounded and continuous in the sense of part i in Theorem 12.1.1, but it is not asymptotically continuous in the sense of part ii in the same theorem. The latter operation is therefore qualitatively robust for finite data dimensionalities $n + l$ only. We will extend the operation in (12.4.26), in like manner as extensions (12.3.31) and (12.3.32) in Section 12.3, to create a filtering operation that is both asymptotically and nonasymptotically robust. We distinguish between causal and noncausal filtering, and we present then two different extensions.

Noncausal Filtering or Smoothing

Consider the Gaussian densities f_{0s} and f_{0N} in Lemma 12.4.1. We then select some ε_N and some finite nonnegative integer m. Let $\{\ldots, X_{-1}, X_0, X_1, \ldots\}$ and

$\{\ldots, W_{-1}, W_0, W_1, \ldots\}$ denote sequences of random variables that are respectively generated by f_{0S} and f_{0N}. Given some integer k and some nonnegative integer n, let $N_{2n+1,k}$ and $M_{2n+1,k}$ denote respectively the autovariance matrices $E\{W_{k-n}^{k+n}(W_{k-n}^{k+n})^T | f_{0N}\}$ and $E\{X_{k-n}^{k+n}(X_{k-n}^{k+n})^T | f_{0S}\}$. Let $a_{2n+1,k}^T$ denote the $(n+1)$th row of the matrix $M_{2n+1,k}$, let $\Lambda_{2n+1,k} = M_{2n+1,k} + N_{2n+1,k}$, and let $g_{kl}^0(x_{k-n}^{k-l}, x_{k+l}^{k+n})$; $n \geq l$, denote the optimal mean-squared interpolation operation at the Gaussian density f_{0S} for the datum x_k, given x_{k-n}^{k-l} and x_{k-l}^{k+n}. Let us then define the sets $\{d_{k,n,l,j}; k-n \leq j \leq k-l, k+l \leq j \leq k+n\}$ and $\{b_{k,n,j}; k-n \leq j \leq k+n\}$ of coefficients as follows, where $\Lambda_{2n+1,k}$ is assumed nonsingular.

$$\{d_{k,n,l,j}\}: g_{kl}^0(x_{k-n}^{k-l}, x_{k+l}^{k+n}) = \sum_{j=k-n}^{k-l} d_{k,n,l,j} x_j + \sum_{j=k+l}^{k+n} d_{k,n,l,j} x_j \quad (12.4.29)$$

$$[b_{k,n,k-n}, \ldots, b_{k,n,k+n}] = a_{2n+1,k}^T \Lambda_{2n+1,k}^{-1}$$

Let us now define

$$g_n^s(x) = \begin{cases} x & \text{if } |x| \leq \lambda_n \\ \lambda_n \text{sgn}(x) & \text{otherwise} \end{cases} \quad (12.4.30)$$

where $c = \lambda_n [a_{2n+1,k}^T \Lambda_{2n+1,k}^{-1} a_{2n+1,k}]^{-1/2}$ is such that

$$\Phi(c) + c^{-1}\phi(c) = 2^{-1}[1 + (1-\varepsilon_N)^{-1}] \quad (12.4.31)$$

Let $\hat{x}_{k,n}^s$ denote the estimate of the signal datum x_k from the observation vector y_{k-n}^{k+n}. Then the estimate $\hat{x}_{k,n}^s$ is designed as

$$\hat{x}_{k,n}^s = \begin{cases} g_n^s(a_{2n+1,k}^T \Lambda_{2n+1,k}^{-1} y_{k-n}^{k+n}) & \text{if } n \leq m \\ g_{kl}^0(\hat{x}_{k-n}^{k-1}, \hat{x}_{k+1}^{k+n}) & n > m \end{cases} \quad (12.4.32)$$

where $\hat{x}_j^i = [\hat{x}_{j,m}^s, \ldots, \hat{x}_{i,m}^s]$; $i > j$ and $g_{kl}^0(\cdot)$ is as in (12.4.29).
Let us define

$$r_n^s(n) = a_{2n+1,k}^T \Lambda_{2n+1,k}^{-1} a_{2n+1,k} \quad (12.4.33)$$

Then, $r_k^s(n)$ represents a variance gain in estimating the signal datum x_k from the observation vector y_{k-n}^{k+n} at the zero mean Gaussian noise density whose autocovariance matrix is as in (12.4.29). Therefore, $r_k^s(n)$ is monotonically nondecreasing with increasing n. Given ε_N, the same monotonicity characterizes the truncation constant λ_n in (12.4.30), whose maximum value λ_∞ equals $c \lim_{n \to \infty} [r_k^s(n)]^{1/2}$, where c is the solution of (12.4.31). If the densities f_{0S} and f_{0N} are both stationary, with respective power spectral densities, $p_{0S}(\lambda)$ and $p_{0N}(\lambda)$; $\lambda \in [-\pi, \pi]$, and if $m \to \infty$, then directly from (12.2.26) we obtain

$$\lambda_\infty = c[E\{X_0^2 | f_{0S}\} - e_F(p_{0S}, p_{0N})]^{1/2}$$
$$= c\{(2\pi)^{-1} \int_{-\pi}^{\pi} p_{0S}(\lambda) d\lambda - (2\pi)^{-1} \int_{-\pi}^{\pi} p_{0S}(\lambda)[p_{0S}(\lambda) + p_{0N}(\lambda)]^{-1} p_{0N}(\lambda) d\lambda\}^{1/2}$$
$$= c\{(2\pi)^{-1} \int_{-\pi}^{\pi} p_{0S}^2(\lambda)[p_{0S}(\lambda) + p_{0N}(\lambda)]^{-1} d\lambda\}^{1/2}$$

12.4 Robust Filtering

Causal Filtering

Given the Gaussian densities f_{0S} and f_{0N} in Lemma 12.4.1, and the sequences $\{\ldots, X_{-1}, X_0, X_1, \ldots\}$ and $\{\ldots, W_{-1}, W_0, W_1, \ldots\}$ of random variables as in the noncausal filtering, let $M_{n,k}$ and $N_{n,k}$ denote respectively the autocovariance matrices $E\{X_{k-n+1}^k (X_{k-n+1}^k)^T | f_{0S}\}$ and $E\{W_{k-n+1}^k (W_{k-n+1}^k)^T | f_{0N}\}$, where $n \geq 0$. Let then $a_{n,k}^T$ denote the first row of the matrix $M_{n,k}$, and let $\Lambda_{n,k} = M_{n,k} + N_{n,k}$. Let $g_{kp}^0(x_{k-n+1}^{k-l})$, $n - 1 \geq l$, denote the optimal mean-squared prediction operation at the Gaussian density f_{0S} for the datum x_k, given x_{k-n+1}^{k-l}. Assuming that $\Lambda_{n,k}$ is nonsingular, let us then define the sets $\{c_{k,n-1,l,j}; k - n + 1 \leq j \leq k - l\}$ and $\{h_{k,n,j}; k - n + 1 \leq j \leq k\}$ of coefficients as

$$\{c_{k,n-1,l,j}\}: g_{kp}^0(x_{k-n+1}^{k-l}) = \sum_{j=k-n+1}^{k-l} c_{k,n-1,l,j} x_j \tag{12.4.34}$$

$$[h_{k,n,k-n+1}, \ldots, h_{k,n,k}] = a_{n,k}^T \Lambda_{n,k}^{-1}$$

Let us now define.

$$g_n^c(x) = \begin{cases} x & \text{if } |x| \leq \mu_n \\ \mu_n \, \text{sgn}(x) & \text{otherwise} \end{cases} \tag{12.4.35}$$

where $c = \mu_n [a_{n,k}^T \Lambda_{n,k}^{-1} a_{n,k}]^{-1/2}$ is such that

$$\Phi(c) + c^{-1}\phi(c) = 2^{-1}[1 + (1 - \varepsilon_N)^{-1}] \tag{12.4.36}$$

Let $\hat{x}_{k,n}^c$ denote the estimate of the signal datum x_k from the observation vector y_{k-n+1}^k. Then, the estimate $\hat{x}_{k,n}^c$ is designed as follows, where ε_N and m are a priori selected.

$$\hat{x}_{k,n}^c = \begin{cases} g_n^c(a_{n,k}^T \Lambda_{n,k}^{-1} y_{k-n+1}^k) & \text{if } n \leq m \\ \sum_{j=k-n+1}^{k-m} c_{k,n-1,m,j} \hat{x}_{j,j+n-k}^c \\ + g_m^c\left(\sum_{j=k-m+1}^{k} h_{k,n,j}[y_j - g_{jp}^0(\hat{x}_{k-n+1}^{j-m})]\right) & \text{if } n > m \end{cases} \tag{12.4.37}$$

where $g_{jp}^0(\cdot)$ is as in (12.4.34), and where $\hat{x}_j^i = [\hat{x}_{j,j+n-k}^c, \ldots, \hat{x}_{i,i+n-k}^c]$.
Let us define,

$$r_k^c(n) = a_{n,k}^T \Lambda_{n,k}^{-1} a_{n,k} \tag{12.4.38}$$

Then, $r_k^c(n)$ represents the variance gain in estimating the datum x_k, from the observation vector y_{k-n+1}^k, at the zero mean Gaussian density, whose autocovariance matrix is as in (12.4.34). Thus, $r_k^c(n)$ is monotonically nondecreasing with increasing n, and so is then the truncation constant μ_n in (12.4.35), where ε_N remains fixed.

It can be shown [Tsaknakis (1986)] that the operations in (12.4.32) and (12.4.37) are both qualitatively robust, in both the asymptotic and the nonasymptotic sense. In both operations, the integer m and the number ε_N represent a tradeoff between optimality at the Gaussian noise density f_{0N} and robustness, and they are both system parameters. As m increases and ε_N decreases, the filtering operations in (12.4.32) and (12.4.37) tend to the optimal at the Gaussian density f_{0N}, linear data operations. The estimates in (12.4.32) and (12.4.37) can be computed recursively in some cases, and their performance can then be found via a combination of analytical and numerical methods. We will consider such a case in Section 12.5.

Let us now consider the case where the Gaussian densities f_{0S} and f_{0N}, in Lemma 12.4.1 are both stationary, with power spectral densities belonging to two disjoint classes $\mathscr{C}_s$ and $\mathscr{C}_N$. Let both those classes be as the class $\mathscr{F}$ in (12.2.33), where the nominal power spectral density $f_0(\lambda)$ and the constant ε in class $\mathscr{F}$ are replaced by $p_{0S}(\lambda)$ and δ_s for class $\mathscr{C}_s$, and $p_{0N}(\lambda)$ and δ_N for class $\mathscr{C}_N$. We now have a second level game, where the quantity $I(f^*)$ in (12.4.28) is asymptotically $(n, l \to \infty)$ minimized on $\mathscr{C}_s \times \mathscr{C}_N$. Equivalently, the limit $\lim_{\substack{n \to \infty \\ l \to \infty}} a_{n+l}^T \Lambda_{n+l}^{-1} a_{n+l}$ must be then minimized on $\mathscr{C}_s \times \mathscr{C}_N$. But if the power spectral densities of the acting signal and noise processes are respectively $p_S(\lambda)$ and $p_N(\lambda)$, then the latter limit equals

$$(2\pi)^{-1} \int_{-\pi}^{\pi} p_s(\lambda)\, d\lambda - (2\pi)^{-1} \int_{-\pi}^{\pi} p_s(\lambda)[p_s(\lambda) + p_N(\lambda)]^{-1} p_N(\lambda)\, d\lambda \quad (12.4.39)$$

Since $(2\pi)^{-1} \int_{-\pi}^{\pi} p_s(\lambda)\, d\lambda = 1 \ \forall\ p_s \in \mathscr{C}_s$, minimization of the expression in (12.4.39) on $\mathscr{C}_s \times \mathscr{C}_N$ corresponds to maximization of the quantity $(2\pi)^{-1} \int_{-\pi}^{\pi} p_s(\lambda)[p_s(\lambda) + p_N(\lambda)]^{-1} p_N(\lambda)\, d\lambda$ on $\mathscr{C}_s \times \mathscr{C}_N$. But this latter problem has been solved in the last part of Section 12.2. Its solution is then used to determine the autocovariance matrices $\Lambda_{2n+1, k}$ and $\Lambda_{n,k}$ in the generation of the robust noncausal and causal filtering operations in, respectively, (12.4.32) and (12.4.37).

12.5 EXAMPLES

In this section, we will concentrate on the important case of filtering, when the information process is zero mean Gaussian and first order Markov with known autocovariance coefficients, and when the nominal additive noise is zero mean Gaussian, white, and independent of the information process. Our main objective is protection against extreme outliers; we will thus adopt the robust operations proposed in Section 12.4. In addition, we will focus on causal filtering and will not necessarily impose stationarity on the Gaussian information process and the Gaussian white nominal noise.

Let us assume that sequences $y_1, y_2, \ldots$ are observed. If X_k denotes the random variable that corresponds to the kth element of the information data sequence, let us denote $r_{kl} = E\{X_k X_l\}$. If W_k denotes the random variable that corresponds to the kth element of the noise data sequence when the nominal Gaussian white noise is present, let us denote $\sigma_k^2 = E\{W_k^2\}$. The optimal at the Gaussian, nominal noise, causal filter is then a linear operation on the observed

12.5 Examples

sequences; it can be computed recursively, as shown by (12.2.27) to (12.2.32). In particular, if $\hat{x}_n^0(y_1^n)$ denotes the estimate of the information datum x_n from the observed sequence y_1^n, as induced by the latter filter, we have

$$\hat{x}_1^0(y_1) = \beta_1 y_1$$
$$\hat{x}_n^0(y_1^n) = \alpha_n \hat{x}_{n-1}^0(y_1^{n-1}) + \beta_n y_n \tag{12.5.1}$$

where

$$\beta_1 = r_{11}[r_{11} + \sigma_1^2]^{-1}$$

$$\beta_n = \frac{r_{nn} - r_{n-1,n}^2 r_{n-1,n-1}^{-1} + \sigma_{n-1}^2 r_{n-1,n}^2 r_{n-1,n-1}^{-2} \beta_{n-1}}{r_{nn} - r_{n-1,n}^2 r_{n-1,n-1}^{-1} + \sigma_{n-1}^2 r_{n-1,n}^2 r_{n-1,nn-1}^{-2} \beta_{n-1} + \sigma_n^2} \quad n \geq 2 \tag{12.5.2}$$

$$\alpha_1 = 0$$

$$\alpha_n = r_{n-1,n} r_{n-1,n-1}^{-1} (1 - \beta_n) \quad n \geq 2$$

Furthermore, the mean-squared error $e_n^0 = E\{[X_n - \hat{x}_n^0(y_1^n)]^2 | f_{ON}\}$ induced by the filtering operation in (12.5.1) and (12.5.2), at the Gaussian nominal noise f_{ON}, is then such that,

$$e_n^0 = \sigma_n^2 \beta_n \tag{12.5.3}$$

Let us now consider the robust operation in (12.4.37), for the present model. The quadratic expression $a_{n,n}^T \Lambda_{n,n}^{-1} y_1^n$ equals the estimate $\hat{x}_n^0(y_1^n)$ in (12.5.1), while the set $\{c_{n,n-1,l,j}\}$ of coefficients in (12.4.34) is here such that $c_{n,n-1,m,n-m} = r_{n-m,n} r_{n-m,n-m}^{-1}$ and $c_{n,n-1,m,j} = 0; j < n - m$. The set $\{h_{n,n,j}; 1 \leq j \leq n\}$ of coefficients in (12.4.34) are here the coefficients of the linear filter $\hat{x}_n^0(y_1^n)$ in (12.5.1). Due to the above, the filtering operation in (12.4.37) is simplified here as follows, where the sequences $\{\alpha_n\}$ and $\{\beta_n\}$ are as in (12.5.2).

$$\hat{x}_n^c = \hat{x}_{n,n}^c = \begin{cases} g_n^c[\hat{x}_n^0(y_1^n)] & n \leq m \\ \dfrac{r_{n-m,n}}{r_{n-m,n-m}} \hat{x}_{n-m}^c + g_m^c\left(\sum_{n-m+1}^{n} b_i\left[y_i - \dfrac{r_{i-m}}{r_{i-m,i-m}} \hat{x}_i^c\right]\right) & n > m \end{cases}$$
$$\tag{12.5.4}$$

where

$$b_n = \beta_n \quad b_{n-l} = \beta_{n-l} \prod_{k=n-l+1}^{n} \alpha_k \quad l \geq 1 \tag{12.5.5}$$

All the quantities included in (12.5.4) can be computed recursively. The coefficients $\{\alpha_n\}$ and $\{\beta_n\}$ are updated as in (12.5.2), and the estimate $\hat{x}_n^0(y_1^n)$ is updated as in (12.5.1). The thresholds $\{\mu_n = c[a_{nn}^T \Lambda_{nn}^{-1} a_{nn}]^{-1/2}; 1 \leq n \leq m\}$ included in the functions $g_n^c(x)$ can be also computed recursively, since $a_{nn}^T \Lambda_{nn}^{-1} a_{nn} = r_{nn} - \sigma_n^2 \beta_n$.

The computations are simplified significantly when the Gaussian information process, and the nominal Gaussian white noise are both stationary. In this case, let us denote

$$\sigma_k^2 = \sigma^2 \qquad \forall k \geq 1$$
$$r_{kk} = r^2 \qquad \forall k \geq 1 \qquad (12.5.6)$$
$$r_{k,k+j} r_{kk}^{-1} = d_j \qquad \forall k \qquad d_j < 1, j \geq 1$$

The recursions in (12.5.2) are then simplified as

$$\beta_1 = [1 + \rho^{-2}]^{-1} \qquad \alpha_1 = 0$$
$$\beta_n = 1 - \frac{1}{\rho^2(1 - d_1^2) + 1 + d_1^2 \beta_{n-1}} \qquad 2 \leq n \leq m \qquad (12.5.7)$$
$$\alpha_n = d_1(1 - \beta_n) \qquad 2 \leq n \leq m$$

where

$$\rho^2 = r^2/\sigma^2 \qquad (12.5.8)$$

Given some ε_N, we find the solution c of the equation in (12.4.36), and we maintain this value in memory. At each point in time n the following recursive operations are then performed.

For $n \leq m$:

$$h_1(y_1) = \beta_1 y_1$$
$$h_k(y_1^k) = \alpha_k h_{k-1}(y_1^{k-1}) + \beta_k y_k \qquad k \geq 2 \qquad (12.5.9)$$
$$\mu_n = c[r^2 - \sigma^2 \beta_n]^{1/2} \qquad (12.5.10)$$
$$\hat{x}_n^c = \hat{x}_n^c(y_1^n) = \begin{cases} h_n(y_1^n) & |h_n(y_1^n)| \leq \mu_n \\ \mu_n \operatorname{sgn}[h_n(y_1^n)] & \text{otherwise} \end{cases} \qquad (12.5.11)$$

For $n > m$:

$$h_{n-m+1} = \beta_{n-m+1}[y_{n-m+1} - \hat{x}_{n-m+1}^c]$$
$$h_{n-m+l} = \alpha_{n-m+l} h_{n-m+l-1} + \beta_{n-m+l}[y_{n-m+l} - \hat{x}_{n-m+l}^c] \qquad 2 \leq l \leq m-1 \qquad (12.5.12)$$

$$\hat{x}_n^c = d_1 \hat{x}_{n-1}^c + g_m^c(\alpha_n h_{n-1} + \beta_n[y_n - d_1 \hat{x}_{n-1}^c]) \qquad (12.5.13)$$

where

$$g_m^c(x) = \begin{cases} x & |x| \leq \mu_m \\ \mu_m \operatorname{sgn} x & |x| > \mu_m \end{cases} \qquad (12.5.14)$$
$$\mu_m = c[r^2 - \sigma^2 \beta_m]^{1/2} \qquad (12.5.15)$$

The expressions in (12.5.11) and (12.5.13) provide the causal estimates of the datum x_n, for respectively $n \leq m$, and $n > m$. We note that at each point in time, the m most recent estimates need to be kept in memory.

12.5 Examples

We will now study the mean-squared performance of the robust estimate both at the nominal Gaussian white noise, and in the presence of a single outlier at the estimated datum. We will compare this performance to the performance of the optimal at the Gaussian noise, linear mean-squared estimate. For simplicity in the computations, we will consider the case where the datum x_m is estimated from the observation sequence y_1^m, where m is the operational integer parameter in (12.5.9). The qualitative characteristics of the performance, when x_n, $n > m$ is estimated instead are similar. Let the outlier at the datum x_m be modeled as follows. With probability $(1 - \zeta)$, the additive to x_m noise is zero mean Gaussian with variance σ^2, and with probability ζ the same noise is a delta function at the value y. Let us denote by e_m^0 the mean-squared error induced by the optimal at the Gaussian noise, linear estimate of x_m when no outlier exists, and let us denote by $e_m^0(\zeta, y)$ the mean-squared error induced by the same estimate, when the above outlier model is present. Let us denote by e_{m,ε_N}^* and $e_{m,\varepsilon_N}^*(\zeta, y)$ the corresponding mean-squared errors induced by the robust estimate, when ε_N has been selected in (12.4.36). Let the nominal Gaussian white noise be stationary, and let the Gaussian information process be first order Markov with autocovariance coefficients δ^k, $k \geq 0$ and $\delta < 1$. Then, given the coefficients β_k; $k \geq 1$ in (12.5.7), where now $\rho^2 = \sigma^{-2}$ and $d_1 = \delta$, we easily obtain

$$e_m^0 = \sigma^2 \beta_m$$
$$e_m^0(\zeta, y) = (1 - \zeta)e_m^0 + \zeta[e_m^0 + \beta_m^2(y^2 - \sigma^2)] = \sigma^2\beta_m + \zeta\beta_m^2(y^2 - \sigma^2) \quad (12.5.16)$$

Via some tedious, but routine, manipulations, we also obtain

$$e_{m,\varepsilon_N}^* = 2[1 - \rho^2(2 + c^2)]\Phi(c\rho) - 2c\rho\phi(c\rho) + \rho^2(2c^2 + 3) - 1 \quad (12.5.17)$$

$$\begin{aligned}e_{m,\varepsilon_N}^*(\zeta, y) = &(1 - \zeta)e_{m,\varepsilon_N}^* \\ &+ \zeta\Bigg\{1 + 2v + 2c^2\rho^2 - m^2 - \lambda^2 + (m^2 + \lambda^2 - c^2\rho^2)\left[\Phi\left(\frac{c\rho - m}{\lambda}\right)\right. \\ &+ \left.\Phi\left(\frac{c\rho + m}{\lambda}\right)\right] - \lambda(c\rho + m)\phi\left(\frac{c\rho - m}{\lambda}\right) - \lambda(c\rho - m)\phi\left(\frac{c\rho + m}{\lambda}\right) \\ &- 2v[\Phi(c\rho - m) + \Phi(c\rho + m)] - 4v(c\rho + m)[1 - v^2(1 - v^2)^{1/2}] \\ &\times \phi([c\rho + m](1 - v^2 + v^4)^{1/2}(1 - v^2)^{-1})\Bigg\} \quad (12.5.18)\end{aligned}$$

where c is the solution of the equation in (12.4.36), where $\phi(x)$ and $\Phi(x)$ denote respectively the density function at the point x and the cumulative distribution at the point x, of the zero mean, unit variance Gaussian random variable, and where

$$m = \beta_m y \qquad \rho^2 = 1 - \sigma^2 \beta_m$$
$$\lambda^2 = 1 - \beta_m(1 - \beta_m)\sigma^2 + \beta_m^2 y^2 \qquad v = \lambda^{-1}\rho^2 \quad (12.5.19)$$

From the above expressions, we can easily find that the error in (12.5.18) is a bounded function of the value y, where the error $e_m^0(\zeta, y)$ in (12.5.16) is clearly not. In fact,

$$\lim_{y \to \infty} e_{m,\varepsilon_N}^*(\zeta, y) = (1 - \zeta)e_{m,\varepsilon_N}^* + \zeta(1 + c^2\rho^2) \qquad (12.5.20)$$

$$\lim_{y \to \infty} e_m^0(\zeta, y) = \infty \qquad (12.5.21)$$

In addition, the difference between the errors e_m^0 in (12.5.16) and e_{m,ε_N}^* in (12.5.17) decreases monotonically with increasing m and with decreasing ε_N. Thus, the robust estimate can combine good performance at the nominal Gaussian noise with excellent protection against infrequent extreme outliers. On the other hand, the optimal at the Gaussian noise, linear estimate offers no protection against outliers, inducing mean-squared error that is quadratically increasing with the outlier value. The largest value ζ^* for which the expression in (12.5.20) does not exceed $r^2 = E\{X_0^2\}$ is called the *breakdown point* of the robust estimate. It can easily be found that ζ^* is strictly positive and strictly less than one. For outlier frequencies above ζ^*, the robust estimate becomes useless, since the induced mean-squared error is then larger than that deduced when no observation data are available. Due to (12.5.21), we conclude that the breakdown point of the linear estimate is zero.

We computed the errors in (12.5.16), (12.5.17), and (12.5.18) for various values of the design parameters ε_N and m, and for $\sigma^2 = 2$, $\delta = 0.5$, and $\zeta = 0.1$. In Figure 12.5.1, we plot the errors e_m^0 and e_{m,ε_N}^* against m for various ε_N values in the latter. In Figures 12.5.2 and 12.5.3, we plot the errors $e_{m,\varepsilon_N}^*(\zeta, y)$ and $e_m^0(\zeta, y)$

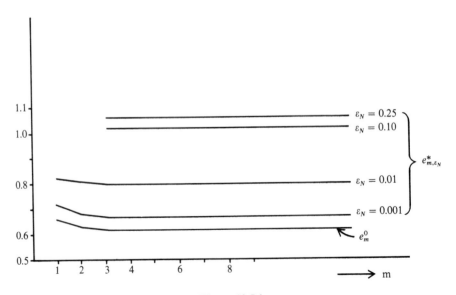

Figure 12.5.1

12.5 Examples

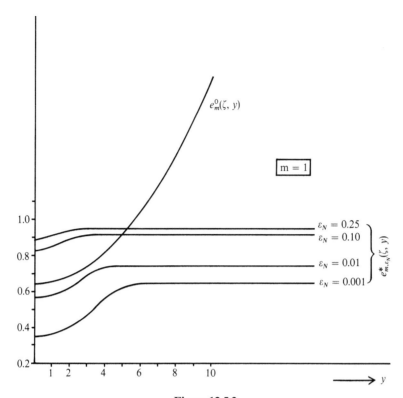

Figure 12.5.2

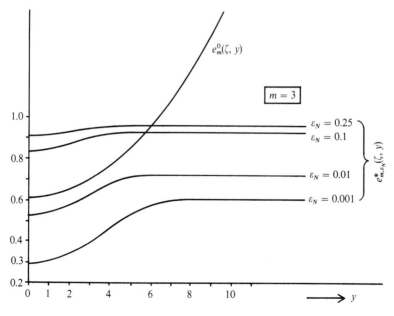

Figure 12.5.3

against y, for respectively $m = 1$ and $m = 3$. From the three figures, we observe that the robust filter with design parameters ε_N and m respectively equal to 0.001 and 3, offers excellent protection against occasional outliers, while its performance at the Gaussian noise is simultaneously very close to the optimal.

12.6 APPENDIX

Proof of (12.2.38): We only need to prove that the power spectral density in (12.2.38) satisfies the left part of the game in (12.2.34). That is,

$$\forall f \in \mathscr{F} \quad e_p(f, g_p^*) \leq e_p(f_p^*, g_p^*)$$
$$= \exp\left\{(2\pi)^{-1} \int_{-\pi}^{\pi} \log[2\pi f_p^*(\lambda)] \, d\lambda\right\} \tag{12.6.1}$$

where, for log being the natural logarithm, we have

$$\|g_p^*(\lambda)\|^2 = k_p [f_p^*(\lambda)]^{-1} \quad \text{a.e. in } [-\pi, \pi] \tag{12.6.2}$$

$$k_p = \exp\left\{(2\pi)^{-1} \int_{-\pi}^{\pi} d\lambda \log f_p^*(\lambda)\right\} \tag{12.6.3}$$

and

$$e_p(f, g_p^*) = (2\pi)^{-1} k_p \int_{-\pi}^{\pi} f(\lambda) [f_p^*(\lambda)]^{-1} \, d\lambda \tag{12.6.4}$$
$$f(\lambda) - (1 - \varepsilon) f_0(\lambda) \geq 0 \quad \forall \lambda \in [-\pi, \pi] \quad \forall f \in \mathscr{F}$$

We now have

$$e_p(f, g_p^*) - e_p(f_p^*, g_p^*)$$
$$= (2\pi)^{-1} k_p \left\{ \int_{(1-\varepsilon)f_0(\lambda) \geq c} [(1-\varepsilon)f_0(\lambda)]^{-1} [f(\lambda) - (1-\varepsilon)f_0(\lambda)] \, d\lambda \right.$$
$$\left. + \int_{(1-\varepsilon)f_0(\lambda) < c} c^{-1} [f(\lambda) - c] \, d\lambda \right\}$$
$$\leq (2\pi)^{-1} k_p \left\{ \int_{(1-\varepsilon)f_0(\lambda) \geq c} c^{-1} [f(\lambda) - (1-\varepsilon)f_0(\lambda)] \, d\lambda \right.$$
$$\left. + \int_{(1-\varepsilon)f_0(\lambda) < c} c^{-1} [f(\lambda) - c] \, d\lambda \right\}$$
$$= (2\pi c)^{-1} k_p \int_{-\pi}^{\pi} [f(\lambda) - f_p^*(\lambda)] \, d\lambda = 0$$

Thus, $e_p(f, g_p^*) \leq e_p(f_p^*, g_p^*) \; \forall f \in \mathscr{F}$, and the proof is complete. ∎

12.6 Appendix

Proof of (12.2.39): First, it can be easily shown that the function $T(\gamma)$ is monotonically increasing with increasing γ, and it is continuous. The inverse $T^{-1}(\gamma)$ is also monotonic and continuous. Thus, there exists d such that

$$(2\pi)^{-1} \int_{-\pi}^{\pi} \max[T^{-1}(d), (1-\varepsilon)f_0(\lambda)] \, d\lambda = 1$$

Now, we only need to show that the left part of (12.2.35) is satisfied by the power spectral density $f_I^*(\lambda)$ in (12.2.39). That is,

$$\forall f \in \mathscr{F} \qquad e_I(f, g_I^*) \leq e_I(f_I^*, g_I^*) = 4\pi^2 \left[\int_{-\pi}^{\pi} f_I^{*-1}(\lambda) \, d\lambda \right]^{-1}$$

where

$$g_I^*(\lambda) = 2\pi \left[\int_{-\pi}^{\pi} f_I^{*-1}(\omega) \, d\omega \right]^{-1} f_I^{*-1}(\lambda) \qquad \text{a.e. in } [-\pi, \pi]$$

We now have

$$e_I(f, g_I^*) - e_I(f_I^*, g_I^*)$$

$$= 2\pi \left[\int_{-\pi}^{\pi} f_I^{*-1}(\omega) \, d\omega \right]^{-2} \left\{ \int_{T^{-1}(d) > (1-\varepsilon)f_0(\lambda)} [T^{-1}(d)]^{-2} [f(\lambda) - T^{-1}(d)] \, d\lambda \right.$$

$$\left. + \int_{T^{-1}(d) \leq (1-\varepsilon)f_0(\lambda)} [(1-\varepsilon)f_0(\lambda)]^{-2} [f(\lambda) - (1-\varepsilon)f_0(\lambda)] \, d\lambda \right\}$$

$$\leq 2\pi d^{-2} [T^{-1}(d)]^2 \left\{ \int_{T^{-1}(d) > (1-\varepsilon)f_0(\lambda)} [T^{-1}(d)]^{-2} [f(\lambda) - T^{-1}(d)] \, d\lambda \right.$$

$$\left. + \int_{T^{-1}(d) \leq (1-\varepsilon)f_0(\lambda)} [T^{-1}(d)]^{-2} [f(\lambda) - (1-\varepsilon)f_0(\lambda)] \, d\lambda \right\}$$

$$= 2\pi d^{-2} \int_{-\pi}^{\pi} [f(\lambda) - (1-\varepsilon)f_0(\lambda)] \, d\lambda = 0$$

Thus, $e_I(f, g_I^*) \leq e_I(f_I^*, g_I^*) \ \forall f \in \mathscr{F}$, and the proof is complete. ∎

Proof of Theorem 12.4.1: The class $\mathscr{F}'$ in (12.4.18) is convex and compact, with respect to the metric $\int_{R^{n+l}} |f_1(x_{-n}^{l-1}) - f_2(x_{-n}^{l-1})| \, dx_{-n}^{l-1}$ and the integral in (12.4.15) is strictly convex with respect to f. Denoting the latter integral $I(f)$, and applying calculus of variations, we conclude that the density f^* in $\mathscr{F}'$ that attains the infimum in (12.4.15), for $\mathscr{F}$ being replaced by $\mathscr{F}'$, exists, it is unique, and it is such that

$$\frac{\partial}{\partial \delta} I\bigl((1-\delta)f^* + \delta h\bigr)\bigg|_{\delta=0} \geq 0 \qquad \forall h \in \mathscr{F}' \qquad (12.6.5)$$

By substitution, and integration by parts, we easily find

$$\frac{\partial}{\partial \delta} I\left((1-\delta)f^* + \delta h\right)\bigg|_{\delta=0}$$

$$= \int_{R^{n+l}} dy_{-n}^{l-1} [h(y_{-n}^{l-1}) - f^*(y_{-n}^{l-1})]\{2Q(\underline{D})[g^*(y_{-n}^{l-1})] - [g^*(y_{-n}^{l-1})]^2\} \quad (12.6.6)$$

where $Q(\underline{D})$ is the operator obtained from the operator $P(\underline{D})$, by changing the signs of the odd terms in the Taylor series expansion of the latter, and where

$$g^*(y_{-n}^{l-1}) = \frac{P(\underline{D})[f^*(y_{-n}^{l-1})]}{f^*(y_{-n}^{l-1})} \quad (12.6.7)$$

The condition $\forall\, h \in \mathscr{F}'$ in (12.6.5) is equivalent to the constraints

$$h: \begin{cases} h(y_{-n}^{l-1}) \geq (1-\varepsilon_N) f_0(y_{-n}^{l-1}) & \forall\, y_{-n}^{l-1} \in R^{n+l} \\ \int_{R^{n+l}} h(y_{-n}^{l-1})\, dy_{-n}^{l-1} = 1 \end{cases} \quad (12.6.8)$$

Thus, we require that the quantity in (12.6.6) be nonnegative, subject to the constraints in (12.6.8). The latter is satisfied if $f^*(y_{-n}^{l-1})$ touches the boundary $(1-\varepsilon_N)f_0(y_{-n}^{l-1})$ for some y_{-n}^{l-1} values in R^{n+l}, and if for the remaining y_{-n}^{l-1} values in R^{n+l} we have

$$2Q(\underline{D})[g^*(y_{-n}^{l-1})] - [g^*(y_{-n}^{l-1})]^2 = c \text{ (constant)} \quad (12.6.9)$$

The condition in (12.6.9) is satisfied if $g^*(y_{-n}^{l-1}) = \lambda$ (constant) for y_{-n}^{l-1} such that $f^*(y_{-n}^{l-1}) > (1-\varepsilon_N)f_0(y_{-n}^{l-1})$. Thus, the statement in the theorem is true. ∎

12.7 PROBLEMS

12.7.1 Given $x = z + n$ with z and n independent. z is uniformly distributed between $-a$ and a; n is uniformly distributed between $-b$ and b. Find the best mean-squared error estimation of z in terms of x.

12.7.2 Find the best linear realizable predictor of δ units ahead for a random signal $s(t)$ which has a power spectral density

$$\Phi_{ss}(\omega) = 1/(\omega^2 + a^2)(\omega^2 + b^2)$$

Find the error of prediction as a function of δ.

12.7.3 A signal $s(t)$ is distorted after passing through a random channel and being corrupted by a random noise as shown in the figure below. The channel is linear and time invariant, that is, the output is the convolution of the input and the impulse response of the channel $h(t, \underline{\alpha})$ where $\underline{\alpha}$ is a set of random parameters. The transfer function

12.7 Problems

of the channel is denoted by $H(\omega, \underline{\alpha})$. Assume the signal, noise, and channel are independent statistically. Find the noncausal optimum m.s. filter $G(\omega)$ to extract the signal.

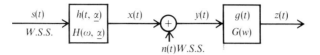

12.7.4 Let $\{s_k\}$ be a stationary time series with mean μ and spectral density

$$p(\lambda) = \frac{1}{2}|1 - e^{i\lambda}|^2$$

Find the best linear estimate (i.e., interpolation) of s_0 based on the set of values $s_{-n}, s_{-n+1}, \ldots, s_{-1}, s_1, s_2, \ldots, s_n$. Evaluate the mean square error. What happens as $n \to \infty$?

12.7.5 Let $X(t)$ be a random process with zero mean and covariance function

$$R(t, s) = \min(t, s)$$

Given observations of the random process at times T, $2T$, and $3T$, find the linear least mean-squared error estimate of $X(3T + \Delta)$ and the corresponding mean-squared error. Reduce both results to their simplest form. What happens as $\Delta \to \infty$?

12.7.6 Consider the following system.

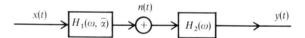

where the linear filter $H_1(\omega, \bar{\alpha})$ contains a set $\bar{\alpha}$ of random parameters; otherwise H_1 is known. Let $y(t)$ be a least-mean-squared estimate of $x(t + T)$ when $x(t)$ and $n(t)$ are wide-sense stationary random processes with zero means.
(a) Find $H_2(\omega)$ ignoring physical realizability (noncausal).
(b) Write a physically realizable expression for $H_2(\omega)$ (causal).

12.7.7 A sequence of random variables $s_1, s_2, \ldots, s_n$ is given with covariance

$$E[s_i s_j] = \text{minimum } (i, j)$$

Using the Gram-Schmidt process, find a linear transformation which takes this set into a set of n uncorrelated variables. Find the minimum variance prediction of s_n given $s_1, s_2, \ldots, s_m$; $m < n$ and evaluate that variance. When would it be possible to say that the best *linear* predictor is also the best predictor from the class of all estimators?

12.7.8 Let $\{s_i\}$ and $\{n_i\}$ be sequences of independent identically distributed random variables. Let $x_i = s_i + n_i$ for all i.
(a) Find a general expression for the least-mean-squared-error filter to recover s_i from x_i.
(b) If s_i is uniformly distributed on $(-a, a)$ and n_i is Gaussian with mean zero and variance σ^2, find an explicit expression for the filter.

12.7.9 The zero mean random process $\{x(t_i)\}$ is said to be wide-sense Markov if the conditional expectation satisfies

$$E\{x(t_n)|x(t_1), x(t_2), \ldots, x(t_{m-1})\} = E\{x(t_n)|x(t_{m-1})\}$$

Let the normalized correlation function be given by

$$R(t_i, t_j) = E\{x(t_i)x(t_j)\}, \quad R(t_i, t_i) = 1$$

Show that the following two equations are equivalent:
(a) The minimum linear mean-square estimate is

$$E\{x(t_m)|x(t_{m-1})\} = R(t_m, t_{m-1})x(t_{m-1})$$

(b) $R(t_i, t_j) = R(t_i, t_k)R(t_k, t_j);\ i > k > j$ or $i < k < j$.

12.7.10 Consider two nonstationary random processes $s(t)$ and $x(t)$, $0 \le t \le T$, and define the correlation functions as

$$R_x(t_1, t_2) = E\{x(t_1)x(t_2)\}$$
$$R_{sx}(t_1, t_2) = E\{s(t_1)x(t_2)\}$$
$$R_s(t_1, t_2) = E\{s(t_1)s(t_2)\}$$

Let ϕ_n and λ_n be the eigenfunctions and eigenvalues associated with R_x

$$\lambda_n \phi_n(t_1) = \int_0^T R_x(t_1, t_2)\phi_n(t_2)\, dt_2$$

As usual, we have:

$$R_x(t_1, t_2) = \sum_{n=0}^{\infty} \lambda_n \phi_n(t_1)\phi_n(t_2)$$

$$x_n = \int_0^T x(t)\phi_n(t)\, dt$$

(a) Suppose you are given the finite sequence of observations $\{x_n/\sqrt{\lambda_n}\}$, $n = 0, 1, \ldots, N$. Find the minimum mean-square estimate of $s(\tau_1)$. Calculate the mean-square error.

(b) Using the results of part (a), show that as $N \to \infty$, the optimum estimate is

$$\hat{s}(\tau_1) = \int_0^T h(\tau_1, u)x(u)\, du$$

where

$$h(\tau, u) = \int_0^T R_{sx}(\tau, t) \sum_{n=0}^{\infty} \frac{\phi_n(t)\phi_n(u)}{\lambda_n}\, dt$$

(c) Show that $h(\tau, u)$ satisfies the usual orthogonality condition for linear estimates.

12.7.11 You are given the amplitude modulated signal $s(t) = A[1 + a(t)] \cdot \cos \omega_0 t$, which is corrupted by an additive noise $n(t)$; $a(t)$ and $n(t)$ are Gaussian random processes that are wide-sense stationary, statistically independent, and with means of zero, autocorrelation functions $R_a(\tau)$ and $R_n(\tau)$, and power spectral densities $\phi_a(\omega)$ and $\phi_n(\omega)$, respectively. Assume that $a(t)$ is lowpass and $n(t)$ is bandpass and centered at ω_0.

(a) The total received signal is demodulated by a synchronous demodulator as shown below

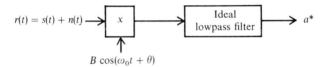

Find the output signal-to-noise ratio.

(b) For any phase insensitive demodulator of your choice, find the output signal-to-noise ratio for the input $r(t)$.

(c) Compare the performance of the demodulators of (a) and (b). When is one superior to the other on a signal-to-noise basis?

(d) Modify, if necessary, the synchronous demodulator to obtain a least-mean-squared-error receiver.

REFERENCES

Blachman, N. M. (1965), "The Convolution Inequality for Entropy Powers", *IEEE Trans. Inform. Th.* IT-11, 267–271.

Boente, G., R. Fraiman, and V. Yohai (1982), "Qualitative Robustness for General Stochastic Processes, *Tech. Report No. 26.* Dept. of Statistics, Univ. of Wash., Seattle.

Bustos, O., R. Fraiman, and V. Yohai (1984), "Asymptotic Behaviour of the Estimates Based on Residual Autocovariances for ARMA Models," Robust and Nonlinear Time Series Analysis, in *Lecture Notes in Statistics*, Springer-Verlag, New York, 26–49.

Cox, D. (1978), "Metrics on Stochastic Processes and Qualitative Robustness," *Tech. Report No. 23.* Dept. of Statistics, Univ. of Wash., Seattle.

Franke, J. and H. W. Poor (1984), "Minimax-Robust filtering and finite-Length Robust Predictors," Robust and Nonlinear Time Series Analysis, in *Lecture Notes in Statistics*, Springer-Verlag, New York, 87–126.

Gelb, A. (1975), *Applied optimal estimation*, M.I.T. Press, Cambridge, MA.

Hannan, E. J. (1970), *Multiple Time Series*, Wiley, New York.

Hampel, F. R. (1971), "A General Qualitative Definition of Robustness," *Ann. Math. Statist.* 42, 1887–1895.

——— (1974), "The Influence Curve and Its Role in Robust Estimation," *J. Amer. Statist. Assoc.* 69, 383–394.

——— (1975), "Beyond Location Parameters: Robust Concepts and Methods," *Proceedings of I.S.I. Meeting, 40th Session*, Warsaw.

——— (1978), "Optimally Bounding the Gross-Error-Sensitivity and the Influence of Position in Factor Space," *1978 Proceedings of the A.S.A. Statistical Computing Section*, Amer. Stat. Assoc., Washington, DC.

Helson, H. and D. Lowdenslager (1958), "Prediction Theory and Fourier Series in Several Variables," *Acta Math.*, 99, 165–202.

Hosoya, Y.,(1978), "Robust Linear Extrapolation of Second Order Stationary Processes," *Ann. Proba.* 6, 574–584.

Huber, P. J. (1973), "Robust Regression: Asymptotics, Conjectures and Monte Carlo", *Ann. Statist.* 1, 799–821.

Huber, P. J. (1981), *Robust Statistics*, Wiley, New York.

Kalman, R. E. (1963), "New methods in Wiener Filtering", *Proc. of the First Symposium on Engineering Applications of Random Function Theory and Probability*, Wiley, New York. Ch. 9.

Kalman, R. E. (1960), "A New Approach to Linear Filtering and Prediction Problems," *Journal of Basic Engineering* (ASME), 82D, 35–45.

Kassam, S. A. and T. L. Lim (1977), "Robust Wiener Filters," *J. Franklin Inst.* 304, 171–185.

Kolmogorov, A. (1941), "Interpolation and Extrapolation," *Bull. Acad. Sci. USSR, Ser. Math.* 5, 3–14.

——— (1941), "Stationary Sequences in Hilbert Space," *Bull. Math. Univ. Moscow*, 2, 40.

Maronna, R. A. and V. J. Yohai (1981), "Asymptotic Behaviour of General M-Estimates for Regression and Scale With Random Carriers," *Z. Wahrscheinlickeitstheorie and Verw. Gebiete* 8, 7–20.

References

Martin, R. D. and G. Debow (1976), "Robust Filtering With Data Dependent covariance," *Proceedings of the 1976 Johns Hopkins Conference on Information Sciences and Systems.*

Martin, R. D. and V. J. Yohai (1984), "Gross-Error Sensitivities of GM and RA-Estimates," Robust and Nonlinear Time Series Analysis, in *Lecture Notes in Statistics*, Springer-Verlag, New York, 198–217.

Martin, R. D. and J. E., Zeh, (1977), "Determining the Character of Time Series Outliers," *Proc. Amer. Stat. Assoc.*

Masreliez, C. J. and R. D. Martin (1977), "Robust Bayesian Estimation for the Linear Model and Robustifying the Kalman Filter," *IEEE Trans. Aut. Control* AC-22, 361–371.

Papantoni-Kazakos, P. (1977), "Robustness in Estimation," IEEE Trans. Inf. Th., Vol. IT-23, 223.

——— (1981a), "Sliding Block Encoders That Are Rho-Bar Continuous Functions of Their Input," IEEE Trans. Inf. Theory, IT-27, 372–376.

——— (1981b), "Stochastic Quantization for Performance Stability," Information and Control, 49, No. 3, 171–198.

——— (1983), "Qualitative Robustness in Time Series," EECS Dept., Univ. of Conn., Tech. Report UCT/DEECS/TR-83-15, Nov.

——— (1984a), "A Game Theoretic Approach to Robust Filtering," Information and Control, 60, Nos. 1-3, 168–191.

——— (1984b), "Some Aspects of Qualitative Robustness in Time Series," Robust and Nonlinear Time Series Analysis, in Lecture Notes in Statistics, Springer-Verlag, New York, 218–230.

Papantoni-Kazakos, P. and R. M. Gray, (1979), "Robustness of Estimators on Stationary Observations", *Ann. Prob.* 7, 989–1002.

Poor, H. V. (1980), "On Robust Wiener Filtering," *IEEE Trans. Automat. Contr.* AC-25, 531–536.

Shannon, C. E. (1948), "A Mathematical Theory of Communication," *Bell Syst. Tech. J.* 27, 623–656.

Stain, A. J. (1959), "Some Inequalities Satisfied by the Quantities of Information of Fisher and Shannon," *Information and Control* 2, 101–112.

Tsaknakis, H. (1986), "Qualitatively Robust Operations in Time Series," Ph.D. Dissertation, EECS Dept., Univ. of Conn., Storrs, CT.

Tsaknakis, H. and P. Papantoni-Kazakos (1986), "Robust Linear Filtering for Multivariable Stationary Time Series," *IEEE Trans. Automat. Cont.* AC-31, 462–466.

Tsaknakis, H., D. Kazakos, and P. Papantoni-Kazakos (1986), "Robust Prediction and Interpolation for Vector Stationary Processes," *Prob. Th. and Related Fields*, Vol. 72 Springer-Verlag, New York, 589–602.

Viterbi, A. J. (1965), "On the Minimum Mean Square Error Resulting From Linear Filtering of Stationary, Signals in White Noise." *IEEE Trans. Inf. Th.* IT-11, 594–595.

Whittle, P. (1953), "The Analysis of Multiple Stationary Time Series," *J. Roy. Statist. Soc.*, Ser. B. 15, 125–139.

Wiener, N. (1949), *Extrapolation, Interpolation and Smoothing of Stationary Time Series*, MIT Press, Cambridge, MA.

Wiener, N. and P. Masani (1957, 1958), "The Prediction Theory of Multivariate Stochastic Processes."
I. "The Regularity Condition," *Acta Math.* 98, 111–150.
II. "The Linear Predictor," *Acta Math.* 99, 93–137.

Zasuhin, V. (1941), "On the Theory of Multidimensional Stationary Random Processes," *Acad. Sci. USSR* 33, 435.

INDEX

Admissible rules, 45, 51, 156
Asymptotic relative efficiency, 199

Bayesian, 47, 51, 61, 260, 261, 388
Bhattacharyya
  coefficient, 342, 345, 350, 351, 356, 360
  distance, 342, 345, 362
Bias, 50
  unbiased estimates, 50
Breakdown point, 175, 324, 325, 368, 369, 388, 416

Causal, 390
Chernoff
  bound, 101, 103
  coefficient, 342, 360
  distance, 341
Consistency, 310

Decision rule, 3, 37, 44, 45
  randomized, 39
Detection, 3, 9, 37, 46
  shift, 7

Efficacy, 199
Efficiency, 291, 292
Empirical mean estimates, 310, 325
Entropy, 306

False alarm, 45, 113, 241
  rate, 48, 112
Filtering, 7, 388, 389, 404
Fisher information, 208, 306

Gram-Schmidt orthogonalization, 71, 73

Hamming distance, 89
Hypothesis testing, 3, 9, 37

Ideal observer, 3, 66
I-divergence distance, 341
Inadmissible rules, 45
Influence function, 324, 325
Interpolation, 388, 389, 397

J-divergence distance, 341, 345, 350, 356, 362

Kalman filter, 394
Karhunen-Loéve expansion, 77
Kullback-Leibler distance, 172, 234, 238, 341, 345, 351, 356

Least favorable, 48, 156, 167, 284
Levy distance, 343, 348
Likelihood ratio, 66
Linear mean squared estimates, 274, 275
Locally most powerful test, 113, 140, 142
Location parameter, 57, 198, 300, 305, 318

Matched filter, 69, 96
Maximum a posteriori estimate, 263
Maximum likelihood, 51, 59, 299
Median estimates, 310, 325, 387
M-estimates, 310, 326, 387

Minimax estimates, 260, 284
Minimax test, 48, 51, 154
Minimum distance decoding, 56, 90
Minimum mean squared estimate, 266

Neyman-Pearson tests, 4, 49, 112
Noncausal, 390
  filtering, 409
Nonparametric
  filtering, 7
  parameter estimation, 49, 52
  tests, 3, 5, 6, 198
Nonparametrically described processes, 2

Orthogonality principle, 276

Parameter estimation, 5, 49
Parametric
  filtering, 7, 388, 389
  interpolation, 388, 389
  parameter estimation, 49, 52
  prediction, 388, 389
  tests, 3, 5
Parametrically known process, 2
Penalty
  coefficients, 42
  conditional expected, 43, 44
  expected, 43
  functions, 42
Prediction, 388, 389, 397
Probability of correct detection, 42
Probability of error, 42
Prohorov distance, 343, 349, 353, 385, 386

Qualitative robustness, 384
Quality control, 7

Radar, 53
Rank tests, 205
Rao-Cramér bound, 292
Rho-Bar distance, 343, 349, 353, 364, 386
Robust
  estimates, 318
  filtering, 404
  interpolation, 397
  prediction, 397
  rule, 167
  test, 6, 154, 166, 168, 233

Sensitivity, 388
Sequential tests, 4, 5, 37, 49, 224
Sign test, 175, 200
Single hypothesis testing, 3, 42
Small sample efficiency, 215
Smoothing, 390, 409
Stopping rule, 37
Sufficient statistics, 68, 299

Test function, 66
Test statistics, 65
Threshold, 66
Transfer function, 392

Unbiased mean squared estimates, 280
Uniformly most powerful tests, 49, 113, 133, 137, 140

Variational distance, 345, 350
Vasershtein distance, 343, 348, 353

Wald stopping variable, 226
Wald test, 226
Wilcoxon rank test, 208